Progress in Mathematics

Volume 242

Jean-Paul Dufour
Nguyen Tien Zung

Poisson Structures and Their Normal Forms

Birkhäuser Verlag
Basel · Boston · Berlin

Authors:

Jean-Paul Dufour
Département de mathématique
Université de Montpellier 2
place Eugène Bataillon
34095 Montpellier
France
e-mail: dufourj@math.univ-montp2.fr

Nguyen Tien Zung
Laboratoire Émile Picard,
UMR 5580 CNRS
Institut de Mathématiques
Université Paul Sabatier
118 route de Narbonne
31062 Toulouse
France
e-mail: tienzung@picard.ups-tlse.fr

2000 Mathematics Subject Classification 53Dxx, 53Bxx, 85Kxx, 70Hxx

A CIP catalogue record for this book is available from the Library of Congress, Washington D.C., USA

Bibliographic information published by Die Deutsche Bibliothek
Die Deutsche Bibliothek lists this publication in the Deutsche Nationalbibliografie;
detailed bibliographic data is available in the Internet at <http://dnb.ddb.de>.

ISBN 3-7643-7334-2 Birkhäuser Verlag, Basel – Boston – Berlin

© 2005 Birkhäuser Verlag, P.O. Box 133, CH-4010 Basel, Switzerland
Part of Springer Science+Business Media
Printed on acid-free paper produced of chlorine-free pulp. TCF ∞
Printed in Germany
ISBN-10: 3-7643-7334-2   e-ISBN: 3-7643-7335-0
ISBN-13: 978-3-7643-7334-4

9 8 7 6 5 4 3 2 1   www.birkhauser.ch

à Hélène
J.-P. D.

à Mai
N. T. Z.

Contents

Preface

"Il ne semblait pas que cette importante théorie pût encore être perfectionnée, lorsque les deux géomètres qui ont le plus contribué à la rendre complète, en ont fait de nouveau le sujet de leurs méditations...". By these words, Siméon Denis Poisson announced in 1809 [293] that he had found an improvement in the theory of Lagrangian mechanics, which was being developed by Joseph-Louis Lagrange and Pierre-Simon Laplace. In that pioneering paper, Poisson introduced (we slightly modernize his writing) the notation

$$(a, b) = \sum_{i=1}^{n} \left(\frac{\partial a}{\partial q_i} \frac{\partial b}{\partial p_i} - \frac{\partial a}{\partial p_i} \frac{\partial b}{\partial q_i} \right), \tag{0.1}$$

where a and b are two functions of the coordinates q_i and the conjugate quantities $p_i = \frac{\partial R}{\partial \dot{q}_i}$ for a mechanical system with Lagrangian function R. He proved that, if a and b are first integrals of the system then (a, b) also is. This (a, b) is nowadays denoted by $\{a, b\}$ and called the Poisson bracket of a and b. Mathematicians of the 19th century already recognized the importance of this bracket. In particular, William Hamilton used it extensively to express his equations in an essay in 1835 [168] on what we now call Hamiltonian dynamics. Carl Jacobi in his "Vorlesungen über Dynamik" around 1842 (see [185]) showed that the Poisson bracket satisfies the famous Jacobi identity:

$$\{\{a, b\}, c\} + \{\{b, c\}, a\} + \{\{c, a\}, b\} = 0. \tag{0.2}$$

This same identity is satisfied by Lie algebras, which are infinitesimal versions of Lie groups, first studied by Sophus Lie and his collaborators in the end of the 19th century [213].

In our modern language, a Poisson structure on a manifold M is a 2-vector field Π (Poisson tensor) on M, such that the corresponding bracket (Poisson bracket) on the space of functions on M, defined by

$$\{f, g\} := \langle \mathrm{d}f \wedge \mathrm{d}g, \Pi \rangle, \tag{0.3}$$

satisfies the Jacobi identity. (M, Π) is then called a Poisson manifold. This notion of Poisson manifolds generalizes both symplectic manifolds and Lie algebras. The

Poisson tensor of the original bracket of Poisson is

$$\Pi = \sum_{i=1}^{n} \frac{\partial}{\partial p_i} \wedge \frac{\partial}{\partial q_i}, \tag{0.4}$$

which is nondegenerate and corresponds to a symplectic 2-form, namely

$$\omega = \sum_{i=1}^{n} \mathrm{d}p_i \wedge \mathrm{d}q_i . \tag{0.5}$$

On the other hand, each finite-dimensional Lie algebra gives rise to a linear Poisson tensor on its dual space and vice versa.

Poisson manifolds play a fundamental role in Hamiltonian dynamics, where they serve as phase spaces. They also arise naturally in other mathematical problems as well. In particular, they form a bridge from the "commutative world" to the "noncommutative world". For example, Lie groupoids give rise to noncommutative operator algebras, while their infinitesimal versions, called Lie algebroids, are nothing but "fiber-wise linear" Poisson structures. Poisson geometry, i.e., the geometry of Poisson structures, which began as an outgrowth of symplectic geometry, has seen rapid growth in the last three decades, and has now become a very large theory, with interactions with many other domains of mathematics, including Hamiltonian dynamics, integrable systems, representation theory, quantum groups, noncommutative geometry, singularity theory, and so on.

This book arises from its authors' efforts to study Poisson structures, and in particular their normal forms. As a result, the book aims to offer a quick introduction to Poisson geometry, and to give an extensive account on known results about the theory of normal forms of Poisson structures and related objects. This theory is relatively young. Though some earlier results may be traced back to V.I. Arnold, it really took off with a fundamental paper of Alan Weinstein in 1983 [346], in which he proved a formal linearization theorem for Poisson structures, a local symplectic realization theorem, and the following splitting theorem: locally any Poisson manifold can be written as the direct product of a symplectic manifold with another Poisson manifold whose Poisson tensor vanishes at a point. Since then, a large number of other results have emerged, many of them very recently.

Here is a brief summary of this book, which only highlights a few important points from each chapter. For a more detailed list of what the book has to offer, the reader may look at the table of contents.

The book consists of eight chapters and some appendices. Chapter 1 is based on lectures given by the authors in Montpellier and Toulouse for graduate students, and is a small self-contained introduction to Poisson geometry. Among other things, we show how Poisson manifolds can be viewed as singular foliations with symplectic leaves, and also as quotients of symplectic manifolds. The reader will also find in this chapter a section about the Schouten bracket of multi-vector fields, which was discovered by Schouten in 1940 [311], and whose importance goes beyond Poisson geometry.

Starting from Chapter 2, the book contains many recent results which have not been previously available in book form. A few results in this book are even original and not published elsewhere.

Chapter 2 is about Poisson cohomology, a natural and important invariant introduced by André Lichnerowicz in 1977 [211]. In particular, we show the role played by this cohomology in normal form problems, and its relations with de Rham cohomology of manifolds and Chevalley–Eilenberg cohomology of Lie algebras. Some known methods for computing Poisson cohomology are briefly discussed, including standard tools from algebraic topology such as the Mayer–Vietoris sequence and spectral sequences, and also tools from singularity theory. Many authors, including Viktor Ginzburg, Johannes Huebschmann, Mikhail Karasev, Jean-Louis Koszul, Izu Vaisman, Ping Xu, etc., contributed to the understanding of Poisson cohomology, and we discuss some of their results in this chapter. However, the computation of Poisson cohomology remains very difficult in general.

Chapter 3 is about a kind of normal form for Poisson structures, which are comparable to Poincaré–Birkhoff normal forms for vector fields, and which are called Levi decompositions because they are analogous to Levi–Malcev decompositions for finite-dimensional Lie algebras. The results of this chapter are due mainly to Aissa Wade [342] (the formal case), the second author and Monnier [369, 263] (the analytic and smooth cases). The proof of the formal case is purely algebraic and relatively simple. The analytic and smooth cases make use of the fast convergence methods of Kolmogorov and Nash–Moser.

Chapter 4 is about linearization of Poisson structures. The results of Chapter 3 are used in this chapter. In particular, Conn's linearization results for Poisson structures with a semi-simple linear part [80, 81] may be viewed as special cases of Levi decomposition. Among results discussed at length in this chapter, we will mention here Weinstein's theorem on the smooth degeneracy of real semisimple Lie algebras of real rank greater than or equal to 2 [348], and our result on the formal and analytic nondegeneracy of the Lie algebra $\mathfrak{aff}(n)$ [120].

In Chapter 5 we explain the links among quadratic Poisson structures, r-matrices, and the theory of Poisson–Lie groups introduced by Drinfeld [107]. So far, all quadratic Poisson structures known to us can be obtained from r-matrices, which have their origins in the theory of integrable systems. Some important contributions of Semenov–Tian–Shansky, Lu, Weinstein and other people can be found in this chapter. We then show how the curl vector field (also known as modular vector field) led the first author and other people to a classification of "nonresonant" quadratic Poisson structures, and quadratization results for Poisson structures which begin with a nonresonant quadratic part. Let us mention that Poisson–Lie groups are classical versions of *quantum groups*, a subject which is beyond the scope of this book.

Chapter 6 is devoted to n-ary generalizations of Poisson structures, which go under the name of Nambu structures. Though originally invented by physicists Nambu [275] and Takhtajan [328], these Nambu structures turn out to be dual to integrable differential forms and play an important role in the theory of singular foliations. A linearization theorem for Nambu structures [119] is given in this chapter. Its proof at one point makes use of Malgrange's "Frobenius with singularities" theorem [233, 234]. Malgrange's theorem is also discussed in this chapter, together with many other results on singular foliations and integrable differential forms. In particular, we present generalizations of Kupka's stability theorem [204], which are due to de Medeiros [244, 245], Camacho and Lins Neto [59], and ourselves.

Chapter 7 deals with Lie groupoids. Among other things, it contains a recent slice theorem due to Weinstein [354] and the second author [370]. This slice theorem is a normal form theorem for proper Lie groupoids near an orbit, and generalizes the classical Koszul–Palais slice theorem for proper Lie group actions. We also discuss symplectic groupoids, an important object of Poisson geometry introduced independently by Karasev [189], Weinstein [349], and Zakrzewski [364] in the 1980s. A local normal form theorem for proper symplectic groupoids is also given.

Chapter 8 is about Lie algebroids, introduced by Pradines [294] in 1967 as infinitesimal versions of Lie groupoids. They correspond to fiber-wise linear Poisson structures, and many results about general Poisson structures, including the splitting theorem and the Levi decomposition, apply to them. Our emphasis is again on their local normal forms, though we also discuss cohomology of Lie algebroids, and the problem of integrability of Lie algebroids, including a recent strong theorem of Crainic and Fernandes [86].

Finally, Appendix A is a collection of discussions which help make the book more self-contained or which point to closely related subjects. It contains, among other things, Vorobjev's description of a neighborhood of a symplectic leaf [340], toric characterization of Poincaré–Birkhoff normal forms of vector fields, a brief introduction to deformation quantization, including a famous theorem of Kontsevich [195] on the existence of deformation quantization for an arbitrary Poisson structure, etc.

The book is biased towards what we know best, i.e., local normal forms. May the specialists in Poisson geometry forgive us for not giving more discussions on other topics, due to our lack of competence. Familiarity with symplectic manifolds is not required, though it will be helpful for reading this book. There are many nice books readily available on symplectic geometry. On the other hand, books on Poisson geometry are relatively rare. The only general introductory reference to date is Vaisman [333]. Some other references are Cannas da Silva and Weinstein [60] (a nice book about geometric models for noncommutative algebras, where Poisson geometry plays a key role), Karasev and Maslov [190] (a book on Poisson manifolds with an emphasis on quantization), Mackenzie [228] (a general reference on Lie groupoids and Lie algebroids), Ortega and Ratiu [288] (a comprehensive book on symmetry and reduction in Poisson geometry), and a book in preparation by Xu [362] (with an emphasis on Poisson groupoids). We hope that our book is complementary to the above books, and will be useful for students and researchers interested in the subject.

Acknowledgements. During the preparation of the book, we benefited from collaboration and discussions with many colleagues. We would like to thank them all, and in particular Rui Fernandes, Viktor Ginzburg, Kirill Mackenzie, Alberto Medina, Philippe Monnier, Michel Nguiffo Boyom, Tudor Ratiu, Pol Vanhaecke, Aissa Wade, Alan Weinstein, Ping Xu, Misha Zhitomirskii. Special thanks go to Rui L. Fernandes who carefully read the preliminary version of this book and made numerous suggestions which greatly helped us to improve the book. N.T.Z. thanks Max-Planck Institut für Mathematik, Bonn, for hospitality during 07-08/2003, when a part of this book was written.

Chapter 1

Generalities on Poisson Structures

1.1 Poisson brackets

Definition 1.1.1. A C^∞-smooth *Poisson structure* on a C^∞-smooth finite-dimensional manifold M is an $\mathbb{R}$-bilinear antisymmetric operation

$$\mathcal{C}^\infty(M) \times \mathcal{C}^\infty(M) \to \mathcal{C}^\infty(M), \ (f,g) \longmapsto \{f,g\} \tag{1.1}$$

on the space $\mathcal{C}^\infty(M)$ of real-valued C^∞-smooth functions on M, which verifies the Jacobi identity

$$\{\{f,g\},h\} + \{\{g,h\},f\} + \{\{h,f\},g\} = 0 \tag{1.2}$$

and the Leibniz identity

$$\{f,gh\} = \{f,g\}h + g\{f,h\}, \ \ \forall f,g,h \in \mathcal{C}^\infty(M). \tag{1.3}$$

In other words, $\mathcal{C}^\infty(M)$, equipped with $\{,\}$, is a Lie algebra whose Lie bracket satisfies the Leibniz identity. This bracket $\{,\}$ is called a *Poisson bracket*. A manifold equipped with such a bracket is called a *Poisson manifold*.

Similarly, one can define real analytic, holomorphic, and formal Poisson manifolds, if one replaces $\mathcal{C}^\infty(M)$ by the corresponding sheaf of local analytic (respectively, holomorphic, formal) functions. In order to define C^k-smooth Poisson structures ($k \in \mathbb{N}$), we will have to express them in terms of 2-vector fields. This will be done in the next section.

Remark 1.1.2. In this book, when we say that something is smooth without making precise its smoothness class, we usually mean that it is C^∞-smooth. However, most of the time, being C^1-smooth or C^2-smooth will also be good enough, though we don't want to go into these details. Analytic means either real analytic or holomorphic. Though we will consider only finite-dimensional Poisson structures in this book, let us mention that infinite-dimensional Poisson structures also appear naturally (especially in problems of mathematical physics), see, e.g., [281, 285] and references therein.

Example 1.1.3. One can define a trivial Poisson structure on any manifold by putting $\{f, g\} = 0$ for all functions f and g.

Example 1.1.4. Take $M = \mathbb{R}^2$ with coordinates (x, y) and let $p : \mathbb{R}^2 \longrightarrow \mathbb{R}$ be an arbitrary smooth function. One can define a smooth Poisson structure on $\mathbb{R}^2$ by putting

$$\{f, g\} = \left(\frac{\partial f}{\partial x} \frac{\partial g}{\partial y} - \frac{\partial f}{\partial y} \frac{\partial g}{\partial x} \right) p . \tag{1.4}$$

Exercise 1.1.5. Verify the Jacobi identity and the Leibniz identity for the above bracket. Show that any smooth Poisson structure of $\mathbb{R}^2$ has the above form.

Definition 1.1.6. A *symplectic manifold* (M, ω) is a manifold M equipped with a nondegenerate closed differential 2-form ω, called the *symplectic form*.

The nondegeneracy of a differential 2-form ω means that the corresponding homomorphism $\omega^\flat : TM \to T^*M$ from the tangent space of M to its cotangent space, which associates to each vector X the covector $i_X \omega$, is an isomorphism. Here $i_X \omega = X \lrcorner \omega$ is the contraction of ω by X and is defined by $i_X \omega(Y) = \omega(X, Y)$.

If $f : M \to \mathbb{R}$ is a function on a symplectic manifold (M, ω), then we can define its *Hamiltonian vector field*, denoted by X_f, as follows:

$$i_{X_f} \omega = -\mathrm{d}f . \tag{1.5}$$

We can also define on (M, ω) a natural bracket, called the Poisson bracket of ω, as follows:

$$\{f, g\} = \omega(X_f, X_g) = -\langle \mathrm{d}f, X_g \rangle = -X_g(f) = X_f(g). \tag{1.6}$$

Proposition 1.1.7. *If (M, ω) is a smooth symplectic manifold, then the bracket $\{f, g\} = \omega(X_f, X_g)$ is a smooth Poisson structure on M.*

Proof. The Leibniz identity is obvious. Let us show the Jacobi identity. Recall the following *Cartan's formula* for the differential of a k-form η (see, e.g., [41]):

$$\mathrm{d}\eta(X_1, \dots, X_{k+1}) = \sum_{i=1}^{k+1} (-1)^{i-1} X_i \left(\eta(X_1, \dots, \widehat{X_i}, \dots, X_{k+1}) \right)$$
$$+ \sum_{1 \leq i < j \leq k+1} (-1)^{i+j} \eta \left([X_i, X_j], X_1, \dots, \widehat{X_i}, \dots, \widehat{X_j}, \dots, X_{k+1} \right), \tag{1.7}$$

where $X_1, \ldots, X_{k+1}$ are vector fields, and the hat means that the corresponding entry is omitted. Applying Cartan's formula to ω and X_f, X_g, X_h, we get:

$$
\begin{aligned}
0 = {}& d\omega(X_f, X_g, X_h) \\
= {}& X_f(\omega(X_g, X_h)) + X_g(\omega(X_h, X_f)) + X_h(\omega(X_f, X_g)) \\
& - \omega([X_f, X_g], X_h) - \omega([X_g, X_h], X_f) - \omega([X_h, X_f], X_g) \\
= {}& X_f\{g, h\} + X_g\{h, f\} + X_h\{f, g\} \\
& + [X_f, X_g](h) + [X_g, X_h](f) + [X_h, X_f](g) \\
= {}& \{f, \{g, h\}\} + \{g, \{h, f\}\} + \{h, \{f, g\}\} + X_f(X_g(h)) - X_g(X_f(h)) \\
& + X_g(X_h(f)) - X_h(X_g(f)) + X_h(X_f(g)) - X_f(X_h(g)) \\
= {}& 3(\{f, \{g, h\}\} + \{g, \{h, f\}\} + \{h, \{f, g\}\}). \qquad \square
\end{aligned}
$$

Thus, any symplectic manifold is also a Poisson manifold, though the inverse is not true.

The classical *Darboux theorem* says that in the neighborhood of every point of (M, ω) there is a local system of coordinates $(p_1, q_1, \ldots, p_n, q_n)$, where $2n = \dim M$, called *Darboux coordinates* or *canonical coordinates*, such that

$$
\omega = \sum_{i=1}^{n} dp_i \wedge dq_i \ . \tag{1.8}
$$

A proof of Darboux's theorem will be given in Section 1.4. In such a Darboux coordinate system one has the following expressions for the Poisson bracket and the Hamiltonian vector fields:

$$
\{f, g\} = \sum_{i=1}^{n} \left(\frac{\partial f}{\partial p_i} \frac{\partial g}{\partial q_i} - \frac{\partial f}{\partial q_i} \frac{\partial g}{\partial p_i} \right), \tag{1.9}
$$

$$
X_h = \sum_{i=1}^{n} \frac{\partial h}{\partial p_i} \frac{\partial}{\partial q_i} - \sum_{i=1}^{n} \frac{\partial h}{\partial q_i} \frac{\partial}{\partial p_i}. \tag{1.10}
$$

The *Hamiltonian equation* of h (also called the *Hamiltonian system* of h), i.e., the ordinary differential equation for the integral curves of X_h, has the following form, which can be found in most textbooks on analytical mechanics:

$$
\dot{q}_i = \frac{\partial h}{\partial p_i}, \quad \dot{p}_i = -\frac{\partial h}{\partial q_i}. \tag{1.11}
$$

In fact, to define the Hamiltonian vector field of a function, what one really needs is not a symplectic structure, but a Poisson structure: The Leibniz identity means that, for a given function f on a Poisson manifold M, the map $g \longmapsto \{f, g\}$ is a *derivation*. Thus, there is a unique vector field X_f on M, called the *Hamiltonian vector field* of f, such that for any $g \in C^\infty(M)$ we have

$$
X_f(g) = \{f, g\} \ . \tag{1.12}
$$

Exercise 1.1.8. Show that, in the case of a symplectic manifold, Equation (1.5) and Equation (1.12) give the same vector field.

Example 1.1.9. If N is a manifold, then its cotangent bundle T^*N has a unique natural symplectic structure, hence T^*N is a Poisson manifold with a natural Poisson bracket. The symplectic form on T^*N can be constructed as follows. Denote by $\pi : T^*N \to N$ the projection which assigns to each covector $p \in T_q^*N$ its base point q. Define the so-called *Liouville 1-form* θ on T^*N by

$$\langle \theta, X \rangle = \langle p, \pi_* X \rangle \ \ \forall \ X \in T_p(T^*N).$$

In other words, $\theta(p) = \pi^*(p)$, where on the left-hand side p is considered as a point of T^*N and on the right-hand side it is considered as a cotangent vector to N. Then $\omega = \mathrm{d}\theta$ is a symplectic form on N: ω is obviously closed; to see that it is nondegenerate take a local coordinate system $(p_1, \ldots, p_n, q_1, \ldots, q_n)$ on T^*N, where $(q_1, \ldots, q_n)$ is a local coordinate system on N and $(p_1, \ldots, p_n)$ are the coefficients of covectors $\sum p_i \mathrm{d}q_i(q)$ in this coordinate system. Then $\theta = \sum p_i \mathrm{d}q_i$ and $\omega = \mathrm{d}\theta = \sum \mathrm{d}p_i \wedge \mathrm{d}q_i$, i.e., $(p_1, \ldots, p_n, q_1, \ldots, q_n)$ is a Darboux coordinate system for ω. In classical mechanics, one often deals with Hamiltonian equations on a cotangent bundle T^*N equipped with the natural symplectic structure, where N is the *configuration space*, i.e., the space of all possible configurations or positions; T^*N is called the *phase space*.

A function g is called a *first integral* of a vector field X if g is constant with respect to X: $X(g) = 0$. Finding first integrals is an important step in the study of dynamical systems. Equation (1.12) means that a function g is a first integral of a Hamiltonian vector field X_f if and only if $\{f, g\} = 0$. In particular, every function h is a first integral of its own Hamiltonian vector field: $X_h(h) = \{h, h\} = 0$ due to the anti-symmetry of the Poisson bracket. This fact is known in physics as the principle of *conservation of energy* (here h is the energy function).

The following classical theorem of Poisson [293] allows one sometimes to find new first integrals from old ones:

Theorem 1.1.10 (Poisson). *If g and h are first integrals of a Hamiltonian vector field X_f on a Poisson manifold M, then $\{g, h\}$ also is.*

Proof. Another way to formulate this theorem is

$$\left. \begin{array}{l} \{g, f\} \ = 0 \\ \{h, f\} \ = 0 \end{array} \right\} \Rightarrow \{\{g, h\}, f\} = 0. \tag{1.13}$$

But this is a corollary of the Jacobi identity.   □

Another immediate consequence of the definition of Poisson brackets is the following lemma:

Lemma 1.1.11. *Given a smooth Poisson manifold $(M, \{,\})$, the map $f \mapsto X_f$ is a homomorphism from the Lie algebra $\mathcal{C}^\infty(M)$ of smooth functions under the*

Poisson bracket to the Lie algebra of smooth vector fields under the usual Lie bracket. In other words, we have the following formula:

$$[X_f, X_g] = X_{\{f,g\}}. \tag{1.14}$$

Proof. For any $f, g, h \in \mathcal{C}^\infty(M)$ we have $[X_f, X_g] h = X_f(X_g h) - X_g(X_f h) = \{f, \{g, h\}\} - \{g, \{f, h\}\} = \{\{f, g\}, h\} = X_{\{f,g\}} h$. Since h is arbitrary, it means that $[X_f, X_g] = X_{\{f,g\}}$. $\qquad\square$

1.2 Poisson tensors

In this section, we will express Poisson structures in terms of 2-vector fields which satisfy some special conditions.

Let M be a smooth manifold and q a positive integer. We denote by $\Lambda^q TM$ the space of tangent q-*vectors* of M: it is a vector bundle over M, whose fiber over each point $x \in M$ is the space $\Lambda^q T_x M = \Lambda^q(T_x M)$, which is the exterior (antisymmetric) product of q copies of the tangent space $T_x M$. In particular, $\Lambda^1 TM = TM$. If $(x_1, \ldots, x_n)$ is a local system of coordinates at x, then $\Lambda^q T_x M$ admits a linear basis consisting of the elements $\dfrac{\partial}{\partial x_{i_1}} \wedge \cdots \wedge \dfrac{\partial}{\partial x_{i_q}}(x)$ with $i_1 < i_2 < \cdots < i_q$. A smooth q-*vector field* Π on M is, by definition, a smooth section of $\Lambda^q TV$, i.e., a map Π from V to $\Lambda^q TM$, which associates to each point x of M a q-vector $\Pi(x) \in \Lambda^q T_x M$, in a smooth way. In local coordinates, Π will have a local expression

$$\Pi(x) = \sum_{i_1 < \cdots < i_q} \Pi_{i_1 \ldots i_q} \frac{\partial}{\partial x_{i_1}} \wedge \cdots \wedge \frac{\partial}{\partial x_{i_q}} = \frac{1}{q!} \sum_{i_1 \ldots i_q} \Pi_{i_1 \ldots i_q} \frac{\partial}{\partial x_{i_1}} \wedge \cdots \wedge \frac{\partial}{\partial x_{i_q}}, \tag{1.15}$$

where the components $\Pi_{i_1 \ldots i_q}$, called the *coefficients* of Π, are smooth functions. The coefficients $\Pi_{i_1 \ldots i_q}$ are antisymmetric with respect to the indices, i.e., if we permute two indices then the coefficient is multiplied by -1. For example, $\Pi_{i_1 i_2 \ldots} = -\Pi_{i_2 i_1 \ldots}$. If $\Pi_{i_1 \ldots i_q}$ are C^k-smooth, then we say that Π is C^k-smooth, and so on.

Smooth q-vector fields are dual objects to differential q-forms in a natural way. If Π is a q-vector field and α is a differential q-form, which in some local system of coordinates are written as $\Pi(x) = \sum_{i_1 < \cdots < i_q} \Pi_{i_1 \ldots i_q} \frac{\partial}{\partial x_{i_1}} \wedge \cdots \wedge \frac{\partial}{\partial x_{i_q}}$ and $\alpha = \sum_{i_1 < \cdots < i_q} a_{i_1 \ldots i_q} dx_{i_1} \wedge \cdots \wedge dx_{i_q}$, then their *pairing* $\langle \alpha, \Pi \rangle$ is a function defined by

$$\langle \alpha, \Pi \rangle = \sum_{i_1 < \cdots < i_q} \Pi_{i_1 \ldots i_q} a_{i_1 \ldots i_q} . \tag{1.16}$$

Exercise 1.2.1. Show that the above definition of $\langle \alpha, \Pi \rangle$ does not depend on the choice of local coordinates.

In particular, smooth q-vector fields on a smooth manifold M can be considered as $\mathcal{C}^\infty(M)$-linear operators from the space of smooth differential q-forms on M to $\mathcal{C}^\infty(M)$, and vice versa.

A C^k-smooth q-vector field Π will define an $\mathbb{R}$-multilinear skewsymmetric map from $\mathcal{C}^\infty(M) \times \cdots \times \mathcal{C}^\infty(M)$ (q times) to $\mathcal{C}^\infty(M)$ by the following formula:

$$\Pi(f_1, \ldots, f_q) := \langle \Pi, \mathrm{d}f_1 \wedge \cdots \wedge \mathrm{d}f_q \rangle . \tag{1.17}$$

Conversely, we have:

Lemma 1.2.2. *An $\mathbb{R}$-multilinear map $\Pi : \mathcal{C}^\infty(M) \times \cdots \times \mathcal{C}^\infty(M) \to \mathcal{C}^k(M)$ arises from a C^k-smooth q-vector field by Formula (1.17) if and only if Π is skewsymmetric and satisfies the Leibniz rule (or condition):*

$$\Pi(fg, f_2, \ldots, f_q) = f\Pi(g, f_2, \ldots, f_q) + g\Pi(f, f_2, \ldots, f_q). \tag{1.18}$$

A map Π which satisfies the above conditions is called a *multi-derivation*, and the above lemma says that multi-derivations can be identified with multi-vector fields.

Proof (sketch). The "only if" part is straightforward. For the "if" part, we have to check that the value of $\Pi(f_1, \ldots, f_q)$ at a point x depends only on the value of $\mathrm{d}f_1, \ldots, \mathrm{d}f_q$ at x. Equivalently, we have to check that if $\mathrm{d}f_1(x) = 0$ then $\Pi(f_1, \ldots, f_q)(x) = 0$. If $\mathrm{d}f_1(x) = 0$ then we can write $f_1 = c + \sum_i x_i g_i$ where c is a constant and x_i and g_i are smooth functions which vanish at x. According to the Leibniz rule we have $\Pi(1 \times 1, f_2, \ldots, f_q) = 1 \times \Pi(1, f_2, \ldots, f_q) + 1 \times \Pi(1, f_2, \ldots, f_q) = 2\Pi(1, f_2, \ldots, f_q)$, hence $\Pi(1, f_2, \ldots, f_q) = 0$. Now according to the linearity and the Leibniz rule we have $\Pi(f_1, \ldots, f_q)(x) = c\Pi(1, f_2, \ldots, f_q)(x) + \sum x_i(x)\Pi(g_i, f_2, \ldots, f_q)(x) + \sum g_i(x)\Pi(x_i, f_2, \ldots, f_q)(x) = 0$. $\qquad\square$

In particular, if Π is a Poisson structure, then it is skewsymmetric and satisfies the Leibniz condition, hence it arises from a 2-vector field, which we will also denote by Π:

$$\{f, g\} = \Pi(f, g) = \langle \Pi, \mathrm{d}f \wedge \mathrm{d}g \rangle . \tag{1.19}$$

A 2-vector field Π, such that the bracket $\{f, g\} := \langle \Pi, \mathrm{d}f \wedge \mathrm{d}g \rangle$ is a Poisson bracket (i.e., satisfies the Jacobi identity $\{\{f, g\}, h\} + \{\{g, h\}, f\} + \{\{h, f\}, g\} = 0$ for any smooth functions f, g, h), is called a *Poisson tensor*, or also a *Poisson structure*. The corresponding Poisson bracket is often denoted by $\{,\}_\Pi$. If the Poisson tensor Π is a C^k-smooth 2-vector field, then we say that we have a C^k-smooth Poisson structure, and so on.

In a local system of coordinates $(x_1, \ldots, x_n)$ we have

$$\Pi = \sum_{i<j} \Pi_{ij} \frac{\partial}{\partial x_i} \wedge \frac{\partial}{\partial x_j} = \frac{1}{2} \sum_{i,j} \Pi_{ij} \frac{\partial}{\partial x_i} \wedge \frac{\partial}{\partial x_j} , \tag{1.20}$$

where $\Pi_{ij} = \langle \Pi, \mathrm{d}x_i \wedge \mathrm{d}x_j \rangle = \{x_i, x_j\}$, and

$$\{f, g\} = \langle \sum_{i<j} \{x_i, x_j\} \frac{\partial}{\partial x_i} \wedge \frac{\partial}{\partial x_j}, \sum_{i,j} \frac{\partial f}{\partial x_i} \frac{\partial g}{\partial x_j} \mathrm{d}x_i \wedge \mathrm{d}x_j \rangle = \sum_{i,j} \Pi_{ij} \frac{\partial f}{\partial x_i} \frac{\partial g}{\partial x_j} . \tag{1.21}$$

Example 1.2.3. The Poisson tensor corresponding to the standard symplectic structure $\omega = \sum_{j=1}^{n} dx_j \wedge dy_j$ on $\mathbb{R}^{2n}$ is $\sum_{j=1}^{n} \frac{\partial}{\partial x_j} \wedge \frac{\partial}{\partial y_j}$.

Notation 1.2.4. In this book, if functions $f_1, \ldots, f_p$ depend on variables $x_1, \ldots, x_p$, and maybe other variables, then we will denote by

$$\frac{\partial(f_1, \ldots, f_p)}{\partial(x_1, \ldots, x_p)} := \det\left(\frac{\partial f_i}{\partial x_j}\right)_{i,j=1}^{p} \qquad (1.22)$$

the *Jacobian determinant* of $(f_1, \ldots, f_p)$ with respect to $(x_1, \ldots, x_p)$. For example,

$$\frac{\partial(f, g)}{\partial(x_i, x_j)} := \frac{\partial f}{\partial x_i}\frac{\partial g}{\partial x_j} - \frac{\partial f}{\partial x_j}\frac{\partial g}{\partial x_i}. \qquad (1.23)$$

With the above notation, we have the following local expression for Poisson brackets:

$$\{f, g\} = \sum_{i,j}\{x_i, x_j\}\frac{\partial f}{\partial x_i}\frac{\partial g}{\partial x_j} = \sum_{i<j}\{x_i, x_j\}\frac{\partial(f, g)}{\partial(x_i, x_j)}. \qquad (1.24)$$

Due to the Jacobi condition, not every 2-vector field will be a Poisson tensor.

Exercise 1.2.5. Show that the 2-vector field $\frac{\partial}{\partial x} \wedge (\frac{\partial}{\partial y} + x\frac{\partial}{\partial z})$ in $\mathbb{R}^3$ is *not* a Poisson tensor.

Exercise 1.2.6. Show that if $X_1, \ldots, X_m$ are pairwise commuting vector fields and a_{ij} are constants, then $\sum_{ij} a_{ij} X_i \wedge X_j$ is a Poisson tensor.

To study the Jacobi identity, we will use the following lemma:

Lemma 1.2.7. *For any C^1-smooth 2-vector field Π, one can associate to it a 3-vector field Λ defined by*

$$\Lambda(f, g, h) = \{\{f, g\}, h\} + \{\{g, h\}, f\} + \{\{h, f\}, g\} \qquad (1.25)$$

where $\{k, l\}$ denotes $\langle \Pi, dk \wedge dl \rangle$ (i.e., the bracket of Π).

Proof. It is clear that the right-hand side of Formula (1.25) is $\mathbb{R}$-multilinear and antisymmetric. To show that it corresponds to a 3-vector field, one has to verify that it satisfies the Leibniz rule with respect to f, i.e.,

$$\{\{f_1 f_2, g\}, h\} + \{\{g, h\}, f_1 f_2\} + \{\{h, f_1 f_2\}, g\}$$
$$= f_1(\{\{f_2, g\}, h\} + \{\{g, h\}, f_2\} + \{\{h, f_2\}, g\})$$
$$+ f_2(\{\{f_1, g\}, h\} + \{\{g, h\}, f_1\} + \{\{h, f_1\}, g\}).$$

This is a simple direct verification, based on the Leibniz rule $\{ab, c\} = a\{b, c\} + b\{a, c\}$ for the bracket of the 2-vector field Π. It will be left to the reader as an exercise. $\qquad\square$

Direct calculations in local coordinates show that

$$\Lambda(f,g,h) = \sum_{ijk}\left(\oint_{ijk}\sum_s \frac{\partial \Pi_{ij}}{\partial x_s}\Pi_{sk}\right)\frac{\partial f}{\partial x_i}\frac{\partial g}{\partial x_j}\frac{\partial h}{\partial x_k}\ , \tag{1.26}$$

where $\oint_{ijk} a_{ijk}$ means the cyclic sum $a_{ijk} + a_{jki} + a_{kij}$. In other words,

$$\Lambda = \sum_{i<j<k}\left(\oint_{ijk}\sum_s \frac{\partial \Pi_{ij}}{\partial x_s}\Pi_{sk}\right)\frac{\partial}{\partial x_i}\wedge\frac{\partial}{\partial x_j}\wedge\frac{\partial}{\partial x_k}\ . \tag{1.27}$$

Clearly, the Jacobi identity for Π is equivalent to the condition that $\Lambda = 0$. Thus we have:

Proposition 1.2.8. *A 2-vector field $\Pi = \sum_{i<j}\Pi_{ij}\frac{\partial}{\partial x_i}\wedge\frac{\partial}{\partial x_j}$ expressed in terms of a given system of coordinates $(x_1,\ldots,x_n)$ is a Poisson tensor if and only if it satisfies the following system of equations:*

$$\oint_{ijk}\sum_s \frac{\partial \Pi_{ij}}{\partial x_s}\Pi_{sk} = 0 \quad (\forall\ i,j,k)\ . \tag{1.28}$$

$\square$

An obvious consequence of the above proposition is that the condition for a 2-vector field to be a Poisson structure is a local condition. In particular, the restriction of a Poisson structure to an open subset of the manifold is again a Poisson structure.

Example 1.2.9. Constant Poisson structures on $\mathbb{R}^n$: Choose arbitrary constants Π_{ij}. Then Equation (1.28) is obviously satisfied. The canonical Poisson structure on $\mathbb{R}^{2n}$, associated to the canonical symplectic form $\omega = \sum dq_i \wedge dp_i$, is of this type.

Example 1.2.10. Any 2-vector field on a two-dimensional manifold is a Poisson tensor. Indeed, the 3-vector field Λ in Lemma 1.2.7 is identically zero because there are no nontrivial 3-vectors on a two-dimensional manifold. Thus the Jacobi identity is nontrivial only starting from dimension 3.

Example 1.2.11. Let V be a finite-dimensional vector space over $\mathbb{R}$ (or $\mathbb{C}$). A *linear Poisson structure* on V is a Poisson structure on V for which the Poisson bracket of two linear functions is again a linear function. Equivalently, in linear coordinates, the components of the corresponding Poisson tensor are linear functions. In this case, by restriction to linear functions, the operation $(f,g)\mapsto\{f,g\}$ gives rise to an operation $[\,,\,]:V^*\times V^*\longrightarrow V^*$, which is a Lie algebra structure on V^*, where V^* is the dual linear space of V.

Conversely, any Lie algebra structure on V^* determines a linear Poisson structure on V. Indeed, consider a finite-dimensional Lie algebra $(\mathfrak{g},[\,,\,])$. For each linear function $f:\mathfrak{g}^*\longrightarrow\mathbb{R}$ we denote by $\tilde{f}$ the element of $\mathfrak{g}$ corresponding to it. If f and g are two linear functions on $\mathfrak{g}^*$, then we put $\{f, g\}(\alpha) = \langle \alpha, [\tilde{f}, \tilde{g}] \rangle$ for every α in $\mathfrak{g}^*$. If we choose a linear basis $e_1, \ldots, e_n$ of $\mathfrak{g}$, with $[e_i, e_j] = \sum c_{ij}^k e_k$, then we have $\{x_i, x_j\} = \sum c_{ij}^k x_k$ where x_l is the function such that $\tilde{x}_l = e_l$. By taking $(x_1, \ldots, x_n)$ as a linear system of coordinates on $\mathfrak{g}^*$, it follows from the Jacobi identity for $[\ ,\ ]$ that the functions $\Pi_{ij} = \{x_i, x_j\}$ verify Equation (1.28). Thus we get a Poisson structure on $\mathfrak{g}^*$. This Poisson structure can be defined intrinsically by the following formula:

$$\{f, g\}(\alpha) = \langle \alpha, [\mathrm{d}f(\alpha), \mathrm{d}g(\alpha)] \rangle \ , \tag{1.29}$$

where $\mathrm{d}f(\alpha)$ and $\mathrm{d}g(\alpha)$ are considered as elements of $\mathfrak{g}$ via the identification $(\mathfrak{g}^*)^* = \mathfrak{g}$. Thus, there is a natural bijection between finite-dimensional linear Poisson structures and finite-dimensional Lie algebras. One can even try to study Lie algebras by viewing them as linear Poisson structures (see, e.g., [61]).

Remark 1.2.12. Multi-vector fields are also known as antisymmetric *contravariant tensors*, because their coefficients change contravariantly under a change of local coordinates. In particular, the local expression of a Poisson bracket will change contravariantly under a change of local coordinates: Let $x = (x_1, \ldots, x_n)$ and $y = (y_1, \ldots, y_n)$ be two local coordinate systems on the same open subset of a Poisson manifold $(M, \{, \})$. Viewing y_i as functions of $(x_1, \ldots, x_n)$, we have

$$\{y_i, y_j\} = \sum_{r < s} \frac{\partial(y_i, y_j)}{\partial(x_r, x_s)} \{x_r, x_s\} \ . \tag{1.30}$$

Denote $\Pi_{rs}(x) = \{x_r, x_s\}(x), \Pi'_{ij}(y) = \{y_i, y_j\}(y)$. Then the above equation can be rewritten as

$$\Pi'_{ij}(y(x)) = \sum_{r < s} \frac{\partial(y_i, y_j)}{\partial(x_r, x_s)}(x)\Pi_{rs}(x). \tag{1.31}$$

Exercise 1.2.13. Consider the Poisson structure on $\mathbb{R}^2$ defined by $\{x, y\} = e^x$. Show that in the new coordinates $(u, v) = (x, ye^{-x})$ the Poisson tensor will have the standard form $\frac{\partial}{\partial u} \wedge \frac{\partial}{\partial v}$.

Exercise 1.2.14. Let $\Pi = \sum \Pi_{ij} \partial/\partial x_i \wedge \partial/\partial x_j$ be a constant Poisson structure on $\mathbb{R}^n$, i.e., the coefficients Π_{ij} are constants. Show that there is a number $p \geq 0$ and a linear coordinate system $(y_1, \ldots, y_n)$ in which the Poisson bracket has the form

$$\{f, g\} = \frac{\partial(f, g)}{\partial(y_1, y_2)} + \frac{\partial(f, g)}{\partial(y_3, y_4)} + \cdots + \frac{\partial(f, g)}{\partial(y_{2p-1}, y_{2p})} \ . \tag{1.32}$$

1.3 Poisson morphisms

Definition 1.3.1. If $(M_1, \{, \}_1)$ and $(M_2, \{, \}_2)$ are two smooth Poisson manifolds, then a smooth map ϕ from M_1 to M_2 is called a smooth *Poisson morphism* or *Poisson map* if the associated pull-back map $\phi^* : \mathcal{C}^\infty(M_2) \to \mathcal{C}^\infty(M_1)$ is a Lie algebra homomorphism with respect to the corresponding Poisson brackets.

In other words, $\phi : (M_1, \{,\}_1) \to (M_2, \{,\}_2)$ is a Poisson morphism if

$$\{\phi^* f, \phi^* g\}_1 = \phi^* \{f, g\}_2 \quad \forall \, f, g \in C^\infty(M_2) . \tag{1.33}$$

Of course, Poisson manifolds together with Poisson morphisms form a category: the composition of two Poisson morphisms is again a Poisson morphism, and so on. Notice that a Poisson morphism which is a diffeomorphism will automatically be a *Poisson isomorphism*: the inverse map is also a Poisson map.

Similarly, a map $\phi : (M_1, \omega_1) \to (M_2, \omega_2)$ is called a *symplectic morphism* if $\phi^* \omega_2 = \omega_1$. Clearly, a symplectic isomorphism is also a Poisson isomorphism. However, a symplectic morphism is *not* a Poisson morphism in general. For example, if M_1 is a point with a trivial symplectic form, and M_2 is a symplectic manifold of positive dimension, then any map $\phi : M_1 \to M_2$ is a symplectic morphism but not a Poisson morphism.

Example 1.3.2. If $\phi : \mathfrak{h} \to \mathfrak{g}$ is a Lie algebra homomorphism, then the linear dual map $\phi^* : \mathfrak{g}^* \to \mathfrak{h}^*$ is a Poisson map, where $\mathfrak{g}^*$ and $\mathfrak{h}^*$ are equipped with their respective linear Poisson structures. The proof of this fact will be left to the reader as an exercise. In particular, if $\mathfrak{h}$ is a Lie subalgebra of $\mathfrak{g}$, then the canonical projection $\mathfrak{g}^* \to \mathfrak{h}^*$ is Poisson.

Example 1.3.3. If ϕ is a diffeomorphism of a manifold N, then it can be lifted naturally to a diffeomorphism $\phi_* : T^*N \to T^*N$ covering ϕ. By definition, ϕ_* preserves the Liouville 1-form θ (see Example 1.1.9), hence it preserves the symplectic form $\mathrm{d}\theta$. Thus, ϕ_* is a Poisson isomorphism.

Example 1.3.4. Direct product of Poisson manifolds. Let $(M_1, \{,\}_1)$ and $(M_2, \{,\}_2)$ be two Poisson manifolds. Then their direct product $M_1 \times M_2$ can be equipped with the following natural bracket:

$$\{f(x_1, x_2), g(x_1, x_2)\} = \{f_{x_2}, g_{x_2}\}_1(x_1) + \{f_{x_1}, g_{x_1}\}_2(x_2) \tag{1.34}$$

where we use the notation $h_{x_1}(x_2) = h_{x_2}(x_1) = h(x_1, x_2)$ for any function h on $M_1 \times M_2$, $x_1 \in M_1$ and $x_2 \in M_2$. Using Equation (1.28), one can verify easily that this bracket is indeed a Poisson bracket on $M_1 \times M_2$. It is called the *product Poisson structure*. With respect to this product Poisson structure, the projection maps $M_1 \times M_2 \to M_1$ and $M_1 \times M_2 \to M_2$ are Poisson maps.

Exercise 1.3.5. Let $M_1 = M_2 = \mathbb{R}^n$ with trivial Poisson structure. Find a nontrivial Poisson structure on $M_1 \times M_2 = \mathbb{R}^{2n}$ for which the two projections $M_1 \times M_2 \to M_1$ and $M_1 \times M_2 \to M_2$ are Poisson maps.

Exercise 1.3.6. Show that any Poisson map from a Poisson manifold to a symplectic manifold is a submersion.

A vector field X on a Poisson manifold (M, Π), is called a *Poisson vector field* if it is an *infinitesimal automorphism* of the Poisson structure, i.e., the Lie derivative of Π with respect to X vanishes:

$$\mathcal{L}_X \Pi = 0 . \tag{1.35}$$

Equivalently, the local flow (φ_X^t) of X, i.e., the one-dimensional pseudo-group of local diffeomorphisms of M generated by X, preserves the Poisson structure: $\forall t \in \mathbb{R}$, (φ_X^t) is a Poisson morphism wherever it is well defined.

By the Leibniz rule we have $\mathcal{L}_X(\{f, g\}) = \mathcal{L}_X(\langle \Pi, df \wedge dg \rangle) = \langle \mathcal{L}_X \Pi, df \wedge dg \rangle + \langle \Pi, d\mathcal{L}_X f \wedge dg \rangle + \langle \Pi, df \wedge d\mathcal{L}_X g \rangle = \langle \mathcal{L}_X \Pi, df \wedge dg \rangle + \{X(f), g\} + \{f, X(g)\}$. So another equivalent condition for X to be a Poisson vector field is the following:

$$\{Xf, g\} + \{f, Xg\} = X\{f, g\} . \tag{1.36}$$

When $X = X_h$ is a Hamiltonian vector field, then Equation (1.36) is nothing but the Jacobi identity. Thus any Hamiltonian vector field is a Poisson vector field. The inverse is not true in general, even locally. For example, if the Poisson structure is trivial, then any vector field is a Poisson vector field, while the only Hamiltonian vector field is the trivial one.

Exercise 1.3.7. Show that on $\mathbb{R}^{2n}$ with the standard Poisson structure $\sum \frac{\partial}{\partial x_i} \wedge \frac{\partial}{\partial y_i}$ any Poisson vector field is also Hamiltonian.

Example 1.3.8. Infinitesimal version of Example 1.3.3. If X is a vector field on a manifold N, then X admits a unique natural lifting to a vector field $\hat{X}$ on T^*N which preserves the Liouville 1-form. In a local coordinate system $(p_1, \ldots, p_n, q_1, \ldots, q_n)$ on T^*N, where $(q_1, \ldots, q_n)$ is a local coordinate system on N and the Liouville 1-form is $\theta = \sum_i p_i dq_i$ (see Example 1.1.9), we have the following expression for $\hat{X}$:

$$\text{If } X = \sum_i \alpha_i(q) \frac{\partial}{\partial q_i} \text{ then } \hat{X} = \sum_i \alpha_i(q) \frac{\partial}{\partial q_i} - \sum_{i,j} \frac{\partial \alpha_i(q)}{\partial q_j} p_i \frac{\partial}{\partial p_j}.$$

The vector field $\hat{X}$ is in fact the Hamiltonian vector field of the function

$$\mathcal{X}(p_1, \ldots, p_n, q_1, \ldots, q_n) = \sum_i \alpha_i(q) p_i$$

on T^*N. This function $\mathcal{X}$ is nothing else than X itself, considered as a fiber-wise linear function on T^*N.

Example 1.3.9. Let G be a connected Lie group, and denote by $\mathfrak{g}$ the Lie algebra of G. By definition, $\mathfrak{g}$ is isomorphic to the Lie algebra of left-invariant tangent vector fields of G (i.e., vector fields which are invariant under left translations $L_g : h \mapsto gh$ on G). Denote by e the neutral element of G. For each $X_e \in T_e G$, there is a unique left-invariant vector field X on G whose value at e is X_e (X obtained from X_e by left translations), so we may identify $T_e G$ with $\mathfrak{g}$ via this association $X_e \mapsto X$. We will write $T_e G = \mathfrak{g}$, and $T_e^* G = \mathfrak{g}^*$ by duality. Consider the left translation map

$$L : T^*G \to \mathfrak{g}^* = T_e^* G, \quad L(p) = (L_g)^* p = L_{g^{-1}} p \;\; \forall p \in T_g^* G, \tag{1.37}$$

where $L_{g^{-1}} p$ means the push-forward $(L_{g^{-1}})_* p$ of p by $L_{g^{-1}}$ (we will often omit the subscript asterisk when writing push-forwards to simplify the notation).

Theorem 1.3.10. *The above left translation map* $L : T^*G \to \mathfrak{g}^*$ *is a Poisson map, where* T^*G *is equipped with the standard symplectic structure, and* $\mathfrak{g}^*$ *is equipped with the standard linear Poisson structure* (*induced from the Lie algebra structure of* $\mathfrak{g}$).

Proof (sketch). It is enough to verify that, if x, y are two elements of $\mathfrak{g}$, considered as linear functions on $\mathfrak{g}^*$, then we have

$$\{L^*x, L^*y\} = L^*([x, y]).$$

Notice that L^*x is nothing else than x itself, considered as a left-invariant vector field on G and then as a left-invariant fiber-wise linear function on T^*G. By the formulas given in Example 1.3.8, the Hamiltonian vector field X_{L^*x} of L^*x is the natural lifting to T^*G of x, considered as a left-invariant vector field on G. Since the process of lifting of vector fields from N to T^*N preserves the Lie bracket for any manifold N, we have

$$[X_{L^*x}, X_{L^*y}] = X_{L^*[x,y]}.$$

It follows from the above equation and Lemma 1.1.11 that $\{L^*x, L^*y\}$ and $L^*([x, y])$ have the same Hamiltonian vector field on T^*G. Hence these two functions differ by a function which vanishes on the zero section of T^*G and whose Hamiltonian vector field is trivial on T^*G. The only such function is 0, so $\{L^*x, L^*y\} = L^*([x, y])$. $\square$

Exercise 1.3.11. Show that the right translation map $R : T^*G \to \mathfrak{g}^* = T_e^*G$, defined by $L(p) = (R_g)_*p \; \forall \, p \in T_g^*G$, is an anti-Poisson map. A map $\phi : (M, \Pi) \to (N, \Lambda)$ is called an *anti-Poisson map* if $\phi : (M, \Pi) \to (N, -\Lambda)$ is a Poisson map.

Given a subspace $V \in T_xM$ of a tangent space T_xM of a symplectic manifold (M, ω), we will denote by $V^\perp$ the *symplectic orthogonal* to V: $V^\perp = \{X \in T_xM \mid \omega(X, Y) = 0 \; \forall \, Y \in V\}$. Clearly, $V = (V^\perp)^\perp$. V is called *Lagrangian* (resp. *isotropic, coisotropic, symplectic*) if $V = V^\perp$ (resp. $V \subset V^\perp$, $V \supset V^\perp$, $V \cap V^\perp = 0$). A submanifold of a symplectic manifold is called *Lagrangian* (resp. *isotropic, coisotropic,* resp. *symplectic*) if its tangent spaces are so. Lagrangian submanifolds play a central role in symplectic geometry, see, e.g., [345, 243]. In particular, we have the following characterization of symplectic isomorphisms in terms of Lagrangian submanifolds:

Proposition 1.3.12. *A diffeomorphism* $\phi : (M, \omega_1) \to (M_2, \omega_2)$ *is a symplectic isomorphism if and only if its graph* $\Delta = \{(x, \phi(x))\} \subset M_1 \times \overline{M_2}$ *is a Lagrangian manifold of* $M_1 \times \overline{M_2}$, *where* $\overline{M_2}$ *means* M_2 *together with the opposite symplectic form* $-\omega_2$.

The proof is almost obvious and is left as an exercise. $\square$

A subspace $V \subset T_xM$ of a Poisson manifold (M, Π) is called *coisotropic* if for any $\alpha, \beta \in T_x^*M$ such that $\langle \alpha, X \rangle = \langle \beta, X \rangle = 0 \; \forall \, X \in V$ we have $\langle \Pi, \alpha \wedge \beta \rangle = 0$.

In other words, $V^\circ \subset (V^\circ)^\perp$, where $V^\circ = \{\alpha \in T_x^*M \mid \langle \alpha, X \rangle = 0 \ \forall \ X \in V\}$ is the annulator of V and $(V^\circ)^\perp = \{\beta \in T_x^*M \mid \langle \Pi, \alpha \wedge \beta \rangle = 0 \ \forall \ \alpha \in V^\circ\}$ is the "Poisson orthogonal" of V°. A submanifold N of a Poisson manifold is called *coisotropic* if its tangent spaces are coisotropic.

Proposition 1.3.13. *A map $\phi : (M_1, \Pi_1) \to (M_2, \Pi_2)$ between two Poisson manifolds is a Poisson map if and only if its graph $\Gamma(\phi) := \{(x, y) \in M_1 \times M_2; \ y = \phi(x)\}$ is a coisotropic submanifold of $(M_1, \Pi_1) \times (M_2, -\Pi_2)$.*

Again, the proof will be left as an exercise.   □

1.4 Local canonical coordinates

In this section, we will prove the *splitting theorem* of Alan Weinstein [346], which says that locally a Poisson manifold is a direct product of a symplectic manifold with another Poisson manifold whose Poisson tensor vanishes at a point. This splitting theorem, together with the Darboux theorem which will be proved at the same time, will give us local canonical coordinates for Poisson manifolds.

Given a Poisson structure Π (or more generally, an arbitrary 2-vector field) on a manifold M, we can associate to it a natural homomorphism

$$\sharp = \sharp_\Pi : T^*M \longrightarrow TM, \tag{1.38}$$

which maps each covector $\alpha \in T_x^*M$ over a point x to a unique vector $\sharp(\alpha) \in T_xM$ such that

$$\langle \alpha \wedge \beta, \Pi \rangle = \langle \beta, \sharp(\alpha) \rangle \tag{1.39}$$

for any covector $\beta \in T_x^*M$. We will call $\sharp = \sharp_\Pi$ the *anchor map* of Π.

The same notations $\sharp$ (or $\sharp_\Pi$) will be used to denote the operator which associates to each differential 1-form α the vector field $\sharp(\alpha)$ defined by $(\sharp(\alpha))(x) = \sharp(\alpha(x))$. For example, if f is a function, then $\sharp(df) = X_f$ is the Hamiltonian vector field of f.

The restriction of $\sharp_\Pi$ to a cotangent space T_x^*M will be denoted by $\sharp_x$ or $\sharp_{\Pi(x)}$. In a local system of coordinates $(x_1, \ldots, x_n)$ we have

$$\sharp\left(\sum_{i=1}^n a_i dx_i\right) = \sum_{ij} \{x_i, x_j\} a_i \frac{\partial}{\partial x_j} = \sum_{ij} \Pi_{ij} a_i \frac{\partial}{\partial x_j}.$$

Thus $\sharp_x$ is a linear operator, given by the matrix $[\Pi_{ij}(x)]$ in the linear bases $(dx_1, \ldots, dx_n)$ and $\left(\frac{\partial}{\partial x_1}, \ldots, \frac{\partial}{\partial x_n}\right)$.

Definition 1.4.1. Let (M, Π) be a Poisson manifold and x a point of M. Then the image $\mathcal{C}_x := \mathrm{Im}\sharp_x$ of $\sharp_x$ is called the *characteristic space* at x of the Poisson structure Π. The dimension $\dim \mathcal{C}_x$ of $\mathcal{C}_x$ is called the *rank* of Π at x, and $\max_{x \in M} \dim \mathcal{C}_x$ is called the *rank* of Π. When rank $\Pi_x = \dim M$ we say that Π

is *nondegenerate* at x. If rank Π_x is a constant on M, i.e., does not depend on x, then Π is called a *regular Poisson structure*.

Example 1.4.2. The constant Poisson structure $\sum_{i=1}^{s} \frac{\partial}{\partial x_i} \wedge \frac{\partial}{\partial x_{i+s}}$ on $\mathbb{R}^m$ ($m \geq 2s$) is a regular Poisson structure of rank $2s$.

Exercise 1.4.3. Show that rank Π_x is always an even number, and that Π is nondegenerate everywhere if and only if it is the associated Poisson structure of a symplectic structure.

The characteristic space $\mathcal{C}_x$ admits a unique natural antisymmetric nondegenerate bilinear scalar product, called the *induced symplectic form*: if X and Y are two vectors of $\mathcal{C}_x$, then we put

$$(X, Y) := \langle \beta, X \rangle = \langle \Pi, \alpha \wedge \beta \rangle = -\langle \Pi, \beta \wedge \alpha \rangle = -\langle \alpha, Y \rangle = -(Y, X) \qquad (1.40)$$

where $\alpha, \beta \in T_x^* M$ are two covectors such that $X = \sharp \alpha$ and $Y = \sharp \beta$.

Exercise 1.4.4. Verify that the above scalar product is anti-symmetric nondegenerate and is well defined (i.e., does not depend on the choice of α and β). When Π is nondegenerate then the above formula defines the corresponding symplectic structure on M.

Theorem 1.4.5 (Splitting theorem [346]). *Let x be a point of rank $2s$ of a Poisson m-dimensional manifold (M, Π): $\dim \mathcal{C}_x = 2s$ where $\mathcal{C}_x$ is the characteristic space at x. Let N be an arbitrary $(m - 2s)$-dimensional submanifold of M which contains x and is transversal to $\mathcal{C}_x$ at x. Then there is a local system of coordinates $(p_1, \ldots, p_s, q_1, \ldots, q_s, z_1, \ldots, z_{m-2s})$ in a neighborhood of x, which satisfy the following conditions:*

  a) *$p_i(N_x) = q_i(N_x) = 0$ where N_x is a small neighborhood of x in N.*
  b) *$\{q_i, q_j\} = \{p_i, p_j\} = 0 \,\forall\, i, j;\ \{p_i, q_j\} = 0$ if $i \neq j$ and $\{p_i, q_i\} = 1 \,\forall\, i$.*
  c) *$\{z_i, p_j\} = \{z_i, q_j\} = 0 \,\forall\, i, j$.*
  d) *$\{z_i, z_j\}(x) = 0 \,\forall\, i, j$.*

A local coordinate system which satisfies the conditions of the above theorem is called a system of local *canonical coordinates*. In such canonical coordinates we have

$$\{f, g\} = \sum_{i,j} \{z_i, z_j\} \frac{\partial f}{\partial z_i} \frac{\partial g}{\partial z_j} + \sum_{i=1}^{s} \frac{\partial(f, g)}{\partial(p_i, q_i)} = \{f, g\}_N + \{f, g\}_S , \qquad (1.41)$$

where

$$\{f, g\}_S = \sum_{i=1}^{s} \frac{\partial(f, g)}{\partial(p_i, q_i)} \qquad (1.42)$$

defines the nondegenerate Poisson structure $\sum \frac{\partial}{\partial p_i} \wedge \frac{\partial}{\partial q_i}$ on the local submanifold $S = \{z_1 = \cdots = z_{m-2s} = 0\}$, and

$$\{f, g\}_N = \sum_{u,v} \{z_i, z_j\} \frac{\partial f}{\partial z_i} \frac{\partial g}{\partial z_j} \qquad (1.43)$$

defines a Poisson structure on a neighborhood of x in N. Notice that, since $\{z_i, p_j\} = \{z_i, q_j\} = 0 \ \forall \ i, j$, the functions $\{z_i, z_j\}$ do not depend on the variables $(p_1, \ldots, p_s, q_1, \ldots, q_s)$. The equality $\{z_i, z_j\}(x) = 0 \ \forall \ i, j$ means that the Poisson tensor of $\{,\}_N$ vanishes at x.

Formula (1.41) means that the Poisson manifold (M, Π) is locally isomorphic (in a neighborhood of x) to the direct product of a symplectic manifold $(S, \sum_1^s dp_i \wedge dq_i)$ with a Poisson manifold $(N_x, \{,\}_N)$ whose Poisson tensor vanishes at x. That's why Theorem 1.4.5 is called the splitting theorem for Poisson manifolds: locally, we can split a Poisson structure in two parts – a regular part and a singular part which vanishes at a point.

Proof of Theorem 1.4.5. If $\Pi(x) = 0$ then $s = 0$ and there is nothing to prove. Suppose that $\Pi(x) \neq 0$. Let p_1 be a local function (defined in a small neighborhood of x in M) which vanishes on N and such that $dp_1(x) \neq 0$. Since C_x is transversal to N, there is a vector $X_g(x) \in C_x$ such that $\langle X_g(x), dp_1(x) \rangle \neq 0$, or equivalently, $X_{p_1}(g)(x) \neq 0$, where X_{p_1} denotes the Hamiltonian vector field of p_1 as usual. Therefore $X_{p_1}(x) \neq 0$. Since $C_x \ni \sharp(dp_1)(x) = X_{p_1}(x) \neq 0$ and is not tangent to N, there is a local function q_1 such that $q_1(N) = 0$ and $X_{p_1}(q_1) = 1$ in a neighborhood of x, or

$$X_{p_1} q_1 = \{p_1, q_1\} = 1 \ . \tag{1.44}$$

Moreover, X_{p_1} and X_{q_1} are linearly independent ($X_{q_1} = \lambda X_{p_1}$ would imply that $\{p_1, q_1\} = -\lambda X_{p_1}(p_1) = 0$), and we have

$$[X_{p_1}, X_{q_1}] = X_{\{p_1, q_1\}} = 0 \ . \tag{1.45}$$

Thus X_{p_1} and X_{q_1} are two linearly independent vector fields which commute. Hence they generate a locally free infinitesimal $\mathbb{R}^2$-action in a neighborhood of x, which gives rise to a local regular two-dimensional foliation. As a consequence, we can find a local system of coordinates $(y_1, \ldots, y_m)$ such that

$$X_{q_1} = \frac{\partial}{\partial y_1}, \qquad X_{p_1} = \frac{\partial}{\partial y_2} \ . \tag{1.46}$$

With these coordinates we have $\{q_1, y_i\} = X_{q_1}(y_i) = 0$ and $\{p_1, y_i\} = X_{p_1}(y_i) = 0$, for $i = 3, \ldots, m$. Poisson's Theorem 1.1.10 then implies that $\{p_1, \{y_i, y_j\}\} = \{q_1, \{y_i, y_j\}\} = 0$ for $i, j \geq 3$, whence

$$\begin{aligned} \{y_i, y_j\} &= \varphi_{ij}(y_3, \ldots, y_n) \quad \forall \ i, j \geq 3 \ , \\ \{p_1, q_1\} &= 1 \ , \\ \{p_1, y_j\} &= \{q_1, y_j\} = 0 \quad \forall \ j \geq 3 \ . \end{aligned} \tag{1.47}$$

We can take $(p_1, q_1, y_3, \ldots, y_n)$ as a new local system of coordinates. In fact, the Jacobian matrix of the map $\varphi : (y_1, y_2, y_3, \ldots, y_m) \mapsto (p_1, q_1, y_3, \ldots, y_m)$ is of the form

$$\begin{pmatrix} 0 & 1 & \\ -1 & 0 & * \\ & 0 & \mathrm{Id} \end{pmatrix} \tag{1.48}$$

(because $\frac{\partial q_1}{\partial y_1} = X_{q_1} q_1 = 0, \frac{\partial q_1}{\partial y_2} = X_{p_1} q_1 = \{q_1, p_1\} = 1, \ldots$), which has a non-zero determinant (equal to 1). In the coordinates $(q_1, p_1, y_3, \ldots, y_m)$, we have

$$\Pi = \frac{\partial}{\partial p_1} \wedge \frac{\partial}{\partial q_1} + \frac{1}{2} \sum_{i,j \geq 3} \Pi'_{ij}(y_3, \ldots, y_m) \frac{\partial}{\partial y_i} \wedge \frac{\partial}{\partial y_j} . \qquad (1.49)$$

The above formula implies that our Poisson structure is locally the product of a standard symplectic structure on a plane $\{(p_1, q_1)\}$ with a Poisson structure on a $(m-2)$-dimensional manifold $\{(y_3, \ldots, y_m)\}$. In this product, N is also the direct product of a point (= the origin) of the plane $\{(p_1, q_1)\}$ with a local submanifold in the Poisson manifold $\{(y_3, \ldots, y_m)\}$. The splitting theorem now follows by induction on the rank of Π at x.   □

Remark 1.4.6. In the above theorem, when $m = 2s$, we recover Darboux's theorem which gives local canonical coordinates for symplectic manifolds. If (M, Π) is a regular Poisson structure, then the Poisson structure of N_x in the above theorem must be trivial, and we get the following generalization of Darboux's theorem: any regular Poisson structure is locally isomorphic to a standard constant Poisson structure.

Exercise 1.4.7. Prove the following generalization of Theorem 1.4.5. Let N be a submanifold of a Poisson manifold (M, Π), and x be a point of N such that $T_x N + \mathcal{C}_x = T_x M$ and $T_x N \cap \mathcal{C}_x$ is a symplectic subspace of $\mathcal{C}_x$, i.e., the restriction of the symplectic form on the characteristic space $\mathcal{C}_x$ to $T_z N \cap \mathcal{C}_x$ is nondegenerate. (Such a submanifold N is sometimes called *cosymplectic*.) Then there is a coordinate system in a neighborhood of x which satisfies the conditions a), b), c) of Theorem 1.4.5, where $2s = \dim M - \dim N = \dim \mathcal{C}_x - \dim(T_x N \cap \mathcal{C}_x)$.

1.5   Singular symplectic foliations

A smooth *singular foliation* in the sense of Stefan–Sussmann [320, 327] on a smooth manifold M is by definition a partition $\mathcal{F} = \{\mathcal{F}_\alpha\}$ of M into a disjoint union of smooth immersed connected submanifolds $\mathcal{F}_\alpha$, called *leaves*, which satisfies the following *local foliation property* at each point $x \in M$: Denote the leaf that contains x by $\mathcal{F}_x$, the dimension of $\mathcal{F}_x$ by d and the dimension of M by m. Then there is a smooth local chart of M with coordinates $y_1, \ldots, y_m$ in a neighborhood U of x, $U = \{-\varepsilon < y_1 < \varepsilon, \ldots, -\varepsilon < y_m < \varepsilon\}$, such that the d-dimensional disk $\{y_{d+1} = \cdots = y_m = 0\}$ coincides with the path-connected component of the intersection of $\mathcal{F}_x$ with U which contains x, and each d-dimensional disk $\{y_{d+1} = c_{d+1}, \ldots, y_m = c_m\}$, where $c_{d+1}, \ldots, c_m$ are constants, is wholly contained in some leaf $\mathcal{F}_\alpha$ of $\mathcal{F}$. If all the leaves $\mathcal{F}_\alpha$ of a singular foliation $\mathcal{F}$ have the same dimension, then one says that $\mathcal{F}$ is a *regular foliation*.

A *singular distribution* on a manifold M is the assignment to each point x of M a vector subspace D_x of the tangent space $T_x M$. The dimension of D_x

may depend on x. For example, if $\mathcal{F}$ is a singular foliation, then it has a natural associated *tangent distribution* $D^{\mathcal{F}}$: at each point $x \in V$, $D_x^{\mathcal{F}}$ is the tangent space to the leaf of $\mathcal{F}$ which contains x.

A singular distribution D on a smooth manifold is called *smooth* if for any point x of M and any vector $X_0 \in D_x$, there is a smooth vector field X defined in a neighborhood U_x of x which is tangent to the distribution, i.e., $X(y) \in D_y \; \forall \; y \in U_x$, and such that $X(x) = X_0$. If, moreover, $\dim D_x$ does not depend on x, then we say that D is a smooth *regular distribution*.

It follows directly from the local foliation property that the tangent distribution $D^{\mathcal{F}}$ of a smooth singular foliation is a smooth singular distribution.

An *integral submanifold* of smooth singular distribution D on a smooth manifold M is, by definition, a connected immersed submanifold W of M such that for every $y \in W$ the tangent space $T_y W$ is a vector subspace of D_y. An integral submanifold W is called *maximal* if it is not contained in any other integral submanifold; it is said to be of *maximum dimension* if its tangent space at every point $y \in W$ is exactly D_y.

We say that a smooth singular distribution D on a smooth manifold M is an *integrable distribution* if every point of M is contained in a maximal integral manifold of maximum dimension of D.

Let C be a family of smooth vector fields on M. Then it gives rise to a smooth singular distribution D^C: for each point $x \in M$, D_x^C is the vector space spanned by the values at x of the vector fields of C. We say that D^C is *generated* by C.

A distribution D is called *invariant* with respect to a family of smooth vector fields C if it is invariant with respect to every element of C: if $X \in C$ and (φ_X^t) denotes the local flow of X, then we have $(\varphi_X^t)_* D_x = D_{\varphi_X^t(x)}$ wherever $\varphi_X^t(x)$ is well defined.

The following result, due to Stefan [320] and Sussmann [327] (see also Dazord [93]), gives an answer to the following question: what are the conditions for a smooth singular distribution to be the tangent distribution of a singular foliation?

Theorem 1.5.1 (Stefan–Sussmann). *Let D be a smooth singular distribution on a smooth manifold M. Then the following three conditions are equivalent:*

a) *D is integrable.*

b) *D is generated by a family C of smooth vector fields, and is invariant with respect to C.*

c) *D is the tangent distribution $D^{\mathcal{F}}$ of a smooth singular foliation $\mathcal{F}$.*

Proof (sketch). a) $\Rightarrow$ b). Suppose that D is integrable. Let C be the family of all smooth vector fields which are tangent to D. The smoothness condition of D implies that D is generated by C. It remains to show that if X is an arbitrary smooth vector field tangent to D, then D is invariant with respect to X. Let x be an arbitrary point in M, and denote by $\mathcal{F}(x)$ the maximal invariant submanifold of maximum dimension which contains x. Then by definition (the condition of maximum dimension), for every point $y \in \mathcal{F}(x)$ we have $T_y\mathcal{F}(x) = D_y$, which implies that the vector field X, when restricted to $\mathcal{F}(x)$, is tangent to $\mathcal{F}(x)$. In particular, the local flow φ_X^t can be restricted to $\mathcal{F}(x)$, i.e., $\mathcal{F}(x)$ is an invariant manifold for this local flow. Moreover, if $\varphi_X^\tau(x)$ is well defined for some $\tau > 0$, then the point $\varphi_X^\tau(x)$ lies on $\mathcal{F}(x)$. This fact follows from the maximality condition on $\mathcal{F}(x)$. (Note that, the union of two invariant submanifolds of maximum dimension is again an invariant submanifold of maximum dimension if it is connected.) Because X is tangent to $\mathcal{F}(x)$, we have $(\varphi_X^\tau)_*(T_x\mathcal{F}(x)) = T_{\varphi_X^\tau(x)}\mathcal{F}(x)$. But $T_x\mathcal{F}(x) = D_x$ and $T_{\varphi_X^\tau(x)}\mathcal{F}(x) = D_{\varphi_X^\tau(x)}$, hence $(\varphi_X^\tau)_*D_x = D_{\varphi_X^\tau(x)}$.

b) $\Rightarrow$ c). Suppose that D is generated by a family C of smooth vector fields, and is invariant with respect to C. Let x be an arbitrary point of M, denote by d the dimension of D_x, and choose d vector fields $X_1, \ldots, X_d$ of C such that $X_1(x), \ldots, X_d(x)$ span D_x. Denote by $\phi_1^t, \ldots, \phi_d^t$ the local flow of $X_1, \ldots, X_d$ respectively. The map

$$(s_1, \ldots, s_d) \mapsto \phi_1^{s_1} \circ \cdots \circ \phi_d^{s_d}(x) \tag{1.50}$$

is a local diffeomorphism from a d-dimensional disk to a d-dimensional submanifold containing x in M. The invariance of D with respect to C implies that this submanifold is an integral submanifold of maximum dimension. Gluing these local integral submanifolds together (wherever they intersect), we obtain a partition of M into a disjoint union of connected immersed integral submanifolds of maximum dimension, called leaves. To see that this partition satisfies the local foliation property of singular foliations, we can proceed by induction on the dimension of D_x: If $\dim D_x = 0$, then the local foliation property at x is empty. If $\dim D_x > 0$, then there is a vector field $X \in C$ such that $X(x) \neq 0$. Then the trajectories of X lie on the leaves, and we can take the quotient of a small neighborhood of x by the trajectories of X to reduce the dimension of M and of the leaves by 1. The invariance with respect to C and the local foliation property does not change under this reduction.

c) $\Rightarrow$ a): If $D = D^{\mathcal{F}}$ is the tangent distribution of a singular foliation $\mathcal{F}$, then the leaves of $\mathcal{F}$ are maximal invariant submanifolds of maximum dimension for D. $\qquad\square$

Definition 1.5.2. An *involutive distribution* is a distribution D such that if X, Y are two arbitrary smooth vector fields which are tangent to D, then their Lie bracket $[X, Y]$ is also tangent to D.

It is clear from Theorem 1.5.1 that if a singular distribution is integrable, then it is involutive. Conversely, for regular distributions we have:

Theorem 1.5.3 (Frobenius). *If a smooth regular distribution is involutive then it is integrable, i.e., it is the tangent distribution of a regular foliation.*

Proof (sketch). One can use Formula (1.50) to construct local invariant submanifolds of maximum dimension and then glue them together, just like in the proof of Theorem 1.5.1. One can also see Theorem 1.5.3 as a special case of Theorem 1.5.1, by first showing that a regular involutive distribution is invariant with respect to the family of all smooth vector fields which are tangent to it. $\qquad\square$

Example 1.5.4. Consider the following singular foliation D on $\mathbb{R}^2$ with coordinates (x, y): $D_{(x,y)} = T_{(x,y)}\mathbb{R}^2$ if $x > 0$, and $D_{(x,y)}$ is spanned by $\frac{\partial}{\partial x}$ if $x \leq 0$. Then D is smooth involutive but not integrable.

The above example shows that, if in Frobenius' theorem we omit the word *regular*, then it is false. The reason is that, though Formula (1.50) still provides us with local invariant submanifolds, they are not necessarily of maximum dimension. However, the situation in the finitely generated case is better. A smooth distribution D on a manifold M is called *locally finitely generated* if for any $x \in M$ there is a neighborhood U of x such that the $\mathcal{C}^\infty(U)$-module of smooth tangent vector fields to D in U is finitely generated: there is a finite number of smooth vector fields $X_1, \ldots, X_n$ in U which are tangent to D, such that any smooth vector field Y in U which is tangent to D can be written as $Y = \sum_{i=1}^n f_i X_i$ with $f_i \in \mathcal{C}^\infty(U)$.

Theorem 1.5.5 (Hermann [171]). *Any locally finitely generated smooth involutive distribution on a smooth manifold is integrable.*

See [366] for a simple proof of Theorem 1.5.5. $\square$

Consider now a smooth Poisson manifold (M, Π). Denote by $\mathcal{C}$ its characteristic distribution. Recall that

$$\mathcal{C}_x = \mathrm{Im}\sharp_x = \{X_f(x), f \in C^\infty(M)\} \ \forall \, x \in M \ . \tag{1.51}$$

Since the Hamiltonian vector fields preserve the Poisson structure, they also preserve the characteristic distribution. Thus, according to Stefan–Sussmann's theorem, the characteristic foliation $\mathcal{C}$ is completely integrable and corresponds to a singular foliation, which we will denote by $\mathcal{F} = \mathcal{F}_\Pi$. For the reasons which will become clear below, this singular foliation is called the *symplectic foliation* of the Poisson manifold (M, Π).

For each point $x \in M$, denote by $\mathcal{F}(x)$ the leaf of $\mathcal{F}$ which contains x. Local charts of $\mathcal{F}(x)$ are readily provided by Theorem 1.4.5: If

$$(p_1, \ldots, p_s, q_1, \ldots, q_s, z_1, \ldots, z_{m-2s})$$

is a local canonical system of coordinates at a point $x \in M$, then the submanifold $\{z_1 = \cdots = z_{m-2s} = 0\}$ is an open subset of $\mathcal{F}(x)$, and it has a natural symplectic structure with Darboux coordinates (p_i, q_i). Notice that this symplectic structure does not depend on the choice of coordinates: at each point of $\{z_1 = \cdots = z_{m-2s} = 0\}$, it coincides with the symplectic form on the characteristic space. Thus, on each leaf $\mathcal{F}(x)$ we have a unique natural symplectic structure, which at each point coincides with the symplectic form on the corresponding characteristic space. It also follows from Assertions b), d) of Theorem 1.4.5 that the injection $i : \mathcal{F} \to M$ is a Poisson morphism: if f, g are two functions on M and $y \in \mathcal{F}(x)$, then

$$\{f, g\}(y) = \{f|_{\mathcal{F}(x)}, g|_{\mathcal{F}(x)}\}_x(y), \tag{1.52}$$

where $\{ , \}_x$ is the Poisson bracket of the symplectic form on $\mathcal{F}(x)$. In other words, we have:

Theorem 1.5.6 ([346]). *Every leaf $\mathcal{F}(x)$ of the symplectic foliation $\mathcal{F}_\Pi$ of a Poisson manifold (M, Π) is an immersed symplectic submanifold, the immersion being a Poisson morphism. The Poisson structure Π is completely determined by the symplectic structures on the leaves of $\mathcal{F}_\Pi$.*

Example 1.5.7. Symplectic foliation of linear Poisson structures. Let G be a connected Lie group, $\mathfrak{g}$ its Lie algebra, and $\mathfrak{g}^*$ the dual of $\mathfrak{g}$. Recall that $\mathfrak{g}^*$ has a natural linear Poisson structure, also known as the *Lie–Poisson structure*, defined by

$$\{f, h\}(\alpha) = \langle \alpha, [\mathrm{d}f(\alpha), \mathrm{d}h(\alpha)] \rangle. \tag{1.53}$$

Denote the neutral element of G by e, and identify $\mathfrak{g}$ with $T_e G$. G acts on $\mathfrak{g}$ by the adjoint action $\mathrm{Ad}_g(x) = (u \mapsto gug^{-1})_{*e}(x)$ and on $\mathfrak{g}^*$ by the coadjoint action (the induced dual action) $\mathrm{Ad}_g^*(\alpha)(x) = \alpha(\mathrm{Ad}_{g^{-1}}x)$, $\alpha \in \mathfrak{g}^*$, $x \in \mathfrak{g}$, $g \in G$. This action is generated infinitesimally by the coadjoint action of $\mathfrak{g}$ on $\mathfrak{g}^*$ defined by $ad_x^*(\alpha)(y) = \langle \alpha, [y, x] \rangle = -\langle \alpha, \mathrm{ad}_x y \rangle$.

Theorem 1.5.8. *The symplectic leaves of the Lie–Poisson structure on the dual of an arbitrary finite-dimensional Lie algebra coincide with the orbits of the coadjoint representation on it.*

Proof. Due to the Leibniz rule, the tangent spaces to the symplectic leaves, i.e., the characteristic spaces, are generated by the Hamiltonian vector fields of linear functions. If f, h are two linear functions on $\mathfrak{g}^*$, also considered as two elements of $\mathfrak{g}$ by duality, and α is a point of $\mathfrak{g}^*$, then we have

$$X_f(h)(\alpha) = \langle \alpha, [f, h] \rangle = -\langle \mathrm{ad}_f^*(\alpha), h \rangle. \tag{1.54}$$

It implies that the tangent spaces of symplectic leaves are the same as the tangent spaces of coadjoint orbits. It follows that coadjoint orbits are open closed subsets of symplectic leaves, so they coincide with symplectic leaves because symplectic leaves are connected by definition. $\square$

A corollary of the above theorem is that the orbits of the coadjoint representation of a finite-dimensional Lie algebra are of even dimension and equipped with a natural symplectic form. This symplectic form is also known as the *Kirillov–Kostant–Souriau form*. Let us mention that coadjoint orbits play a very important role in the theory of unitary representations of Lie groups (the so-called *orbit method*), see, e.g., [193].

Exercise 1.5.9. Describe the symplectic leaves of $so^*(3)$ and $sl^*(2)$.

Remark 1.5.10. A direct way to define the symplectic foliation of a Poisson manifold (M, Π) is as follows: two points x, y are said to belong to the same leaf if they can be connected by a piecewise-smooth curve consisting of integral curves of Hamiltonian vector fields. Then it is a direct consequence of the splitting Theorem 1.4.5 that the corresponding partition of M into leaves satisfies the local foliation property. Thus, in fact, we can use the splitting Theorem 1.4.5 instead of the Stefan–Sussmann Theorem 1.5.1 in order to show that on (M, Π) there is a natural associated foliation whose tangent distribution is the characteristic distribution.

1.6 Transverse Poisson structures

Let N be a smooth local (i.e., sufficiently small) disk of dimension $m - 2s$ of an m-dimensional Poisson manifold (M, Π), which intersects transversally a $2s$-dimensional leaf $\mathcal{F}(x)$ of the symplectic foliation $\mathcal{F}$ of (M, Π) at a point x. In other words, N contains x and is transversal to the characteristic space $\mathcal{C}_x$. Then according to the splitting Theorem 1.4.5, there are local canonical coordinates in a neighborhood of x, which will define on N a Poisson structure. This Poisson structure on N is called the *transverse Poisson structure* at x of the Poisson manifold (M, Π).

To justify the above definition of transverse Poisson structures, we must show that the Poisson structure on N given by Theorem 1.4.5 does not depend on the choice of local canonical coordinates, nor on the choice of N itself, modulo local Poisson diffeomorphisms.

Theorem 1.6.1. *With the above notations, we have:*

a) *The local Poisson structure on N given by Theorem 1.4.5 does not depend on the choice of local canonical coordinates.*

b) *If x_0 and x_1 are two points on the symplectic leaf $\mathcal{F}(x)$, and N_0 and N_1 are two smooth local disks of dimension $m - 2s$ which intersect $\mathcal{F}(x)$ transversally at x_0 and x_1 respectively, then there is a smooth local Poisson diffeomorphism from (N_0, x_0) to (N_1, x_1).*

Proof. a) Theorem 1.4.5 implies that the local symplectic leaves near point x are direct products of the symplectic leaves of a neighborhood of x in N with the local symplectic manifold $\{(p_1, \ldots, p_s, q_1, \ldots, q_s)\}$. In particular, the symplectic leaves of N are connected components of intersections of the symplectic leaves of M with N, and the symplectic form on the symplectic leaves of N is the restriction of the symplectic form of the leaves of M to those intersections. This geometric characterization of the symplectic leaves of N and their corresponding symplectic forms shows that they do not depend on the choice of local canonical coordinates. Hence, according to Theorem 1.5.6, the Poisson structure of N does not depend on the choice of local canonical coordinates.

b) Let N_0 and N_1 be two local disks which intersect a symplectic leaf $\mathcal{F}(x)$ transversally at x_0 and x_1 respectively. Then there is a smooth one-dimensional family of local submanifolds N_t $(0 \leq t \leq 1)$, connecting N_0 to N_1, such that N_t intersects $\mathcal{F}_\alpha$ transversely at a point x_t. Point x_t depends smoothly on t. According to (a parameterized version of) Theorem 1.4.5, there is a smooth family of local functions

$$(p_1^t, \ldots, p_s^t, q_1^t, \ldots, q_s^t, z_1^t, \ldots, z_{m-2s}^t),$$

such that for each $t \in [0,1]$, $(p_1^t, \ldots, p_s^t, q_1^t, \ldots, q_s^t, z_1^t, \ldots, z_{m-2s}^t)$ is a local canonical system of coordinates for (M, Π) in a neighborhood of x_t such that $p_i^t(N_t) = q_i^t(N_t) = 0$.

For each point $y \in N_t$ close enough to x_t, define a tangent vector $Y_t(y) \in T_y M$ as follows: For each τ near t, the disk N_τ intersects the local submanifold $\{z_1^t = z_1^t(y), \ldots, z_{m-2s}^t = z_{m-2s}^t(y)\}$ transversally at a unique point y_τ. The map $\tau \mapsto y_\tau$ is smooth. Vector $Y_t(y)$ is defined to be the derivation of this map at $\tau = t$. In particular, $Y_t(x_t)$ is the derivation of the map $\tau \mapsto x_\tau$ at $\tau = t$.

There is a unique cotangent vector $\beta_t(y) \in T_y^* M$ such that $\beta_t(y)$ annulates $T_y N_t$ (the tangent space of N_t at y), and $\sharp \beta_t(y) = Y_t(y)$. For each $t \in [0,1]$ we can choose a function f_t defined in a neighborhood of N_t, in such a way that f_t depends smoothly on t, and that $\mathrm{d}f_t(y) = \beta_t(y) \; \forall \; y \in N_t$. It implies that we have $X_t(y) = Y_t(y) \; \forall \; y \in N_t$, where $X_t = X_{f_t}$ denotes the Hamiltonian vector field of f_t.

Denote by φ_t the local flow of the time-dependent Hamiltonian vector field X_t (where t is considered as the time variable). Then of course φ_t preserves the Poisson structure of V (wherever φ_t is defined). From the construction of X_t, we also see that φ_t moves x_0 to x_t, and it moves a sufficiently small neighborhood of x_0 in N_0 into N_t. In particular, φ_1 defines a local diffeomorphism from N_0 to N_1. Since φ_1 preserves the Poisson structure of M, it also preserves the Poisson structure of N_0. (As explained in the first part of this theorem, the Poisson structure of N_0 depends only on Π and N_0, and does not depend on other things like local canonical coordinates.) In other words, φ_1 defines a local Poisson diffeomorphism from (N_0, x_0) to (N_1, x_1). $\qquad\square$

In practice, the transverse Poisson structure may be calculated by the following so-called Dirac's constrained bracket formula, or *Dirac's formula*[1] for short.

Proposition 1.6.2 (Dirac's formula). *Let N be a local submanifold of a Poisson manifold (M, Π) which intersects a symplectic leaf transversely at a point z. Let $\psi_1, \ldots, \psi_{2s}$, where $2s = \mathrm{rank}\,\Pi(z)$, be functions in a neighborhood U of z such that*

$$N = \{x \in U \mid \psi_i(x) = \text{constant}\}. \tag{1.55}$$

Denote by $P_{ij} = \{\psi_i, \psi_j\}$ and by (P^{ij}) the inverse matrix of $(P_{ij})_{i,j=1}^{2s}$. Then the bracket formula for the transverse Poisson structure on N is given as follows:

$$\{f, g\}_N(x) = \{\widetilde{f}, \widetilde{g}\}(x) - \sum_{i,j=1}^{2s} \{\widetilde{f}, \psi_i\}(x) P^{ij}(x) \{\psi_j, \widetilde{g}\}(x) \quad \forall \; x \in N, \tag{1.56}$$

where f, g are functions on N and $\widetilde{f}, \widetilde{g}$ are extensions of f and g to U. The above formula is independent of the choice of extensions $\widetilde{f}$ and $\widetilde{g}$.

---

[1] According to Weinstein [347], Dirac's formula was actually found by T. Courant and R. Montgomery, who generalized a constraint procedure of Dirac.

Proof (sketch). If one replaces $\widetilde{f}$ by ψ_k ($\forall\ k = 1, \ldots, 2s$) in the above formula, then the right-hand side vanishes. If $\widetilde{f}$ and $\widehat{f}$ are two extensions of f, then we can write $\widehat{f} = \widetilde{f} + \sum_{i=1}^{2s}(\psi_i - \psi_i(z))h_i$. Using the Leibniz rule, one verifies that the right-hand side in Formula (1.56) does not depend on the choice of $\widetilde{f}$. By anti-symmetricity, the right-hand side does not depend on the choice of $\widetilde{g}$ either. Finally, we can choose $\widetilde{f}$ and $\widetilde{g}$ to be independent of p_i, q_i in a canonical coordinate system $(p_1, \ldots, p_s, q_1, \ldots, q_s, z_1, \ldots, z_{m-2s})$ provided by the splitting Theorem 1.4.5. For that particular choice we have $\{\widetilde{f}, \psi_i\}(x) = 0$ and $\{\widetilde{f}, \widetilde{g}\}(x) = \{f, g\}_N(x)$. $\qquad\square$

1.7 Group actions and reduction

Recall that, a *left action* of a Lie group G on manifold M is a (smooth) map $\varrho : G \times M \to M$ such that

$$\varrho(e, z) = z \quad \forall\ z \in M, \tag{1.57}$$

where e denotes the neutral element of G, and

$$\varrho(gh, z) = \varrho(g, \varrho(h, z)) \ \forall\ g, h \in G, z \in M. \tag{1.58}$$

Similarly, a *right action* of a Lie group G on a manifold M is a map $\rho : M \times G \to M$ such that $\rho(z, e) = z \ \forall\ z \in M$ and

$$\rho(z, gh) = \rho(\rho(z, g), h) \ \forall\ g, h \in G, z \in M. \tag{1.59}$$

If we write the right action $\rho(g)(z)$ of g on z as $z.g$, then we have $z.(gh) = (z.g).h$. Sometimes, for convenience, a right action will also be denoted as $\rho : G \times M \to M$.

If ρ is a right action, then $\widehat{\rho}(g, z) := \rho(z, g^{-1})$ is a left action, and vice versa. So left actions and right actions are essentially the same. A (left) action of G on M may also be defined as a homomorphism $\rho : G \to Diff(M)$, $\rho(g) := \rho(g, .)$, from G to the group of diffeomorphisms of M.

Recall that, a *Lie algebra action* of a Lie algebra $\mathfrak{g}$ on a manifold M is a linear map $\xi : \mathfrak{g} \to \mathcal{V}^1(M)$ from $\mathfrak{g}$ to the space of vector fields on M, which preserves the Lie bracket, i.e., $[\xi(x), \xi(y)] = \xi([x, y]) \ \forall\ x, y \in \mathfrak{g}$. In other words, ξ is a Lie homomorphism from $\mathfrak{g}$ to $\mathcal{V}^1(M)$.

Example 1.7.1. An action of a Lie algebra $\mathfrak{g}$ on a vector space V is a map $\eta : \mathfrak{g} \times V \to V$ such that

$$\eta([x, y])(v) = \eta(x)(\eta(y)(v)) - \eta(y)(\eta(x)(v)).$$

If η is an action of $\mathfrak{g}$ on a vector space V, then the corresponding action of $\mathfrak{g}$ on V, considered as a manifold, is given by

$$-\eta : (x, v) \mapsto -\eta(x)(v) , \tag{1.60}$$

where $\eta(x)(v)$ is considered as a tangent vector at v to $T_v V$. (Notice the minus sign there.)

Let $\rho : M \times G \to M$ be a smooth *right* action of a Lie group G on a manifold M. For each $x \in \mathfrak{g}$, define ξ_x to be the generator of the one-dimensional group action $\rho(\exp(tx))$ on M. In other words, $\rho(\exp(tx))$ is the time-t flow of the vector field ξ_x: $\rho(\exp(tx)) = \exp(t\xi_x)$. Then the map $\xi : \mathfrak{g} \to \mathcal{V}^1(M), \xi(x) = \xi_x$, is a Lie algebra action of $\mathfrak{g}$ on M.

If a *left* action $\varrho : G \times M \to M$ is said to be generated by a Lie algebra action $\xi : \mathfrak{g} \to \mathcal{V}^1(M)$ of $\mathfrak{g}$, then it means that $\varrho(\exp(x), z) = \exp(-\xi(x))(z) \ \forall \ x \in \mathfrak{g}$, $z \in M$.

Motivated in particular by the fact that Hamiltonian systems and their phase spaces often admit symmetries, we are interested in Lie group and Lie algebra actions on Poisson manifolds M. If a connected Lie group G acts on a Poisson manifold (M, Π), in such a way that the action preserves the Poisson structure Π, then the space $\mathcal{C}^\infty(M)^G$ of G-invariant functions on M is a Poisson algebra in the following natural sense:

Definition 1.7.2. An associative commutative algebra A together with an anti-symmetric bracket $\{,\} : A \times A \to$ is called a *Poisson algebra* if $\{,\}$ satisfies the Leibniz identity $\{f, gh\} = \{f, g\}h + g\{f, h\}$ and the Jacobi identity $\{f, \{g, h\}\} + \{g, \{h, f\}\} + \{h, \{f, g\}\} = 0 \ \forall \ f, g, h \in A$.

Recall that, an action of a Lie group G on a manifold M is called a *proper action* if the corresponding map $\rho : G \times M \to M$ is a *proper map*, i.e., for any compact $K \subset M$ its preimage $\rho^{-1}(K)$ is also compact. In particular, if G is a compact Lie group, then any action of G is automatically proper. Assume that we have a free and proper action of a Lie group G on a manifold M. Then the quotient M/G of M by the action of G is a manifold, and $\mathcal{C}^\infty(M)$ can be naturally identified with $\mathcal{C}^\infty(M)^G$ via the pull-back of the projection map $M \to M/G$. Hence M/G has a natural *reduced Poisson structure*, for which the projection map $M \to M/G$ is a Poisson map. Note that $\dim M/G = \dim M - \dim G$ in this case. If $h \in \mathcal{C}^\infty(M)^G$ is a G-invariant function on M, and $\widetilde{h}$ its projection on M/G, then the Hamiltonian vector field X_h projects to the Hamiltonian vector field $X_{\widetilde{h}}$ under the projection $M \to M/G$. The Hamiltonian system of $\widetilde{h}$ on M/G is called the *reduced Hamiltonian system* of the Hamiltonian system of h on M.

Example 1.7.3. Consider the left action of a connected Lie group G on T^*G. Then, according to Theorem 1.3.10, the corresponding reduced Poisson manifold is isomorphic to $\mathfrak{g}^*$, with the left translation map $L : T^*G \to \mathfrak{g}^*$ as the projection map. For example, consider the *Euler top*, i.e., the rotational movement of a rigid body around its fixed center of gravity. The configuration space is $SO(3)$ (the space of all possible rotational positions), and the corresponding Hamiltonian system is a system on $T^*SO(3)$, which can be written in reduced form as a Hamiltonian system on $so^*(3)$. A Hamiltonian system on the dual $\mathfrak{g}^*$ of a Lie algebra with the corresponding linear Poisson structure is often called an *Euler equation*.

A particularly interesting class of actions on Poisson manifolds consists of the so-called Hamiltonian actions.

Definition 1.7.4. A *Hamiltonian action* of a Lie algebra $\mathfrak{g}$ on a Poisson manifold (M, Π) is a Lie algebra action $\xi : \mathfrak{g} \to \mathcal{V}^1(M)$ which is induced from a Poisson morphism $\mu : (M, \Pi) \to \mathfrak{g}^*$ (where $\mathfrak{g}^*$ is equipped with the standard Lie–Poisson structure) by the following formula:

$$\xi(x) = X_{\mu^*(x)} \quad \forall \, x \in \mathfrak{g}, \tag{1.61}$$

where on the right-hand side of the above equation x is considered as a linear function on $\mathfrak{g}^*$, and $X_{\mu^*(x)}$ denotes the Hamiltonian vector field of $\mu^*(x)$. This Poisson morphism $\mu : (M, \Pi) \to \mathfrak{g}^*$ is called an *equivariant momentum map*, or *momentum map*[2] for short. A left (resp. right) *Hamiltonian action* of a connected Lie group G on a Poisson manifold (M, Π) is a left (resp. right) action of G on M whose corresponding Lie algebra action is a Hamiltonian action.

Remark 1.7.5. If $\mu : (M, \Pi) \to \mathfrak{g}^*$ is a Poisson morphism, then it automatically gives rise to a Hamiltonian action of $\mathfrak{g}$ on M by Formula (1.61). The adjective *equivariant* in the above definition is due to the fact that the momentum map $\mu : (M, \Pi) \to \mathfrak{g}^*$ is a Poisson map if and only if it intertwines the (left) action of G on M with the coadjoint action of G on $\mathfrak{g}^*$. The proof of this fact will be left to the reader as an exercise.

Remark 1.7.6. It may happen that a map $\mu : (M, \Pi) \to \mathfrak{g}^*$ is *not* Poisson, but still defines a Lie algebra action $\xi : \mathfrak{g} \to \mathcal{V}^1(M)$ by Formula (1.61). In that case ξ is called a weakly Hamiltonian action, and μ a *non-equivariant momentum map*.

Example 1.7.7. Consider the right action of a connected Lie group G on itself. Then its associated Lie algebra action of $\mathfrak{g}$ on G is given by left-invariant vector fields. The right action of G on itself can be lifted to a right Hamiltonian action of G on T^*G, whose momentum map is the natural Poisson morphism $L : T^*G \to \mathfrak{g}^*$ given in Theorem 1.3.10. Thus, the map $L : T^*G \to \mathfrak{g}^*$ is at the same time the projection map for the left action of G and the momentum map for the right action of G on T^*G.

Example 1.7.8. If a connected Lie group G acts on a manifold N on the right, then its natural lifting to T^*N is a Hamiltonian action on T^*N, whose momentum map $\mu : T^*N \to \mathfrak{g}^*$ is defined as follows:

$$\langle \mu(p), x \rangle = \langle p, \xi_x(\pi(p)) \rangle, \tag{1.62}$$

where $p \in T^*N$, $x \in \mathfrak{g}$, $\pi : T^*N \to N$ is the projection, ξ_x is the vector field on N which generates the action of the one-dimensional group $\exp(tx)$. This example generalizes the previous example.

If the action of G on M is Hamiltonian with a given equivariant momentum map $\mu : M \to \mathfrak{g}^*$, then $\mu^{-1}(0)$ is invariant under the action of G, and we can form another reduced Poisson space, denoted by $M//G$, as follows:

$$M//G = \mu^{-1}(0)/G. \tag{1.63}$$

---

[2]The terminology *momentum map* is the English translation of the French term *application de moment* introduced by Souriau [319]. Some people prefer to call it *moment map*.

Theorem 1.7.9. *With the above notations, if the action of G on $\mu^{-1}(0)$ is free and proper, then $M//G = \mu^{-1}(0)/G$ has a unique natural Poisson structure coming from the Poisson structure of M, called the* reduced Poisson structure.

Proof. Since the action of G on $\mu^{-1}(0)$ is free, its infinitesimal action is also free, i.e., $\xi_x(z) \neq 0 \; \forall \; 0 \neq x \in \mathfrak{g}, z \in \mu^{-1}(0)$, where $\xi : \mathfrak{g} \to \mathcal{V}^{-1}(M)$ denotes the corresponding action of $\mathfrak{g}$. It implies that $\mathrm{d}f\mu(z) : T_z M \to T_{\mu(z)}\mathfrak{g}^*$ is surjective for any $z \in \mu^{-1}(0)$. In particular, 0 is a regular value of the momentum map, and $\mu^{-1}(0)$ is a closed submanifold of M. Let f, h be any two functions on $\mu^{-1}(0)/G$, viewed also as G-invariant functions on $\mu^{-1}(0)$. Extend them to two G-invariant functions $\tilde{f}, \tilde{h}$ on a neighborhood of $\mu^{-1}(0)$ in M. Since the Poisson structure on M is G-invariant, the Poisson bracket $\{\tilde{f}, \tilde{h}\}$ is also G-invariant. Similarly to the proof of Dirac's formula (1.56), it is easy to see that the restriction of $\{\tilde{f}, \tilde{h}\}$ to $\mu^{-1}(0)$ depends only on f, h but not on the extensions $\tilde{f}, \tilde{h}$. We can define the Poisson bracket of f, h on $\mu^{-1}(0)/G$ to be the projection of $\{\tilde{f}, \tilde{h}\}|_{\mu^{-1}(0)}$ to $\mu^{-1}(0)/G$. $\qquad \square$

Exercise 1.7.10. Show that (assuming that the action of G is Hamiltonian, proper and free) the inclusion map $M//G \to M/G$ is a Poisson morphism. When M is a symplectic manifold, then $M//G$ is a symplectic leaf of M/G.

Remark 1.7.11. The Poisson reduction $M//G$ is also known as the *Marsden–Ratiu reduction* [238]. When M is symplectic, it is known as the *Marsden–Weinstein– Meyer reduction* [237, 247].

Example 1.7.12 ([348]). If G is a Lie group, B is a principal G-bundle over a manifold X (i.e., G acts on B freely on the right and $B/G = X$), and M is a Poisson manifold together with a Hamiltonian G-action, then the triple (G, B, M) is called a *classical Yang–Mills–Higgs setup*. Given a classical Yang–Mills–Higgs setup (G, B, M), the product Hamiltonian action of G on $T^*B \times M$ is free and proper, so by Theorem 1.7.9, we can form the reduced Poisson manifold $\mathcal{Y}(G, B, M) = (T^*B \times M)//G$. This Poisson manifold $\mathcal{Y}(G, B, M)$ is called the *Yang–Mills–Higgs phase space* for a classical particle with configuration space X, gauge group G and internal phase space M.

When the action of G is proper but *not* free, the reduced spaces M/G and $M//G$ are no longer manifolds in general, but they are still Hausdorff spaces with the structure of a *stratified manifold*. We will not go into the details of this so-called singular reduction theory (see, e.g., [317, 23, 287, 90] and references therein). We refer the reader to the book by Ortega and Ratiu [288] for a comprehensive treatment of reduction theory.

1.8 The Schouten bracket

1.8.1 Schouten bracket of multi-vector fields

Recall that, if $A = \sum_i a_i \frac{\partial}{\partial x_i}$ and $B = \sum_i b_i \frac{\partial}{\partial x_i}$ are two vector fields written in a local system of coordinates $(x_1, \ldots, x_n)$, then the Lie bracket of A and B is

$$[A, B] = \sum_i a_i \left(\sum_j \frac{\partial b_j}{\partial x_i} \frac{\partial}{\partial x_j} \right) - \sum_i b_i \left(\sum_j \frac{\partial a_j}{\partial x_i} \frac{\partial}{\partial x_j} \right). \qquad (1.64)$$

We will redenote $\frac{\partial}{\partial x_i}$ by ζ_i and consider them as *formal*, or *odd variables*[3] (formal in the sense that they don't take values in a field, but still form an algebra, and odd in the sense that $\zeta_i \zeta_j = -\zeta_j \zeta_i$, i.e., $\frac{\partial}{\partial x_j} \wedge \frac{\partial}{\partial x_i} = -\frac{\partial}{\partial x_j} \wedge \frac{\partial}{\partial x_i}$). We can write $A = \sum_i a_i \zeta_i$ and $B = \sum_i b_i \zeta_i$ and consider them formally as functions of variables (x_i, ζ_i) which are linear in the odd variables (ζ_i). We can write $[A, B]$ formally as

$$[A, B] = \sum_i \frac{\partial A}{\partial \zeta_i} \frac{\partial B}{\partial x_i} - \sum_i \frac{\partial B}{\partial \zeta_i} \frac{\partial A}{\partial x_i}. \qquad (1.65)$$

The above formula makes the Lie bracket of two vector fields look pretty much like the Poisson bracket of two functions in a Darboux coordinate system.

Now if $\Pi = \sum_{i_1 < \cdots < i_p} \Pi_{i_1 \ldots i_p} \frac{\partial}{\partial x_{i_1}} \wedge \cdots \wedge \frac{\partial}{\partial x_{i_p}}$ is a p-vector field, then we will consider it as a homogeneous polynomial of degree p in the odd variables (ζ_i):

$$\Pi = \sum_{i_1 < \cdots < i_p} \Pi_{i_1 \ldots i_p} \zeta_{i_1} \ldots \zeta_{i_p}. \qquad (1.66)$$

It is important to remember that the variables ζ_i do not commute. In fact, they anti-commute among themselves, and commute with the variables x_i:

$$\zeta_i \zeta_j = -\zeta_j \zeta_i; \ x_i \zeta_j = \zeta_j x_i; \ x_i x_j = x_j x_i. \qquad (1.67)$$

Due to the anti-commutativity of (ζ_i), one must be careful about the signs when dealing with multiplications and differentiations involving these odd variables. The differentiation rule that we will adopt is as follows:

$$\frac{\partial(\zeta_{i_1} \ldots \zeta_{i_p})}{\partial \zeta_{i_p}} := \zeta_{i_1} \ldots \zeta_{i_{p-1}}. \qquad (1.68)$$

Equivalently, $\frac{\partial(\zeta_{i_1} \ldots \zeta_{i_p})}{\partial \zeta_{i_k}} = (-1)^{p-k} \zeta_{i_1} \ldots \widehat{\zeta_{i_k}} \ldots \zeta_{i_p}$, where the hat means that ζ_{i_k} is missing in the product $(1 \leq k \leq p)$.

---

[3]The name *odd variable* comes from the theory of supermanifolds, though it is not necessary to know what a supermanifold is in order to understand this section.

If

$$A = \sum_{i_1 < \cdots < i_a} A_{i_1,\ldots,i_a} \frac{\partial}{\partial x_{i_1}} \wedge \cdots \wedge \frac{\partial}{\partial x_{i_a}} = \sum_{i_1,\ldots,i_a} A_{i_1,\ldots,i_a} \zeta_{i_1} \cdots \zeta_{i_a} \qquad (1.69)$$

is an a-vector field, and

$$B = \sum_{i_1 < \cdots < i_b} B_{i_1,\ldots,i_b} \frac{\partial}{\partial x_{i_1}} \wedge \cdots \wedge \frac{\partial}{\partial x_{i_b}} = \sum_{i_1,\ldots,i_a} B_{i_1,\ldots,i_a} \zeta_{i_1} \cdots \zeta_{i_b} \qquad (1.70)$$

is a b-vector field, then generalizing Formula (1.65), we can define a bracket of A and B as follows:

$$[A, B] = \sum_i \frac{\partial A}{\partial \zeta_i} \frac{\partial B}{\partial x_i} - (-1)^{(a-1)(b-1)} \sum_i \frac{\partial B}{\partial \zeta_i} \frac{\partial A}{\partial x_i} \ . \qquad (1.71)$$

Clearly, the bracket $[A, B]$ of A and B as defined above is a homogeneous polynomial of degree $a + b - 1$ in the odd variables (ζ_i), so it is a $(a+b-1)$-vector field.

Theorem 1.8.1 (Schouten–Nijenhuis). *The bracket defined by Formula (1.71) satisfies the following properties:*

a) *Graded anti-commutativity: if A is an a-vector field and B is a b-vector field then*

$$[A, B] = -(-1)^{(a-1)(b-1)}[B, A] \ . \qquad (1.72)$$

b) *Graded Leibniz rule: if A is an a-vector field, B is a b-vector field and C is a c-vector field then*

$$[A, B \wedge C] = [A, B] \wedge C + (-1)^{(a-1)b} B \wedge [A, C] \ , \qquad (1.73)$$

$$[A \wedge B, C] = A \wedge [B, C] + (-1)^{(c-1)b}[A, C] \wedge B \ . \qquad (1.74)$$

c) *Graded Jacobi identity:*

$$(-1)^{(a-1)(c-1)}[A, [B, C]] + (-1)^{(b-1)(a-1)}[B, [C, A]]$$
$$+ (-1)^{(c-1)(b-1)}[C, [A, B]] = 0 \ . \qquad (1.75)$$

d) *If $A = X$ is a vector field then*

$$[X, B] = \mathcal{L}_X B \ , \qquad (1.76)$$

where $\mathcal{L}_X$ denotes the Lie derivative by X. In particular, if A and B are two vector fields, then the Schouten bracket of A and B coincides with their Lie bracket. If $A = X$ is a vector field and $B = f$ is a function (i.e., a 0-vector field), then we have

$$[X, f] = X(f) = \langle \mathrm{d}f, X \rangle \ . \qquad (1.77)$$

Proof. Assertion a) follows directly from the definition.

b) The differentiation rule (1.68) implies that

$$\frac{\partial(B \wedge C)}{\partial \zeta_i} = B\frac{\partial C}{\partial \zeta_i} + (-1)^c \frac{\partial B}{\partial \zeta_i}C.$$

Hence we have

$$
\begin{aligned}
[A, B \wedge C] &= \sum \frac{\partial A}{\partial \zeta_i}\frac{\partial(B \wedge C)}{\partial x_i} - (-1)^{(a-1)(b+c-1)}\sum \frac{\partial(B \wedge C)}{\partial \zeta_i}\frac{\partial A}{\partial x_i} \\
&= \sum \frac{\partial A}{\partial \zeta_i}\frac{\partial B}{\partial x_i}C + \sum \frac{\partial A}{\partial \zeta_i}B\frac{\partial C}{\partial x_i} - (-1)^{(a-1)(b+c-1)}\sum B\frac{\partial C}{\partial \zeta_i}\frac{\partial A}{\partial x_i} \\
&\quad -(-1)^{(a-1)(b+c-1)+c}\sum \frac{\partial B}{\partial \zeta_i}C\frac{\partial A}{\partial x_i} \\
&= \sum \frac{\partial A}{\partial \zeta_i}\frac{\partial B}{\partial x_i}C - (-1)^{(a-1)(b+c-1)+c+ac}\sum \frac{\partial B}{\partial \zeta_i}\frac{\partial A}{\partial x_i}C \\
&\quad +(-1)^{(a-1)b}\left(-(-1)^{(a-1)(c-1)}\sum B\frac{\partial C}{\partial \zeta_i}\frac{\partial A}{\partial x_i} + \sum B\frac{\partial A}{\partial \zeta_i}\frac{\partial C}{\partial x_i}\right) \\
&= [A, B] \wedge C + (-1)^{(a-1)b}B \wedge [A, C].
\end{aligned}
$$

The proof of Formula (1.74) is similar.

c) By direct calculations we have

$$(-1)^{(a-1)(c-1)}[A, [B, C]] = S_1 + S_2 + S_3 + S_4 ,$$

where

$$S_1 = (-1)^{(a-1)(c-1)}\sum_{i,j} \frac{\partial A}{\partial \zeta_j}\frac{\partial^2 B}{\partial x_j\partial \zeta_i}\frac{\partial C}{\partial x_i} - (-1)^{(a-1)(b-1)}\sum_{i,j}\frac{\partial B}{\partial \zeta_i}\frac{\partial^2 C}{\partial x_i\partial \zeta_j}\frac{\partial A}{\partial x_j} ,$$

$$S_2 = (-1)^{(a-1)(c-1)}\sum_{i,j} \frac{\partial A}{\partial \zeta_j}\frac{\partial B}{\partial \zeta_i}\frac{\partial^2 C}{\partial x_i\partial x_j} - (-1)^{(c-1)(b-1)}\sum_{i,j}\frac{\partial C}{\partial \zeta_i}\frac{\partial A}{\partial \zeta_j}\frac{\partial^2 B}{\partial x_i\partial x_j} ,$$

$$S_3 = (-1)^{(b-1)(a-1)}\sum_{i,j} \frac{\partial^2 B}{\partial \zeta_j\partial x_i}\frac{\partial C}{\partial \zeta_i}\frac{\partial A}{\partial x_j} - (-1)^{(c-1)(b-1)}\sum_{i,j}\frac{\partial^2 C}{\partial \zeta_i\partial x_j}\frac{\partial A}{\partial \zeta_j}\frac{\partial B}{\partial x_i} ,$$

$$
\begin{aligned}
S_4 &= (-1)^{(b-1)(a+c)+b}\sum_{i,j} \frac{\partial^2 C}{\partial \zeta_j\partial \zeta_i}\frac{\partial B}{\partial x_i}\frac{\partial A}{\partial x_j} - (-1)^{(a-1)(b-1)+c}\sum_{i,j}\frac{\partial^2 B}{\partial \zeta_j\partial \zeta_i}\frac{\partial C}{\partial x_i}\frac{\partial A}{\partial x_j} \\
&= (-1)^{(b-1)(c-1)+a}\sum_{i,j} \frac{\partial^2 C}{\partial \zeta_j\partial \zeta_i}\frac{\partial A}{\partial x_i}\frac{\partial B}{\partial x_j} - (-1)^{(a-1)(b-1)+c}\sum_{i,j}\frac{\partial^2 B}{\partial \zeta_j\partial \zeta_i}\frac{\partial C}{\partial x_i}\frac{\partial A}{\partial x_j}
\end{aligned}
$$

(because $\frac{\partial^2 C}{\partial \zeta_i\partial \zeta_j} = -\frac{\partial^2 C}{\partial \zeta_j\partial \zeta_i}$).

Each of the summands S_1, S_2, S_3, S_4 will become zero when adding similar terms from $(-1)^{(b-1)(a-1)}[B, [C, A]]$ and $(-1)^{(c-1)(b-1)}[C, [A, B]]$.

d) If f is a function and $X = \sum_i \xi_i \frac{\partial}{\partial x_i}$ is a vector field, then $\frac{\partial f}{\partial \zeta_i} = 0$, and $[X, f] = \sum \frac{\partial X}{\partial \zeta_i} \frac{\partial f}{\partial x_i} = \sum \xi_i \frac{\partial f}{\partial x_i} = X(f)$. When A and B are vector fields, Formula (1.71) clearly coincides with Formula 1.65. When B is a multi-vector field, Assertion d) can be proved by induction on the degree of B, using the Leibniz rules given by Assertion b).   □

A priori, the bracket of an a-vector field A with a b-vector field B, as defined by Formula (1.71), may depend on the choice of local coordinates $(x_1, \ldots, x_n)$. However, the Leibniz rules (1.73) and (1.74) show that the computation of $[A, B]$ can be reduced to the computation of the Lie brackets of vector fields. Since the Lie bracket of vector fields does not depend on the choice of local coordinates, it follows that the bracket $[A, B]$ is in fact a well-defined $(a+b-1)$-vector field which does not depend on the choice of local coordinates.

Definition 1.8.2. If A is an a-vector field and B is a b-vector field, then the uniquely defined $(a+b-1)$-vector field $[A, B]$, given by Formula (1.71) in each local system of coordinates, is called the *Schouten bracket* of A and B.

Remark 1.8.3. Our sign convention in the definition of the Schouten bracket is the same as Koszul's [201], but different from Vaisman's [333] and some other authors.

The Schouten bracket was first discovered by Schouten [311, 312]. Theorem 1.8.1 is essentially due to Schouten [311, 312] and Nijenhuis [280]. The graded Jacobi identity (1.75) means that the Schouten bracket is a graded Lie bracket: the space $\mathcal{V}^\star(M) = \bigoplus_{p \geq 0} \mathcal{V}^p(M)$, where $\mathcal{V}^p(M)$ is the space of smooth p-vector fields on a manifold M, is a graded Lie algebra, also known as Lie super-algebra, under the Schouten bracket, if we define the grade of $\mathcal{V}^p(M)$ to be $p - 1$. In other words, we have to shift the natural grading by -1 for $\mathcal{V}(M)$ together with the Schouten bracket to become a graded Lie algebra in the usual sense.

Another equivalent definition of the Schouten bracket, due to Lichnerowicz [211], is as follows. If A is an a-vector field, B is a b-vector field, and η is an $(a + b - 1)$-form then

$$\langle \eta, [A, B] \rangle = (-1)^{(a-1)(b-1)} \langle d(i_B \eta), A \rangle - \langle d(i_A \eta), B \rangle + (-1)^a \langle d\eta, A \wedge B \rangle. \quad (1.78)$$

In this formula, $i_A : \Omega^\star(M) \to \Omega^\star(M)$ denotes the *inner product* of differential forms with A, i.e., $\langle i_A \beta, C \rangle = \langle \beta, A \wedge C \rangle$ for any k-form β and $(k - a)$-vector field C. If $k < a$ then $i_A \beta = 0$.

More generally, we have the following useful formula, due to Koszul [201][4].

Lemma 1.8.4. *For any $A \in \mathcal{V}^a(M), B \in \mathcal{V}^b(M)$ we have*

$$i_{[A,B]} = (-1)^{(a-1)(b-1)} i_A \circ d \circ i_B - i_B \circ d \circ i_A$$

$$+ (-1)^a i_{A \wedge B} \circ d + (-1)^b d \circ i_{A \wedge B}. \quad (1.79)$$

---

[4]Koszul [201] wrote (1.79) as $i_{[A,B]} = [[i_A, d], i_B]$. But the brackets on the right-hand side must be understood as graded commutators of graded endomorphisms of $\Omega^\star(M)$, not usual commutators.

Proof. By induction, using the Leibniz rule.   □

Yet another equivalent definition of the Schouten bracket, via the so-called curl operator, will be given in Section 2.6.

The Schouten bracket offers a very convenient way to characterize Poisson structures and Hamiltonian vector fields:

Theorem 1.8.5. *A 2-vector field Π is a Poisson tensor if and only if the Schouten bracket of Π with itself vanishes:*

$$[\Pi, \Pi] = 0 . \tag{1.80}$$

If Π is a Poisson tensor and f is a function, then the corresponding Hamiltonian vector field X_f satisfies the equation

$$X_f = -[\Pi, f] . \tag{1.81}$$

Proof. It follows directly from Formula (1.71) that Equation (1.80), when expressed in local coordinates, is the same as Equation (1.28). Thus the first part of the above theorem is a consequence of Proposition 1.2.8. The second part also follows directly from Formula (1.71) and the definition of X_f: $-[\Pi, f] = -[\sum_{i<j} \Pi_{ij} \frac{\partial}{\partial x_i} \wedge \frac{\partial}{\partial x_j}, f] = -\sum_{i,j} \Pi_{ij} \frac{\partial}{\partial x_i} \frac{\partial f}{\partial x_j} = \sum_{i,j} \frac{\partial f}{\partial x_i} \Pi_{ij} \frac{\partial}{\partial x_j} = X_f$.   □

By abuse of language, we will call Equation (1.80) the *Jacobi identity*, because it is equivalent to the usual Jacobi identity (1.2).

Exercise 1.8.6. Let Π be a smooth Poisson tensor on a manifold M. Using Theorem 1.8.5, show that the following two statements are equivalent: a) $\operatorname{rank} \Pi \leq 2$; b) $f\Pi$ is a Poisson tensor for every smooth function f on M.

Exercise 1.8.7. Show that, if Λ is a p-vector field on a Poisson manifold (M, Π), then the Schouten bracket $[\Pi, \Lambda]$ can be given, in terms of multi-derivations, as follows:

$$[\Pi, \Lambda](f_1, \ldots, f_{p+1}) = \sum_{i=1}^{p+1} (-1)^{i+1} \{f_i, \Lambda(f_1, \ldots \hat{f}_i \ldots, f_{p+1})\}$$

$$+ \sum_{i<j} (-1)^{i+j} \Lambda(\{f_i, f_j\}, f_1, \ldots \hat{f}_i \ldots \hat{f}_j \ldots, f_{p+1}), \quad (1.82)$$

where the hat over f_i and f_j means that these terms are missing in the expression. (Hint: one can use Formula (1.78)).

1.8.2 Schouten bracket on Lie algebras

The Schouten bracket on $\mathcal{V}^*(M)$ extends the Lie bracket on $\mathcal{V}^1(M)$ by the graded Leibniz rule. Similarly, by the graded Leibniz rule (1.73,1.74), we can extend the Lie bracket on any Lie algebra $\mathfrak{g}$ to a natural graded Lie bracket on $\wedge^\star \mathfrak{g} = \bigoplus_{k=0}^\infty \wedge^k \mathfrak{g}$, where $\wedge^k \mathfrak{g}$ means $\mathfrak{g} \wedge \cdots \wedge \mathfrak{g}$ (k times), which will also be called the *Schouten bracket*. More precisely, we have:

Lemma 1.8.8. *Given a Lie algebra $\mathfrak{g}$ over $\mathbb{K}$, there is a unique bracket on $\wedge^\star \mathfrak{g} = \bigoplus_{k=0}^\infty \wedge^k \mathfrak{g}$ which extends the Lie bracket on $\mathfrak{g}$ and which satisfies the following properties, $\forall\ A \in \wedge^a \mathfrak{g}, B \in \wedge^b \mathfrak{g}, C \in \wedge^c \mathfrak{g}$:*

a) *Graded anti-commutativity:*

$$[A, B] = -(-1)^{(a-1)(b-1)}[B, A]. \tag{1.83}$$

b) *Graded Leibniz rule:*

$$[A, B \wedge C] = [A, B] \wedge C + (-1)^{(a-1)b} B \wedge [A, C], \tag{1.84}$$

$$[A \wedge B, C] = A \wedge [B, C] + (-1)^{(c-1)b}[A, C] \wedge B. \tag{1.85}$$

c) *Graded Jacobi identity:*

$$(-1)^{(a-1)(c-1)}[A, [B, C]] + (-1)^{(b-1)(a-1)}[B, [C, A]] \\ +(-1)^{(c-1)(b-1)}[C, [A, B]] = 0. \tag{1.86}$$

d) *The bracket of any element in $\wedge^\star \mathfrak{g}$ with an element in $\wedge^0 \mathfrak{g} = \mathbb{K}$ is zero.*

Proof. The proof is straightforward and is left to the reader as an exercise. Remark that, another equivalent way to define the Schouten bracket on $\wedge^\star \mathfrak{g}$ is to identify $\wedge^\star \mathfrak{g}$ with the space of left-invariant multi-vector fields on G, where G is the simply-connected Lie group whose Lie algebra is $\mathfrak{g}$, then restrict the Schouten bracket on $\mathcal{V}^\star(G)$ to these left-invariant multi-vector fields. $\square$

If $\xi : \mathfrak{g} \to \mathcal{V}^1(M)$ is an action of a Lie algebra G on a manifold M, then it can be extended in a unique way by wedge product to a map

$$\wedge \xi : \wedge^\star \mathfrak{g} \to \mathcal{V}^\star(M).$$

For example, if $x, y \in \mathfrak{g}$ then $\wedge \xi(x \wedge y) = \xi(x) \wedge \xi(y)$.

Lemma 1.8.9. *If $\xi : \mathfrak{g} \to \mathcal{V}^1(M)$ is a Lie algebra homomorphism, then its extension $\wedge \xi : \wedge^\star \mathfrak{g} \to \mathcal{V}^\star(M)$ preserves the Schouten bracket, i.e.,*

$$\wedge \xi([\alpha, \beta]) = [\wedge \xi(\alpha), \wedge \xi(\beta)] \quad \forall\ \alpha, \beta \in \wedge \mathfrak{g}.$$

Proof. The proof is straightforward, by induction, based on the Leibniz rule. $\square$

Notation 1.8.10. For an element $\alpha \in \wedge^\star \mathfrak{g}$, we will denote by α^+ the left-invariant multi-vector field on G whose value at the neutral element e of G is α, i.e., $\alpha^+(g) = L_g \alpha$, where L_g means the left translation by g. Similarly, α^- denotes the right-invariant multi-vector field $\alpha^-(g) = R_g \alpha$, where R_g means the right translation by g.

As a direct consequence of Lemma 1.8.9, we have:

Theorem 1.8.11. *For an element $r \in \mathfrak{g} \wedge \mathfrak{g}$, where $\mathfrak{g}$ is the Lie algebra of a connected Lie group G, the following three conditions are equivalent:*

a) *r satisfies the equation $[r, r] = 0$,*
b) *r^+ is a left-invariant Poisson structure on G,*
c) *r^- is a right-invariant Poisson structure on G.*

Proof. Obvious. $\qquad\qquad\qquad\qquad\qquad\qquad\qquad\qquad\qquad\qquad\qquad\qquad\qquad$ $\square$

The equation $[r, r] = 0$ is called the *classical Yang–Baxter equation*[5] [141], or *CYBE* for short. This equation will be discussed in more detail in Chapter 5.

Example 1.8.12. If $x, y \in \mathfrak{g}$ such that $[x, y] = 0$ and $x \wedge y \neq 0$, then $r = x \wedge y$ satisfies the classical Yang–Baxter equation, and the corresponding left- and right-invariant Poisson structures on G have rank 2.

1.8.3 Compatible Poisson structures

Definition 1.8.13. Two Poisson tensors Π_1 and Π_2 are called *compatible* if their Schouten bracket vanishes:

$$[\Pi_1, \Pi_2] = 0. \tag{1.87}$$

Another equivalent definition is: two Poisson structures Π_1 and Π_2 are compatible if $\Pi_1 + \Pi_2$ is also a Poisson structure. Indeed, we have $[\Pi_1 + \Pi_2, \Pi_1 + \Pi_2] = [\Pi_1, \Pi_1] + [\Pi_2, \Pi_2] + 2[\Pi_1, \Pi_2] = 2[\Pi_1, \Pi_2]$, provided that $[\Pi_1, \Pi_1] = [\Pi_2, \Pi_2] = 0$. So Equation (1.87) is equivalent to $[\Pi_1 + \Pi_2, \Pi_1 + \Pi_2] = 0$.

If Π_1 and Π_2 are two compatible Poisson structures, then we have a whole two-dimensional family of compatible Poisson structures (or projective one-dimensional family): for any scalars c_1 and c_2, $c_1\Pi_1 + c_2\Pi_2$ is a Poisson structure. Such a family of Poisson structures is often called a *pencil of Poisson structures*.

Example 1.8.14. The linear Poisson structure $x_1 \frac{\partial}{\partial x_2} \wedge \frac{\partial}{\partial x_3} + x_2 \frac{\partial}{\partial x_3} \wedge \frac{\partial}{\partial x_1} + x_3 \frac{\partial}{\partial x_1} \wedge \frac{\partial}{\partial x_2}$ on $so^*(3) = \mathbb{R}^3$ can be decomposed into the sum of two compatible linear Poisson structures $(x_1 \frac{\partial}{\partial x_2} - x_2 \frac{\partial}{\partial x_1}) \wedge \frac{\partial}{\partial x_3}$ and $x_3 \frac{\partial}{\partial x_1} \wedge \frac{\partial}{\partial x_2}$.

Example 1.8.15. If $r_1, r_2 \in \mathfrak{g} \wedge \mathfrak{g}$ are solutions of the CYBE $[r, r] = 0$, then r_1^+ and r_2^- form a pair of compatible Poisson structures on G, where G is a Lie group whose Lie algebra is $\mathfrak{g}$.

Example 1.8.16 ([253]). On the dual $\mathfrak{g}^*$ of a Lie algebra $\mathfrak{g}$, besides the standard Lie–Poisson structure $\{f, g\}_{LP}(x) = \langle [\mathrm{d}f(x), \mathrm{d}g(x)], x \rangle$, consider the following *constant* Poisson structure:

$$\{f, g\}_a(x) = \langle [\mathrm{d}f(x), \mathrm{d}g(x)], a \rangle, \tag{1.88}$$

---

[5]The Yang–Baxter equation has its origins in integrable models in statistical mechanics, and is one of the main tools in the study of integrable systems (see, e.g., [187]).

where a is a fixed element of $\mathfrak{g}^*$. This constant Poisson structure $\{,\}_a$ and the Lie–Poisson structure $\{,\}_{LP}$ are compatible. In fact, their sum is the affine (i.e., nonhomogeneous linear) Poisson structure

$$\{f, g\}(x) = \langle [\mathrm{d}f(x), \mathrm{d}g(x)], x + a \rangle, \tag{1.89}$$

which can be obtained from the linear Poisson structure $\{,\}_{LP}$ by the pull-back of the translation map $x \mapsto x + a$ on $\mathfrak{g}^*$.

Exercise 1.8.17. Suppose that Π_1 is a nondegenerate Poisson structure, i.e., it corresponds to a symplectic form ω_1. For a Poisson structure Π_2, denote by ω_2 the differential 2-form defined as follows:

$$\omega_2(X, Y) = \langle \Pi_2, i_X \omega_1 \wedge i_Y \omega_1 \rangle \ \ \forall \, X, Y \in \mathcal{V}^1(M).$$

Show that $[\Pi_1, \Pi_2] = 0$ if and only if $\mathrm{d}\omega_2 = 0$.

Exercise 1.8.18 ([37]). Consider a complex pencil of holomorphic Poisson structures $\lambda_1 \Pi_1 + \lambda_2 \Pi_2$, $\gamma_1, \gamma_2 \in \mathbb{C}$. Let S be the set of points $(\gamma_1, \gamma_2) \in \mathbb{C}^2$ such that the rank of $\gamma_1 \Pi_1 + \gamma_2 \Pi_2$ is smaller than the rank of a generic Poisson structure in the pencil. Show that if $(\gamma_1, \gamma_2), (\delta_1, \delta_2) \in \mathbb{C}^2 \setminus S$ are two arbitrary "regular" points of the pencil (which may coincide), f is a Casimir function for $\gamma_1 \Pi_1 + \gamma_2 \Pi_2$ and g is a Casimir function for $\delta_1 \Pi_1 + \delta_2 \Pi_2$, then $\{f, g\}_{\Pi_1} = \{f, g\}_{\Pi_2} = 0$.

Remark 1.8.19. A vector field X on a manifold is called a *bi-Hamiltonian system* if it is Hamiltonian with respect to two compatible Poisson structures: $X = X_{H_1}^{\Pi_1} = X_{H_2}^{\Pi_2}$. Bi-Hamiltonian systems often admit large sets of first integrals, which make them into integrable Hamiltonian systems. Conversely, a vast majority of known integrable systems turn out to be bi-Hamiltonian. The theory of bi-Hamiltonian systems starts with Magri [232] and Mischenko–Fomenko [253], and there is now a very large amount of articles on the subject. See, e.g., [2, 19, 20, 38, 102] for an introduction to the theory of integrable Hamiltonian systems.

1.9   Symplectic realizations

We have seen in Section 1.5 that Poisson manifolds can be viewed as singular foliations by symplectic manifolds. In this section, we will discuss another way to look at Poisson manifolds, namely as quotients of symplectic manifolds.

Definition 1.9.1. A *symplectic realization* of a Poisson manifold (P, Π) is a symplectic manifold (M, ω) together with a surjective Poisson submersion $\Phi : (M, \omega) \to (P, \Pi)$ (i.e., a submersion which is a Poisson map).

For example, Theorem 1.3.10 says that if G is a Lie group, then T^*G together with the left translation map $L : T^*G \to \mathfrak{g}^*$ is a symplectic realization for $\mathfrak{g}^*$.

The existence of symplectic realization for arbitrary Poisson manifolds is an important result due to Karasev [189] and Weinstein [349]:

Theorem 1.9.2 (Karasev–Weinstein). *Any smooth Poisson manifold of dimension n admits a symplectic realization of dimension $2n$.*

In fact, the result of Karasev and Weinstein is stronger: any Poisson manifold can be realized by a local symplectic groupoid (see Section 8.8). In this section, we will give a pedestrian proof of Theorem 1.9.2. First let us show a local version of it, which can be proved by an explicit formula. We will say that $\Phi : (M, \omega, L) \to (P, \Pi)$ is a *marked symplectic realization* of (P, Π), where L is a Lagrangian submanifold of M, if it is a symplectic realization such that $\Phi|_L : L \to P$ is a diffeomorphism. Note that in this case we automatically have $\dim M = 2 \dim L = 2 \dim P$.

Theorem 1.9.3 ([346]). *Any point z of a smooth Poisson manifold (P, Π) has an open neighborhood U such that (U, Π) admits a marked symplectic realization.*

Proof. Denote by $(x_1, \ldots, x_n)$ a local system of coordinates at z. We will look for functions $w_i(x, y)$, $i = 1, \ldots, n$, $x = (x_1, \ldots, x_n)$, viewed as functions in a neighborhood of z which depend smoothly on n parameters $y = (y_1, \ldots, y_n)$, such that $w_i(x, 0) = x_i$, and if we denote by $x_i = x_i(w, y)$ ($w = (w_1, \ldots, w_n)$) the inverse functions, then the map $\Theta : (w, y) \mapsto x(w, y)$ is a Poisson submersion from a symplectic manifold M with coordinates (w, y) and standard symplectic structure $\omega = \sum_i dw_i \wedge dy_i$ to a neighborhood (U, Π) of z. We may also view (x, y) as a local coordinate system on M. The condition that Θ be a Poisson map can be written as:

$$\{x_i, x_j\}_\omega(x, y) = \{x_i, x_j\}_\Pi(x) \quad (\forall\, i, j = 1, \ldots, n), \tag{1.90}$$

or

$$\sum_{h=1}^{n} \Big(\frac{\partial x_i}{\partial w_h} \frac{\partial x_j}{\partial y_h} - \frac{\partial x_i}{\partial y_h} \frac{\partial x_j}{\partial w_h} \Big) = \{x_i, x_j\}_\Pi \quad (\forall\, i, j = 1, \ldots, n). \tag{1.91}$$

Viewing the above equation as a matrix equation, and multiplying it by $(\frac{\partial w_k}{\partial x_i})_{k,i}$ on the left and $(\frac{\partial w_l}{\partial x_j})_{j,l}$ on the right of each side, we get

$$\Big(\frac{\partial w_k}{\partial x_i} \Big)_{k,i} \Big(\sum_{h=1}^{n} \Big(\frac{\partial x_i}{\partial w_h} \frac{\partial x_j}{\partial y_h} - \frac{\partial x_i}{\partial y_h} \frac{\partial x_j}{\partial w_h} \Big) \Big)_{ij} \Big(\frac{\partial w_l}{\partial x_j} \Big)_{j,l}$$
$$= \Big(\frac{\partial w_k}{\partial x_i} \Big)_{k,i} (\{x_i, x_j\}_\Pi)_{ij} \Big(\frac{\partial w_l}{\partial x_j} \Big)_{j,l}, \tag{1.92}$$

which means

$$\frac{\partial w_l}{\partial y_k} - \frac{\partial w_k}{\partial y_l} = \{w_k, w_l\}_\Pi \quad (\forall\, k, l = 1, \ldots, n). \tag{1.93}$$

Equation (1.93) with the initial condition $w_i(x, 0) = x_i$ has the following explicit local solution: denote by φ_y^t the local time-t flow of the local Hamiltonian vector field X_{f_y} of the local function $f_y = \sum_i y_i x_i$ on (P, Π). Then put (noting that φ_y^1 is well defined in a neighborhood of z when y is small enough)

$$w_i(x, y) = \int_0^1 x_i \circ \varphi_y^t dt .\qquad (1.94)$$

A straightforward computation, which will be left as an exercise (see [333, 346]), shows that this is a solution of (1.93). The local Lagrangian submanifold in question can be given by $L = \{y = 0\}$.   □

Proposition 1.9.4. *If* $\Phi_1 : (M_1, \omega_1, L_1) \to (P, \Pi)$ *and* $\Phi_2 : (M_2, \omega_2, L_2) \to (P, \Pi)$ *are two marked symplectic realizations of a Poisson manifold* (P, Π), *then there is a unique symplectomorphism* $\Psi : U(L_1) \to U(L_2)$ *from a neighborhood* $U(L_1)$ *of* L_1 *in* (M_1, ω_1) *to a neighborhood* $U(L_2)$ *of* L_2 *in* (M_2, ω_2), *which sends* L_1 *to* L_2 *and such that* $\Phi_1|_{U(L_1)} = \Phi_2|_{U(L_2)} \circ \Psi$.

Proof (sketch). Clearly, $\psi = (\Phi_2|_{L_2})^{-1} \circ \Phi_1|_{L_1} : L_1 \to L_2$ is a diffeomorphism. We want to extend it to a symplectomorphism Ψ from a neighborhood of L_1 to a neighborhood of L_2 which satisfies the conditions of the theorem. Let f be a function on P. Then Ψ must send $\Phi_1^* f$ to $\Phi_2^* f$, hence it sends the Hamiltonian vector field $X_{\Phi_1^* f}$ to $X_{\Phi_2^* f}$. If $x_1 \in M_1$ is a point close enough to L_1, then there is a point $y_1 \in L_1$ and a function f on P such that $x_1 = \phi_{\Phi_1^* f}^1(y_1)$, where ϕ_g^t denotes the time-t flow of the Hamiltonian vector field X_g of the function g, and we must have

$$\Psi(x_1) = \phi_{\Phi_2^* f}^1(\psi(y_1)).\qquad (1.95)$$

This formula shows the uniqueness of Ψ (if it can be defined) in a neighborhood of L_1. To show that this formula also defines Ψ unambiguously, we will find the graph of Ψ in $M_1 \times M_2$. Consider the distribution D on $M_1 \times M_2$, generated at each point $(x_1, x_2) \in M_1 \times M_2$ by the tangent vectors of the type $(X_{\Phi_1^* f}(x_1), X_{\Phi_2^* f}(x_2))$. The fact that Φ_1 and Φ_2 are Poisson submersions imply that this distribution is regular involutive of dimension $n = \dim P$, so we have an n-dimensional foliation. The graph of Ψ is nothing but the union of the leaves which go through the n-dimensional submanifold $\{(y_1, \psi(y_1)) \mid y_1 \in L_1\}$.

It remains to show that Φ is symplectic, and sends $\Phi_1^* f$ to $\Phi_2^* f$ for any function f on P. To show that Φ is symplectic, it suffices to show that its graph in $M_1 \times \overline{M_2}$ is Lagrangian (see Proposition 1.3.12). Since the involutive distribution D is generated by Hamiltonian vector fields $(X_{\Phi_1^* f}, X_{\Phi_2^* f})$, and the property of being Lagrangian is invariant under Hamiltonian flow, it is enough to show that the tangent spaces to the graph of Ψ at points $(y_1, \psi(y_1))$, $y_1 \in L_1$, are Lagrangian. But this last fact can be verified immediately.

Since Ψ is symplectic and $\Psi_* X_{\Phi_1^* f} = X_{\Phi_2^* f}$ by construction, it means that $\Psi_*(\Phi_1^* f)$ is equal to $\Phi_2^* f$ up to a constant. But this constant is zero, because these two functions coincide on L_2 by the construction of Ψ. Thus $\Psi_*(\Phi_1^* f) = \Phi_2^* f$ for any function f on P, implying that $\Phi_1 = \Phi_2 \circ \Psi$.   □

Theorem 1.9.2 is now a direct consequence of the local realization Theorem 1.9.3 and the uniqueness Proposition 1.9.4: there is a unique way to glue local marked symplectic realizations together, which glues the marked Lagrangian submanifolds together on their overlaps, to get a marked symplectic realization of a given Poisson manifold.   □

Remark 1.9.5. Of course, (non-marked) symplectic realizations of a Poisson manifold (P, Π) of dimension n are not necessarily of dimension $2n$. For example, if (M, ω) is a symplectic realization of (P, Π) and (N, σ) is a symplectic manifold, then $M \times N$ is also a symplectic realization of P. And if (P, Π) is symplectic then it is a symplectic realization of itself. Proposition 1.9.4 can be generalized to an "essential uniqueness" result for non-marked local symplectic realizations (see [346]).

An important notion in symplectic geometry, directly related to symplectic realizations, is the following:

Definition 1.9.6 ([209]). A foliation $\mathcal{F}$ on a symplectic manifold (M, ω) is called a *symplectically complete foliation* if the symplectically orthogonal distribution $(T\mathcal{F})^{\perp}$ to $\mathcal{F}$ is integrable.

In other words, $\mathcal{F}$ is a symplectically complete foliation if there is another foliation $\mathcal{F}'$ such that $T_x\mathcal{F} = (T_x\mathcal{F}')^{\perp} \ \forall \ x \in M$. In this case, the pair $(\mathcal{F}, \mathcal{F}')$ is called a *dual pair*. For example, any Lagrangian foliation is a symplectically complete foliation which is dual to itself.

Theorem 1.9.7 (Libermann [209]). *Let $\Phi : (M, \omega) \to P$ be a surjective submersion from a symplectic manifold (M, ω) to a manifold P, such that the level sets of Φ are connected. Denote by $\mathcal{F}$ the foliation whose leaves are level sets of Φ. Then there is a (unique) Poisson structure Π on P such that $\Phi : (M, \omega) \to (P, \Pi)$ is Poisson if and only if $\mathcal{F}$ is symplectically complete.*

Proof (sketch). The symplectically orthogonal distribution $(T\mathcal{F})^{\perp}$ to $\mathcal{F}$ is generated by Hamiltonian vector fields of the type $X_{\Phi^* f}$ where f is a function on P. The integrability of $(T\mathcal{F})^{\perp}$ is equivalent to the fact that $[X_{\Phi^* f}, X_{\Phi^* g}]$ is tangent to $(T\mathcal{F})^{\perp}$ for any functions f, g on P. In other words, $X_{\{\Phi^* f, \Phi^* g\}}$ is tangent to $(T\mathcal{F})^{\perp}$, i.e., $\{\Phi^* f, \Phi^* g\}$ is constant on the leaves of $\mathcal{F}$. Since the leaves of $\mathcal{F}$ are level sets of Φ, it means that there is a function h on P such that $\{\Phi^* f, \Phi^* g\} = \Phi^* h$. In other words, $\{f, g\} := h$ is a Poisson bracket on P such that Φ is Poisson.   □

Chapter 2

Poisson Cohomology

2.1 Poisson cohomology

2.1.1 Definition of Poisson cohomology

Poisson cohomology was introduced by Lichnerowicz [211]. Its existence is based on the following simple lemma.

Lemma 2.1.1. *If Π is a Poisson tensor, then for any multi-vector field A we have*

$$[\Pi, [\Pi, A]] = 0 \ . \tag{2.1}$$

Proof. By the graded Jacobi identity (1.75) for the Schouten bracket, if Π is a 2-vector field and A is an a-vector field, then

$$(-1)^{a-1}[\Pi, [\Pi, A]] - [\Pi, [A, \Pi]] + (-1)^{a-1}[A, [\Pi, \Pi]] = 0 \ .$$

Moreover, $[A, \Pi] = -(-1)^{a-1}[\Pi, A]$ due to the graded anti-commutativity, hence $[\Pi, [\Pi, A]] = -\frac{1}{2}[A, [\Pi, \Pi]]$. Now if Π is a Poisson structure, then $[\Pi, \Pi] = 0$, and therefore $[\Pi, [\Pi, A]] = 0$. $\qquad\square$

Let (M, Π) be a smooth Poisson manifold. Denote by $\delta = \delta_\Pi : \mathcal{V}^\star(M) \longrightarrow \mathcal{V}^\star(M)$ the $\mathbb{R}$-linear operator on the space of multi-vector fields on M, defined as follows:

$$\delta_\Pi(A) = [\Pi, A]. \tag{2.2}$$

Then Lemma 2.1.1 says that δ_Π is a *differential operator* in the sense that $\delta_\Pi \circ \delta_\Pi = 0$. The corresponding differential complex $(\mathcal{V}^\star(M), \delta)$, i.e.,

$$\cdots \longrightarrow \mathcal{V}^{p-1}(M) \xrightarrow{\delta} \mathcal{V}^p(M) \xrightarrow{\delta} \mathcal{V}^{p+1}(M) \longrightarrow \cdots \ , \tag{2.3}$$

will be called the *Lichnerowicz complex*. The cohomology of this complex is called *Poisson cohomology*.

By definition, Poisson cohomology groups of (M, Π), i.e., the cohomology groups of the Lichnerowicz complex (2.3), are the quotient groups

$$H_\Pi^p(M) = \frac{\ker(\delta : \mathcal{V}^p(M) \longrightarrow \mathcal{V}^{p+1}(M))}{\operatorname{Im}(\delta : \mathcal{V}^{p-1}(M) \longrightarrow \mathcal{V}^p(M))} . \qquad (2.4)$$

The above Poisson cohomology groups are also denoted by $H^p(M, \Pi)$, or also $H_{LP}^p(M, \Pi)$, where LP stands for Lichnerowicz–Poisson.

Remark 2.1.2. Poisson cohomology groups can be very big, infinite-dimensional. For example, when $\Pi = 0$ then $H_\Pi^\star(M) := \bigoplus_k H_\Pi^k(M) = \mathcal{V}^\star(M)$. Poisson cohomology groups of smooth Poisson manifolds have a natural induced topology from the Fréchet spaces of multi-vector fields, which make them into not-necessarily-separated locally convex topological vector spaces (see Ginzburg [143, 144]).

2.1.2   Interpretation of Poisson cohomology

The zeroth Poisson cohomology group $H_\Pi^0(M)$ is the group of functions $f \in \mathcal{C}^\infty(M)$ such that $X_f = -[\Pi, f] = 0$. In other words, $H_\Pi^0(M)$ is the space of *Casimir functions* of Π, i.e., the space of first integrals of the associated symplectic foliation.

The first Poisson cohomology group $H_\Pi^1(M)$ is the quotient of the space of Poisson vector fields (i.e., vector fields X such that $[\Pi, X] = 0$) by the space of Hamiltonian vector fields (i.e., vector fields of the type $[\Pi, f] = X_{-f}$). Poisson vector fields are infinitesimal automorphisms of the Poisson structures, while Hamiltonian vector fields may be interpreted as *inner* infinitesimal automorphisms. Thus $H_\Pi^1(M)$ may be interpreted as the space of *outer infinitesimal automorphisms* of Π.

The second Poisson cohomology group $H_\Pi^2(M)$ is the quotient of the space of 2-vector fields Λ which satisfy the equation $[\Pi, \Lambda] = 0$ by the space of 2-vector fields of the type $\Lambda = [\Pi, Y]$. If $[\Pi, \Lambda] = 0$ and ε is a formal (infinitesimal) parameter, then $\Pi + \varepsilon \Lambda$ satisfies the Jacobi identity up to terms of order ε^2:

$$[\Pi + \varepsilon\Lambda, \Pi + \varepsilon\Lambda] = \varepsilon^2[\Lambda, \Lambda] = 0 \mod \varepsilon^2. \qquad (2.5)$$

So one may view $\Pi + \varepsilon\Lambda$ as an infinitesimal deformation of Π in the space of Poisson tensors. On the other hand, up to terms of order ε^2, $\Pi + \varepsilon[\Pi, Y]$ is equal to $(\varphi_Y^\varepsilon)_*\Pi$, where φ_Y^ε denotes the time-ε flow of Y. Therefore $\Pi + \varepsilon[\Pi, Y]$ is a trivial infinitesimal deformation of Π up to an infinitesimal diffeomorphism. Thus, $H_\Pi^2(M)$ is the quotient of the space of all possible infinitesimal deformations of Π by the space of trivial deformations. In other words, $H_\Pi^2(M)$ may be interpreted as the moduli space of formal *infinitesimal deformations* of Π. For this reason, the second Poisson cohomology group plays a central role in the study of normal forms of Poisson structures.

The third Poisson cohomology group $H_\Pi^3(M)$ may be interpreted as the space of *obstructions to formal deformation*. Suppose that we have an infinitesimal deformation $\Pi + \varepsilon\Lambda$, i.e., $[\Pi, \Lambda] = 0$. Then a priori, $\Pi + \varepsilon\Lambda$ satisfies the Jacobi identity only modulo ε^2. To make it satisfy the Jacobi identity modulo ε^3, we have to add a term $\varepsilon^2 \Lambda_2$ such that

$$[\Pi + \varepsilon\Lambda + \varepsilon^2\Lambda_2, \Pi + \varepsilon\Lambda + \varepsilon^2\Lambda_2] = 0 \mod \varepsilon^3. \tag{2.6}$$

The equation to solve is $2[\Pi, \Lambda_2] = -[\Lambda, \Lambda]$. This equation can be solved if and only if the cohomology class of $[\Lambda, \Lambda]$ in $H_\Pi^3(M)$ is trivial. Similarly, if (2.6) is already satisfied, to find a term $\varepsilon^3 \Lambda_3$ such that

$$[\Pi + \varepsilon\Lambda + \varepsilon^2\Lambda_2 + \varepsilon^3\Lambda_3, \Pi + \varepsilon\Lambda + \varepsilon^2\Lambda_2 + \varepsilon^3\Lambda_3] = 0 \mod \varepsilon^4, \tag{2.7}$$

we have to make sure that the cohomology class of $[\Lambda, \Lambda_2]$ in $H_\Pi^3(M)$ vanishes, and so on.

The Poisson tensor Π is itself a cocycle in the Lichnerowicz complex. If the cohomology class of Π in $H_\Pi^2(M)$ vanishes, i.e., there is a vector field Y such that $\Pi = [\Pi, Y]$, then Π is called an *exact Poisson structure*.

2.1.3 Poisson cohomology versus de Rham cohomology

Recall that, the Poisson structure Π gives rise to a homomorphism

$$\sharp = \sharp_\Pi : T^*M \longrightarrow TM, \tag{2.8}$$

which associates to each covector α a unique vector $\sharp(\alpha)$ such that

$$\langle \alpha \wedge \beta, \Pi \rangle = \langle \beta, \sharp(\alpha) \rangle \tag{2.9}$$

for any covector β. This homomorphism is an isomorphism if and only if Π is nondegenerate, i.e., is a symplectic structure. By taking exterior powers of the above map, we can extend it to a homomorphism

$$\sharp : \Lambda^p T^*M \longrightarrow \Lambda^p TM, \tag{2.10}$$

and hence a $\mathcal{C}^\infty(M)$-linear homomorphism

$$\sharp : \Omega^p(M) \longrightarrow \mathcal{V}^p(M), \tag{2.11}$$

where $\Omega^p(M)$ denotes the space of smooth differential forms of degree p on M. Recall that $\sharp$ is called the *anchor map* of Π.

Lemma 2.1.3. *For any smooth differential form η on a given smooth Poisson manifold (M, Π) we have*

$$\sharp(d\eta) = -[\Pi, \sharp(\eta)] = -\delta_\Pi(\sharp(\eta)) . \tag{2.12}$$

Proof. By induction on the degree of η, using the Leibniz rule. If η is a function then $\sharp(\eta) = \eta$ and $\sharp(d\eta) = -[\Pi, \eta] = X_\eta$, the Hamiltonian vector field of η. If $\eta = df$ is an exact 1-form then $\sharp(d\eta) = 0$ and $[\Pi, \sharp(\eta)] = [\Pi, X_f] = 0$, hence Equation

(2.12) is satisfied. If Equation (2.12) is satisfied for a differential p-form η and a differential q-form μ, then its also satisfied for their exterior product $\eta \wedge \mu$. Indeed, we have $\sharp(d(\eta \wedge \mu)) = \sharp(d\eta \wedge \mu + (-1)^p \eta \wedge d\mu) = \sharp(d\eta) \wedge \sharp(\mu) + (-1)^p \sharp(\eta) \wedge \sharp(d\mu) = -[\Pi, \sharp(\eta)] \wedge \sharp(\mu) - (-1)^p \sharp(\eta) \wedge [\Pi, \sharp(\mu)] = -[\Pi, \sharp(\eta) \wedge \sharp(\mu)] = -[\Pi, \sharp(\eta \wedge \mu)]$. $\square$

The above lemma means that, up to a sign, the operator $\sharp$ intertwines the usual differential operator d of the *de Rham complex*

$$\cdots \longrightarrow \Omega^{p-1}(M) \xrightarrow{\;d\;} \Omega^p(M) \xrightarrow{\;d\;} \Omega^{p+1}(M) \longrightarrow \cdots \tag{2.13}$$

with the differential operator δ_Π of the Lichnerowicz complex. In particular, it induces a linear homomorphism of the corresponding cohomologies. In other words, we have:

Theorem 2.1.4 ([211]). *For every smooth Poisson manifold (M, Π), there is a natural homomorphism*

$$\sharp^* : H^\star_{dR}(M) = \bigoplus_p H^p_{dR}(M) \longrightarrow H^\star_\Pi(M) = \bigoplus_p H^p_\Pi(M) \tag{2.14}$$

from its de Rham cohomology to its Poisson cohomology, induced by the map $\sharp = \sharp_\Pi$. If M is a symplectic manifold, then this homomorphism is an isomorphism.

When M is symplectic, $\sharp$ is an isomorphism, and that's why $\sharp^*$ is also an isomorphism. $\square$

Remark 2.1.5. The de Rham cohomology has a graded Lie algebra structure, given by the cap product (induced from the exterior product of differential forms). So does the Poisson cohomology. The Lichnerowicz homomorphism $\sharp^* : H^\star_{dR}(M) \longrightarrow H^\star_\Pi(M)$ in the above theorem is not only a linear homomorphism, but also an algebra homomorphism.

Remark 2.1.6. If (M, Π) is not symplectic then the map $\sharp^* : H^\star_{dR}(M) \to H^\star_\Pi(M)$ is not an isomorphism in general. In particular, while de Rham cohomology groups of manifolds of "finite type" (e.g., compact manifolds) are of finite dimensions, Poisson cohomology groups may have infinite dimension in general. An interesting and largely open question is: what are the conditions for the Lichnerowicz homomorphism to be injective or surjective?

2.1.4   Other versions of Poisson cohomology

If, in the Lichnerowicz complex, instead of *smooth* multi-vector fields, we consider other classes of multi-vector fields, then we arrive at other versions of Poisson cohomology. For example, if Π is an analytic Poisson structure, and one considers analytic multi-vector fields, then one gets *analytic Poisson cohomology*.

Recall that, a *germ* of an object (e.g., a function, a differential form, a Riemannian metric, etc.) at a point z is an object defined in a neighborhood of z. Two germs at z are considered to be the same if there is a neighborhood of z in which they coincide. When considering a germ of smooth (resp. analytic) Poisson structure Π at a point z, it is natural to talk about *germified Poisson cohomology*: the space $\mathcal{V}^\star(M)$ in the Lichnerowicz complex is replaced by the space of germs of smooth (resp. analytic) multi-vector fields. More generally, given any subset N of a Poisson manifold (M, Π), one can define germified Poisson cohomology at N. Similarly, one can talk about *formal Poisson cohomology*. By convention, the germ of a formal multi-vector field is itself. Viewed this way, formal Poisson cohomology is the formal version of germified Poisson cohomology.

If M is not compact, then one may be interested in *Poisson cohomology with compact support*, by restricting one's attention to multi-vector fields with compact support. Remark that Theorem 2.1.4 also holds in the case with compact support: if (M, Π) is a symplectic manifold then its de Rham cohomology with compact support is isomorphic to its Poisson cohomology with compact support.

If one considers only multi-vector fields which are tangent to the characteristic distribution, then one gets *tangential Poisson cohomology*. (A multi-vector field λ is said to be *tangent* to a distribution $\mathcal{D}$ on a manifold M if at each point $x \in M$ one can write $\Lambda(x) = \sum a_i v_{i1} \wedge \cdots \wedge v_{is}$ where v_{ij} are vectors lying in $\mathcal{D}$.) It is easy to see that the homomorphism $\sharp^*$ in Theorem 2.1.4 also makes sense for tangential Poisson cohomology (and tangential de Rham cohomology).

The above versions of Poisson cohomology also have a natural interpretation, similar to the one given for smooth Poisson cohomology.

2.1.5 Computation of Poisson cohomology

If a Poisson structure Π on a manifold M is nondegenerate (i.e., symplectic), then Poisson cohomology of Π is the same as de Rham cohomology of M. There are many tools for computing de Rham cohomology groups, and these groups have probably been computed for most "familiar" manifolds, see, e.g., [41, 138]. However, when Π is not symplectic, $H_\Pi^\star(M)$ is much more difficult to compute than $H_{dR}^\star(M)$ in general, and at the moment of writing of this book, there are few Poisson (non-symplectic) manifolds for which Poisson cohomology has been computed. For one thing, $H_\Pi^\star(M)$ can have infinite dimension even when M is compact, and the problem of determining whether $H_\Pi^\star(M)$ is finite dimensional or not is already a difficult open problem for most Poisson structures that we know of.

Nevertheless, various tools from algebraic topology and homological algebra can be adapted to the problem of computation of Poisson cohomology. One of them is the classical Mayer–Vietoris sequence (see, e.g., [41]). The following Poisson cohomology version of Mayer–Vietoris sequence is absolutely analogous to its de Rham cohomology version.

Proposition 2.1.7 ([333]). *Let U and V be two open subsets of a smooth Poisson manifold (M, Π). Then*

$$0 \longrightarrow \mathcal{V}^\star(U \cup V) \xrightarrow{\alpha} \mathcal{V}^\star(U) \oplus \mathcal{V}^\star(V) \xrightarrow{\beta} \mathcal{V}^\star(U \cap V) \longrightarrow 0, \qquad (2.15)$$

where $\alpha(\Lambda) = (\Lambda|_U, \Lambda|_V)$ is the restriction map, and $\beta(\Lambda_1, \Lambda_2) = \Lambda_1|_{U \cap V} - \Lambda_2|_{U \cap V}$ is the difference map, is an exact sequence of smooth Lichnerowicz complexes, and the corresponding cohomological long exact sequence (called the Mayer–Vietoris *sequence) has the form*

$$\cdots \longrightarrow H^k(U \cup V, \Pi) \xrightarrow{\alpha_*} H^k(U, \Pi) \oplus H^k(V, \Pi) \xrightarrow{\beta_*}$$
$$H^k(U \cap V, \Pi) \longrightarrow H^{k+1}(U \cup V, \Pi) \xrightarrow{\alpha_*} \cdots . \quad (2.16)$$

The proof of Proposition 2.1.7 is also absolutely similar to the proof of its de Rham version. The above Mayer–Vietoris sequence reduces the computation of Poisson cohomology on a manifold to the computation of Poisson cohomology on small open sets (which contain singularities of the Poisson structure). To study (germified) Poisson cohomology of singularities of Poisson structures, one can try to use the tools from singularity theory. This will be done in Section 2.5 for Poisson structures in dimension 2.

Another standard tool is the spectral sequence, which will be discussed in Section 2.4.

In the case of linear Poisson structures, Poisson cohomology is intimately related to Lie algebra cohomology, also known as Chevalley–Eilenberg cohomology, which will be discussed in Section 2.3.

In Chapter 8, we will interpret Poisson cohomology as a particular case of cohomology of Lie algebroids. This leads to a definition and study of Poisson cohomology from a purely algebraic point of view, as was done by Huebschmann [180].

Some other methods for computing and studying Poisson cohomology include: the use of *symplectic groupoids*[1] to reduce the computation of Poisson cohomology of certain Poisson manifolds to the computation of de Rham cohomology of other manifolds [359]; the *van Est map* which relates Lie algebroid cohomology with differentiable cohomology of Lie groupoids [355, 85]; comparison of Poisson cohomology of Poisson manifolds which are *Morita equivalent* [147, 146, 145, 85]; equivariant Poisson cohomology [144].

2.2 Normal forms of Poisson structures

Consider a Poisson structure Π on a manifold M. In a given system of coordinates $(x_1, \ldots, x_m)$, Π has the expression

$$\Pi = \sum_{i<j} \Pi_{ij} \frac{\partial}{\partial x_i} \wedge \frac{\partial}{\partial x_j} = \frac{1}{2} \sum_{i,j} \Pi_{ij} \frac{\partial}{\partial x_i} \wedge \frac{\partial}{\partial x_j}. \quad (2.17)$$

A priori, the coefficients Π_{ij} of Π may be very complicated, non-polynomial functions. The idea of normal forms is to simplify these coefficients in the expression of Π.

---

[1]Symplectic groupoids will be introduced in Section 7.5.

A (local) *normal form* of Π is a Poisson structure

$$\Pi' = \sum_{i<j} \Pi'_{ij} \frac{\partial}{\partial x'_i} \wedge \frac{\partial}{\partial x'_j} = \frac{1}{2} \sum_{i,j} \Pi'_{ij} \frac{\partial}{\partial x'_i} \wedge \frac{\partial}{\partial x'_j} \qquad (2.18)$$

which is (locally) isomorphic to Π, i.e., there is a (local) diffeomorphism $\varphi : (x_i) \mapsto (x'_i)$ called a *normalization* such that $\varphi_* \Pi = \Pi'$, such that the functions Π'_{ij} are "simpler" than the functions Π_{ij}. The ideal would be that Π'_{ij} were constant functions. According to Remark 1.4.6, such a local normal form exists when Π is a (locally) regular Poisson structure.

Near a singular point of Π, we can use the splitting Theorem 1.4.5 to write Π as the direct sum of a constant symplectic structure with a Poisson structure which vanishes at a point. The local normal form problem for Π is then reduced to the problem of local normal forms for a Poisson structure which vanishes at a point.

Having this in mind, we now assume that Π vanishes at the origin 0 of a given local coordinate system $(x_1, \ldots, x_m)$. Denote by

$$\Pi = \Pi^{(k)} + \Pi^{(k+1)} + \cdots + \Pi^{(k+n)} + \cdots \qquad (k \geq 1) \qquad (2.19)$$

the Taylor expansion of Π in the coordinate system $(x_1, \ldots, x_m)$, where for each $h \in \mathbb{N}$, $\Pi^{(h)}$ is a 2-vector field whose coefficients $\Pi^{(h)}_{ij}$ are homogeneous polynomial functions of degree h. $\Pi^{(k)}$, assumed to be nontrivial, is the term of lowest degree in Π, and is called the *homogeneous part*, or *principal part* of Π. If $k = 1$ then $\Pi^{(1)}$ is called the *linear part* of Π, and so on. This homogeneous part can be defined intrinsically, i.e., it does not depend on the choice of local coordinates.

At the formal level, the Jacobi identity for Π can be written as

$$\begin{aligned}
0 &= [\Pi, \Pi] = [\Pi^{(k)} + \Pi^{(k+1)} + \cdots, \Pi^{(k)} + \Pi^{(k+1)} + \cdots] \\
&= [\Pi^{(k)}, \Pi^{(k)}] + 2[\Pi^{(k)}, \Pi^{(k+1)}] + 2[\Pi^{(k)}, \Pi^{(k+2)}] + [\Pi^{(k+1)}, \Pi^{(k+1)}] + \cdots,
\end{aligned}$$

which leads to (by considering terms of the same degree):

$$\begin{aligned}
&[\Pi^{(k)}, \Pi^{(k)}] = 0, \\
&2[\Pi^{(k)}, \Pi^{(k+1)}] = 0, \\
&2[\Pi^{(k)}, \Pi^{(k+2)}] + [\Pi^{(k+1)}, \Pi^{(k+1)}] = 0, \\
&\cdots
\end{aligned} \qquad (2.20)$$

In particular, the homogeneous part $\Pi^{(k)}$ of Π is a Poisson structure, and Π may be viewed as a deformation of $\Pi^{(k)}$. A natural homogenization question arises: is this deformation trivial? In other words, is Π locally (or formally) isomorphic to its homogeneous part $\Pi^{(k)}$? That's where Poisson cohomology comes in, because, as explained in Subsection 2.1.2, Poisson cohomology governs (formal) deformations of Poisson structures.

When $k = 1$, one talks about the linearization problem, and when $k = 2$ one talks about the quadratization problem, and so on. These problems, for Poisson structures and related structures like Nambu structures, Lie algebroids and Lie groupoids, will be studied in detail in the subsequent chapters of this book. Here we will discuss, at the formal level, a more general problem of *quasi-homogenization*.

Denote by

$$Z = \sum_{i=1}^{n} w_i x_i \frac{\partial}{\partial x_i}, \quad w_i \in \mathbb{N} \tag{2.21}$$

a given diagonal linear vector field with the following special property: its coefficients w_i are positive integers. Such a vector field is called a *quasi-radial vector field*. (When $w_i = 1 \, \forall i$, we get the usual *radial vector field*.)

A multi-vector field Λ is called *quasi-homogeneous* of degree d ($d \in \mathbb{Z}$) with respect to Z if

$$\mathcal{L}_Z \Lambda = d\Lambda. \tag{2.22}$$

For a function f, it means $Z(f) = df$. For example, a monomial k-vector field

$$\left(\prod_{i=1}^{n} x_i^{a_i}\right) \frac{\partial}{\partial x_{j_1}} \wedge \cdots \wedge \frac{\partial}{\partial x_{j_k}}, \quad a_i \in \mathbb{Z}_{\geq 0}, \tag{2.23}$$

is quasi-homogeneous of degree $\sum_{i=1}^{n} a_i w_i - \sum_{s=1}^{k} w_{j_s}$. In particular, monomial terms of high degree in the usual sense (i.e., with large $\sum a_i$) also have high quasi-homogeneous degree. As a consequence, quasi-homogeneous (smooth, formal or analytic) multi-vector fields are automatically polynomial in the usual sense. Note that the quasi-homogeneous degree of a monomial multi-vector field can be negative, though it is always greater than or equal to $-\sum_{i=1}^{n} w_i$.

Given a Poisson structure Π with $\Pi(0) = 0$, by abuse of notation, we will now denote by

$$\Pi = \Pi^{(d_1)} + \Pi^{(d_2)} + \cdots, \quad d_1 < d_2 < \cdots \tag{2.24}$$

the quasi-homogeneous Taylor expansion of Π with respect to Z, where each term $\Pi^{(d_i)}$ is quasi-homogeneous of degree d_i. The term $\Pi^{(d_1)}$, assumed to be nontrivial, is called the *quasi-homogeneous part* of Π. Similarly to the case with usual homogeneous Taylor expansion, the Jacobi identity for Π implies the Jacobi identity for $\Pi^{(d_1)}$, which means that $\Pi^{(d_1)}$ is a quasi-homogeneous Poisson structure, and Π may be viewed as a deformation of $\Pi^{(d_1)}$. The quasi-homogenization problem is the following: is there a transformation of coordinates which sends Π to $\Pi^{(d_1)}$, i.e., which kills all the terms of quasi-homogeneous degree $> d_1$ in the expression of Π?

In order to treat this quasi-homogenization problem at the formal level, we will need the quasi-homogeneous graded version of Poisson cohomology.

Let $\Pi^{(d)}$ be a Poisson structure on an n-dimensional space $V = \mathbb{K}^n$, which is quasi-homogeneous of degree d with respect to a given quasi-radial vector field

$Z = \sum_{i=1}^{n} w_i x_i \frac{\partial}{\partial x_i}$. For each $r \in \mathbb{Z}$, denote by $\mathcal{V}_{(r)}^k = \mathcal{V}_{(r)}^k(\mathbb{K}^n)$ the space of quasi-homogeneous polynomial k-vector fields on $\mathbb{K}^n$ of degree r with respect to Z. Of course, we have

$$\mathcal{V}^k = \oplus_r \mathcal{V}_{(r)}^k, \tag{2.25}$$

where $\mathcal{V}^k = \mathcal{V}^k(\mathbb{K}^n)$ is the space of all polynomial vector fields on $\mathbb{K}^n$. Note that, if $\Lambda \in \mathcal{V}_{(r)}^k$ then

$$\mathcal{L}_Z[\Pi^{(d)}, \Lambda] = [\mathcal{L}_Z \Pi^{(d)}, \Lambda] + [\Pi^{(d)}, \mathcal{L}_Z \Lambda] = (d+r)[\Pi^{(d)}, \Lambda],$$

i.e., $\delta_{\Pi^{(d)}} \Lambda = [\Pi^{(d)}, \Lambda] \in \mathcal{V}_{(r+d)}^{k+1}$. The group

$$H_{(r)}^k(\Pi^{(d)}) = \frac{\ker(\delta_{\Pi^{(d)}} : \mathcal{V}_{(r)}^k \longrightarrow \mathcal{V}_{(r+d)}^{k+1})}{\operatorname{Im}(\delta_{\Pi^{(d)}} : \mathcal{V}_{(r-d)}^{k-1} \longrightarrow \mathcal{V}_{(r)}^k)} \tag{2.26}$$

is called the kth quasi-homogeneous of degree r Poisson cohomology group of $\Pi^{(d)}$. Of course, there is a natural injection from $H_{(r)}^k(\Pi^{(d)})$ to the usual (formal, analytic or smooth) Poisson cohomology group $H^k(\Pi^{(d)})$ of $\Pi^{(d)}$ over $\mathbb{K}^n$. While $H^k(\Pi^{(d)})$ may be of infinite dimension, $H_{(r)}^k(\Pi^{(d)})$ is always of finite dimension (for each r).

Return now to the quasi-homogeneous Taylor series $\Pi = \Pi^{(d_1)} + \Pi^{(d_2)} + \cdots$. The Jacobi identity for Π implies that $[\Pi^{(d_1)}, \Pi^{(d_2)}] = 0$, i.e., $\Pi^{(d_2)}$ is a quasi-homogeneous cocycle in the Lichnerowicz complex of $\Pi^{(d_1)}$. If this term $\Pi^{(d_2)}$ is a coboundary, i.e., $\Pi^{(d_2)} = [\Pi^{(d_1)}, X^{(d_2 - d_1)}]$ for some quasi-homogeneous vector field $X^{(d_2 - d_1)} = X_i^{(d_2 - d_1)} \partial/\partial x_i$, then the coordinate transformation $x_i' = x_i - X_i^{(d_2 - d_1)}$ will kill the term $\Pi^{(d_2)}$ in the expression of Π. More generally, we have:

Proposition 2.2.1. *With the above notations, suppose that* $\Pi^{(d_k)} = [\Pi^{(d_1)}, X] + \Lambda^{(d_k)}$ *for some $k > 1$, where $X = X_i \partial/\partial x_i$ is a quasi-homogeneous vector field of degree $d_k - d_1$. Then the diffeomorphism (coordinate transformation) $\phi : (x_i) \mapsto (x_i') = (x_i - X_i)$ transforms Π into*

$$\phi_* \Pi = \Pi^{(d_1)} + \cdots + \Pi^{(d_{k-1})} + \Lambda^{(d_k)} + \widetilde{\Pi}^{(d_{k+1})} \cdots . \tag{2.27}$$

In other words, this transformation suppresses the term $[\Pi^{(d_1)}, X]$ without changing the terms of degree strictly smaller than d_k.

Proof. Denote by $\Gamma = \phi_* \Pi$. For the Poisson structure Π we have

$$\{x_i', x_j'\} = \sum_{uv} \frac{\partial x_i'}{\partial x_u} \frac{\partial x_j'}{\partial x_v} \{x_u, x_v\} = \sum_{uv} \frac{\partial x_i'}{\partial x_u} \frac{\partial x_j'}{\partial x_v} \Pi_{uv}$$

$$= \sum_{uv} (\delta_i^u - \frac{\partial X_i}{\partial x_u})(\delta_j^v - \frac{\partial X_j}{\partial x_v})(\Pi^{(d_1)} + \Pi^{(d_2)} + \cdots)_{uv},$$

where δ_i^u is the Kronecker symbol, and the terms of degree smaller than or equal to d_k in this expression give

$$(\Pi^{(d_1)} + \cdots + \Pi^{(d_k)})_{ij} - \sum_u \frac{\partial X_i}{\partial x_u} \Pi_{uj}^{(d_1)} - \sum_v \frac{\partial X_j}{\partial x_v} \Pi_{iv}^{(d_1)}.$$

On the other hand, by definition, $\{x_i', x_j'\}$ is equal to $\Gamma_{ij} \circ \phi$. But the terms of degree smaller than or equal to d_k in the expansion of $\Gamma_{ij} \circ \phi$ are

$$(\Gamma^{(d_1)} + \cdots + \Gamma^{(d_k)})_{ij} - \sum_s X_s \frac{\partial \Gamma_{ij}^{(d_1)}}{\partial x_s}.$$

Comparing the terms of degree $d_1, \ldots, d_{k-1}$, we get $\Gamma_{ij}^{(d_1)} = \Pi_{ij}^{(d_1)}, \ldots, \Gamma_{ij}^{(d_{k-1})} = \Pi_{ij}^{(d_{k-1})}$. As for the terms of degree d_k, they give

$$\Gamma_{ij}^{(d_k)} - \sum_s X_s \frac{\partial \Pi_{ij}^{(d_1)}}{\partial x_s} = \Pi_{ij}^{(d_k)} - \sum_u \frac{\partial X_i}{\partial x_u} \Pi_{uj}^{(d_1)} - \sum_v \frac{\partial X_j}{\partial x_v} \Pi_{iv}^{(d_1)}.$$

As we have

$$[X, \Pi^{(d_1)}]_{ij} = \sum_s X_s \frac{\partial \Pi_{ij}^{(d_1)}}{\partial x_s} - \sum_u \frac{\partial X_i}{\partial x_u} \Pi_{uj}^{(d_1)} - \sum_v \frac{\partial X_j}{\partial x_v} \Pi_{iv}^{(d_1)},$$

it follows that

$$\Gamma_{ij}^{(d_k)} = \Pi_{ij}^{(d_k)} + [X, \Pi^{(d_1)}]_{ij} = \Pi_{ij}^{(d_k)} - [\Pi^{(d_1)}, X]_{ij} = \Lambda_{ij}^{(d_k)}.$$

The proposition is proved. $\square$

Theorem 2.2.2. *If the quasi-homogeneous Poisson cohomology groups* $H_{(r)}^2(\Pi^{(d)})$ *of a quasi-homogeneous Poisson structure* $\Pi^{(d)}$ *of degree d are trivial for all $r > d$, then any Poisson structure admitting a formal quasi-homogeneous expansion* $\Pi = \Pi^{(d)} + \Pi^{(d_2)} + \cdots$ *is formally isomorphic to its quasi-homogeneous part* $\Pi^{(d)}$.

Proof. Use Proposition 2.2.1 to kill the terms of degree strictly greater than d in Π consecutively. $\square$

Example 2.2.3. One can use Theorem 2.2.2 to show that any Poisson structure on $\mathbb{K}^2$ of the form $\Pi = f \frac{\partial}{\partial x} \wedge \frac{\partial}{\partial y}$, where $f = x^2 + y^3 +$ higher-order terms, is formally isomorphic to $(x^2 + y^3) \frac{\partial}{\partial x} \wedge \frac{\partial}{\partial y}$. (This is a simple singularity studied by Arnold, see Theorem 2.5.2.) The quasi-radial vector field in this case is $Z = 3x \frac{\partial}{\partial x} + 2y \frac{\partial}{\partial y}$.

2.3 Cohomology of Lie algebras

Let $\Pi^{(1)}$ be a linear Poisson structure on a vector space $\mathbb{K}^n$. Denote by $\mathfrak{g} = ((\mathbb{K}^n)^*, \{,\}_{\Pi^{(1)}})$ the Lie algebra corresponding to $\Pi^{(1)}$. We will see in this section that Poisson cohomology groups of $\Pi^{(1)}$ are special cases of Lie algebra cohomology of $\mathfrak{g}$.

2.3.1 Chevalley–Eilenberg complexes

Let W be a $\mathfrak{g}$-module, i.e., a vector space together with a Lie algebra homomorphism $\rho : \mathfrak{g} \to End(W)$ from $\mathfrak{g}$ to the Lie algebra of endomorphisms of W. In other words, ρ is a linear map such that $\rho([x,y]) = \rho(x).\rho(y) - \rho(y).\rho(x)\ \forall x, y \in \mathfrak{g}$. The action of an element $x \in \mathfrak{g}$ on a vector $v \in W$ is defined by

$$x.v = \rho(x)(v). \tag{2.28}$$

One associates to W the following complex, called *Chevalley–Eilenberg complex* of $\mathfrak{g}$ with coefficients in W [77]:

$$\cdots \xrightarrow{\delta} C^{k-1}(\mathfrak{g}, \rho) \xrightarrow{\delta} C^k(\mathfrak{g}, \rho) \xrightarrow{\delta} C^{k+1}(\mathfrak{g}, \rho) \xrightarrow{\delta} \cdots, \tag{2.29}$$

where

$$C^k(\mathfrak{g}, \rho) = (\wedge^k \mathfrak{g}^*) \otimes W \tag{2.30}$$

$(k \geq 0)$ is the space of k-multilinear antisymmetric maps from $\mathfrak{g}$ to W: an element $\theta \in C^k(\mathfrak{g}, \rho)$ may be presented as a k-multilinear antisymmetric map from $\mathfrak{g}$ to W, or a linear map from $\wedge^k \mathfrak{g}$ to W:

$$\theta(x_1, \ldots, x_k) = \theta(x_1 \wedge \cdots \wedge x_k) \in W,\ x_i \in \mathfrak{g}. \tag{2.31}$$

The operator $\delta = \delta_{CE} : C^k(\mathfrak{g}, \rho) \to C^{(k+1)}(\mathfrak{g}, \rho)$ in the Chevalley–Eilenberg complex is defined as follows:

$$(\delta\theta)(x_1, \ldots, x_{k+1}) = \sum_i (-1)^{i+1} \rho(x_i)(\theta(x_1, \ldots, \widehat{x_i}, \ldots, x_{k+1}))$$
$$+ \sum_{i<j} (-1)^{i+j} \theta([x_i, x_j], x_1, \ldots, \widehat{x_i} \ldots \widehat{x_j}, \ldots, x_{k+1}), \tag{2.32}$$

the symbol $\widehat{\ }$ above a variable means that this variable is missing in the list.

It is a classical result [77], which follows directly from the Jacobi identity of $\mathfrak{g}$, that $\delta_{CE} \circ \delta_{CE} = 0$. It means that the Chevalley–Eilenberg complex is a differential complex with differential operator $\delta = \delta_{CE}$. Its cohomology groups

$$H^k(\mathfrak{g}, \rho) = H^k(\mathfrak{g}, W) = \frac{\ker(\delta : C^k(\mathfrak{g}, \rho) \longrightarrow C^{k+1}(\mathfrak{g}, \rho))}{\mathrm{Im}(\delta : C^{k-1}(\mathfrak{g}, \rho) \longrightarrow C^k(\mathfrak{g}, \rho))} \tag{2.33}$$

are called *cohomology groups of* $\mathfrak{g}$ with coefficients in W (or with respect to the representation ρ).

Remark 2.3.1. Formula (2.32) is absolutely analogous to Cartan's formula (1.7). This construction of differential operators is sometimes referred to as *Cartan–Chevalley–Eilenberg construction*. If G is a connected Lie group with Lie algebra $\mathfrak{g}$, then the space $\Omega_L^\star(G)$ of left-invariant differential forms on G is a subcomplex of the de Rham complex of G which is naturally isomorphic to the Chevalley–Eilenberg $C^\star(\mathfrak{g}, \mathbb{R})$ for the trivial action of $\mathfrak{g}$ on $\mathbb{R}$, which implies that their cohomologies are also isomorphic:

$$H_L^\star(G) \cong H^\star(\mathfrak{g}, \mathbb{R}). \tag{2.34}$$

(The isomorphism from $\Omega_L^\star(G)$ to $C^\star(\mathfrak{g}, \mathbb{R})$ associates to each left-invariant differential form on G its value at the neutral element e of G, after the identification of $\mathfrak{g}^*$ with T_e^*G.) In particular, when G is compact, the averaging process $\alpha \mapsto \int_G L_g^* \alpha \, dg$ (where α denotes a differential form on G, and L_g denotes the left translation by $g \in G$) induces an isomorphism from $H_{dR}^\star(G)$ to $H_L^\star(G)$, and we have $H_{dR}^\star(G) \cong H^\star(g, \mathbb{R})$.

Exercise 2.3.2. Show that, given a smooth Poisson manifold (M, Π), its Lichnerowicz complex can be identified with a subcomplex of the Chevalley–Eilenberg complex of the (infinite-dimensional) Lie algebra $C^\infty(M)$ with coefficients in $C^\infty(M)$ (with respect to the adjoint action given by the Poisson bracket), which consists of cochains which are multi-derivations. (Hint: use Formula (1.82)).

In general, the problem of computation of $H(\mathfrak{g}, W)$ for a finite-dimensional $\mathfrak{g}$-module W of a finite-dimensional Lie algebra $\mathfrak{g}$ is a problem of linear algebra: one simply has to deal with finite-dimensional systems of linear equations. However, even for low-dimensional Lie algebras, these systems of linear equations often have high dimensions and require thousands or millions of computations, so it is not easy to do it by hand.

Fortunately, cohomology of semisimple Lie algebras is relatively simple, due in part to the following results, known as *Whitehead's lemmas*.

Theorem 2.3.3 (Whitehead). *If $\mathfrak{g}$ is semisimple, and W is a finite-dimensional $\mathfrak{g}$-module, then $H^1(\mathfrak{g}, W) = 0$ and $H^2(\mathfrak{g}, W) = 0$.*

Theorem 2.3.4 (Whitehead). *If $\mathfrak{g}$ is semisimple, and W is a finite-dimensional $\mathfrak{g}$-module such that $W^\mathfrak{g} = 0$, where $W^\mathfrak{g} = \{w \in W \mid x.w = 0 \ \forall \ x \in \mathfrak{g}\}$ denotes the set of elements in W which are invariant under the action of $\mathfrak{g}$, then $H^k(\mathfrak{g}, W) = 0 \ \forall \ k \geq 0$.*

See, e.g., [186] for the proof of Whitehead's lemmas. A refined (normed) version of Theorem 2.3.3 will be proved in Chapter 3. Let us also mention that if $\mathfrak{g}$ is simple then $\dim H^3(\mathfrak{g}, \mathbb{K}) = 1$. Combining the two Whitehead's lemmas with the fact that any finite-dimensional module W of a semisimple Lie algebra $\mathfrak{g}$ is completely reducible, one gets the following formula:

$$H^\star(\mathfrak{g}, W) = H^\star(\mathfrak{g}, \mathbb{K}) \otimes W^\mathfrak{g} = \bigoplus_{k \neq 1,2} H^k(\mathfrak{g}, \mathbb{K}) \otimes W^\mathfrak{g}. \tag{2.35}$$

Remark 2.3.5. If W is a smooth Fréchet module of a compact Lie group G and $\mathfrak{g}$ is the Lie algebra of G, then the formula $H^\star(\mathfrak{g}, W) = H^\star(\mathfrak{g}, \mathbb{R}) \otimes W^\mathfrak{g}$ is still valid, see Ginzburg [144]. In particular, if a compact Lie group G acts on a smooth manifold M, then $C^\infty(M)$ is a smooth Fréchet G-module, and we have

$$H^\star(\mathfrak{g}, C^\infty(M)) = H^\star(\mathfrak{g}, \mathbb{R}) \otimes (C^\infty(M))^\mathfrak{g}. \tag{2.36}$$

Remark 2.3.6. Cohomology of Lie algebras is closely related to differentiable (or continuous) cohomology of Lie groups, via the so-called Van Est map and Van Est spectral sequence. See, e.g., [40, 157].

2.3.2 Cohomology of linear Poisson structures

Consider now the case $W = \mathcal{S}^q \mathfrak{g}$, the q-symmetric power of $\mathfrak{g}$ together with the adjoint action of $\mathfrak{g}$:

$$\rho(x)(x_{i_1} \ldots x_{i_q}) = \sum_{s=1}^{q} x_{i_1} \ldots [x, x_{i_s}] \ldots x_{i_q}. \tag{2.37}$$

Since $\mathfrak{g} = ((\mathbb{K}^n)^*, \{,\}_{\Pi^{(1)}})$, the space $W = \mathcal{S}^q \mathfrak{g}$ can be naturally identified with the space of homogeneous polynomials of degree q on $\mathbb{K}^n$, and we can write

$$\rho(x).f = \{x, f\}, \tag{2.38}$$

where $f \in \mathcal{S}^q \mathfrak{g}$, and $\{x, f\}$ denotes the Poisson bracket of x with f with respect to $\Pi^{(1)}$.

Denote by $\mathcal{V}^p_{(q)} = \mathcal{V}^p_{(q)}(\mathbb{K}^n)$ the space of homogeneous p-vector fields of degree q. (It is the same as the space of quasi-homogeneous p-vector fields of quasi-homogeneous degree $q - p$ with respect to the radial vector field $\sum x_i \partial/\partial x_i$.) $\mathcal{V}^p_{(q)}$ can be identified with $C^p(\mathfrak{g}, \mathcal{S}^q \mathfrak{g})$ as follows: For

$$A = \sum_{i_1 < \cdots < i_p} A_{i_1,\ldots,i_p} \frac{\partial}{\partial x_{i_1}} \wedge \cdots \wedge \frac{\partial}{\partial x_{i_p}} \in \mathcal{V}^p_{(q)} \tag{2.39}$$

define $\theta_A \in C^p(\mathfrak{g}, \mathcal{S}^q \mathfrak{g})$ by

$$\theta_A(x_{i_1}, \ldots, x_{i_p}) = A_{i_1,\ldots,i_p}. \tag{2.40}$$

Lemma 2.3.7. *With the above identification $A \leftrightarrow \theta_A$, the Lichnerowicz differential operator $\delta_{LP} = [\Pi^{(1)}, .] : \mathcal{V}^p_{(q)} \longrightarrow \mathcal{V}^{p+1}_{(q)}$ is identified with the Chevalley–Eilenberg differential operator $\delta_{CE} : C^p(\mathfrak{g}, \mathcal{S}^q \mathfrak{g}) \longrightarrow C^{p+1}(\mathfrak{g}, \mathcal{S}^q \mathfrak{g})$.*

Proof. We must show that $\theta_{[\Pi^{(1)}, A]} = \delta_{CE} \theta_A$ for $A \in \mathcal{V}^p_{(q)}$. Write

$$\Pi^{(1)} = \tfrac{1}{2} \sum_{i,j,k} c^k_{ij} x_k \frac{\partial}{\partial x_i} \wedge \frac{\partial}{\partial x_j}, \quad \text{and} \quad A = \tfrac{1}{p!} \sum_{i_1,\ldots,i_p} A_{i_1,\ldots,i_p} \frac{\partial}{\partial x_{i_1}} \wedge \cdots \wedge \frac{\partial}{\partial x_{i_p}}.$$

Denote the Poisson bracket of $\Pi^{(1)}$ by $\{,\}$. By the Leibniz rule, we have

$$[\Pi^{(1)}, A] = E_1 + E_2$$

where

$$
\begin{aligned}
E_1 &= \frac{1}{p!} \sum [\Pi^{(1)}, A_{i_1,\dots,i_p}] \wedge \frac{\partial}{\partial x_{i_1}} \wedge \cdots \wedge \frac{\partial}{\partial x_{i_p}} \\
&= \frac{1}{p!} \sum_{i_1,\dots,i_p,i} \{x_i, A_{i_1,\dots,i_p}\} \frac{\partial}{\partial x_i} \wedge \frac{\partial}{\partial x_{i_1}} \wedge \cdots \wedge \frac{\partial}{\partial x_{i_p}}
\end{aligned}
$$

and

$$
\begin{aligned}
E_2 &= \frac{1}{p!} \sum A_{i_1,\dots,i_p} \left[\Pi^{(1)}, \frac{\partial}{\partial x_{i_1}} \wedge \cdots \wedge \frac{\partial}{\partial x_{i_p}}\right] \\
&= \frac{1}{2p!} \sum_{i_1,\dots,i_p,i,j,s} (-1)^s A_{i_1,\dots,i_p} c_{ij}^{i_s} \frac{\partial}{\partial x_i} \wedge \frac{\partial}{\partial x_j} \wedge \frac{\partial}{\partial x_{i_1}} \wedge \cdots \wedge \widehat{\frac{\partial}{\partial x_{i_s}}} \cdots \wedge \frac{\partial}{\partial x_{i_p}}.
\end{aligned}
$$

It means that

$$E_1(dx_{i_1},\dots,dx_{i_{p+1}}) = \sum_s (-1)^{s+1} \{x_{i_s}, A_{i_1,\dots\hat{i_s}\dots,i_{p+1}}\}$$

and

$$E_2(dx_{i_1},\dots,dx_{i_{p+1}}) = \sum_{u<v;k} (-1)^{u+v} A_{k,i_1,\dots\hat{i_u}\dots\hat{i_v}\dots,i_{p+1}} c_{i_u i_v}^k.$$

On the other hand, we have

$$
\begin{aligned}
\delta\theta_A & (x_{i_1},\dots,x_{i_{p+1}}) \\
&= \sum_u (-1)^{u+1} \rho(x_{i_u}) A(x_{i_1},\dots \widehat{x_{i_u}} \dots, x_{i_{p+1}}) \\
&\quad + \sum_{u<v} (-1)^{u+v} A([x_{i_u}, x_{i_v}], x_{i_1}, \dots \widehat{x_{i_u}} \dots \widehat{x_{i_v}} \dots, x_{i_{p+1}}) \\
&= \sum_u (-1)^{u+1} \{x_{i_u}, A_{i_1,\dots\hat{i_u}\dots,i_{p+1}}\} \\
&\quad + \sum_{u<v;k} (-1)^{u+v} c_{i_u i_v}^k A(x_k, x_{i_1}, \dots \widehat{x_{i_u}} \dots \widehat{x_{i_v}} \dots, x_{i_{p+1}}).
\end{aligned}
$$

It remains to compare the above formulas. $\square$

An immediate consequence of Lemma 2.3.7 and Theorem 2.2.2 is the following:

Theorem 2.3.8 ([346]). *If $\mathfrak{g}$ is a finite-dimensional Lie algebra such that*

$$H^2(\mathfrak{g}, \mathcal{S}^k \mathfrak{g}) = 0 \quad \forall\, k \geq 2,$$

then any formal Poisson structure Π which vanishes at a point and whose linear part $\Pi^{(1)}$ at that point corresponds to $\mathfrak{g}$ is formally linearizable. In particular, it is the case when $\mathfrak{g}$ is semisimple.

Remark 2.3.9. In Lemma 2.3.7, the fact that A is homogeneous is not so important. What is important is that the module W in question can be identified with a subspace of the space of functions on $\mathbb{K}^n$, where the action of $\mathfrak{g}$ is given by the Poisson bracket, i.e., by Formula (2.38). The following smooth (as compared to homogeneous) version of Lemma 2.3.7 is also true, with a similar proof (see, e.g., [144, 148, 221, 222]): if U is an Ad^*-invariant open subset of the dual $\mathfrak{g}^*$ of the Lie algebra $\mathfrak{g}$ of a connected Lie group G (or more generally, an open subset of a dual Poisson–Lie group G^* which is invariant under the dressing action of G – Poisson–Lie groups will be introduced later in the book), then

$$H_\Pi^\star(U) \cong H^\star(\mathfrak{g}, C^\infty(U)), \tag{2.41}$$

where the action of $\mathfrak{g}$ on $C^\infty(U)$ is induced by the coadjoint (or dressing) action, and a natural isomorphism exists already at the level of cochain complexes. In particular, if G is compact semisimple, then this formula together with the Fréchet-module version of Whitehead's lemmas (Remark 2.3.5) leads to the following formula (see [148]):

$$H_\Pi^\star(U) = H^\star(\mathfrak{g}) \otimes (C^\infty(U))^G = \bigoplus_{k \neq 1,2} H^k(\mathfrak{g}) \otimes (C^\infty(U))^G. \tag{2.42}$$

2.3.3 Rigid Lie algebras

Theorem 2.3.8 means that the second cohomology group

$$H^2(\mathfrak{g}, \mathcal{S}_{\geq 2}\mathfrak{g}) = \bigoplus_{k \geq 2} H^2(\mathfrak{g}, \mathcal{S}^k \mathfrak{g})$$

governs nonlinear deformations of the linear Poisson structure of $\mathfrak{g}^*$. Meanwhile, the group $H^2(\mathfrak{g}, \mathfrak{g})$ governs deformations of $\mathfrak{g}$ itself (or equivalently, linear deformations of the Poisson structure on $\mathfrak{g}^*$ associated to $\mathfrak{g}$).

An n-dimensional Lie algebra $\mathfrak{g}$ over $\mathbb{K}$ can be determined by its structure constants c_{ij}^k with respect to a given basis (x_i): $[x_i, x_j]_{\mathfrak{g}} = \sum c_{ij}^k x_k$. The n^3-tuple of coefficients (c_{ij}^k) is called a Lie algebra structure of dimension n. The set $\mathcal{A}(n, \mathbb{K}) \subset \mathbb{K}^{n^3}$ of all Lie algebra structures of dimension n is an algebraic variety (the Jacobi identity and the anti-commutativity give the system of algebraic equations which determine this set). The full linear group $GL(n, \mathbb{K})$ acts naturally on $\mathcal{A}(n, \mathbb{K})$ by changes of basis, and two Lie algebra structures are isomorphic if and only if they lie on the same orbit of $GL(n, \mathbb{K})$. An n-dimensional Lie algebra $\mathfrak{g}$ is called *rigid* if the orbit of its structure is an open subset of $\mathcal{A}(n, \mathbb{K})$ with respect to the usual topology induced from the Euclidean topology of $\mathbb{K}^{n^3}$; equivalently, any Lie algebra $\mathfrak{g}'$ close enough to $\mathfrak{g}$ is isomorphic to $\mathfrak{g}$.

Theorem 2.3.10 (Nijenhuis–Richardson [279]). *If $\mathfrak{g}$ is a finite-dimensional Lie algebra such that $H^2(\mathfrak{g}, \mathfrak{g}) = 0$, then $\mathfrak{g}$ is rigid. In particular, semisimple Lie algebras are rigid.*

Remark 2.3.11. The condition $H^2(\mathfrak{g}, \mathfrak{g}) = 0$ is a sufficient but not a necessary condition for the rigidity of a Lie algebra. For example, Richardson [300] showed that, for any odd integer $n > 5$, the semi-direct product $\mathfrak{l}_n = sl(2, \mathbb{K}) \ltimes W^{2n+1}$, where W^{2n+1} is the $(2n + 1)$-dimensional irreducible $sl(2, \mathbb{K})$-module, is rigid but has $H^2(\mathfrak{l}_n, \mathfrak{l}_n) \neq 0$. In fact, $H^2(\mathfrak{g}, \mathfrak{g}) \neq 0$ means that there are nontrivial infinitesimal deformations, but not every infinitesimal deformation can be made into a true deformation. See, e.g., [62, 63, 151].

2.4 Spectral sequences

2.4.1 Spectral sequence of a filtered complex

Spectral sequences are one of the main tools for computing cohomology groups. The general idea is as follows.

Let $(C = \oplus_{k \in \mathbb{Z}_+} C^k, \delta)$ be a differential complex. It means that C^k $(k \geq 0)$ are vector spaces (or more generally, Abelian groups), and $\delta : C^k \to C^{k+1}$ are linear operators such that $\delta \circ \delta = 0$.

Assume that (C, δ) admits a *filtration* $(C_h)_{h \in \mathbb{N}}$. It means that each C^k is filtered by subspaces

$$C^k = C_0^k \supset C_1^k \supset C_2^k \supset \cdots , \tag{2.43}$$

such that $\delta C_h^k \subset C_h^{k+1} \; \forall \; k, h$. In other words, $C_h = \bigoplus_k C_h^k$ is a differential subcomplex of C_{h-1} for $h \geq 1$, and $C_0 = C$. Put $C_h = C$ if $h < 0$ by convention. Using this filtration, one decomposes cohomology groups

$$H^k(C) = \frac{Z^k}{B^k} = \frac{\ker(\delta : C^k \longrightarrow C^{k+1})}{\mathrm{Im}(\delta : C^{k-1} \longrightarrow C^k)} \tag{2.44}$$

into smaller pieces $H_h^k(C)/H_{h+1}^k(C)$, where $H_h^k(C)$ consists of the elements of $H^k(C)$ which can be represented by cocycles lying in C_h^k. The group

$$\bigoplus_{h \geq 0} H_h^k(C)/H_{h+1}^k(C) \tag{2.45}$$

is called the graded version of $H^k(C)$; it is linearly isomorphic to $H^k(C)$ if, say, the filtration is finite, i.e., $C_n = 0$ for some $n \in \mathbb{N}$.

A way to compute $H_h^k(C)/H_{h+1}^k(C)$ and $H^k(C)$ is to use the *spectral sequence* $(E_r^{p,q})_{r\geq 0}$ of the above filtered complex. By definition,

$$E_r^{p,q} = \frac{Z_r^{p,q} + C_{p+1}^{p+q}}{B_r^{p,q} + C_{p+1}^{p+q}} \tag{2.46}$$

where

$$Z_r^{p,q} = \left\{ y \in C_p^{p+q} \mid \delta y \in C_{p+r}^{p+q+1} \right\} \tag{2.47}$$

and

$$B_r^{p,q} = \left\{ y \in C_p^{p+q} \mid y = \delta z, z \in C_{p-r+1}^{p+q-1} \right\}. \tag{2.48}$$

Clearly, $B_r^{p,q} \subset Z_r^{p,q}$, $B_r^{p,q} \subset B_{r+1}^{p,q}$ and $Z_r^{p,q} \supset Z_{r+1}^{p,q}$. Hence $E_r^{p,q}$ is well defined, and is bigger than $E_{r+1}^{p,q}$. (There is a surjection from a subgroup of $E_r^{p,q}$ to $E_{r+1}^{p,q}$.) As r tends to ∞, the group $E_r^{p,q}$ gets smaller and smaller, and it approximates better and better the group $H_p^{p+q}(C)/H_{p+1}^{p+q}(C)$. In fact, if the filtration is of finite length, i.e., $C_n = 0$ for some $n \in \mathbb{N}$, then

$$E_r^{p,q} = \frac{Z^{p+q} \cap C_p^{p+q} + C_{p+1}^{p+q}}{B^{p+q} \cap C_p^{p+q} + C_{p+1}^{p+q}} \cong H_p^{p+q}(C)/H_{p+1}^{p+q}(C) \quad \forall\, r \geq n, p. \tag{2.49}$$

In general, one says that $(E_r^{p,q})$ converges if its limit

$$E_\infty^{p,q} = \lim_{r\to\infty} E_r^{p,q} \tag{2.50}$$

is isomorphic to $H_p^{p+q}(C)/H_{p+1}^{p+q}(C)$.

The terms $E_r^{p,q}$ of the spectral sequence can be computed inductively on r (that's why they are useful for computing $H_p^{p+q}(C)/H_{p+1}^{p+q}(C)$). The zeroth term is:

$$E_0^{p,q} = C_p^{p+q}/C_{p+1}^{p+q}. \tag{2.51}$$

In other words, $E_0 = \oplus E_0^{p,q}$ is just the graded version of the complex C. For $r \geq 0$, the differential operator δ induces an operator on $(E_r^{p,q})$, denoted by δ_r:

$$\delta_r : E_r^{p,q} \to E_r^{p+r,q-r+1}. \tag{2.52}$$

(The image of $y \in Z_r^{p,q}$ mod $B_r^{p,q} + C_{p+1}^{p+q}$ under δ_r is $\delta y \in Z_r^{p+r,q-r+1}$ mod $B_r^{p+r,q-r+1} + C_{p+r+1}^{p+q}$. One verifies directly that δ_r is well defined.)

Since $\delta \circ \delta = 0$, we also have $\delta_r \circ \delta_r = 0$, i.e., δ_r is a differential operator. It turns out that $E_{r+1}^{p,q}$ is nothing but the cohomology of $E_r^{p,q}$ with respect to δ_r:

$$E_{r+1}^{p,q} = \frac{\ker(\delta_r : E_r^{p,q} \longrightarrow E_r^{p+r,q-r+1})}{\mathrm{Im}(\delta_r : E_r^{p-r,q+r-1} \longrightarrow E_r^{p,q})}. \tag{2.53}$$

Exercise 2.4.1. Verify Formula (2.53), starting from (2.46), (2.47) and (2.48).

Remark 2.4.2. In the literature, Formula (2.53) is often used in the definition of spectral sequences, and Formulas (2.46), (2.47) and (2.48) only show up after or not at all.

2.4.2 Leray spectral sequence

As a first example of spectral sequences, let us consider a locally trivial fibration $\pi : M \to N$ of a manifold M over a connected manifold N with fibers diffeomorphic to F. The de Rham complex $\Omega^{\star}(M)$ of differential forms on M admits a natural filtration with respect to this fibration: $\Omega_h^k(M)$ $(h \geq 0)$ is the subspace of $\Omega^k(M)$ consisting of k-forms ω which satisfy the following condition:

$$\omega_x(X_1, \ldots, X_k) = 0 \; \forall x \in M, X_1, \ldots, X_k \in T_x M \; s.t. \; \pi_* X_1 = \cdots = \pi_* X_{k-h+1} = 0. \tag{2.54}$$

The associated spectral sequence of this filtration is known as the *Leray spectral sequence*. Its zeroth term $E_0^{p,q}$ can be written as follows:

$$E_0^{p,q} \cong \Omega^p(N, \Omega^q(F)). \tag{2.55}$$

More precisely, $E_0^{p,q} = \Omega_p^{p+q}(M)/\Omega_{p+1}^{p+q}(M)$ is naturally isomorphic to the space of vector-valued p-forms on N with values in the vector bundle over N whose fiber over a point $y \in N$ is the space of q-forms on the fiber $F_y = \pi^{-1}(y)$ of the fibration of M over N. The first and second terms are

$$E_1^{p,q} = \Omega^p(N, H^q(F)) \tag{2.56}$$

and

$$E_2^{p,q} = H^p(N, H^q(F)). \tag{2.57}$$

In the above formulas, $H^q(F)$ must be understood as a vector bundle over N whose fiber over $y \in N$ is $H_{dR}^q(F_y)$, i.e., it is a local system of coefficients. If N is simply connected then this bundle is automatically trivial and we can write

$$E_2^{p,q} = H_{dR}^p(N) \otimes H_{dR}^q(F). \tag{2.58}$$

Example 2.4.3. The de Rham cohomology of the special unitary groups $SU(n)$ can be computed inductively on n with the help of the Leray spectral sequence associated to the natural fibration of $SU(n)$ over S^{2n-1} with fiber $SU(n-1)$ (this fibration is obtained via the natural action of $SU(n)$ on the unit sphere S^{2n-1} in $\mathbb{C}^n$). When $n = 2$, $SU(2)$ is diffeomorphic to the three-dimensional sphere S^3, so we will simply write $H_{dR}^{\star}(SU(2)) = H_{dR}^{\star}(S^3)$. When $n = 3$, the second terms of the Leray spectral sequence of the fibration $SU(2) \to SU(3) \to S^5$ are as follows:

$$E_2^{0,3} = \mathbb{R}, \quad E_2^{5,3} = \mathbb{R},$$
$$E_2^{0,0} = \mathbb{R}, \quad E_2^{5,0} = \mathbb{R},$$

(the other second terms are zero). The differential δ_2 is automatically trivial, because, for example, it maps the nontrivial term $E_2^{0,3}$ to the trivial term $E_2^{2,2}$. Similarly, all the other differentials δ_r, $r \geq 2$, are trivial, because there are only four nontrivial cells $E^{0,3}, E^{5,3}, E^{0,0}, E^{5,0}$, and no differential δ_r connects two of these cells. It means that the Leray spectral sequence degenerates at E_2, implying that

$$H^\star_{dR}(SU(3)) \cong H^\star_{dR}(SU(2)) \otimes H^\star_{dR}(S^5) \cong H^\star_{dR}(S^3 \times S^5). \qquad (2.59)$$

This isomorphism between $H^\star_{dR}(SU(3))$ and $H^\star_{dR}(S^3 \times S^5)$ is actually an algebra isomorphism, because the Leray spectral sequence is compatible with the product structure of the de Rham cohomology in a natural sense. Using this compatibility, one can show inductively on n that the Leray spectral sequence for the fibration $SU(n-1) \to SU(n) \to S^{n-1}$ degenerates at the second term E_2 for any $n \geq 3$, leading to the following algebra isomorphism:

$$H^\star_{dR}(SU(n)) \cong H^\star_{dR}(S^3 \times S^5 \times \cdots \times S^{2n-1}). \qquad (2.60)$$

See, e.g., [41, 138] for details and other applications of Leray spectral sequences and other spectral sequences in topology.

2.4.3 Hochschild–Serre spectral sequence

Given a Lie algebra $\mathfrak{l}$, a Lie subalgebra $\mathfrak{r} \subset \mathfrak{l}$, and an $\mathfrak{l}$-module W, the Chevalley–Eilenberg complex $C(\mathfrak{l}, W)$ has the following natural filtration with respect to $\mathfrak{r}$:

$$C^k_h(\mathfrak{l}, W) = \left\{ f \in C^k(\mathfrak{l}, W) \mid f(x_1, \ldots, x_k) = 0 \text{ if } x_1, \ldots, x_{k-h+1} \in \mathfrak{r} \right\}. \qquad (2.61)$$

By convention, $C^k_0(\mathfrak{l}, W) = C^k(\mathfrak{l}, W)$, and $C^k_h(\mathfrak{l}, W) = 0$ if $h > k$. One checks directly that $\delta C_h(\mathfrak{l}, W) \subset C_h(\mathfrak{l}, W)$, i.e., it is really a filtered complex. The corresponding spectral sequence is known as the *Hochschild–Serre spectral sequence* [178].

We will be mainly interested in a special case of this spectral sequence, when $\mathfrak{r}$ is an ideal of $\mathfrak{l}$, and the quotient Lie algebra $\mathfrak{g} = \mathfrak{l}/\mathfrak{r}$ is semisimple. (This is the case, for example, when $\mathfrak{r}$ is the *radical*, i.e., the maximal solvable ideal of $\mathfrak{l}$.) In this case, the Hochschild–Serre spectral sequence leads to the following theorem.

Theorem 2.4.4 (Hochschild–Serre [178]). *Let $\mathfrak{l}$ be a finite-dimensional Lie algebra over $\mathbb{K}$ ($\mathbb{K} = \mathbb{R}$ or $\mathbb{C}$), $\mathfrak{r}$ be an ideal of $\mathfrak{l}$ such that $\mathfrak{g} = \mathfrak{l}/\mathfrak{r}$ is semisimple, and W be a finite-dimensional $\mathfrak{l}$-module. Then*

$$H^k(\mathfrak{l}, W) \cong \bigoplus_{i+j=k} H^i(\mathfrak{g}, \mathbb{K}) \otimes H^j(\mathfrak{r}, W)^{\mathfrak{g}} \quad \forall \, k \geq 0. \qquad (2.62)$$

In the above theorem, $H^j(\mathfrak{r}, W)$ has a natural structure of $\mathfrak{g}$-module which will be explained below, and $H^j(\mathfrak{r}, W)^{\mathfrak{g}}$ is the subspace of $H^j(\mathfrak{r}, W)$ consisting of elements which are invariant under the action of $\mathfrak{g}$.

Proof (sketch). Since $\mathfrak{g}$ is semisimple, by the classical Levi–Malcev theorem there is a Lie algebra injection $\mathfrak{g} \to \mathfrak{l}$ whose composition with the projection map $\mathfrak{l} \to \mathfrak{l}/\mathfrak{r} = \mathfrak{g}$ is identity. (See, e.g., [42, 335], and the beginning of Chapter 3.) In other words, we may assume that $\mathfrak{g}$ is a Lie subalgebra of $\mathfrak{l}$. As a vector space, $\mathfrak{l}$ is the direct sum of $\mathfrak{g}$ with $\mathfrak{r}$. As a Lie algebra, $\mathfrak{l}$ can be written a semi-direct product $\mathfrak{l} = \mathfrak{g} \ltimes \mathfrak{r}$.

By definition, the zeroth term $E_0^{p,q}$ of the spectral sequence is

$$E_0^{p,q} = C_p^{p+q}(\mathfrak{l}, W)/C_{p+1}^{p+q}(\mathfrak{l}, W). \tag{2.63}$$

This space can be naturally identified with $C^p(\mathfrak{g}, C^q(\mathfrak{r}, W))$. Indeed, if we denote by $(x_1, \ldots, x_m, y_1, \ldots, y_{n-m})$ a basis of $\mathfrak{l}$ such that $(x_1, \ldots, x_m)$ span $\mathfrak{g}$ and $(y_1, \ldots, y_{n-m})$ span $\mathfrak{r}$, then an element $f \in C_p^{p+q}(\mathfrak{l}, W)$ mod $C_{p+1}^{p+q}(\mathfrak{l}, W)$ is completely determined by its value on elements of the type

$$x_{i_1} \wedge \cdots \wedge x_{i_p} \wedge y_{j_1} \wedge \cdots \wedge y_{j_q}.$$

The map

$$\theta_f : x_{i_1} \wedge \cdots \wedge x_{i_p} \mapsto \left(y_{j_1} \wedge \cdots \wedge y_{j_q} \mapsto f(x_{i_1} \wedge \cdots \wedge x_{i_p} \wedge y_{j_1} \wedge \cdots \wedge y_{j_q}) \right)$$

is a linear map from $\wedge^p \mathfrak{g}$ to $C^q(\mathfrak{r}, W)$, i.e., $\theta_f \in C^p(\mathfrak{g}, C^q(\mathfrak{r}, W))$. Note that $C^q(\mathfrak{r}, W) = \wedge^q \mathfrak{r} \otimes W$ is a $\mathfrak{g}$-module: $\mathfrak{g}$ acts on W by the restriction of the action of $\mathfrak{l}$; it acts on $\mathfrak{r}$ by the adjoint action of $\mathfrak{g}$ in $\mathfrak{l}$, and on $\mathfrak{r}^*$ by the dual action. It is clear that the correspondence $f \leftrightarrow \theta_f$ is one-to-one. Thus, we can write

$$E_0^{p,q} \cong C^p(\mathfrak{g}, C^q(\mathfrak{r}, W)). \tag{2.64}$$

The next step is to look at the first spectral term $E_1^{p,q}$, which is the cohomology of $E_0^{p,q}$ with respect to $\delta_0 : E_0^{p,q} \to E_0^{p,q+1}$. Using the identification $E_0^{p,q} \cong C^p(\mathfrak{g}, C^q(\mathfrak{r}, W))$, we can write δ_0 as

$$\delta_0 : C^p(\mathfrak{g}, C^q(\mathfrak{r}, W)) \to C^p(\mathfrak{g}, C^{q+1}(\mathfrak{r}, W)). \tag{2.65}$$

It follows that

$$E_1^{p,q} \cong C^p(\mathfrak{g}, H^q(\mathfrak{r}, W)). \tag{2.66}$$

Similarly, we have

$$E_2^{p,q} \cong H^p(\mathfrak{g}, H^q(\mathfrak{r}, W)). \tag{2.67}$$

Whitehead's lemma (see Formula (2.35)) implies that

$$E_2^{p,q} \cong H^p(\mathfrak{g}, H^q(\mathfrak{r}, W)^{\mathfrak{g}}) \cong H^p(\mathfrak{g}, \mathbb{K}) \otimes H^q(\mathfrak{r}, W)^{\mathfrak{g}}. \tag{2.68}$$

If $f \in \wedge^p \mathfrak{g}^*$ and $g \in \wedge^q \mathfrak{r}^* \otimes W$ are cocycles, and moreover g is invariant under the action of $\mathfrak{g}$ on $\wedge^q \mathfrak{r}^* \otimes W$, then their product

$$f \otimes g \in \wedge^p \mathfrak{g}^* \otimes \wedge^q \mathfrak{r}^* \otimes W \subset \wedge^{p+q} \mathfrak{l}^* \otimes W \tag{2.69}$$

is a cocycle. Equation (2.68) means that elements of $E_2^{p,q}$ can be written as linear combinations of such cocycles $f \otimes g$. In particular, any element of $E_2^{p,q}$ can be represented by a cocycle in $Z^{p+q}(\mathfrak{l}, W)$. It implies that $\delta_2 = \delta_3 = \cdots = 0$, and the Hochschild–Serre spectral sequence degenerates (stabilizes) at E_2, i.e.,

$$E_2^{p,q} = E_3^{p,q} = \cdots = E_\infty^{p,q}.$$

Since the filtration is clearly of finite length, we have

$$H_p^{p+q}(\mathfrak{l}, W)/H_{p+1}^{p+q}(\mathfrak{l}, W) \cong E_\infty^{p,q} \cong H^p(\mathfrak{g}, \mathbb{K}) \otimes H^q(\mathfrak{r}, W)^{\mathfrak{g}}$$

and

$$H^k(\mathfrak{l}, W) \cong \bigoplus_p \frac{H_p^k(\mathfrak{l}, W)}{H_{p+1}^k(\mathfrak{l}, W)} \cong \bigoplus_{p+q=k} H^p(\mathfrak{g}, \mathbb{K}) \otimes H^q(\mathfrak{r}, W)^{\mathfrak{g}}. \qquad \square$$

2.4.4   Spectral sequence for Poisson cohomology

Given a smooth Poisson manifold (M, Π), there is a natural filtration of the Lichnerowicz complex, induced by the characteristic distribution as follows. Denote by $\mathcal{V}_k^q(M, \Pi)$ the space of smooth q-vector fields Λ on M with the following property: $\Lambda(x) \in \wedge^k(\mathrm{Im}\sharp_x) \wedge \wedge^{q-k} T_x M \ \forall \ x \in M$. In other words, $\Lambda(x)$ is a linear combination of q-vectors of the type $Y_1 \wedge \cdots \wedge Y_q$ with $Y_1, \ldots, Y_k \in \mathrm{Im}\sharp_x$. It is clear that $\Pi \in \mathcal{V}_2^2(M, \Pi)$, $\mathcal{V}^\star(M) = \mathcal{V}_0^\star(M, \Pi) \supset \mathcal{V}_1^\star(M, \Pi) \supset \cdots \supset \mathcal{V}_{(\mathrm{rank}\,\Pi)}^\star(M, \Pi) \supset \mathcal{V}_{(\mathrm{rank}\,\Pi)+1}^\star(M, \Pi) = 0$, and $[\Pi, \Lambda] \in \mathcal{V}_k^{p+1}(M, \Pi)$ if $\Lambda \in \mathcal{V}_k^q(M, \Pi)$. So we have a filtration of finite length. The corresponding spectral sequence was first written down explicitly by Vaisman [332, 333], and by Karasev and Vorobjev [338, 339].

Let us mention that the zeroth column $E_0^{p,0}$ of the zeroth term of the above spectral sequence consists of multi-vector fields which are tangent to the characteristic distribution. Consequently, the zeroth column $E_1^{p,0}$ of the first term of the above spectral sequence consists of tangential Poisson cohomology groups, mentioned in Subsection 2.1.4.

The use of the above spectral sequence in the computation of Poisson cohomology has yielded only limited success so far, mainly in the case when the Poisson structure is regular and the characteristic symplectic foliation is a fibration [338, 332]. For this reason, we will not write down explicitly the above spectral sequence in the general case (the reader may try to do it as an exercise). Instead, we will give here a concrete simple example.

Example 2.4.5. Let $M = P \times B^n$, where P is a closed manifold such that $H_{dR}^1(P) = 0$, and B^n is an open ball of dimension $n = \dim H_{dR}^2(P)$. Let Π be a regular Poisson structure on M, whose symplectic leaves are $P \times \{y\}, y \in B^n$, such that the map $B^n \to H_{dR}^2(P), y \mapsto [\omega_y]$, where ω_y is the symplectic form of the symplectic leaf $P \times \{y\}$ of Π, is a diffeomorphism from B to its image. Then the second Poisson cohomology of M vanishes: $H_\Pi^2(M) = 0$. To see this, decompose any 2-vector field Λ such that $[\Lambda, \Pi] = 0$ into the sum of three parts, $\Lambda = \Lambda_{xx} + \Lambda_{xy} + \Lambda_{yy}$, where Λ_{xx} is tangent to the symplectic leaves, $\Lambda_{xy} = \sum_{i=1}^n X_i \wedge \partial/\partial y_i$ with X_i being vector fields tangent to the symplectic leaves and (y_i) being a system of coordinates on B^n, and $\Lambda_{yy} = \sum f_{ij} \partial/\partial y_i \wedge \partial/\partial y_j$. The condition $[\Lambda, \Pi] = 0$ is equivalent to the following system of equations:

$$[\Lambda_{xx}, \Pi] = -\sum_i \alpha_i \wedge [\partial/\partial y_i, \Pi], \tag{2.70}$$

$$\sum_i [X_i, \Pi] \wedge \partial/\partial y_i = \sum_{i,j} f_{ij} [\partial/\partial y_j, \Pi] \wedge \partial/\partial y_i, \tag{2.71}$$

$$\sum_{i,j} [f_{ij}, \Pi] \wedge \partial/\partial y_i \wedge \partial/\partial y_j = 0. \tag{2.72}$$

The second equation means that $[X, \Pi] = \sum_j f_{ij} [\partial/\partial y_j, \Pi] \ \forall\ i$. If we fix a symplectic leaf $\{y = \text{constant}\}$, then $[X_i, \Pi]$ is exact on that leaf while $\sum_j f_{ij} [\partial/\partial y_j, \Pi]$ is not exact unless $f_{ij} = 0$ because of the hypothesis that $y \mapsto [\omega_y]$ is a diffeomorphism. Thus the equation $[\Lambda, \Pi] = 0$ implies that $f_{ij} = 0$, i.e., $\Lambda_{yy} = 0$, and $[X_i, \Pi] = 0 \ \forall\ i$. It follows from the hypothesis $H^1_{dR}(P) = 0$ that X_i is exact on each symplectic leaf, hence we can write $X_i = [g_i, \Pi]$. The 2-vector field $\Lambda' = \Lambda + \sum [g_i \partial/\partial y_i, \Pi]$ is tangent to the symplectic leaves (i.e., $\Lambda' = \Lambda'_{xx}$), and $[\Lambda', \Pi] = 0$. It follows again from the hypothesis that $y \mapsto [\omega_y]$ is a diffeomorphism that Λ' is exact, $\Lambda' = [Z, \Pi]$, and so is Λ. Thus, any 2-cocycle is a coboundary, and $H^2_\Pi(M) = 0$. In terms of spectral sequences, the decomposition $\Lambda = \Lambda_{xx} + \Lambda_{xy} + \Lambda_{yy}$ corresponds to the decomposition $H^2_\Pi(M) \cong E^{0,2}_\infty \oplus E^{1,1}_\infty \oplus E^{2,0}_\infty$, and we showed that each of the three summands in this cohomology decomposition is trivial.

Exercise 2.4.6. Write down more explicitly the spectral sequence for the Poisson cohomology of the above example.

Remark 2.4.7. There are some other natural filtrations of the Lichnerowicz complex, e.g., the filtration associated to a momentum map, studied by Viktor Ginzburg [143, 144], and the filtration given by the powers of an ideal (usually the maximal ideal) of functions at a point where the Poisson structure vanishes). This last filtration is a general one, appearing in the study of local normal forms of many different objects – we already used it in Section 2.2, without even mentioning the spectral sequence.

2.5   Poisson cohomology in dimension 2

Consider a Poisson structure Π on a two-dimensional surface Σ. For simplicity, assume that Σ is orientable. Then Π can be written as

$$\Pi = f\Lambda \tag{2.73}$$

where Λ is a given 2-vector field on Σ which does not vanish anywhere.

If the zero-level set $f^{-1}(0)$ of f is empty, then Π is a symplectic structure, and (smooth) Poisson cohomology of Π is the same as the de Rham cohomology of Σ. The more interesting and difficult case is when $f^{-1}(0) \neq \emptyset$. Using Mayer–Vietoris sequences, the computation of $H_\Pi(\Sigma)$ can be reduced to the computation of Poisson cohomology of Π over small neighborhoods of points $x \in f^{-1}(0)$. Let $U \ni x$ be a small open disk containing a point x such that $f(x) = 0$. Under "reasonable assumptions" (for example, when x is either a regular point or an isolated singular point of f), one can show that $H_\Pi(U)$ is isomorphic to the germified (smooth) cohomology of Π at x. Thus, the computation of $H_\Pi(\Sigma)$ can often be reduced to the computation of germified cohomology.

In this section, we will present the results of Monnier [262] on germified Poisson cohomology of a large class of two-dimensional Poisson structures. This class consists of Poisson tensors of the form

$$\Pi = f(1+h)\frac{\partial}{\partial x} \wedge \frac{\partial}{\partial y}, \tag{2.74}$$

where f is a quasi-homogeneous polynomial of degree D with respect to a quasi-radial vector field $Z = w_1 x \frac{\partial}{\partial x} + w_2 \frac{\partial}{\partial y}$ $(w_1, w_2 \in \mathbb{N})$, i.e., $Z(f) = Df$, and either $h = 0$ or h is a non-constant quasi-homogeneous polynomial of degree $D - w_1 - w_2$. (If $D - w_1 - w_2$ can't be written as $D - w_1 - w_2 = m_1 w_1 + m_2 w_2$ with $m_1, m_2 \in \mathbb{Z}, m_1, m_2 \geq 0, m_1 + m_2 > 0$, then $h = 0$.)

One of the reasons why this class of Poisson structures is interesting is that it contains *all* simple singularities of two-dimensional Poisson structures in the sense of Arnold [15]. Before showing the cohomology of these Poisson structures, we will first discuss briefly Arnold's classification of simple singularities.

2.5.1 Simple singularities

Two germs f_1, f_2 of (smooth or analytic) functions on $\mathbb{K}^2$, where $\mathbb{K} = \mathbb{R}$ or $\mathbb{C}$, are called R-equivalent if there is a germ of diffeomorphism $\phi : (\mathbb{K}^2, 0) \to (\mathbb{K}^2, 0)$ such that $f_1 = f_2 \circ \phi$. A germ g is called *adjacent* to f if there is a sequence of germs g_n, $n \in \mathbb{N}$, such that g_n is R-equivalent to g and $\lim_{n \to \infty} g_n = f$. This limit may be understood in a formal sense, i.e., for any $k \in \mathbb{N}$ the k-jet of g_n at 0 tends to the k-jet of f when $n \to \infty$.

A germ of function f such that $f(0) = 0$ and $\mathrm{d}f(0) = 0$ is called a *simple singularity* if up to R-equivalence there are only a finite number of function germs adjacent to it. Simple singularities of functions of many variables are defined similarly.

Theorem 2.5.1 (Arnold [13]). *The following is a complete list of simple singularities, up to R-equivalence, of functions of two variables:*

$A_k^\pm$ $(k \geq 1)$	$D_k^\pm$ $(k \geq 4)$	$E_6^\pm$	E_7	E_8
$x^2 \pm y^{k+1}$	$x^2 y \pm y^{k-1}$	$x^3 \pm y^4$	$x^3 + xy^3$	$x^3 + y^5$

If $\mathbb{K} = \mathbb{C}$, or if k is even in the real A_k case, then the symbols $\pm$ can be omitted or replaced by $+$.

The above classification also holds in higher dimensions: to get any simple singularity with n variables, one simply adds a Morse term $\sum_{i=3}^{n} \pm x_i^2$ to a simple singularity with two variables. The letters A, D, E in the above classification are due to natural relations between these simple singularities and Dynkin diagrams of the same names. See, e.g., [16], for details.

Two Poisson germs $\Pi_1 = f_1 \frac{\partial}{\partial x} \wedge \frac{\partial}{\partial y}$ and $\Pi_2 = f_2 \frac{\partial}{\partial x} \wedge \frac{\partial}{\partial y}$ are *equivalent* if there exists a germ of diffeomorphism $\varphi : (\mathbb{K}^2, 0) \to (\mathbb{K}^2, 0)$ such that $\varphi_* \Pi_1 = \Pi_2$. This condition means that

$$f_2 \circ \varphi = Jac(\varphi) f_1, \tag{2.75}$$

where $Jac(\varphi)$ denotes the Jacobian of φ. Due to the additional term $Jac(\varphi)$, Poisson tensors in dimension 2 do not behave exactly like functions under diffeomorphisms, but quite similarly.

Theorem 2.5.2 (Arnold [15]). *Let $f \in \mathcal{F}(\mathbb{K}^2, 0)$ be a simple singularity. Then, the Poisson germ $\Pi = f \frac{\partial}{\partial x} \wedge \frac{\partial}{\partial y}$ is equivalent, up to a multiplicative constant, to a Poisson germ of the type $g \frac{\partial}{\partial x} \wedge \frac{\partial}{\partial y}$, where g is in the following list (where λ is a constant):*

$$
\begin{array}{lll}
A_{2p} & : & x^2 + y^{2p+1}, \quad p \geq 1 \\
A_{2p-1}^{\pm} & : & (x^2 \pm y^{2p})(1 + \lambda y^{p-1}), \quad p \geq 1 \\
D_{2p}^{\pm} & : & (x^2 y \pm y^{2p-1})(1 + \lambda y^{p-1}), \quad p \geq 2 \\
D_{2p+1} & : & (x^2 y + y^{2p})(1 + \lambda x), \quad p \geq 2 \\
E_6 & : & x^3 + y^4 \\
E_7 & : & (x^3 + xy^3)(1 + \lambda y^2) \\
E_8 & : & x^3 + y^5
\end{array}
\tag{2.76}
$$

When $\mathbb{K} = \mathbb{C}$, the symbols $\pm$ can be omitted or replaced by $+$.

Remark 2.5.3. A simple case excluded from the above theorem is when $f(0) = 0$ but $df(0) \neq 0$. It is easy to see that in this case Π is isomorphic to the linear Poisson structure $x \frac{\partial}{\partial x} \wedge \frac{\partial}{\partial y}$ (see Theorem 4.2.1).

The proof of Theorem 2.5.2 is based on Theorem 2.5.1 and the following

Theorem 2.5.4 ([15]). *Let $\Pi = fa \frac{\partial}{\partial x} \wedge \frac{\partial}{\partial y}$, where f is a quasi-homogeneous polynomial of degree $D > 0$ with respect to a given quasi-radial vector field $Z = w_1 x \frac{\partial}{\partial x} + w_2 y \frac{\partial}{\partial y}$ ($w_1, w_2 \in \mathbb{N}$), and $a(0) \neq 0$. Then up to a multiplicative constant, the germ of Π at 0 is equivalent to a germ of Poisson structure of the type $f(1 + h) \frac{\partial}{\partial x} \wedge \frac{\partial}{\partial y}$ (with the same f), where either $h = 0$ or h is a non-constant quasi-homogeneous polynomial of degree $D - w_1 - w_2$.*

Proof (sketch). In order to preserve f, one uses only coordinate transformations generated by vector fields of the type αZ, where α is a function. In the formal category, one shows easily that all the quasi-homogeneous terms in a, except the constant term and the terms of quasi-homogeneous degree $D - w_1 - w_2$, can be killed consecutively by such coordinate transformations. To prove the result in analytic and smooth categories, one first kills the terms of degree $< D - w_1 - w_2$

as above to arrive at a Poisson germ of the type $f(1 + h + R)\frac{\partial}{\partial x} \wedge \frac{\partial}{\partial y}$, where $\deg(h) = D - w_1 - w_2$ and the Taylor expansion of R contains only terms of degree $> D - w_1 - w_2$. One then kills R by Moser's path method. For details, see [15, 262]. $\qquad\qquad\qquad\qquad\qquad\qquad\qquad\qquad\qquad\qquad\qquad\qquad\qquad\qquad\qquad\square$

2.5.2 Cohomology of Poisson germs

In this subsection, we will denote by

$$\Pi_0 = f\frac{\partial}{\partial x} \wedge \frac{\partial}{\partial y} \tag{2.77}$$

a Poisson structure on $\mathbb{K}^2$, where f is a quasi-homogeneous polynomial function of degree D with respect to a given quasi-radial vector field $Z = w_1 x\frac{\partial}{\partial x} + w_2 y\frac{\partial}{\partial y}$, where $w_1, w_2 \in \mathbb{N}$. Recall that it means that f is polynomial and $\mathcal{L}_Z f = Df$. We will also denote by

$$\Pi = f(1 + h)\frac{\partial}{\partial x} \wedge \frac{\partial}{\partial y} \tag{2.78}$$

a deformation of Π, where h is a quasi-homogeneous polynomial function of degree $D - w_1 - w_2$ with respect to Z. (If $D - w_1 - w_2 \leq 0$ then $h = 0$.)

We will denote by $\mathcal{F}(\mathbb{K}^2, 0), \mathcal{V}^1(\mathbb{K}^2, 0)$ and $\mathcal{V}^2(\mathbb{K}^2, 0)$ the space of germs of functions, vector fields and 2-vector fields on $\mathbb{K}^2$, respectively. We will view Π_0 and Π as germs of Poisson structures at 0, and denote their germified cohomology by $H_{\text{germ}}(\Pi_0)$ and $H_{\text{germ}}(\Pi)$. The spaces of germified k-cocycles and k-coboundaries of the Lichnerowicz complex of Π (resp. Π_0) will de denoted by $Z^k_{\text{germ}}(\Pi)$ and $B^k_{\text{germ}}(\Pi)$ (resp. $Z^k_{\text{germ}}(\Pi_0)$ and $B^k_{\text{germ}}(\Pi_0)$).

Actually, by a germ, we will mean either a C^∞-smooth germ ($\mathbb{K} = \mathbb{R}$), a real analytic germ ($\mathbb{K} = \mathbb{R}$), a holomorphic germ ($\mathbb{K} = \mathbb{C}$), or a formal series ($\mathbb{K} = \mathbb{R}$ or $\mathbb{C}$). So there are several versions of this germified cohomology. However, the resulting cohomology groups (for the Poisson structures (2.77) and (2.78)) will be the same, so we consider these different versions together.

Recall that f is said to have *finite codimension* (or equivalently, the germ of f at 0 is an *isolated singularity*, if indeed it is a singularity), if its *multiplicity* $c = \dim Q_f$, where $Q_f = \mathcal{F}(\mathbb{K}^2, 0)/I_f$, is finite. Here $\mathcal{F}(\mathbb{K}^2, 0)$ is the space of germs of functions on $\mathbb{K}^2$, and I_f denotes the ideal generated by $\frac{\partial f}{\partial x}$ and $\frac{\partial f}{\partial y}$. This ideal consists of germs of functions of the type $Y(f), Y \in \mathcal{V}^1(\mathbb{K}^2, 0)$.

Theorem 2.5.5 (Monnier [262]). *With the above notations, assume that f has finite codimension. Then we have:*

a) $H^0_{\text{germ}}(\Pi) \cong H^0_{\text{germ}}(\Pi_0) \cong \mathbb{K}$ *and* $H^k_{\text{germ}}(\Pi) = H^k_{\text{germ}}(\Pi_0) = 0$ *if* $k \geq 3$.

b) $H^1_{\text{germ}}(\Pi) \cong H^1_{\text{germ}}(\Pi_0) \cong \mathbb{K}^{r+1}$, *where r is the dimension of the space of polynomials which are quasi-homogeneous of degree $D - w_1 - w_2$ with respect to Z.*

c) $H^2_{\text{germ}}(\Pi) \cong H^2_{\text{germ}}(\Pi_0) \cong \mathbb{K}^{r+c}$, *where r is defined as above, and c is the multiplicity of f at 0.*

Assertion a) in the above theorem is clear, because the only Casimir functions are constants, and the dimension of the manifold is 2. The proof of b) and c) consists of several lemmas.

In this subsection, for a function g on $\mathbb{K}^2$, we will denote by H_g the Hamiltonian vector field of g with respect to the standard symplectic form $\mathrm{d}x \wedge \mathrm{d}y$:

$$H_g = \frac{\partial g}{\partial x}\frac{\partial}{\partial y} - \frac{\partial g}{\partial y}\frac{\partial}{\partial x}. \tag{2.79}$$

Lemma 2.5.6. *Let X be in $\mathcal{V}^1(\mathbb{K}^2, 0)$. Then we have:*

  i) *If* Div $X = 0$, *where* Div X *denotes the divergence of X with respect to the volume form $\mathrm{d}x \wedge \mathrm{d}y$, then there exists $g \in \mathcal{F}(\mathbb{K}^2, 0)$ such that $X = H_g$.*

  ii) *If $X(f) = 0$, then $X = \alpha H_f$ for some $\alpha \in \mathcal{F}(\mathbb{K}^2, 0)$.*

Proof. Write $X = A\frac{\partial}{\partial x} + B\frac{\partial}{\partial y}$. Consider the 1-form $\theta = -B\mathrm{d}x + A\mathrm{d}y = i_X(\mathrm{d}x \wedge \mathrm{d}y)$.

  i) If Div $X = 0$ then $d\omega = 0$, which implies that $\theta = -\mathrm{d}g$ for some $g \in \mathcal{F}(\mathbb{K}^2)$, and so $X = H_g$.

  ii) If $X(f) = 0$ then $\mathrm{d}f \wedge \theta = 0$. Since f has finite codimension, de Rham's division theorem [98] implies that $\theta = -\alpha \mathrm{d}f$ for some $\alpha \in \mathcal{F}(\mathbb{K}^2, 0)$. $\qquad\square$

Lemma 2.5.7. *Let $X \in Z^1(\Pi_0)$. Then there exists $\alpha \in \mathcal{F}(\mathbb{K}^2)$ such that*

$$X = \alpha H_f + \frac{\mathrm{Div}\, X}{D} Z.$$

Proof. $0 = \mathcal{L}_X(f\frac{\partial}{\partial x} \wedge \frac{\partial}{\partial y}) = (X(f) - (\mathrm{Div}\, X)f)\frac{\partial}{\partial x} \wedge \frac{\partial}{\partial y}$, implying that $X(f) = (\mathrm{Div}\, X)f = \frac{\mathrm{Div}\, X}{D}Z(f)$. Now apply Lemma 2.5.6. $\qquad\square$

Lemma 2.5.8. *Let $X \in Z^1(\Pi_0)$ be such that $\mathrm{ord}(X) > D - w_1 - w_2$, where $\mathrm{ord}(X)$ denotes the order of X at 0 with respect to Z (if X_0 is the quasi-homogeneous part of X with respect to Z, then $\mathcal{L}_Z X_0 = \mathrm{ord}(X)X_0$). Then $X \in B^1(\Pi_0)$.*

Proof. Suppose for the moment that Div $X = 0$. Then $X(f) = (\mathrm{Div}\, X)f = 0$, hence $X = \gamma H_f$ by Lemma 2.5.6. Since $\mathrm{ord}(X) > D - w_1 - w_2$ and $\mathrm{ord}(H_f) = D - w_1 - w_2$, we have $\mathrm{ord}(\gamma) > 0$, so $\gamma(0) = 0$. It is easy to see that, because $\gamma(0) = 0$, there exists $\mu \in \mathcal{F}(\mathbb{K}^2, 0)$ such that $Z(\mu) = \gamma$. We have $\mathrm{ord}(\mu) = \mathrm{ord}(\gamma) > 0$. Hence $\mathrm{ord}(H_f(\mu)) \geq \mathrm{ord}(\mu) + \mathrm{ord}(H_f) > \mathrm{ord}(H_f) = D - w_1 - w_2$. On the other hand, we have

$$0 = \mathrm{Div}\, X = H_f(\gamma) = H_f(Z(\mu)) = Z(H_f(\mu)) + [H_f, Z](\mu)$$
$$= Z(H_f(\mu)) - (D - w_1 - w_2)H_f(\mu),$$

i.e., $H_f(\mu)$ either vanishes or is quasi-homogeneous of degree $D - w_1 - w_2$. Therefore $H_\mu(f) = -H_f(\mu) = 0$. By Lemma 2.5.6, there exists $\nu \in \mathcal{F}(\mathbb{K}^2, 0)$ such that $\frac{\partial \mu}{\partial x} = \nu\frac{\partial f}{\partial x}$ and $\frac{\partial \mu}{\partial y} = \nu\frac{\partial f}{\partial y}$. Therefore, $Z(\mu) = \nu Z(f)$, which means that $\gamma = D\nu f$. We deduce that $X = fY$ with $Y \in \mathcal{V}^1(\mathbb{K}^2, 0)$. Finally, since $X \in Z^1(\Pi_0)$, we have Div $Y = 0$, hence $Y = H_g$ for some $g \in \mathcal{F}(\mathbb{K}^2)$ by Lemma 2.5.6. Thus $X = fH_g = X_g$ (the Hamiltonian vector field of g with respect to Π_0).

Consider now the case Div $X \neq 0$. If we find $\beta \in \mathcal{F}(\mathbb{K}^2, 0)$ such that Div $X =$ Div X_β, then the 1-cocycle $X - X_\beta$ satisfies Div$(X - X_\beta) = 0$, which implies that $X - X_\beta = X_\varepsilon$ and $X = X_{\beta+\varepsilon}$ for some $\varepsilon \in \mathcal{F}(\mathbb{K}^2)$. Since Div $X_\beta = H_\beta(f) = -H_f(\beta)$, we are looking for β such that $H_f(\beta) = -$ Div X.

By Lemma 2.5.7, we have $X = \alpha H_f + \frac{\text{Div } X}{D} Z$ for some $\alpha \in \mathcal{F}(\mathbb{K}^2, 0)$. Taking the divergence of the two sides of this equation, we get:

$$Z(\text{Div } X) - (D - w_1 - w_2)\,\text{Div } X = -DH_f(\alpha).$$

Note that, because ord$(X) > D - w_1 - w_2$, we have ord$(\alpha) > 0$, which implies that there is $\beta \in \mathcal{F}(\mathbb{K}^2)$ such that $Z(\beta) = D\alpha$ (and moreover, ord$(\beta) > 0$). Simple calculations show that $Z(\text{Div } X + H_f(\beta)) = (D - w_1 - w_2)(\text{Div } X + H_f(\beta))$, i.e., Div $X + H_f(\beta)$ is either 0 or quasi-homogeneous of degree $D - w_1 - w_2$. On the other hand, by evaluating directly the order, we get ord$(\text{Div } X + H_f(\beta)) > D - w_1 - w_2$. Therefore Div $X = -H_f(\beta)$. $\square$

Denote by $\{e_1, \ldots, e_r\}$ a linear basis of the space of quasi-homogeneous polynomials of degree $D - w_1 - w_2$. For $X \in Z^1_{\text{germ}}(\Pi_0)$ (resp. $Z^1_{\text{germ}}(\Pi_0)$), we will denote by $[X]_{\Pi_0}$ (resp. $[X]_\Pi$) its cohomological class in $H^1_{\text{germ}}(\Pi_0)$ (resp. $H^1_{\text{germ}}(\Pi)$).

Lemma 2.5.9. *The family* $\{[H_f]_{\Pi_0}, [e_1Z]_{\Pi_0}, \ldots, [e_rZ]_{\Pi_0}\}$ *is a linear basis of* $H^1_{\text{germ}}(\Pi_0)$. *In particular,* $\dim H^1(\Pi_0) = r + 1 < \infty$.

Proof. First we prove that $H^1_{\text{germ}}(\Pi_0)$ is spanned by this family.

Lemma 2.5.8 implies that every X in $Z^1_{\text{germ}}(\Pi_0)$ is cohomologous to a polynomial vector field of quasi-homogeneous degree $\leq D - w_1 - w_2$. (Decompose X as $X = X_1 + X_2$ where ord$(X_1) > D - w_1 - w_2$ and X_2 is a polynomial vector field of quasi-homogeneous degree at most $D - w_1 - w_2$. Then both X_1 and X_2 belong to $Z^1_{\text{germ}}(\Pi_0)$, and moreover X_1 is a coboundary by Lemma 2.5.8.)

On the other hand, if $X \in Z^1_{\text{germ}}(\Pi_0)$ is quasi-homogeneous of degree $\deg X < D - w_1 - w_2$, then $X = 0$. Indeed, according to Lemma 2.5.7, we have $X = \frac{\text{Div } X}{D} Z$, and so

$$\text{Div } X = \frac{\text{Div } X}{D}\,\text{Div } Z + Z\left(\frac{\text{Div } X}{D}\right),$$

which implies that $(D - w_1 - w_2 - \deg X)\,\text{Div } X = 0$.

The above arguments show that $H^1_{\text{germ}}(\Pi_0)$ is spanned by quasi-homogeneous 1-cocycles of degree $D - w_1 - w_2$. If $X \in Z^1(\Pi_0)$ is such a cocycle, then by Lemma 2.5.7 we have $X = \alpha H_f + \frac{\text{Div } X}{D} Z$, where $\alpha \in \mathbb{K}$ and Div X is a quasi-homogeneous polynomial of degree $D - w_1 - w_2$. Therefore the family $\{[H_f]_{\Pi_0}, [e_1Z]_{\Pi_0}, \ldots, [e_rZ]_{\Pi_0}\}$ generates $H^1_{\text{germ}}(\Pi_0)$.

Now let us prove that this family is free. Suppose that $\sum_i \lambda_i e_i Z + \alpha H_f \in B^1_{\text{germ}}(\Pi_0)$, where $\alpha, \lambda_1, \ldots, \lambda_r$ are scalars. Then $\sum_i \lambda_i e_i Z + \alpha H_f = 0$. Indeed, if g is a quasi-homogeneous polynomial, then $\deg X_g = \deg H_f + \deg g = D - w_1 -$

$w_2 + \deg g$, which is strictly larger than $D - w_1 - w_2$ as soon as $g \neq$ constant. Consequently, $\text{Div}\left(\sum_i \lambda_i e_i Z + \alpha H_f\right) = 0$, i.e., $\sum_i \lambda_i e_i = 0$. We deduce that $\lambda_1, \ldots, \lambda_r$ are 0, and so $\alpha = 0$. $\qquad\square$

Lemma 2.5.10. $\left\{[(1+h)H_f]_\Pi, [(1+h)e_1 W]_\Pi, \ldots, [(1+h)e_r W]_\Pi\right\}$ *is a basis of* $H^1_{\text{germ}}(\Pi)$. *In particular,* $H^1_{\text{germ}}(\Pi) \cong H^1_{\text{germ}}(\Pi_0)$.

Proof. Notice that $X \in Z^1_{\text{germ}}(\Pi)$ (resp. $X \in B^1_{\text{germ}}(\Pi)$) if and only if $\frac{X}{1+h} \in Z^1_{\text{germ}}(\Pi_0)$ (resp. $B^1_{\text{germ}}(\Pi_0)$). $\qquad\square$

Lemma 2.5.11. *Let* $g \in \mathcal{F}(\mathbb{K}^2, 0)$.

  i) *If* $g\frac{\partial}{\partial x} \wedge \frac{\partial}{\partial y} \in B^2_{\text{germ}}(\Pi_0)$, *then* $g \in I_f$.
  ii) *If the* ∞-*jet at* 0 *of* g *does not contain components of quasi-homogeneous degree* $2D - w_1 - w_2$, *then* $g\frac{\partial}{\partial x} \wedge \frac{\partial}{\partial y} \in B^2_{\text{germ}}(\Pi_0)$ *if and only if* $g \in I_f$.

Proof. i) For $X \in \mathcal{V}^1(\mathbb{K}^2, 0)$, we have

$$[\Pi_0, X] = \left((\text{Div } X)f - X(f)\right)\frac{\partial}{\partial x} \wedge \frac{\partial}{\partial y}. \qquad (2.80)$$

So $g\frac{\partial}{\partial x} \wedge \frac{\partial}{\partial y} \in B^2_{\text{germ}}(\Pi_0)$ if and only if g can be written as $g = X(f) - (\text{Div } X)f = Y(f) \in I_f$, where $Y = X - \frac{\text{Div } X}{D}Z$.

ii) Assume that $g = Y(f) \in I_f$, and that the ∞-jet of g does not contain a component of degree $2D - w_1 - w_2$. We will find $X \in \mathcal{V}^1(\mathbb{K}, 0)$ such that $g = X(f) - (\text{Div } X)f$. In the analytic (or formal) case, write $g = \sum_{i \geq 0} g^{(i)}$ and $Y = \sum_{i \geq D - \max(w_1, w_2)} Y^{(i-D)}$, where $g^{(i)}$ is quasi-homogeneous of degree i and $Y^{(i-D)}$ is quasi-homogeneous of degree $i - D$. Note that $Y^{(D-w_1-w_2)} = 0$. Direct calculations show that X can be given by

$$X = Y + \sum_{i \neq 2D - w_1 - w_2} \frac{\text{Div } Y^{(i-D)}}{2D - w_1 - w_2 - i} Z. \qquad (2.81)$$

In the C^∞-smooth case, using the result from the formal case and Borel's theorem about existence of smooth functions with an arbitrary given Taylor series, we can write $g = Y(f) = Y_1(f) + X_1(f) - (\text{Div } X_1)f$, where X_1, Y_1 are smooth, and Y_1 is flat at 0. The equation

$$Z(\alpha) - (D - w_1 - w_2)\alpha = -\text{Div } Y_1 \qquad (2.82)$$

has a smooth (actually flat) solution α, and we can put $X = \alpha Z + X_1$. $\qquad\square$

Remark 2.5.12. Lemma 2.5.11 holds true even when f is not of finite codimension, and implies that the map $g\frac{\partial}{\partial x} \wedge \frac{\partial}{\partial y} \mapsto g$ induces a surjection from $H^2_{\text{germ}}(\Pi_0) = \mathcal{V}^2(\mathbb{K}, 0)/B^2_{\text{germ}}(\Pi_0)$ to $Q_f = \mathcal{F}(\mathbb{K}, 0)/I_f$. Therefore, if f is of infinite codimension, i.e., $\dim Q_f = \infty$, then $\dim H^2_{\text{germ}}(\Pi_0) = \infty$.

Denote by $\{u_1, \ldots, u_c\}$ a monomial linear basis of $Q_f = \mathcal{F}(\mathbb{K}^2, 0)/I_f$, $c = \dim Q_f$. In other words, $u_1, \ldots, u_c$ are monomial functions which span $\mathcal{F}(\mathbb{K}^2, 0)$ mod I_f. We will need the following result from the singularity theory of functions (see [16]): such a monomial linear basis always exists, and moreover $\deg(u_i) \leq 2(D - w_1 - w_2)$ $\forall i$, where deg denotes the quasi-homogeneous degree. In particular, any quasi-homogeneous function of degree $> 2(D - w_1 - w_2)$ belongs to I_f.

Denote by $\{e_1, \ldots, e_r\}$ a basis of the space of quasi-homogeneous polynomials of degree $D - w_1 - w_2$. For a function $g \in \mathcal{F}(\mathbb{K}^2)$, denote by $[g]_{\Pi_0}$ the cohomology class of $g\frac{\partial}{\partial x} \wedge \frac{\partial}{\partial y}$ in $H^2_{\mathrm{germ}}(\Pi_0)$.

Lemma 2.5.13. *The family* $\{[e_1 f]_{\Pi_0}, \ldots, [e_r f]_{\Pi_0}, [u_1]_{\Pi_0}, \ldots, [u_c]_{\Pi_0}\}$ *is a basis of* $H^2_{\mathrm{germ}}(\Pi_0)$. *In particular,* $\dim H^2_{\mathrm{germ}}(\Pi_0) = r + c$.

Proof. First let us show that this family generates $\dim H^2_{\mathrm{germ}}(\Pi_0)$. Let $g \in \mathcal{F}(\mathbb{K}^2, 0)$. Write $g = \sum_{i=1}^{c} \lambda_i u_i + \xi$ where $\lambda_i \in \mathbb{K}$ and $\xi \in I_f$. Invoking Lemma 2.5.11, we can write

$$g = \sum_{i=1}^{c} \lambda_i u_i + g' \quad \mathrm{mod}\, B^2_{\mathrm{germ}}(f),$$

where g' is a quasi-homogeneous polynomial of degree $2D - w_1 - w_2$, and

$$B^2_{\mathrm{germ}}(f) = \left\{ \psi \in \mathcal{F}(\mathbb{K}^2, 0) \mid \psi \frac{\partial}{\partial x} \wedge \frac{\partial}{\partial y} \in B^2_{\mathrm{germ}}(\Pi_0) \right\}. \qquad (2.83)$$

Since $\deg(g') = 2D - w_1 - w_2 > 2(D - w_1 - w_2)$, g' belongs to I_f, i.e., $g' = X(f)$ for some quasi-homogeneous vector field X of degree $D - w_1 - w_2$. So we can write $g' = (\mathrm{Div}\, X)f + \big(X(f) - (\mathrm{Div}\, X)f\big) \equiv (\mathrm{Div}\, X)f \;\mathrm{mod}\, B^2_{\mathrm{germ}}(\Pi_0)$. Div X is quasi-homogeneous of degree $D - w_1 - w_2$, so $[(\mathrm{Div}\, X)f]_{\Pi_0}$ is in the linear hull of $[e_1 f]_{\Pi_0}, \ldots, [e_r f]_{\Pi_0}$.

Now let us show that this family is free. Let $g_1 = \sum_{i=1}^{r} \lambda_i e_i$ and $g_2 = \sum_{j=1}^{c} \mu_j u_j$, with $\lambda_i, \mu_j \in \mathbb{K}$, such that $g_1 f + g_2 \in B^2_{\mathrm{germ}}(f) \subset I_f$. Since $\deg(g_1 f) = 2D - w_1 - w_2 > 2(D - w_1 - w_2)$, we have $g_1 f \in I_f$, so g_2 is also in I_f. It is possible only if $g_2 = 0$, and $\mu_j = 0$ $\forall j$. We are left with $g_1 f \in B^2_{\mathrm{germ}}(f)$, i.e., $g_1 f = X(f) - (\mathrm{Div}\, X)f$ for some quasi-homogeneous vector field X of degree $D - w_1 - w_2$. Denote by $Y = \frac{g_1 + \mathrm{Div}\, X}{D} Z$, then $(X - Y)(f) = 0$. Therefore, by Lemma 2.5.6, $X = Y + \alpha H_f$ with $\alpha \in \mathbb{K}$. Note that $Y \in Z^1_{\mathrm{germ}}(\Pi_0)$ and $H_f \in B^1_{\mathrm{germ}}(\Pi_0)$. Hence $X \in Z^1_{\mathrm{germ}}(\Pi_0)$, i.e., $g_1 = X(f) - (\mathrm{Div}\, X)f = 0$, which implies that $\lambda_1 = \cdots = \lambda_r = 0$. $\qquad \square$

Lemma 2.5.14. *The family* $\{[e_1 f]_\Pi, \ldots, [e_r f]_\Pi, [u_1]_\Pi, \ldots, [u_c]_\Pi\}$ *is a basis of* $H^2_{\mathrm{germ}}(\Pi)$. *In particular,* $H^2_{\mathrm{germ}}(\Pi) \cong H^2_{\mathrm{germ}}(\Pi_0) \cong \mathbb{K}^{r+c}$.

The proof of Lemma 2.5.14 is similar to the proof of Lemma 2.5.13, though somewhat more complicated. We will leave it as an exercise (see [262]). $\qquad \square$

2.5.3   Some examples and remarks

Example 2.5.15. The case when $f = x$ is regular: $\Pi = \Pi_0 = x\frac{\partial}{\partial x} \wedge \frac{\partial}{\partial y}$. In this case, we have $w_1 = w_2 = D = 1$, $I_f = \mathcal{F}(\mathbb{K}^2, 0)$, $r = c = 0$, $H^1_{\text{germ}}(\Pi) \cong \mathbb{K}\frac{\partial}{\partial y}$, and $H^2_{\text{germ}}(\Pi) = 0$.

Example 2.5.16. Morse singularity (A_1): Suppose that $\Pi = (x^2 + y^2)\frac{\partial}{\partial x} \wedge \frac{\partial}{\partial y}$. Here we have $w_1 = w_2 = 1$ and $D = 2$. The only polynomials of degree $D - w_1 - w_2$ are the scalars. In particular, $r = 1$ Moreover, $Q_{x^2+y^2} \cong \mathbb{K}.1$, implying that $c = \dim Q_{x^2+y^2} = 1$. Hence $\dim H^1_{\text{germ}}(\Pi) = r + 1 = 2$ and $\dim H^2_{\text{germ}}(\Pi) = r + c = 2$. More precisely, these cohomology groups can be written as follows:

$$H^1_{\text{germ}}(\Pi) \cong \mathbb{K}\left(y\frac{\partial}{\partial x} - x\frac{\partial}{\partial y}\right) \oplus \mathbb{K}\left(x\frac{\partial}{\partial x} + y\frac{\partial}{\partial y}\right),$$

$$H^2_{\text{germ}}(\Pi) \cong \mathbb{K}\left(\frac{\partial}{\partial x} \wedge \frac{\partial}{\partial y}\right) \oplus \mathbb{K}\left(f\frac{\partial}{\partial x} \wedge \frac{\partial}{\partial y}\right).$$

The origin is the only singular point of $f = x^2 + y^2$, and when $\mathbb{K} = \mathbb{R}$ the above germified cohomology groups are naturally isomorphic to the smooth Poisson cohomology groups of $(x^2 + y^2)\frac{\partial}{\partial x} \wedge \frac{\partial}{\partial y}$ on $\mathbb{R}^2$. These groups were computed earlier in [143] and [273]. Vaisman [333] also mentioned them, without actually computing them. Similarly, the Poisson structure $(x^2 - y^2)\frac{\partial}{\partial x} \wedge \frac{\partial}{\partial y}$ has the same cohomology.

Example 2.5.17. Singularity D_{2p+1}, $p \geq 2$: Suppose that

$$\Pi = (x^2 y + y^{2p})(1 + x)\frac{\partial}{\partial x} \wedge \frac{\partial}{\partial y}.$$

Then $w_1 = 2p - 1, w_2 = 2$ and $D = 4p$. The the family $\{1, x, y, y^2, \ldots, y^{2p-1}\}$ is a monomial basis of $Q_{x^2 y + y^{2p}}$, and space of polynomials of quasi-homogeneous degree $D - w_1 - w_2$ is spanned by x. In particular, $r = 1, c = 2p+1, \dim H^1_{\text{germ}}(\Pi) = 2$, and $\dim H^2_{\text{germ}}(\Pi) = 2p + 2$.

Example 2.5.18. The algebraic Poisson cohomology of homogeneous Poisson structures on $\mathbb{K}^2$ was computed by Roger and Vanhaecke in [302] (algebraic means that one considers only cochains with algebraic coefficients), by direct algebraic calculations instead of using singularity theory. The result is as follows: Let $\Pi = f\frac{\partial}{\partial x} \wedge \frac{\partial}{\partial y}$ be a homogeneous Poisson structure of degree n, where $f = \prod_{i=1}^{n}(x - a_i y)$, $a_i \in \mathbb{K}$. Then $\dim H^2_\Pi(\mathbb{K}^2) = \infty$ if ϕ is not reduced, i.e., if $a_i = a_j$ for some $i \neq j$. If $a_i \neq a_j \ \forall i \neq j$ then $H^1_\Pi(\mathbb{K}^2) \cong \mathbb{K}^n$ and $H^2_\Pi(\mathbb{K}^2) \cong \mathbb{K}^{n(n-1)}$. They are isomorphic to germified Poisson cohomology groups of Π: $r = n-1$ and $c = (n-1)^2$ in this case.

In principle, cohomology of Poisson structures on closed two-dimensional surfaces can be computed via local results and Mayer–Vietoris sequences. For example, consider a Poisson structure of the type $\Lambda = f\Lambda_0$ on an orientable closed surface Σ of genus g, where Π_0 is nondegenerate, f is a function such that the zero level set $S = \{f = 0\}$ of f is regular and consists of n connected components (n circles on which Λ vanishes). Then Poisson cohomology of such a Λ was computed by Radko [296], based on some earlier computations of Roytenberg [305]. The result is: $H^0(\Sigma, \Lambda) = \mathbb{R}^1$, $H^1(\Sigma, \Lambda) \cong \mathbb{R}^n \oplus H^1_{dR}(\Sigma) \cong \mathbb{R}^{n+2g}$, and $H^2(\Sigma, \Lambda) \cong \mathbb{R}^n \oplus H^2_{dR}(\Sigma) \cong \mathbb{R}^{n+1}$ (the Lichnerowicz map is injective in this case). The dimension $n + 1$ of $H^2(\Sigma, \Lambda)$ corresponds to $n + 1$ numerical invariants of Λ, which may be described as follows:

- Near each circle on which Λ vanishes (connected component of the set S), there is a coordinate system (θ, x), where $\theta \in \mathbb{R}/\mathbb{Z}$ is a periodic coordinate, in which $\Lambda = \frac{x}{c} \frac{\partial}{\partial \theta} \wedge \frac{\partial}{\partial x}$ for some positive constant c. This number c is an invariant of Λ, and is the period of a curl vector field of Λ on the circle (see Section 2.6): any curl vector field of Λ will have S as a union of periodic orbits and gives an orientation of S, and this orientation is also an invariant of λ. Since there are n circles on which Λ vanishes, we have n periods $c_1, \ldots, c_n$.

- Let U be a neighborhood of S, such that each connected component of U is an annulus U_i with canonical coordinates (θ_i, x_i) such as above ($U_i = \{-\varepsilon_i < x_i < \varepsilon_i\}$ for some positive ε_i). Λ is nondegenerate on $\Sigma \setminus U$, and the number

$$Vol(\Lambda) = \int_{\Sigma \setminus U} \omega,$$

where ω is the symplectic form dual to Λ, does not depend on the choice of U. This number is called the *regularized volume* of (Σ, Λ) and is an invariant of Λ.

Theorem 2.5.19 (Radko [296]). *With the above notations and assumptions, the Poisson structure $\Lambda = f\Lambda_0$ on Σ is uniquely determined, up to Poisson isomorphisms, by the topological configuration of the oriented singular set S, the periods $c_1, \ldots, c_n$, and the regularized volume $Vol(\Lambda)$.*

The proof of the above theorem, based on Moser's path method (see Appendix A.1), is relatively simple. See [296, 303] for some generalizations of the above results to the case when Σ is not necessarily closed and S may be a singular level of f, and [240] for a generalization of Theorem 2.5.19 to the case of multi-vector fields of top degree on a manifold.

2.6 The curl operator

2.6.1 Definition of the curl operator

Recall that, if A is an a-vector field and ω is a differential p-form with $p \geq a$, then the *inner product* of ω by A is a unique $(p - a)$-form, denoted by $i_A\omega$ or $A \lrcorner \omega$, such that

$$\langle i_A\omega, B \rangle = \langle \omega, A \wedge B \rangle \tag{2.84}$$

for any $(p - a)$-vector field B. If $p < a$ then we put $i_A\omega = 0$ by convention.

For example, if X is a vector field then $i_X\omega(X_1,\ldots,X_{p-1}) = \langle i_X\omega, X_1 \wedge \cdots \wedge X_{p-1}\rangle = \langle \omega, X \wedge X_1 \wedge \cdots \wedge X_{p-1}\rangle = \omega(X, X_1,\ldots,X_{p-1})$.

Similarly, when $a \geq p$, then we can define the *inner product of an a-vector field A by a p-form η* to be a unique $(a-p)$-vector field, denoted by $i_\eta A$ or $\eta \lrcorner A$, such that

$$\langle \beta, i_\eta A \rangle = \langle \beta \wedge \eta, A \rangle \tag{2.85}$$

for any $(a-p)$-form β.

Warning: Due to the noncommutativity of the wedge product, one must be careful with the signs when dealing with inner products. Also, our sign convention may be different from some other authors.

Exercise 2.6.1. If f is a function and A a multi-vector field then

$$i_{\mathrm{d}f} A = [A, f] . \tag{2.86}$$

In particular, the Hamiltonian vector field of f with respect to a given Poisson structure Π is $X_f = \sharp_\Pi(df) = -i_{\mathrm{d}f}\Pi$.

Let Ω be a smooth *volume form* on an m-dimensional manifold M, i.e., a nowhere vanishing differential m-form. Then for every $p = 0, 1, \ldots, m$, the map

$$\Omega^\flat : \mathcal{V}^p(M) \longrightarrow \Omega^{m-p}(M) \tag{2.87}$$

defined by $\Omega^\flat(A) = i_A\Omega$, is a $\mathcal{C}^\infty(M)$-linear isomorphism from the space $\mathcal{V}^p(M)$ of smooth p-vector fields to the space $\Omega^{m-p}(M)$ of smooth $(m-p)$-forms. The inverse map of $\Omega^\flat$ is denoted by $\Omega^\sharp : \Omega^{n-p}(M) \longrightarrow \mathcal{V}^p(M)$, which can be defined by $\Omega^\sharp(\eta) = i_\eta\widehat{\Omega}$, where $\widehat{\Omega}$ is the dual m-vector field of Ω, i.e., $\langle \Omega, \widehat{\Omega}\rangle = 1$.

Exercise 2.6.2. Prove the formula $(\eta\lrcorner\widehat{\Omega})\lrcorner\Omega = \eta$.

Denote by $D_\Omega : \mathcal{V}^p(M) \longrightarrow \mathcal{V}^{p-1}(M)$ the linear operator defined by $D_\Omega = \Omega^\sharp \circ \mathrm{d} \circ \Omega^\flat$. Then we have the following commutative diagram:

$$\begin{array}{ccc}
\mathcal{V}^p(M) & \xrightarrow{\ \Omega^\flat\ } & \Omega^{m-p}(M) \\
{\scriptstyle D_\Omega}\Big\downarrow & & \Big\downarrow{\scriptstyle \mathrm{d}} \\
\mathcal{V}^{p-1}(M) & \xrightarrow{\ \Omega^\flat\ } & \Omega^{m-p+1}(M)
\end{array} \tag{2.88}$$

Since $\mathrm{d} \circ \mathrm{d} = 0$, we also have $D_\Omega \circ D_\Omega = 0$.

Definition 2.6.3. The above operator D_Ω is called the *curl operator* (with respect to the volume form Ω). If A is an a-vector field then $D_\Omega A$ is called the *curl of A* (with respect to Ω).

Example 2.6.4. The curl $D_\Omega X$ of a vector field X is nothing but the divergence of X with respect to the volume form Ω: $(D_\Omega X)\Omega = \Omega^\flat(D_\Omega X) = \mathrm{d}i_X\Omega = \mathcal{L}_X\Omega = (\mathrm{Div}_\Omega X)\Omega$, which implies that $D_\Omega X = \mathrm{Div}_\Omega X$.

In a local system of coordinates $(x_1, \ldots, x_n)$ with $\Omega = dx_1 \wedge \cdots \wedge dx_n$, and denoting $\frac{\partial}{\partial x_i}$ by ζ_i as in Section 1.8, we have the following convenient formal formula for the curl operator:

$$D_\Omega A = \sum_i \frac{\partial^2 A}{\partial x_i \partial \zeta_i}. \tag{2.89}$$

The following proposition shows what happens to the curl when we change the volume form.

Proposition 2.6.5. *If f is a non-vanishing function, then we have*

$$D_{f\Omega} A = D_\Omega A + [A, \ln|f|] . \tag{2.90}$$

Proof. We have $D_{f\Omega} A - D_\Omega A = \Omega^\# \Omega^\flat (D_{f\Omega} A - D_\Omega A) = \Omega^\# (\frac{1}{f} d i_A(f\Omega) - d i_A \Omega)$
$= \Omega^\# (d\ln|f| \wedge i_A \Omega) = i_{d\ln|f|} \Omega^\# (i_A \Omega) = i_{d\ln|f|} A = [A, \ln|f|]$. $\square$

Remark 2.6.6. It follows from the above proposition that, if we multiply the volume form by a non-zero constant, then the curl operator does not change. In particular, the curl operator D_Ω can be defined on non-orientable manifolds as well. Non-orientable manifolds don't admit global volume forms in the sense of non-vanishing differential forms of top degree, but they do admit measure-theoretic volume forms with smooth positive distribution. Such a measure-theoretic volume form is a non-oriented (or absolute) version of differential volume forms, and is also called a *density*. Proposition 2.6.5 implies that one can replace a volume form by a density in the definition of the curl operator.

2.6.2 Schouten bracket via curl operator

Theorem 2.6.7 (Koszul [201]). *If A is an a-vector field, B is a b-vector field and Ω is a volume form then*

$$[A, B] = (-1)^b D_\Omega(A \wedge B) - (D_\Omega A) \wedge B - (-1)^b A \wedge (D_\Omega B). \tag{2.91}$$

Proof. By Formula (2.89) and Formula (1.71) we have:
$(-1)^b D_\Omega(A \wedge B) = (-1)^b \sum \frac{\partial^2 (A \wedge B)}{\partial x_i \partial \zeta_i} = \sum \frac{\partial}{\partial x_i} \left(\frac{\partial A}{\partial \zeta_i} B + (-1)^b A \frac{\partial B}{\partial \zeta_i} \right)$
$= \sum \frac{\partial^2 A}{\partial x_i \partial \zeta_i} B + (-1)^b A \sum \frac{\partial^2 B}{\partial x_i \partial \zeta_i} + \frac{\partial A}{\partial \zeta_i} \frac{\partial B}{\partial x_i} + (-1)^b \sum \frac{\partial A}{\partial x_i} \frac{\partial B}{\partial \zeta_i}$
$= (D_\Omega A) B + (-1)^b A (D_\Omega B) + \left(\frac{\partial A}{\partial \zeta_i} \frac{\partial B}{\partial x_i} - (-1)^{(b-1)(a-1)} \sum \frac{\partial B}{\partial \zeta_i} \frac{\partial A}{\partial x_i} \right)$
$= (D_\Omega A) B + (-1)^b A (D_\Omega B) + [A, B]$. $\square$

The curl operator is, up to a sign, a derivation of the Schouten bracket. More precisely, we have the following formula:

$$D_\Omega[A, B] = [A, D_\Omega B] + (-1)^{b-1} [D_\Omega A, B]. \tag{2.92}$$

Exercise 2.6.8. Prove the above formula, either by direct calculations, or by using Theorem 2.6.7 and the fact that $D_\Omega \circ D_\Omega = 0$.

2.6.3   The modular class

A particularly important application of the curl operator in Poisson geometry is the *curl vector field* $D_\Omega \Pi$, also called *modular vector field*, of a Poisson structure Π with respect to a volume form Ω. This curl vector field is an infinitesimal automorphism of the Poisson structure, i.e., it is a Poisson vector field. Moreover, it also preserves the volume form:

Lemma 2.6.9. *If* Π *is a Poisson tensor and* Ω *a volume form, then*

$$[D_\Omega \Pi, \Pi] = 0 \quad and \quad \mathcal{L}_{(D_\Omega \Pi)} \Omega = 0. \tag{2.93}$$

Proof. It follows from Formula (2.92) and the fact that $[\Pi, \Pi] = 0$ that we have $0 = D_\Omega[\Pi, \Pi] = [\Pi, D_\Omega \Pi] - [D_\Omega \Pi, \Pi] = -2[D_\Omega \Pi, \Pi]$. Hence we have $[D_\Omega \Pi, \Pi] = 0$.

To prove the second equality, we don't even need the fact that Π is a Poisson structure. Indeed, we have $\mathcal{L}_{(D_\Omega \Pi)} \Omega = i_{(D_\Omega \Pi)} \mathrm{d}\Omega + \mathrm{d} i_{(D_\Omega \Pi)} \Omega = \mathrm{d}(\mathrm{d}(\Pi \lrcorner \Omega)) = 0$. □

Exercise 2.6.10. Show that the curl vector field of the linear Poisson structure $\Pi = y \frac{\partial}{\partial x} \wedge \frac{\partial}{\partial y}$ with respect to the volume form $\mathrm{d}x \wedge \mathrm{d}y$ is $\frac{\partial}{\partial x}$, and it is *not* a Hamiltonian vector field.

Lemma 2.6.9 means that the curl vector field $D_\Omega \Pi$ is a 1-cocycle in the Lichnerowicz complex. Proposition 2.6.5 implies that if we change the volume form (or more precisely, the density, see Remark 2.6.6) then this cocycle changes by a coboundary. Thus the cohomology class of the curl vector field $D_\Omega \Pi$ in $H^1(M, \Pi)$ does not depend on the choice of the volume form Ω.

Definition 2.6.11. If (M, Π) is a Poisson manifold and Ω a smooth density on M, then the cohomology class of the curl vector field $D_\Omega \Pi$ in $H^1(M, \Pi)$ is called the *modular class* of (M, Π). If this class is trivial, then (M, Π) is called a *unimodular Poisson manifold*.

Definition 2.6.12. A density Ω on a Poisson manifold (M, Π) is called an *invariant density* if it is preserved by all Hamiltonian vector fields on M: $\mathcal{L}_{X_f} \Omega = 0 \ \forall \ f \in \mathcal{C}^\infty(M)$.

Lemma 2.6.13. *If* Π *is a Poisson structure and* Ω *is a smooth density, then* $D_\Omega \Pi = 0$ *if and only if* Ω *is an invariant density. In particular, a Poisson manifold is unimodular if and only if it admits a smooth invariant density.*

We will leave the proof of the above lemma as an exercise. □

Exercise 2.6.14. Show that, if (M^{2n}, ω) is a symplectic manifold of dimension $2n$, then it is unimodular as a Poisson manifold. Up to multiplication by a constant, the only invariant volume form on M is the so-called *Liouville form*

$$\Omega = \frac{1}{n!} \wedge^n \omega. \tag{2.94}$$

(Don't confuse this Liouville volume form with the Liouville 1-form mentioned in Example 1.1.9.)

Exercise 2.6.15. A *unimodular Lie algebra* is a Lie algebra $\mathfrak{g}$ such that for any $x \in \mathfrak{g}$, the linear operator $\mathrm{ad}_x : \mathfrak{g} \to \mathfrak{g}$ is traceless. In other words, $\mathfrak{g}$ is called unimodular if its adjoint action preserves a standard volume form. Show that a linear Poisson structure is unimodular if and only if its corresponding Lie algebra is unimodular.

For more about the modular class, see, e.g., [1, 127, 130, 146, 181]. For the theory of (secondary) characteristic classes of Poisson manifolds (and Lie algebroids), of which the modular class is a particular case, see Fernandes [130] and Crainic [85].

2.6.4 The curl operator of an affine connection

Recall that a *linear connection* on a vector bundle E over a manifold M is an $\mathbb{R}$-bilinear map

$$\nabla : \mathcal{V}^1(M) \times \Gamma(E) \to \Gamma(E), \ (X, \xi) \mapsto \nabla_X \xi, \qquad (2.95)$$

(where $\Gamma(E)$ denotes the space of sections of E), which is $C^\infty(M)$-linear with respect to X, i.e., $\nabla_{fX}\xi = f\nabla_X\xi \ \forall f \in C^\infty(M)$, and which satisfies the Leibniz rule with respect to ξ, i.e., $\nabla_X(f\xi) = f\nabla_X\xi + X(f)\xi$. A linear connection is also called a *covariant derivation* on E.

Let ∇ be an *affine connection* on a manifold M, i.e., a linear connection on the tangent bundle TM of M. By the Leibniz rule, one can extend ∇ to a map

$$\nabla : \mathcal{V}^1(M) \times \mathcal{V}^\star(M) \to \mathcal{V}^\star(M) \qquad (2.96)$$

(and more generally, to a covariant derivation on all kinds of tensor fields on M). For example, $\nabla_X(Y \wedge Z) = (\nabla_X Y) \wedge Z + Y \wedge (\nabla_X Z)$. The operator

$$D_\nabla = \sum_k i_{\mathrm{d}x_k} \circ \nabla_{\partial/\partial x_k} : \mathcal{V}^\star(M) \to \mathcal{V}^\star(M), \qquad (2.97)$$

where $(x_1, \ldots, x_m)$ denotes a system of coordinates on M, is called the *curl operator* of ∇.

Exercise 2.6.16. Show that the above definition of D_∇ does not depend on the choice of local coordinates.

Recall that, an affine connection ∇ on a manifold M is called *torsionless* if $\nabla_X Y - \nabla_Y X = [X, Y]$ for all $X, Y \in \mathcal{V}^1(M)$. We have the following statement, similar to Theorem 2.6.7:

Theorem 2.6.17 (Koszul [201]). *If A is an a-vector field, B is a b-vector field and ∇ is a torsionless affine connection, then*

$$[A, B] = (-1)^b D_\nabla(A \wedge B) - (D_\nabla A) \wedge B - (-1)^b A \wedge (D_\nabla B). \qquad (2.98)$$

Proof. By induction, using the Leibniz identity.   □

If ∇ is a flat torsionless connection with $(x_1, \ldots, x_m)$ as a trivializing co-ordinate system, i.e., $\nabla_{\partial/\partial x_i} \partial/\partial x_j = 0$ $\forall i, j$, then Formula (2.97) coincides with Formula (2.89).

2.7   Poisson homology

Given a Poisson manifold (M, Π), there is another differential complex associated to it, with the following differential operator:

$$\delta_{KB} = [i_\Pi, d] = i_\Pi \circ d - d \circ i_\Pi : \Omega^k(M) \to \Omega^{k-1}(M). \qquad (2.99)$$

Here KB stands for Koszul–Brylinski. This operator was introduced by Koszul [201], with a more explicit expression given by Brylinski [51]:

$$\delta_{KB}(f_0 df_1 \wedge \cdots \wedge df_k) = \sum_i (-1)^{i+1}\{f_0, f_i\} df_1 \wedge \cdots \widehat{df_i} \cdots \wedge df_k +$$

$$+ \sum_{i<j} (-1)^{i+j} f_0 d\{f_i, f_j\} \wedge df_1 \wedge \cdots \widehat{df_i} \cdots \widehat{df_j} \cdots \wedge df_k. \quad (2.100)$$

The fact that

$$\delta_{KB} \circ \delta_{KB} = 0 \qquad (2.101)$$

follows easily from Formula (2.100) and the Jacobi identity for the Poisson bracket. The homology[2] groups

$$H_k(M, \Pi) = \frac{\ker(\delta_{KB} : \Omega^k(M) \longrightarrow \Omega^{k-1}(M))}{\mathrm{Im}(\delta_{KB} : \Omega^{k+1}(M) \longrightarrow \Omega^k(M))} \qquad (2.102)$$

of the differential complex

$$\cdots \longrightarrow \Omega^{k+1}(M) \xrightarrow{\delta_{KB}} \Omega^k(M) \xrightarrow{\delta_{KB}} \Omega^{k-1}(M) \longrightarrow \cdots \qquad (2.103)$$

are called *Poisson homology* groups.

Poisson homology is a natural and important invariant of Poisson manifolds. However, for the study of normal forms of Poisson structures, it is the Poisson cohomology which is more relevant, so we will only mention here briefly some main features of Poisson homology.

It follows directly from Formula (2.99) that

$$\delta_{KB} \circ d + d \circ \delta_{KB} = 0. \qquad (2.104)$$

---

[2]The word homology is used instead of cohomology because δ_{KB} is of degree -1, i.e., it sends Ω^k to Ω^{k-1}.

This equality implies that the space of differential forms on M together with (d, δ_{KB}) is a double complex, and there is a spectral sequence associated to this double complex [51]. In analogy with Riemannian geometry, one says that a differential form β is a *Poisson harmonic form* if $d\beta = \delta_{KB}\beta = 0$ [51]. On a compact Riemannian manifold, every de Rham cohomology class can be represented by a unique harmonic form (see, e.g., [41]). The analogue of this fact in the Poisson case does not hold in general, even for symplectic manifolds [133, 241]. In fact, Mathieu [241] proved that, on a compact symplectic manifold (M^{2n}, ω), every de Rham cohomology class has a symplectic harmonic representative if and only if (M^{2n}, ω) satisfies the so-called strong Lefschetz theorem about Kähler manifolds, i.e., the cup product map $\beta \mapsto \wedge^k \omega \wedge \beta$ induces an isomorphism $[\omega]^k : H^{n-k}(M) \to H^{n+k}(M)$ for any $k \leq n$. Another proof of this fact is given by Yan [363].

Bhaskara and Viswanath [30, 31] showed that there is a natural pairing between Poisson cohomology and Poisson homology, with values in the zeroth Poisson homology group. In fact, it follows directly from Formula (1.8.1) that if $\Lambda \in \mathcal{V}^{k-1}(M)$ and $\beta \in \Omega^k(M)$ then

$$\langle \beta, [\Pi, \Lambda] \rangle - \langle \delta_{KB}\beta, \Lambda \rangle = (-1)^k \delta_{KB}(i_\Lambda \beta). \qquad (2.105)$$

This formula implies that the usual pairing between k-vector fields and k-forms induces a pairing

$$H_k(M, \Pi) \times H^k(M, \Pi) \to H_0(M, \Pi). \qquad (2.106)$$

When (M^{2n}, Π) is a compact symplectic manifold, there is a natural involution

$$\star : \Omega^k(M) \to \Omega^{2n-k}(M) \qquad (2.107)$$

(defined by $\star \alpha = i_{\sharp \alpha} \Omega$, where $\Omega = (1/n!) \wedge^n \omega$ is the standard symplectic form and $\sharp = \sharp_\Pi$ is the anchor map of Π), and δ_{KB} can be written as

$$\delta = \pm \star \circ d \circ \star. \qquad (2.108)$$

It follows that, in this case, $H_k(M, \Pi)$ is naturally isomorphic to $H^{2n-k}_{dR}(M)$ [51], and Formula (2.105) gives the usual Poincaré duality if we identify $H^k(M, \Pi)$ with $H^k_{dR}(M)$.

Huebschmann [180] gave a purely algebraic definition of Poisson homology in terms of the *Tor* functor (and Poisson cohomology in terms of the *Ext* functor). Xu [360] showed that if (M, Π) is an orientable unimodular Poisson manifold of dimension n then $H_k(M, \Pi) \cong H^{n-k}(M, \Pi)$. The duality between Poisson homology and Poisson cohomology was also studied in [181, 127]. There are natural relations between Poisson homology and Hochschild homology, see, e.g., [51, 129, 27]. The paper [129] also contains explicit computations of Poisson homology for some Poisson–Lie groups. Similarly to Poisson cohomology, it is very hard to compute Poisson homology in general.

Chapter 3

Levi Decomposition

In this chapter, we will discuss a type of local normal forms for Poisson structures which vanish at a point, called Levi normal forms, or Levi decompositions. A Levi normal form is a kind of partial linearization of a Poisson structure, and in "good" cases this leads to a true linearization. The name *Levi decomposition* comes from the analogy with the classical Levi decomposition for finite-dimensional Lie algebras. Let us briefly recall here the classical theory (see, e.g., [42, 335]):

Let $\mathfrak{l}$ be a finite-dimensional Lie algebra. Denote by $\mathfrak{r}$ the radical of $\mathfrak{l}$, i.e., the maximal solvable ideal of $\mathfrak{l}$. Then the quotient Lie algebra $\mathfrak{g} = \mathfrak{l}/\mathfrak{r}$ is semi-simple, and we have the following exact sequence:

$$0 \to \mathfrak{r} \to \mathfrak{l} \to \mathfrak{g} \to 0. \tag{3.1}$$

The classical *Levi–Malcev theorem* says that the above sequence splits, i.e., there is an injective Lie algebra homomorphism $\imath : \mathfrak{g} \to \mathfrak{l}$ such that its composition with the projection map $\mathfrak{l} \to \mathfrak{g}$ is identity. The image $\imath(\mathfrak{g})$ of $\mathfrak{g}$ in $\mathfrak{l}$ is called a *Levi factor* of $\mathfrak{l}$. Up to conjugations in $\mathfrak{l}$, the Levi factor of $\mathfrak{l}$ is unique. We will identify $\mathfrak{g}$ with $\imath(\mathfrak{g})$. Then $\mathfrak{g}$ acts on $\mathfrak{r}$ by the adjoint action in $\mathfrak{l}$, and $\mathfrak{l}$ can be decomposed into a semi-direct product of $\mathfrak{g}$ with $\mathfrak{r}$:

$$\mathfrak{l} = \mathfrak{g} \ltimes \mathfrak{r}. \tag{3.2}$$

The above decomposition is called the *Levi decomposition* of $\mathfrak{l}$.

In the study of Poisson structures or other structures involving Lie brackets, we often have *infinite*-dimensional Lie algebras. So the idea is to find analogs of the Levi–Malcev theorem which hold for these infinite-dimensional Lie algebras. These analogs will give interesting information about Poisson structures.

In Section 3.1 we will give a formal infinite-dimensional analog of the Levi–Malcev theorem, and illustrate its use in the example of singular foliations. Then in the rest of this chapter, we will discuss Levi decomposition for Poisson structures.

3.1   Formal Levi decomposition

Let $\mathcal{L}$ be a Lie algebra of infinite dimension. Suppose that $\mathcal{L}$ admits a filtration

$$\mathcal{L} = \mathcal{L}_0 \supset \mathcal{L}_1 \supset \mathcal{L}_2 \supset \cdots, \tag{3.3}$$

such that $\forall i, j \geq 0$, $[\mathcal{L}_i, \mathcal{L}_j] \subset \mathcal{L}_{i+j}$ and $\dim(\mathcal{L}_i/\mathcal{L}_{i+1}) < \infty$. Then we say that $\mathcal{L}$ is a *pro-finite Lie algebra*, and call the inverse limit

$$\widehat{\mathcal{L}} = \lim_{\infty \leftarrow i} \mathcal{L}/\mathcal{L}_i \tag{3.4}$$

the *formal completion* of $\mathcal{L}$ (with respect to a given pro-finite filtration).

Example 3.1.1. Let $\mathcal{L}$ be the Lie algebra of smooth vector fields on $\mathbb{R}^n$ which vanish at the origin 0, and $\mathcal{L}_k$ be the ideal of $\mathcal{L}$ consisting of vector fields with zero k-jet at 0. Then $\mathcal{L}$ is pro-finite, and its formal completion is the algebra of formal vector fields at 0.

Given a pro-finite Lie algebra $\mathcal{L}$ as above, denote by $\mathfrak{r}$ the radical of $\mathfrak{l} = \mathcal{L}/\mathcal{L}_1$ and by $\mathfrak{g}$ the semisimple quotient $\mathfrak{l}/\mathfrak{r}$. Denote by $\mathcal{R}$ the preimage of $\mathfrak{r}$ under the projection $\mathcal{L} \to \mathfrak{l} = \mathcal{L}/\mathcal{L}_1$. Then $\mathcal{R}$ is an ideal of $\mathcal{L}$, called the *pro-solvable radical*, and we have $\mathcal{L}/\mathcal{R} \cong \mathfrak{l}/\mathfrak{r} = \mathfrak{g}$. Denote by $\widehat{\mathcal{R}} = \lim_{\leftarrow} \mathcal{R}/\mathcal{L}_i$ the formal completion of $\mathcal{R}$. Then we have the following exact sequences:

$$0 \to \mathcal{R} \to \mathcal{L} \to \mathfrak{g} \to 0, \tag{3.5}$$

$$0 \to \widehat{\mathcal{R}} \to \widehat{\mathcal{L}} \to \mathfrak{g} \to 0. \tag{3.6}$$

The exact sequence (3.5) does not necessarily split, but its formal completion (3.6) always does:

Theorem 3.1.2. *With the above notations, there is a Lie algebra injection $\imath : \mathfrak{g} \to \widehat{\mathcal{L}}$ whose composition with the projection map $\widehat{\mathcal{L}} \to \mathfrak{g}$ is the identity map. Up to conjugations in $\widehat{\mathcal{L}}$, such an injection is unique.*

Proof. By induction, for each $k \in \mathbb{N}$ we will construct an injection $\imath_k : \mathfrak{g} \to \mathcal{L}/\mathcal{L}_k$, whose composition with the projection map $\mathcal{L}/\mathcal{L}_k \to \mathfrak{g}$ is identity, and moreover the following compatibility condition is satisfied: the diagram

$$
\begin{array}{ccc}
\mathfrak{g} & \xrightarrow{\ \imath_{k+1}\ } & \mathcal{L}/\mathcal{L}_{k+1} \\
{\scriptstyle\mathrm{Id}}\big\downarrow & & \big\downarrow {\scriptstyle\mathrm{proj.}} \\
\mathfrak{g} & \xrightarrow{\ \imath_k\ } & \mathcal{L}/\mathcal{L}_k
\end{array}
\tag{3.7}
$$

is commutative. Then $\imath = \lim_{\leftarrow} \imath_k$ will be the required injection. When $k = 1$, $\imath_1$ is given by the Levi–Malcev theorem. If we forget about the compatibility condition, then the other $\imath_k$, $k > 1$, can also be provided by the Levi–Malcev theorem. But to achieve the compatibility condition, we will construct $\imath_{k+1}$ directly from $\imath_k$.

Assume that $\imath_k$ has been constructed. Denote by $\rho : \mathfrak{g} \to \mathcal{L}/\mathcal{L}_{k+1}$ an arbitrary linear map which lifts the injective Lie algebra homomorphism $\imath_k : \mathfrak{g} \to \mathcal{L}/\mathcal{L}_k$. We will modify ρ into a Lie algebra injection.

Note that $\mathcal{L}_k/\mathcal{L}_{k+1}$ is a $\mathfrak{g}$-module. The action of $\mathfrak{g}$ on $\mathcal{L}_k/\mathcal{L}_{k+1}$ is defined as follows: for $x \in \mathfrak{g}, v \in \mathcal{L}_k/\mathcal{L}_{k+1}$, put $x.v = [\rho(x), v] \in \mathcal{L}_k/\mathcal{L}_{k+1}$. If $x, y \in \mathfrak{g}$ then $[\rho(x), \rho(y)] - \rho([x, y]) \in \mathcal{L}_k/\mathcal{L}_{k+1} \subset \mathcal{L}_1/\mathcal{L}_{k+1}$, and therefore $[[\rho(x), \rho(y)] - \rho([x, y]), v] = 0$ because $[\mathcal{L}_1/\mathcal{L}_{k+1}, \mathcal{L}_k/\mathcal{L}_{k+1}] = 0$. The Jacobi identity in $\mathcal{L}/\mathcal{L}_{k+1}$ then implies that $x.(y.v) - y.(x.v) = [x, y].v$, so $\mathcal{L}_k/\mathcal{L}_{k+1}$ is a $\mathfrak{g}$-module.

Define the following 2-cochain $f : \mathfrak{g} \wedge \mathfrak{g} \to \mathcal{L}_k/\mathcal{L}_{k+1}$:

$$x \wedge y \in \mathfrak{g} \wedge \mathfrak{g} \mapsto f(x, y) = [\rho(x), \rho(y)] - \rho([x, y]) \in \mathcal{L}_k/\mathcal{L}_{k+1}. \tag{3.8}$$

One verifies directly that f is a 2-cocycle of the corresponding Chevalley–Eilenberg complex: denoting by $\oint_{xyz}$ the cyclic sum in (x, y, z), we have

$$
\begin{aligned}
\delta f(x, y, z) &= \oint_{xyz} \left(x.f(y, z) - f([y, z], x) \right) \\
&= \oint_{xyz} \left([\rho(x), [\rho(y), \rho(z)] - \rho([y, z])] - [\rho[y, z], \rho(x)] + \rho([[y, z], x]) \right) \\
&= \oint_{xyz} [\rho(x), [\rho(y), \rho(z)]] + \oint_{xyz} \rho([[y, z], x]) = 0 + 0 = 0.
\end{aligned}
$$

Since $\mathfrak{g}$ is semisimple, by Whitehead's lemma every 2-cocycle of $\mathfrak{g}$ is a 2-coboundary. In particular, there is a 1-cochain $\phi : \mathfrak{g} \to \mathcal{L}_k/\mathcal{L}_{k+1}$ such that $\delta\phi = f$, i.e.,

$$[\rho(x), \phi(y)] - [\rho(y), \phi(x)] - \phi([x, y]) = [\rho(x), \rho(y)] - \rho([x, y]). \tag{3.9}$$

It implies that the linear map $\imath_{k+1} = \rho - \phi$ is a Lie algebra homomorphism from $\mathfrak{g}$ to $\mathcal{L}/\mathcal{L}_{k+1}$. Since the image of ϕ lies in $\mathcal{L}_k/\mathcal{L}_{k+1}$, it is clear that $\imath_{k+1}$ is a lifting of $\imath_k$. Thus $\imath_{k+1}$ satisfies our requirements. By induction, the existence of $\imath$ is proved.

The uniqueness of $\imath$ up to conjugations in $\widehat{\mathcal{L}}$ is proved similarly. Suppose that $\imath_{k+1}, \imath'_{k+1} : \mathfrak{g} \to \mathcal{L}/\mathcal{L}_{k+1}$ are two different injections which lift $\imath_k$. Then $\imath'_{k+1} - \imath_{k+1}$ is a 1-cocycle, and therefore a 1-coboundary by Whitehead's lemma. Denote by α an element of $\mathcal{L}_k/\mathcal{L}_{k+1}$ such that $\delta\alpha$ is this 1-coboundary. Then the inner automorphism of $\mathcal{L}/\mathcal{L}_{k+1}$ given by

$$v \in \mathcal{L}/\mathcal{L}_{k+1} \mapsto \mathrm{Ad}_{\exp \alpha} v = v + [\alpha, v] \tag{3.10}$$

(because the other terms vanish) is a conjugation in $\mathcal{L}/\mathcal{L}_{k+1}$ which intertwines $\imath_{k+1}$ and $\imath'_{k+1}$, and which projects to the identity map on $\mathcal{L}/\mathcal{L}_k$. $\qquad\square$

The image $\imath(\mathfrak{g})$ of $\mathfrak{g}$ in $\widehat{\mathcal{L}}$, where $\imath$ is given by Theorem 3.1.2, is called a *formal Levi factor* of $\mathcal{L}$.

Remark 3.1.3. The above proof can be modified slightly to yield a proof of the classical Levi–Malcev theorem, pretty close to the one given in [335]. (Put $\mathcal{L}_1 =$ the radical of $\mathcal{L}$ in the finite-dimensional case.)

Remark 3.1.4. Every semisimple subalgebra of a finite-dimensional Lie algebra is contained in a Levi factor. Similarly, each semisimple subalgebra of a pro-finite Lie algebra is formally contained in a formal Levi factor. These facts can also be proved by a slight modification of the uniqueness part of the proof of Theorem 3.1.2.

Relations between Levi decomposition and linearization problems were observed, for example, by Flato and Simon [135] in their work on linearization of field equations. Here we will show a simple example of such relations, involving singular foliations.

Let $\mathcal{F}$ be a singular holomorphic foliation in a neighborhood of 0 in $\mathbb{C}^n$. Holomorphic means that $\mathcal{F}$ is generated by holomorphic vector fields. We will assume that the rank of $\mathcal{F}$ at 0 is 0, i.e., $X(0) = 0$ for any tangent vector field X tangent to $\mathcal{F}$. Denote by $\mathcal{X}(\mathcal{F})$ the Lie algebra of germs at 0 of holomorphic vector fields tangent to $\mathcal{F}$. Denote by $\mathcal{X}^{(1)}(\mathcal{F})$ the Lie algebra consisting of linear parts of elements of $\mathcal{X}(\mathcal{F})$ at 0. Then $\mathcal{X}^{(1)}(\mathcal{F})$ is a Lie algebra of linear vector fields. Denote by $\mathcal{F}^{(1)}$ the singular foliation generated by $\mathcal{X}^{(1)}(\mathcal{F})$ and call it the linear part of $\mathcal{F}$.

Theorem 3.1.5 (Cerveau [71]). *With the above notations, if $\mathcal{X}^{(1)}(\mathcal{F})$ is semisimple and $\dim \mathcal{F} = \dim \mathcal{F}^{(1)}$, then $\mathcal{F}$ is formally linearizable at 0, i.e., it is formally isomorphic to $\mathcal{F}^{(1)}$.*

Proof. $\mathcal{X}(\mathcal{F})$ is a pro-finite Lie algebra with the standard filtration given by the order of vanishing of vector fields at 0, hence it admits a formal Levi factor $\mathfrak{g}$. When $\mathcal{X}^{(1)}(\mathcal{F})$ is semisimple, then $\mathfrak{g}$ is isomorphic to $\mathcal{X}^{(1)}(\mathcal{F})$. Since $\mathfrak{g}$ is semisimple, its formal action on $\mathbb{C}^n$ is formally linearizable by a classical theorem of Hermann (Theorem 3.1.6). Suppose that the action of $\mathfrak{g}$ has been linearized. It means that $\mathfrak{g}$ consists of linear vector fields, hence it coincides with $\mathcal{X}^{(1)}(\mathcal{F})$. In other words, after the formal linearization, we have an inclusion $\mathcal{X}^{(1)}(\mathcal{F}) \subset \mathcal{X}(\mathcal{F})$, hence $\mathcal{F}^{(1)} \subset \mathcal{F}$. But $\mathcal{F}$ and $\mathcal{F}^{(1)}$ have the same dimension by assumptions, hence they must coincide. $\square$

Theorem 3.1.6 (Hermann [172]). *If $\mathfrak{g} \subset \mathcal{V}^1_{formal,0}(\mathbb{K}^n)$ is a finite-dimensional semisimple subalgebra of the Lie algebra $\mathcal{V}^1_{formal,0}(\mathbb{K}^n)$ of formal vector fields on $\mathbb{K}^n$ which vanish at 0, where $\mathbb{K} = \mathbb{R}$ or $\mathbb{C}$, then there is a formal coordinate system $(z_1, \ldots, z_n)$ of $\mathbb{K}^n$ at 0, with respect to which the elements of $\mathfrak{g}$ have linear coefficients.*

Proof (sketch). The proof follows the usual formal normalization procedure, and is based on Whitehead's lemma $H^1(\mathfrak{g}, W) = 0$. Let $X_1, \ldots, X_d$ be a basis of $\mathfrak{g}$. Suppose that, in a coordinate system $(z_1, \ldots, z_n)$, we have

$$X_i = X_i^{(1)} + X_i^{(s)} + X_i^{(s+1)} + \cdots \tag{3.11}$$

with $s \geq 2$, where $X_i^{(s)}$ is a vector field whose coefficients are homogeneous of degree s, and so on. We want to kill the term $X_i^{(s)}$ in the expression of X_i by a coordinate transformation of the type $z_i' = z_i +$ terms of degree $\geq s$. Due to the Jacobi identity, the map $X_i \mapsto X_i^{(s)}$ is a 1-cocycle of $\mathfrak{g}$ with coefficients in the $\mathfrak{g}$-module of homogeneous vector fields of degree s. By Whitehead's lemma, this 1-cocycle is a coboundary, i.e., we can write

$$X_i^{(s)} = [X_i^{(1)}, Y], \tag{3.12}$$

where $Y = \sum_j f_j \partial / \partial z_j$ is homogeneous of degree s. Put $z_i' = z_i - f_i$. This coordinate transformation will kill the term of degree s in the Taylor expansion of X_i. $\qquad\square$

3.2 Levi decomposition of Poisson structures

Let Π be a Poisson structure in a neighborhood of 0 in $\mathbb{K}^n$, where $\mathbb{K} = \mathbb{R}$ or $\mathbb{C}$, which vanishes at 0: $\Pi(0) = 0$. Denote by $\Pi^{(1)}$ the linear part of Π at 0, and by $\mathfrak{l}$ the Lie algebra of linear functions on $\mathbb{K}^n$ under the linear Poisson bracket of $\Pi^{(1)}$. Let $\mathfrak{g} \subset \mathfrak{l}$ be a semisimple subalgebra of $\mathfrak{l}$. If Π is formal or analytic, we will assume that $\mathfrak{g}$ is a Levi factor of $\mathfrak{l}$. If Π is smooth (but not analytic), we will assume that $\mathfrak{g}$ is a maximal compact semisimple subalgebra of $\mathfrak{l}$, and we will call such a subalgebra a *compact Levi factor*. Denote by $(x_1, \ldots, x_m, y_1, \ldots, y_{n-m})$ a linear basis of $\mathfrak{l}$, such that $x_1, \ldots, x_m$ span $\mathfrak{g}$ ($\dim \mathfrak{g} = m$), and $y_1, \ldots, y_{n-m}$ span a complement $\mathfrak{r}$ of $\mathfrak{g}$ with respect to the adjoint action of $\mathfrak{g}$ on $\mathfrak{l}$, i.e., $[\mathfrak{g}, \mathfrak{r}] \subset \mathfrak{r}$. (In the formal and analytic cases, $\mathfrak{r}$ is the radical of $\mathfrak{l}$; in the smooth case it is not the radical in general.) Denote by c_{ij}^k and a_{ij}^k the structural constants of $\mathfrak{g}$ and of the action of $\mathfrak{g}$ on $\mathfrak{r}$ respectively: $[x_i, x_j] = \sum_k c_{ij}^k x_k$ and $[x_i, y_j] = \sum_k a_{ij}^k y_k$.

Definition 3.2.1. With the above notations, we will say that Π admits a formal (resp. analytic, resp. smooth) *Levi decomposition* or *Levi normal form* at 0, with respect to the (compact) Levi factor $\mathfrak{g}$, if there is a formal (resp. analytic, resp. smooth) coordinate system

$$(x_1^\infty, \ldots, x_m^\infty, y_1^\infty, \ldots, y_{n-m}^\infty),$$

with $x_i^\infty = x_i +$ higher-order terms and $y_i^\infty = y_i +$ higher-order terms, such that in this system of coordinates we have

$$\Pi = \sum_{i<j} c_{ij}^k x_k^\infty \frac{\partial}{\partial x_i^\infty} \wedge \frac{\partial}{\partial x_j^\infty} + \sum a_{ij}^k y_k^\infty \frac{\partial}{\partial x_i^\infty} \wedge \frac{\partial}{\partial y_j^\infty} + \sum_{i<j} P_{ij} \frac{\partial}{\partial y_i^\infty} \wedge \frac{\partial}{\partial y_j^\infty}, \tag{3.13}$$

where P_{ij} are formal (resp. analytic, resp. smooth) functions.

Remark 3.2.2. Another way to express Equation (3.13) is as follows:

$$\{x_i^\infty, x_j^\infty\} = \sum c_{ij}^k x_k^\infty \quad \text{and} \quad \{x_i^\infty, y_j^\infty\} = \sum a_{ij}^k y_k^\infty. \tag{3.14}$$

In other words, the Poisson brackets of x-coordinates with x-coordinates, and of x-coordinates with y-coordinates, are linear. Yet another way to say it is that the Hamiltonian vector fields of x_i^∞ are linear:

$$X_{x_i^\infty} = \sum c_{ij}^k x_k^\infty \frac{\partial}{\partial x_j^\infty} + \sum a_{ij}^k y_k^\infty \frac{\partial}{\partial y_j^\infty}. \tag{3.15}$$

In particular, the vector fields $X_{x_1^\infty}, \ldots, X_{x_m^\infty}$ form a Lie algebra isomorphic to $\mathfrak{g}$, and we have an infinitesimal linear Hamiltonian action of $\mathfrak{g}$ on $(\mathbb{K}^n, \Pi)$, whose momentum map $\mu : \mathbb{K}^n \to \mathfrak{g}^*$ is defined by $\langle \mu(z), x_i \rangle = x_i(z)$.

Theorem 3.2.3 (Wade [342]). *Any formal Poisson structure Π in $\mathbb{K}^n$ ($\mathbb{K} = \mathbb{R}$ or $\mathbb{C}$) which vanishes at 0 admits a formal Levi decomposition.*

Proof. Denote by $\mathcal{L}$ the algebra of formal functions in $\mathbb{K}^n$ which vanish at 0, under the Lie bracket of Π. Then it is a pro-finite Lie algebra, whose completion is itself. The Lie algebra $\mathcal{L}/\mathcal{L}_1$, where $\mathcal{L}_1$ is the ideal of $\mathcal{L}$ consisting of functions which vanish at 0 together with their first derivatives, is isomorphic to the Lie algebra $\mathfrak{l}$ of linear functions on $\mathbb{K}^n$ whose Lie bracket is given by the linear Poisson structure $\Pi^{(1)}$. By Theorem 3.1.2, $\mathcal{L}$ admits a Levi factor, which is isomorphic to the Levi factor $\mathfrak{g}$ of $\mathfrak{l}$. Denote by $x_1^\infty, \ldots, x_m^\infty$ a linear basis of a Levi factor of $\mathcal{L}$, $\{x_i^\infty, x_j^\infty\} = \sum_k c_{ij}^k x_k^\infty$ where c_{ij}^k are structural constants of $\mathfrak{g}$. Then the Hamiltonian vector fields $X_{x_1^\infty}, \ldots, X_{x_m^\infty}$ give a formal action of $\mathfrak{g}$ on $\mathbb{K}^n$. By Hermann's formal linearization Theorem 3.1.6, this formal action can be linearized formally, i.e., there is a formal coordinate system $(x_1^0, \ldots, y_{n-m}^0)$ in which we have

$$X_{x_i^\infty} = \sum c_{ij}^k x_k^0 \frac{\partial}{\partial x_j^0} + \sum a_{ij}^k y_k^0 \frac{\partial}{\partial y_j^0}. \tag{3.16}$$

A priori, it may happen that $x_i^0 \neq x_i^\infty$, but in any case we have $x_i^0 = x_i^\infty +$ higher-order terms, and $X_{x_i^\infty}(x_j^0) = \sum_k c_{ij}^k x_k^\infty$, $X_{x_i^\infty}(y_j^0) = \sum_k a_{ij}^k y_k^0$. Renaming y_j^0 by y_j^∞, we get a formal coordinate system $(x_1^\infty, \ldots, y_{n-m}^\infty)$ which puts Π in formal Levi normal form. $\qquad\square$

Remark 3.2.4. A particular case of Theorem 3.2.3 is the following formal linearization theorem of Weinstein [346] mentioned in Chapter 2: if the linear part of Π at 0 is semisimple (i.e., it corresponds to a semisimple Lie algebra $\mathfrak{l} = \mathfrak{g}$), then Π is formally linearizable at 0.

Remark 3.2.5. As observed by Chloup [79], Theorem 3.2.3 may also be viewed as a consequence of Hochschild–Serre's Theorem 2.4.4. Indeed, according to Theorem 2.4.4 and Whitehead's lemma, we have

$$H^2(\mathfrak{l}, \mathcal{S}^p\mathfrak{l}) \cong H^0(\mathfrak{g}, \mathbb{K}) \otimes H^2(\mathfrak{r}, \mathcal{S}^p\mathfrak{l})^\mathfrak{g} \cong H^2(\mathfrak{r}, \mathcal{S}^p\mathfrak{l})^\mathfrak{g} \quad \forall\, p. \tag{3.17}$$

It means that any nonlinear term in the Taylor expansion of Π, which is represented by a 2-cocycle of $\mathfrak{l}$ with values in $\mathcal{S}\mathfrak{l} = \oplus_p \mathcal{S}^p\mathfrak{l}$, can be "pushed to $\mathfrak{r}$", i.e., pushed to the "y-part" (consisting of terms $P_{ij}\partial/\partial y_i \wedge \partial/\partial y_j$) of Π.

In the analytic case, we have:

Theorem 3.2.6 ([369]). *Any analytic Poisson structure Π in a neighborhood of 0 in $\mathbb{K}^n$, where $\mathbb{K} = \mathbb{R}$ or $\mathbb{C}$, which vanishes at 0, admits an analytic Levi decomposition.*

Remark 3.2.7. If in the above theorem, $\mathfrak{l} = ((\mathbb{K}^n)^*, \{.,.\}_{\Pi^{(1)}})$ is a semi-simple Lie algebra, i.e., $\mathfrak{g} = \mathfrak{l}$, then we recover the following analytic linearization theorem of Conn [80]: any analytic Poisson structure with a semi-simple linear part is locally analytically linearizable. When $\mathfrak{l} = \mathfrak{g} \oplus \mathbb{K}$, then a Levi decomposition of Π is still automatically a linearization (because $\{y_1, y_1\} = 0$), and Theorem 3.2.6 implies the following result of Molinier [258] and Conn (unpublished): If the linear part of an analytic Poisson structure Π which vanishes at 0 corresponds to $\mathfrak{l} = \mathfrak{g} \oplus \mathbb{K}$, where $\mathfrak{g}$ is semisimple, then Π is analytically linearizable in a neighborhood of 0.

Remark 3.2.8. The existence of a local analytic Levi decomposition of Π is essentially equivalent to the existence of a Levi factor (and not just a formal Levi factor) for the Lie algebra $\mathcal{O}$ of germs at 0 of analytic functions under the Poisson bracket of Π. Indeed, if Π is in analytic Levi normal form with respect to a coordinate system $(x_1, \ldots, y_{n-m})$, then the functions $x_1, \ldots, x_m$ form a linear basis of a Levi factor of $\mathcal{O}$. Conversely, suppose that $\mathcal{O}$ admits a Levi factor with a linear basis $x_1, \ldots, x_m$. Then $X_{x_1}, \ldots, X_{x_m}$ generate a local analytic action of $\mathfrak{g}$ on $\mathbb{K}^n$. According to the Kushnirenko–Guillemin–Sternberg analytic linearization theorem for analytic actions of semisimple Lie algebras [160, 205], we may assume that

$$X_{x_i} = \sum c_{ij}^k x_k^0 \frac{\partial}{\partial x_j^0} + \sum a_{ij}^k y_k^0 \frac{\partial}{\partial y_j^0} \tag{3.18}$$

in a local analytic system of coordinates $(x_1^0, \ldots, y_{n-m}^0)$, where $x_i^0 = x_i +$ higher-order terms. Renaming y_i^0 by y_i, we get a local analytic system of coordinates $(x_1, \ldots, y_{n-m})$ which puts Π in Levi normal form.

In the smooth case, we have:

Theorem 3.2.9 (Monnier–Zung [263]). *For any $n \in \mathbb{N}$ and $p \in \mathbb{N} \cup \{\infty\}$ there is $p' \in \mathbb{N} \cup \{\infty\}$, $p' < \infty$ if $p < \infty$, such that the following statement holds: Let Π be a $C^{p'}$-smooth Poisson structure in a neighborhood of 0 in $\mathbb{R}^n$. Denote by $\mathfrak{l}$ the Lie algebra of linear functions in $\mathbb{R}^n$ under the Lie–Poisson bracket Π_1 which is the linear part of Π, and by $\mathfrak{g}$ a compact Levi factor of $\mathfrak{l}$. Then there exists a C^p-smooth Levi decomposition of Π with respect to $\mathfrak{g}$ in a neighborhood of 0 .*

Remark 3.2.10. The condition that $\mathfrak{g}$ be compact in Theorem 3.2.9 is in a sense necessary, already in the case when $\mathfrak{l} = \mathfrak{g}$. (See Section 4.3.)

Remark 3.2.11. Remark 3.2.7 and Remark 3.2.8 also apply to the smooth case (provided that $\mathfrak{g}$ is compact). In particular, when $\mathfrak{l} = \mathfrak{g}$, one recovers from Theorem 3.2.9 the following smooth linearization theorem of Conn [81]: any smooth Poisson structure whose linear part is compact semisimple is locally smoothly linearizable. When $\mathfrak{l} = \mathfrak{g} \oplus \mathbb{R}$ with $\mathfrak{g}$ compact semisimple, Theorem 3.2.9 still gives a smooth linearization. And the existence of a local smooth Levi decomposition is equivalent to the existence of a compact Levi factor.

In the rest of this chapter, we will give a full proof of Theorem 3.2.6, and then a sketch of the proof of Theorem 3.2.9, which is similar but more technical. These proofs of Theorem 3.2.6 and Theorem 3.2.9 are inspired by and based on Conn's work [80, 81], and use a normed version of Whitehead's lemma (on vanishing cohomology of semisimple Lie algebras) and the fast convergence method (of Kolmogorov in the analytic case and Nash–Moser in the smooth case) in order to show the convergence of a formal coordinate transformation putting the Poisson structure in Levi normal form.

3.3   Construction of Levi decomposition

In this section we will construct, by a recurrence process, a formal system of coordinates $(x_1^\infty, \ldots, x_m^\infty, y_1^\infty, \ldots, y_{n-m}^\infty)$ which satisfy Relations (3.14) for a given local analytic Poisson structure Π. We will later use analytic estimates to show that our construction actually yields a local analytic system of coordinates.

Each step in our recurrence process consists of two substeps: the first substep is to find an *almost Levi factor*. The second substep consists of *almost linearizing* this almost Levi factor.

We begin the first step with the original linear coordinate system

$$(x_1^0, \ldots, x_m^0, y_1^0, \ldots, y_{n-m}^0) = (x_1, \ldots, x_m, y_1, \ldots, y_{n-m}) \ . \tag{3.19}$$

For each positive integer l, after Step l we will find a local coordinate system $(x_1^l, \ldots, x_m^l, y_1^l, \ldots, y_{n-m}^l)$ with the following properties (3.20), (3.21), (3.24):

$$(x_1^l, \ldots, x_m^l, y_1^l, \ldots, y_{n-m}^l) = (x_1^{l-1}, \ldots, x_m^{l-1}, y_1^{l-1}, \ldots, y_{n-m}^{l-1}) \circ \phi_l \ , \tag{3.20}$$

where ϕ_l is a local analytic diffeomorphism of $(\mathbb{K}^n, 0)$ of the type

$$\phi_l(z) = z + \text{terms of order} \geq 2^{l-1} + 1 \ . \tag{3.21}$$

The space $(\mathbb{K}^n, 0)$ above is fixed (our local Poisson manifold). The functions x_1^{l-1}, x_1^l, etc. are local functions on that fixed space.

Denote by

$$X_i^l = X_{x_i^l} \ (i = 1, \ldots, m) \tag{3.22}$$

the Hamiltonian vector field of x_i^l with respect to our Poisson structure Π. Then we have

$$X_i^l = \hat{X}_i^l + Y_i^l \ , \tag{3.23}$$

where

$$\hat{X}_i^l = \sum_{jk} c_{ij}^k x_k^l \frac{\partial}{\partial x_j^l} + \sum_{jk} a_{ij}^k y_k^l \frac{\partial}{\partial y_j^l} \ , \ Y_i^l \in O(|z|^{2^l+1}) \ , \tag{3.24}$$

i.e., $\hat{X}_i^l$ is the linear part of $X_i^l = X_{x_i^l}$ in the coordinate system $(x_1^l, \ldots, y_{n-m}^l)$, c_{ij}^k and a_{ij}^k are structural constants as appeared in Theorem 3.2.6, and $Y_i^l = X_i^l - \hat{X}_i^l$ does not contain terms of order $\leq 2^l$.

Condition (3.24) may be rewritten as

$$\{x_i^l, x_j^l\} = \sum_k c_{ij}^k x_k^l \text{ modulo terms of order } \geq 2^l + 1 , \tag{3.25}$$

$$\{x_i^l, y_j^l\} = \sum_k a_{ij}^k y_k^l \text{ modulo terms of order } \geq 2^l + 1 . \tag{3.26}$$

So we may say that the functions $(x_1^l, \ldots, x_m^l)$ form an *almost Levi factor*, and their corresponding Hamiltonian vector fields are *almost linearized*, up to terms of order $2^l + 1$.

Of course, when $l = 0$, then Relation (3.24) is satisfied by the assumptions of Theorem 3.2.6. Let us show how to construct the coordinate system $(x_1^{l+1}, \ldots, y_{n-m}^{l+1})$ from the coordinate system $(x_1^l, \ldots, y_{n-m}^l)$. Denote

$$\mathcal{O}_l = \{\text{local analytic functions in } (\mathbb{K}^n, 0) \text{ without terms of order } \leq 2^l\} . \tag{3.27}$$

Due to Relations (3.20) and (3.21), it doesn't matter if we use the original coordinate system $(x_1, \ldots, x_m, y_1, \ldots, y_{n-m})$ or the new one $(x_1^l, \ldots, x_m^l, y_1^l, \ldots, y_{n-m}^l)$ in the above definition of $\mathcal{O}_l$. It follows from Relation (3.24) that

$$f_{ij}^l := \{x_i^l, x_j^l\} - \sum_k c_{ij}^k x_k^l = Y_i^l(x_j^l) \in \mathcal{O}_l . \tag{3.28}$$

Denote by $(\xi_1, \ldots, \xi_m)$ a fixed basis of the semi-simple algebra $\mathfrak{g}$, with

$$[\xi_i, \xi_j] = \sum_k c_{ij}^k \xi_k . \tag{3.29}$$

Then $\mathfrak{g}$ acts on $\mathcal{O}$ via vector fields $\hat{X}_1^l, \ldots, \hat{X}_m^l$, and this action induces the following linear action of $\mathfrak{g}$ on the finite-dimensional vector space $\mathcal{O}_l/\mathcal{O}_{l+1}$: if $g \in \mathcal{O}_l$, considered modulo $\mathcal{O}_{l+1}$, then we put

$$\xi_i \cdot g := \hat{X}_i^l(g) = \sum_{jk} c_{ij}^k x_k^l \frac{\partial g}{\partial x_j^l} + \sum_{jk} a_{ij}^k y_k^l \frac{\partial g}{\partial y_j^l} \quad \text{mod } \mathcal{O}_{l+1} . \tag{3.30}$$

Notice that if $g \in \mathcal{O}_l$ then $Y_i^l(g) \in \mathcal{O}_{l+1}$, and hence we have

$$\xi_i \cdot g = X^l(g) \mod \mathcal{O}_{l+1} = \{x_i^l, g\} \mod \mathcal{O}_{l+1} . \tag{3.31}$$

The functions f_{ij}^l in (3.28) form a 2-cochain f^l of $\mathfrak{g}$ with values in the $\mathfrak{g}$-module $\mathcal{O}_l/\mathcal{O}_{l+1}$:

$$f^l : \mathfrak{g} \wedge \mathfrak{g} \to \mathcal{O}_l/\mathcal{O}_{l+1}$$

$$f^l(\xi_i \wedge \xi_j) := f_{ij}^l \mod \mathcal{O}_{l+1} = \{x_i^l, x_j^l\} - \sum_k c_{ij}^k x_k^l \mod \mathcal{O}_{l+1} . \tag{3.32}$$

In other words, if we denote by $\mathfrak{g}^*$ the dual space of $\mathfrak{g}$, and by $(\xi_1^*, \ldots, \xi_m^*)$ the basis of $\mathfrak{g}^*$ dual to $(\xi_1, \ldots, \xi_m)$, then we have

$$f^l = \sum_{i<j} \xi_i^* \wedge \xi_j^* \otimes (f_{ij}^l \mod \mathcal{O}_{l+1}) \in \wedge^2 \mathfrak{g}^* \otimes \mathcal{O}_l/\mathcal{O}_{l+1} . \tag{3.33}$$

It follows from (3.28), and the Jacobi identity for the Poisson bracket of Π and the algebra $\mathfrak{g}$, that the above 2-cochain is a 2-cocycle. Because $\mathfrak{g}$ is semi-simple, we have $H^2(\mathfrak{g}, \mathcal{O}_l/\mathcal{O}_{l+1}) = 0$, i.e., the second cohomology of $\mathfrak{g}$ with coefficients in $\mathfrak{g}$-module $\mathcal{O}_l/\mathcal{O}_{l+1}$ vanishes, and therefore the above 2-cocycle is a coboundary. In other words, there is a 1-cochain

$$w^l \in \mathfrak{g}^* \otimes \mathcal{O}_l/\mathcal{O}_{l+1} \tag{3.34}$$

such that

$$f^l(\xi_i \wedge \xi_j) = \xi_i \cdot w^l(\xi_j) - \xi_j \cdot w^l(\xi_i) - w^l\left(\sum_k c_{ij}^k \xi_k\right) . \tag{3.35}$$

Denote by w_i^l the element of $\mathcal{O}_l$ which is a polynomial of order $\leq 2^{l+1}$ in variables $(x_1^l, \ldots, x_m^l, y_1^l, \ldots, y_{n-m}^l)$ such that the projection of w_i^l in $\mathcal{O}_l/\mathcal{O}_{l+1}$ is $w^l(\xi_i)$.

Remark 3.3.1. Remember that w_i^l are local functions on our fixed space $(\mathbb{K}^n, 0)$. They are not functions of variables $(x_1^l, \ldots, x_m^l, y_1^l, \ldots, y_{n-m}^l)$ per se, but when expressed in terms of these variables they become polynomial functions.

Define x_i^{l+1} as follows:

$$x_i^{l+1} = x_i^l - w_i^l \ \forall \ i = 1, \ldots, m . \tag{3.36}$$

Then it follows from (3.28) and (3.35) that we have

$$\{x_i^{l+1}, x_j^{l+1}\} - \sum_k c_{ij}^k x_k^{l+1} \in \mathcal{O}_{l+1} \ for \ i, j \leq m . \tag{3.37}$$

This concludes our first substep (the (x_i^{l+1}) form a better "almost Levi factor" than (x_i^l)). Let us now proceed to the second substep.

Denote by $\mathcal{Y}^l$ the space of local analytic vector fields of the type $u = \sum_{i=1}^{n-m} u_i \partial/\partial y_i^l$ (with respect to the coordinate system $(x_1^l, \ldots, y_{n-m}^l)$), with u_i being local analytic functions. For each natural number k, denote by $\mathcal{Y}_k^l$ the following subspace of $\mathcal{Y}^l$:

$$\mathcal{Y}_k^l = \left\{ u = \sum_{i=1}^{n-m} u_i \partial/\partial y_i^l \ \Big| \ u_i \in \mathcal{O}_k \right\} . \tag{3.38}$$

Then $\mathcal{Y}^l$, as well as $\mathcal{Y}_l^l/\mathcal{Y}_{l+1}^l$, are $\mathfrak{g}$-modules under the following action:

$$\xi_i \cdot \sum_j u_j \partial/\partial y_j^l := [\hat{X}_i^l, u] = \left[\sum_{jk} c_{ij}^k x_k^l \frac{\partial}{\partial x_j^l} + \sum_{jk} a_{ij}^k y_k^l \frac{\partial}{\partial y_j^l} , \ \sum_j u_j \partial/\partial y_j^l\right] . \tag{3.39}$$

The above linear action of $\mathfrak{g}$ on $\mathcal{Y}_l/\mathcal{Y}_{l+1}$ can also be written as

$$\xi_i \cdot \sum_j u_j \partial/\partial y_j^l = \sum_j (\{x_i^l, u_j\} - \sum_k a_{ij}^k u_k)\partial/\partial y_j^l \mod \mathcal{Y}_{l+1}^l . \qquad (3.40)$$

Define the following 1-cochain of $\mathfrak{g}$ with values in $\mathcal{Y}_l^l/\mathcal{Y}_{l+1}^l$:

$$\sum_{i=1}^{m} \left(\xi_i^* \otimes \left(\sum_{j=1}^{n-m} (\{x_i^{l+1}, y_j^l\} - \sum_k a_{ij}^k y_k^l)\partial/\partial y_j^l \mod \mathcal{Y}_{l+1}^l \right) \right) \in \mathfrak{g}^* \otimes \mathcal{Y}_l^l/\mathcal{Y}_{l+1}^l . \quad (3.41)$$

Due to Relation (3.37), the above 1-cochain is a 1-cocycle. Since $\mathfrak{g}$ is semi-simple, we have $H^1(\mathfrak{g}, \mathcal{Y}_l^l/\mathcal{Y}_{l+1}^l) = 0$, and the above 1-cocycle is a 1-coboundary. In other words, there exists a vector field

$$\sum_{j=1}^{n-m} v_j^l \partial/\partial y_j^l \in \mathcal{Y}_l^l , \qquad (3.42)$$

with v_j^l being polynomial functions of degree $\leq 2^{l+1}$ in variables $(x_1^l, \ldots, y_{n-m}^l)$, such that for every $i = 1, \ldots, m$ we have

$$\sum_j \left(\{x_i^{l+1}, y_j^l\} - \sum a_{ij}^k y_k^l \right)\partial/\partial y_j^l = \sum_j \left(\{x_i^l, v_j^l\} - \sum a_{ij}^k v_k^l \right)\partial/\partial y_j^l \mod \mathcal{Y}_{l+1}^l .$$
$$(3.43)$$

We now define the new system of coordinates as

$$\begin{aligned} x_i^{l+1} &= x_i^l - w_i^l \ (i = 1, \ldots, m), \\ y_i^{l+1} &= y_i^l - v_i^l \ (i = 1, \ldots, n-m), \end{aligned} \qquad (3.44)$$

where functions $w_i^l, v_i^l \in \mathcal{O}_l$ are chosen as above. In particular, Relations (3.37) and (3.43) are satisfied, which means that

$$\begin{aligned} \{x_i^{l+1}, x_j^{l+1}\} - \sum c_{ij}^k x_k^{l+1} &\in \mathcal{O}_{l+1} , \\ \{x_i^{l+1}, y_j^{l+1}\} - \sum a_{ij}^k y_k^{l+1} &\in \mathcal{O}_{l+1} , \end{aligned} \qquad (3.45)$$

i.e., Relation (3.24) is satisfied with l replaced by $l+1$. Of course, Relations (3.20) and (3.21) are also satisfied with l replaced by $l+1$, and with ϕ_{l+1} being the map which when written in the variables $(x_1^l, \ldots, y_{n-m}^l)$ has the form

$$\phi_{l+1} = \mathrm{Id} + \psi_{l+1} , \qquad (3.46)$$

where

$$\psi_{l+1} = -(w_1^l, \ldots, w_m^l, v_1^l, \ldots, v_{n-m}^l) \in (O_l)^n . \qquad (3.47)$$

Remark 3.3.2. We stress here the fact that Formula (3.46) is valid with respect to the coordinate system $(x_1^l, \ldots, y_{n-m}^l)$ only. In particular, the sum there is taken with respect to the local linear structure given by the coordinate system $(x_1^l, \ldots, y_{n-m}^l)$ and not by the original coordinate system $(x_1, \ldots, y_{n-m})$. If we want to express ϕ_{l+1} in terms of the original coordinate system then it will be much more complicated.

Recall that, by the above construction, $w_1^l, \ldots, w_m^l, v_1^l, \ldots, v_{n-m}^l$ are polynomial functions of degree $\leq 2^{l+1}$ in variables $(x_1^l, \ldots, y_{n-m}^l)$, which do not contain terms of degree $\leq 2^l$.

Define the limits

$$(x_1^\infty, \ldots, y_{n-m}^\infty) = \lim_{l \to \infty} (x_1^l, \ldots, y_{n-m}^l) \,,$$

$$\Phi_\infty = \lim_{l \to \infty} \Phi_l \quad \text{where } \Phi_l = \phi_1 \circ \cdots \circ \phi_l \,. \tag{3.48}$$

It is clear that the above limits exist in the formal category,

$$(x_1^\infty, \ldots, y_{n-m}^\infty) = (x_1^0, \ldots, y_{n-m}^0) \circ \Phi_\infty, \tag{3.49}$$

and the formal coordinate system $(x_1^\infty, \ldots, y_{n-m}^\infty)$ satisfies Relation (3.14).

The above construction works not only for local analytic Poisson structures, but also for formal Poisson structures, so it gives us another proof of Theorem 3.2.3. To prove Theorem 3.2.6, it remains to show that, when Π is analytic, we can choose functions w_i^l, v_i^l in such a way that $(x_1^\infty, \ldots, y_{n-m}^\infty)$ is in fact a local analytic system of coordinates.

Remark 3.3.3. The above construction of formal Levi decomposition differs from the construction of Wade [342] and Weinstein [353]. Their construction is simpler (they don't almost linearize the almost Levi factor at each step, and they kill only one term at each step), and is good enough to show the existence of a formal Levi decomposition. However, in order to prove the existence of an *analytic* Levi decomposition, using Kolmogorov's fast convergence method, one needs to kill a bunch of terms at each step, and that's why the second substep (almost linearizing an almost Levi factor) is important.

3.4 Normed vanishing of cohomology

In this section, using *normed vanishing* of first and second cohomology groups of $\mathfrak{g}$, we will obtain some estimates on $w_i^l = x_i^l - x_i^{l+1}$ and $v_i^l = y_i^l - y_i^{l+1}$. For some basic results on semi-simple Lie algebras and their representations which will be used below, one may consult a book on Lie algebras, e.g., [186, 335].

We will denote by $\mathfrak{g}_\mathbb{C}$ the algebra $\mathfrak{g}$ if $\mathbb{K} = \mathbb{C}$, and the complexification of $\mathfrak{g}$ if $\mathbb{K} = \mathbb{R}$. So $\mathfrak{g}_\mathbb{C}$ is a complex semi-simple Lie algebra of dimension m. Denote by $\mathfrak{g}_0$ the compact real form of $\mathfrak{g}_\mathbb{C}$, and identify $\mathfrak{g}_\mathbb{C}$ with $\mathfrak{g}_0 \otimes_\mathbb{R} \mathbb{C}$. Fix an orthonormal basis $(e_1, \ldots, e_m)$ of $\mathfrak{g}_\mathbb{C}$ with respect to the Killing form: $\langle e_i, e_j \rangle = \delta_{ij}$. We may assume that $e_1, \ldots, e_m \in \sqrt{-1}\mathfrak{g}_0$. Denote by $\Gamma = \sum_i e_i^2$ the *Casimir element* of $\mathfrak{g}_\mathbb{C}$: Γ lies in the center of the universal enveloping algebra $\mathcal{U}(\mathfrak{g}_\mathbb{C})$ and does not depend on the choice of the basis (e_i). When $\mathbb{K} = \mathbb{R}$ then Γ is real, i.e., $\Gamma \in \mathcal{U}(\mathfrak{g})$.

Let W be a finite-dimensional complex linear space endowed with a Hermitian metric denoted by $\langle, \rangle$. If $v \in W$ then its norm is denoted by $\|v\| = \sqrt{\langle v, v \rangle}$. Assume that W is a Hermitian $\mathfrak{g}_0$-module. In other words, the linear action of $\mathfrak{g}_0$ on W is via infinitesimal unitary (i.e., skew-adjoint) operators. W is a $\mathfrak{g}_\mathbb{C}$-module via the identification $\mathfrak{g}_\mathbb{C} = \mathfrak{g}_0 \otimes_\mathbb{R} \mathbb{C}$. We have the decomposition $W = W_0 + W_1$, where $W_1 = \mathfrak{g}_\mathbb{C} \cdot W$ (the image of the representation), and $\mathfrak{g}_\mathbb{C}$ acts trivially on W_0. Since W_1 is a $\mathfrak{g}_\mathbb{C}$-module, it is also a $\mathcal{U}(\mathfrak{g}_\mathbb{C})$-module. The action of Γ on W_1 is invertible: $\Gamma \cdot W_1 = W_1$, and we will denote by Γ^{-1} the inverse mapping.

Denote by $\mathfrak{g}_\mathbb{C}^*$ the dual of $\mathfrak{g}_\mathbb{C}$, and by $(e_1^*, \ldots, e_m^*)$ the basis of $\mathfrak{g}_\mathbb{C}^*$ dual to $(e_1, \ldots, e_m)$. If $w \in \mathfrak{g}_\mathbb{C}^* \otimes W$ is a 1-cochain and $f : \wedge^2 \mathfrak{g}_\mathbb{C}^* \otimes W$ is a 2-cochain with values in W, then we will define the norm of f and w as follows:

$$\|w\| = \max_i \|w(e_i)\| \ , \ \|f\| = \max_{i,j} \|f(e_i \wedge e_j)\| \ . \tag{3.50}$$

Since $H^2(\mathfrak{g}, \mathbb{K}) = 0$, there is a (unique) linear map $h_0 : \wedge^2 \mathfrak{g}^* \to \mathfrak{g}^*$ such that if $u \in \wedge^2 \mathfrak{g}^*$ is a 2-cocycle for the trivial representation of $\mathfrak{g}$ in $\mathbb{K}$ (i.e., $u([x,y], z) + u([y,z], x) + u([z,x], y) = 0$ for any $x, y, z \in \mathfrak{g}$), then $u = \delta h_0(u)$, i.e., $u(x,y) = h_0(u)([x,y])$. By complexifying h_0 if $\mathbb{K} = \mathbb{R}$, and taking its tensor product with the projection map $P_0 : W \to W_0$, we get a map

$$h_0 \otimes P_0 : \wedge^2 \mathfrak{g}_\mathbb{C}^* \otimes W \to \mathfrak{g}_\mathbb{C}^* \otimes W_0 \ . \tag{3.51}$$

Define another map

$$h_1 : \wedge^2 \mathfrak{g}_\mathbb{C}^* \otimes W \to \mathfrak{g}_\mathbb{C}^* \otimes W_1 \tag{3.52}$$

as follows: if $f \in \wedge^2 \mathfrak{g}_\mathbb{C}^* \otimes W$ then we put

$$h_1(f) = \sum_i e_i^* \otimes \left(\Gamma^{-1} \cdot \sum_j (e_j \cdot f(e_i \wedge e_j)) \right) \ . \tag{3.53}$$

Then the map

$$h = h_0 \otimes P_0 + h_1 : \wedge^2 \mathfrak{g}_\mathbb{C}^* \otimes W \to \mathfrak{g}_\mathbb{C}^* \otimes W \tag{3.54}$$

is an explicit *homotopy operator*, in the sense that if $f \in \wedge^2 \mathfrak{g}_\mathbb{C}^* \otimes W$ is a 2-cocycle (i.e., $\delta f = 0$ where δ denotes the differential of the Chevalley–Eilenberg complex $\cdots \to \wedge^k \mathfrak{g}_\mathbb{C}^* \otimes W \to \wedge^{k+1} \mathfrak{g}_\mathbb{C}^* \otimes W \to \cdots$), then $f = \delta(h(f))$.

Similarly, the map $h : \mathfrak{g}_\mathbb{C}^* \otimes W \to W$ defined by

$$h(w) = \Gamma^{-1} \cdot \left(\sum_i e_i \cdot w(e_i) \right) \tag{3.55}$$

is also a homotopy operator, in the sense that if $w \in \mathfrak{g}_\mathbb{C}^* \otimes W$ is a 1-cocycle then $w = \delta(h(w))$.

When $\mathbb{K} = \mathbb{R}$, i.e., when $\mathfrak{g}_{\mathbb{C}}$ is the complexification of $\mathfrak{g}$, then the above homotopy operators h are real, i.e., they map real cocycles into real cochains.

The above formulas make it possible to control the norm of a primitive of a 1-cocycle w or a 2-cocycle f in terms of the norm of w or f. More precisely, we have the following lemma about *normed vanishing of cohomology*, which is a normed version of Whitehead's lemma which says that $H^1(\mathfrak{g}, W) = H^2(\mathfrak{g}, W) = 0$.

Lemma 3.4.1 (Conn). *There is a positive constant D (which depends on $\mathfrak{g}$ but does not depend on W) such that with the above notations we have*

$$\|h(f)\| \leq D\|f\| \text{ and } \|h(w)\| \leq D\|w\| \qquad (3.56)$$

for any 1-cocycle w and any 2-cocycle f of $\mathfrak{g}_{\mathbb{C}}$ with values in W.

Remark 3.4.2. The above lemma is essentially due to Conn (see Proposition 2.1 of [80]). Conn stated the result only for some particular modules that he needed, but his proof, which we give below, works without any change for other Hermitian modules.

Proof (sketch). We can decompose W into an orthogonal sum (with respect to the Hermitian metric of W) of irreducible modules of $\mathfrak{g}_0$. The above homotopy operators decompose correspondingly, so it is enough to prove the above lemma for the case when W is a nontrivial irreducible module, which we will now suppose. Let $\lambda \neq 0$ denote the highest weight of the irreducible $\mathfrak{g}_0$-module W, and by δ one-half the sum of positive roots of $\mathfrak{g}_0$ (with respect to a fixed Cartan subalgebra and Weyl chamber). Then Γ acts on W by multiplication by the scalar $\langle \lambda, \lambda + 2\delta \rangle$, which is greater than or equal to $\|\lambda\|^2$. Denote by $\mathcal{J}$ the weight lattice of $\mathfrak{g}_0$, and $D = m(\min_{\gamma \in \mathcal{J}} \|\gamma\|)^{-1}$. Then $D < \infty$ does not depend on W, and $\|\lambda\|^2 > \frac{m\|\lambda\|}{D}$, which implies that the norm of the inverse of the action of Γ on W is smaller than or equal to $\frac{D}{m\|\lambda\|}$. On the other hand, the norm of the action of e_i on W is smaller than or equal to $\|\lambda\|$ for each $i = 1, \ldots, m$ (recall that $\sqrt{-1}e_i \in \mathfrak{g}_0$ and $\langle e_i, e_i \rangle = 1$), hence the norm of the operator $\sum_{i=1}^{m} e_i \cdot \Gamma^{-1} : W \to W$ is smaller than or equal to D. Now apply Formulas (3.53) and (3.55). The lemma is proved. $\square$

Let us now apply the above lemma to $\mathfrak{g}$-modules $\mathcal{O}_l/\mathcal{O}_{l+1}$ and $\mathcal{Y}_l^l/\mathcal{Y}_{l+1}^l$ introduced in the previous section. Recall that $\mathfrak{g}$ is a Levi factor of $\mathfrak{l}$, the space of linear functions in $\mathbb{K}^n$, which is a Lie algebra under the linear Poisson bracket $\Pi^{(1)}$. $\mathfrak{g}$ acts on $\mathfrak{l}$ by the (restriction of the) adjoint action, and on $\mathbb{K}^n$ by the coadjoint action. By complexifying these actions if necessary, we get a natural action of $\mathfrak{g}_{\mathbb{C}}$ on $(\mathbb{C}^n)^*$ (the dual space of $\mathbb{C}^n$) and on $\mathbb{C}^n$. The elements $x_1, \ldots, x_m, y_1, \ldots, y_{n-m}$ of the original linear coordinate system in $\mathbb{K}^n$ may be view as a basis of $(\mathbb{C}^n)^*$. Notice that the action of $\mathfrak{g}_{\mathbb{C}}$ on $(\mathbb{C}^n)^*$ preserves the subspace spanned by $(x_1, \ldots, x_m)$ and the subspace spanned by $(y_1, \ldots, y_{n-m})$. Fix a basis $(z_1, \ldots, z_n)$ of $(\mathbb{C}^n)^*$, such that the Hermitian metric of $(\mathbb{C}^n)^*$ for which this basis is orthonormal is preserved by the action of $\mathfrak{g}_0$, and such that

$$z_i = \sum_{j \leq m} A_{ij} x_j + \sum_{j \leq n-m} A_{i,j+m} y_j , \qquad (3.57)$$

with the constant transformation matrix (A_{ij}) satisfying the following condition:

$$A_{ij} = 0 \text{ if } (i \leq m < j \text{ or } j \leq m < i) . \qquad (3.58)$$

Such a basis $(z_1, \ldots, z_n)$ always exists, and we may view $(z_1, \ldots, z_n)$ as a linear coordinate system on $\mathbb{C}^n$. We will also define local complex analytic coordinate systems $(z_1^l, \ldots, z_n^l)$ as follows:

$$z_i^l = \sum_{j \leq m} A_{ij} x_j^l + \sum_{j \leq n-m} A_{i,j+m} y_j^l . \qquad (3.59)$$

Let l be a natural number, ρ a positive number, and f a local complex analytic function of n variables. Define the following ball $B_{l,\rho}$ and L^2-norm $\|f\|_{l,\rho}$, whenever it makes sense:

$$B_{l,\rho} = \left\{ x \in \mathbb{C}^n \mid \sqrt{\sum |z_i^l(x)|^2} \leq \rho \right\} , \qquad (3.60)$$

$$\|f\|_{l,\rho} = \sqrt{\frac{1}{V_\rho} \int_{S_{l,\rho}} |f(x)|^2 \mathrm{d}\mu_l} , \qquad (3.61)$$

where $\mathrm{d}\mu_l$ is the standard volume form on the boundary $S_{l,\rho} = \partial B_{l,\rho}$ of the complex ball $B_{l,\rho}$ with respect to the coordinate system $(z_1^l, \ldots, z_n^l)$, and V_ρ is the volume of $S_{l,\rho}$, i.e., of a $(2n-1)$-dimensional sphere of radius ρ.

We will say that the ball $B_{l,\rho}$ is well defined if it is analytically diffeomorphic to the standard ball of radius ρ via the coordinate system $(z_1^l, \ldots, z_n^l)$, and will use $\|f\|_{l,\rho}$ only when $B_{l,\rho}$ is well defined. When $B_{l,\rho}$ is not well defined we simply put $\|f\|_{l,\rho} = \infty$. We will write B_ρ and $\|f\|_\rho$ for $B_{0,\rho}$ and $\|f\|_{0,\rho}$ respectively. If f is a real analytic function (the case when $\mathbb{K} = \mathbb{R}$), we will complexify it before taking the norms.

It is well known (see, e.g., Chapter 1 of [306]) that the L^2-norm $\|f\|_\rho$ is given by a Hermitian metric, in which the monomial functions form an orthogonal basis: if $f = \sum_{\alpha \in \mathbb{N}^n} a_\alpha \prod_i z_i^{\alpha_i}$ and $g = \sum_{\alpha \in \mathbb{N}^n} b_\alpha \prod_i z_i^{\alpha_i}$ then the scalar product $\langle f, g \rangle_\rho$ is given by

$$\langle f, g \rangle_\rho = \sum_{\alpha \in \mathbb{N}^n} \frac{\alpha!(n-1)!}{(|\alpha|+n-1)!} \rho^{2|\alpha|} a_\alpha \bar{b}_\alpha , \qquad (3.62)$$

(where $\alpha! = \prod_i \alpha_i!$, $|\alpha| = \sum \alpha_i$, and $\bar{b}$ is the complex conjugate of b), and the norm $\|f\|_\rho$ is given by

$$\|f\|_\rho = \left(\sum_{\alpha \in \mathbb{N}^n} \frac{\alpha!(n-1)!}{(|\alpha|+n-1)!} |a_\alpha|^2 \rho^{2|\alpha|} \right)^{1/2} . \qquad (3.63)$$

The above scalar product turns $\mathcal{O}_l/\mathcal{O}_{l+1}$ into a Hermitian space, if we consider elements of $\mathcal{O}_l/\mathcal{O}_{l+1}$ as polynomial functions of degree less than or equal to 2^{l+1} and which do not contain terms of order $\leq 2^l$. Of course, when $\mathbb{K} = \mathbb{R}$ we will have to complexify $\mathcal{O}_l/\mathcal{O}_{l+1}$, but will redenote $(\mathcal{O}_l/\mathcal{O}_{l+1})_{\mathbb{C}}$ by $\mathcal{O}_l/\mathcal{O}_{l+1}$, for simplicity.

For the space $\mathcal{Y}^l$ of local vector fields of the type $u = \sum_{i=1}^{n-m} u_i \partial/\partial z_{i+m}^l$ (due to (3.58) and (3.59)), this is the same as the space of vector fields of the type $\sum_{i=1}^{n-m} u_i' \partial/\partial y_i^l$ defined in the previous section, up to a complexification if $\mathbb{K} = \mathbb{R}$), we define the L^2-norms in a similar way:

$$\|u\|_{l,\rho} = \sqrt{\frac{1}{V_\rho} \int_{S_{l,\rho}} \sum_{i=1}^{n-m} |u_i(x)|^2 \mathrm{d}\mu_l} \ . \tag{3.64}$$

These L^2-norms are given by Hermitian metrics similar to (3.62), which make $\mathcal{Y}_l^l/\mathcal{Y}_{l+1}^l$ into Hermitian spaces.

Remark that if $u = (u_1, \ldots, u_{n-m})$ then

$$\sum_i \|u_i\|_{l,\rho} \geq \|u\|_{l,\rho} \geq \max_i \|u_i\|_{l,\rho} \ . \tag{3.65}$$

It is an important observation that, since the action of $\mathfrak{g}_0$ on $\mathbb{C}^n$ preserves the Hermitian metric of $\mathbb{C}^n$, its actions on $\mathcal{O}_l/\mathcal{O}_{l+1}$ and $\mathcal{Y}_l^l/\mathcal{Y}_{l+1}^l$, as given in the previous section, also preserve the Hermitian metrics corresponding to the norms $\|f\|_{l,\rho}$ and $\|u\|_{l,\rho}$ (with the same l). Thus, applying Lemma 3.4.1 to these $\mathfrak{g}_{\mathbb{C}}$-modules, we get:

Lemma 3.4.3. *There is a positive constant D_1 such that for any $l \in \mathbb{N}$ and any positive number ρ there exist local analytic functions $w_1^l, \ldots, w_m^l, v_1^l, \ldots, v_{n-m}^l$, which satisfy the relations of the previous section (in particular Relation (3.35) and Relation (3.43)), and which have the following additional property whenever $B_{l,\rho}$ is well defined:*

$$\max_i \|w_i^l\|_{l,\rho} \leq D_1 . \max_{i,j} \left\| \{x_i^l, x_j^l\} - \sum_k c_{ij}^k x_k^l \right\|_{l,\rho} \tag{3.66}$$

and

$$\max_i \|v_i^l\|_{l,\rho} \leq D_1 . \max_{i,j} \left\| \{x_i^l - w_i^l, y_j^l\} - \sum_k a_{ij}^k y_k^l \right\|_{l,\rho} \ . \tag{3.67}$$

$\square$

3.5 Proof of analytic Levi decomposition theorem

Besides the L^2-norms defined in the previous section, we will need the following L^∞-norms: If f is a local function then put

$$|f|_{l,\rho} = \sup_{x \in B_{l,\rho}} |f(x)| \ , \tag{3.68}$$

where the complex ball $B_{l,\rho}$ is defined by (3.60). Similarly, if $g = (g_1, \ldots, g_N)$ is a vector-valued local map then put $|g|_{l,\rho} = \sup_{x \in B_{l,\rho}} \sqrt{\sum_i |g_i(x)|^2}$. For simplicity, we will write $|f|_\rho$ for $|f|_{0,\rho}$.

For the Poisson structure Π, we will use the following norms:

$$|\Pi|_{l,\rho} := \max_{i,j=1,\ldots,n} |\{z_i^l, z_j^l\}|_{l,\rho} \, . \tag{3.69}$$

Due to the following lemma, we will be able to use the norms $|f|_\rho$ and $\|f\|_\rho$ interchangeably for our purposes, and control the norms of the derivatives:

Lemma 3.5.1. *For any $\varepsilon > 0$ there is a finite number $K < \infty$ depending on ε such that for any integer $l > K$, positive number ρ, and local analytic function $f \in \mathcal{O}_l$ we have*

$$|f|_{(1+\varepsilon/l^2)\rho} \geq \exp(2^{l/2})|f|_{(1+\varepsilon/2l^2)\rho} \geq \rho|df|_\rho \, , \tag{3.70}$$

and

$$|f|_{(1-\varepsilon/l^2)\rho} \leq \|f\|_\rho \leq |f|_\rho \, . \tag{3.71}$$

We will postpone the proof of Lemma 3.5.1 a little bit. Now we want to show a key proposition which, together with a simple lemma, will imply Theorem 3.2.6.

Proposition 3.5.2. *Under the assumptions of Theorem 3.2.6, there exists a constant C, such that for any positive number $\varepsilon < 1/4$, there is a natural number $K = K(\varepsilon)$ and a positive number $\rho = \rho(\varepsilon)$, such that for any $l \geq K$ we can construct a local analytic coordinate system $(x_1^l, \ldots, y_{n-m}^l)$ as in the previous sections, with the following additional properties (using the previous notations):*

(i)$_l$ *(Chains of balls) The ball $B_{l,\exp(1/l)\rho}$ is well defined, and if $l > K$ we have*

$$B_{l-1,\exp(\frac{1}{l}-\frac{2\varepsilon}{l^2})\rho} \subset B_{l,\exp(1/l)\rho} \subset B_{l-1,\exp(\frac{1}{l}+\frac{2\varepsilon}{l^2})\rho} \, . \tag{3.72}$$

(ii)$_l$ *(Norms of changes) If $l > K$ then we have*

$$|\psi_l|_{l-1,\exp(\frac{1}{l-1}-\frac{\varepsilon}{(l-1)^2})\rho} < \rho \, . \tag{3.73}$$

(iii)$_l$ *(Norms of the Poisson structure):*

$$|\Pi|_{l,\exp(1/l)\rho} \leq C.\exp(-1/\sqrt{l})\rho \, . \tag{3.74}$$

Theorem 3.2.6 follows immediately from the first part of Proposition 3.5.2 and the following lemma:

Lemma 3.5.3. *If there is a finite number K such that Condition (i)$_l$ of Proposition 3.5.2 is satisfied for all $l \geq K$, then the formal coordinate system*

$$(x_1^\infty, \ldots, x_m^\infty, y_1^\infty, \ldots, y_{n-m}^\infty)$$

is convergent (i.e., locally analytic).

The main idea behind Lemma 3.5.3 is that, if Condition (i)$_l$ is true for any $l \geq K$, then the infinite intersection $\bigcap_{l=K}^{\infty} B_{l,\exp(1/l)\rho}$ contains an open neighborhood of 0, implying a positive radius of convergence.

The second and third parts of Proposition 3.5.2 are needed for the proof of the first part. Proposition 3.5.2 will be proved by recurrence: By taking ρ small enough, we can obviously achieve Conditions (iii)$_K$ and (i)$_K$ (Condition (ii)$_K$ is void). Then provided that K is large enough, when $l \geq K$ we have that Condition (ii)$_l$ implies Conditions (i)$_l$ and (iii)$_l$, and Condition (iii)$_l$ in turn implies Condition (ii)$_{l+1}$. In other words, Proposition 3.5.2 is a direct consequence of the following three technical lemmas:

Lemma 3.5.4. *There exists a finite number K (depending on ε) such that if Condition (ii)$_{l+1}$ is satisfied and $l \geq K$, then Condition (i)$_{l+1}$ is also satisfied.*

Lemma 3.5.5. *There exists a finite number K (depending on ε) such that if Condition (iii)$_l$ (of Proposition 3.5.2) is satisfied and $l \geq K$, then Condition (ii)$_{l+1}$ is also satisfied.*

Lemma 3.5.6. *There exists a finite number K (depending on ε) such that if Conditions (ii)$_{l+1}$ and (iii)$_l$ are satisfied and $l \geq K$, then Condition (iii)$_{l+1}$ is also satisfied.*

The lemmas of this section will be proved now, one by one. But first let us mention here the main ingredients behind the last three ones: The proof of Lemma 3.5.4 and Lemma 3.5.6 is straightforward and uses only the first part of Lemma 3.5.1. Lemma 3.5.5 (the most technical one) follows from the estimates on the primitives of cocycles as provided by Lemma 3.4.3.

Proof of Lemma 3.5.1. Let f be a local analytic function in $(\mathbb{C}^n, 0)$. To make an estimate on df, we use the Cauchy integral formula. For $z \in B_\rho$, denote by γ_i the following circle: $\gamma_i = \{v \in \mathbb{C}^n \mid v_j = z_j \text{ if } j \neq i \,, \ |v_i - z_i| = \varepsilon\rho/2l^2\}$. Then $\gamma_i \subset B_{(1+\varepsilon/l^2)\rho}$, and we have

$$\left| \frac{\partial f}{\partial z_i}(z) \right| = \frac{1}{2\pi} \left| \oint_{\gamma_i} \frac{f(v)dv}{(v-z)^2} \right| \leq \frac{2l^2}{\varepsilon\rho} |f|_{(1+\varepsilon/2l^2)\rho} \,,$$

which implies that $\exp(2^{l/2})|f|_{(1+\varepsilon/2l^2)\rho} \geq \rho|df|$ when l is large enough.

Now let $f \in \mathcal{O}_l$ such that $|f|_{(1+\varepsilon/l^2)\rho} < \infty$. We want to show that if $x \in B_{(1+\varepsilon/2l^2)\rho}$ then $|f(x)| \leq \exp(2^{l/2})|f|_{(1+\varepsilon/l^2)\rho}$ (provided that l is large enough compared to $1/\varepsilon$). Fix a point $x \in B_{(1+\varepsilon/2l^2)\rho}$ and consider the following holomorphic function of one variable: $g(z) = f(\frac{x}{|x|}z)$. This function is holomorphic in the complex one-dimensional disk $B_{(1+\varepsilon/l^2)\rho}^1$ of radius $(1+\varepsilon/l^2)\rho$, and is bounded by $|f|_{(1+\varepsilon/l^2)\rho}$ in this disk. Because $f \in \mathcal{O}_l$, we have that $g(z)$ is divisible by z^{2^l}, that is $g(z)/z^{2^l}$ is holomorphic in $B_{(1+\varepsilon/l^2)\rho}^1$. By the maximum principle we have

$$\frac{|f(x)|}{|x|^{2^l}} = \left| \frac{g(|x|)}{|x|^{2^l}} \right| \leq \max_{|z|=(1+\varepsilon/l^2)\rho} \left| \frac{g(z)}{z^{2^l}} \right| \leq \frac{|f|_{(1+\varepsilon/l^2)\rho}}{((1+\varepsilon/l^2)\rho)^{2^l}} \,,$$

which implies that

$$|f(x)| \leq \left(\frac{1+\varepsilon/2l^2}{1+\varepsilon/l^2}\right)^{2^l}|f|_{(1+\varepsilon/l^2)\rho} \approx \exp(-\frac{2^l}{2\varepsilon l^2})|f|_{(1+\varepsilon/l^2)\rho}$$

$$\leq \exp(-2^{l/2})|f|_{(1+\varepsilon/l^2)\rho}$$

(when l is large enough). Thus we have proved that there is a finite number K depending on ε such that

$$|f|_{(1+\varepsilon/l^2)\rho} \geq \exp(2^{l/2})|f|_{(1+\varepsilon/2l^2)\rho}$$

for any $l > K$ and any $f \in \mathcal{O}_l$.

To compare the norms of f, we use Cauchy–Schwartz inequality: for $f = \sum_{\alpha\in\mathbb{N}^k} c_\alpha \prod_i z_i^{\alpha_i}$ and $|z| = (1-\varepsilon/2l^2)\rho$ we have

$$|f(z)| \leq \sum_{\alpha\in\mathbb{N}^k} |c_\alpha| \prod_i |z_i|^{\alpha_i}$$

$$\leq \left(\sum_\alpha |c_\alpha|^2 \frac{\alpha!(n-1)!}{(|\alpha|+n-1)!}\rho^{2|\alpha|}\right)^{1/2}\left(\sum_\alpha \frac{(|\alpha|+n-1)!}{\alpha!(n-1)!}\rho^{-2|\alpha|}\prod_i |z_i|^{2\alpha}\right)^{1/2}$$

$$= \|f\|_\rho\left(1 - \sum_i \frac{|z_i|^2}{\rho^2}\right)^{-n/2} = \|f\|_\rho(1-(1-\varepsilon/2l^2)^2)^{-n/2} \leq \frac{(2l)^n}{\varepsilon^{n/2}}\|f\|_\rho .$$

It means that for any local analytic function f we have

$$|f|_{(1-\varepsilon/2l^2)\rho} \leq \frac{(2l)^n}{\varepsilon^{n/2}}\|f\|_\rho .$$

Now if $f \in \mathcal{O}_l$, we can apply Inequality (3.70) to get

$$|f|_{(1-\varepsilon/l^2)\rho} \leq \exp(-2^{l/2})|f|_{(1-\varepsilon/2l^2)\rho} \leq \frac{(2l)^n}{\varepsilon^{n/2}}\exp(-2^{l/2})\|f\|_\rho \leq \|f\|_\rho ,$$

provided that l is large enough compared to $1/\varepsilon$. Lemma 3.5.1 is proved. $\square$

Proof of Lemma 3.5.3. The main point is to show that the limit $\bigcap_{l=K}^\infty B_{l,\rho}$ contains a ball B_r of positive radius centered at 0. Then for $x \in B_r$, we have $x \in B_{l,\rho}$, implying $\|(z_1^l(x),\dots,z_n^l(x))\| < \rho$ is uniformly bounded, which in turn implies that the formal functions $z_i^\infty = \lim_{l\to\infty} z_i^l$ are analytic functions inside B_r (recall that $(z_1^l,\dots,z_n^l)$ is obtained from $(x_1^l,\dots,y_{n-m}^l)$ by a constant linear transformation (A_{ij}) which does not depend on l).

Recall the following fact of complex analysis, which is a consequence of the maximum principle: if g is a complex analytic map from a complex ball of radius ρ to some linear Hermitian space such that $g(0) = 0$ and $|g(x)| \leq C$ for all $|x| < \rho$

and some constant C, then we have $|g(x)| \leq C|x|/\rho$ for all x such that $|x| < \rho$. If $l_1, l_2 \in \mathbb{N}$ and $r_1, r_2 > 0, s > 1$, then applying this fact we get:

$$\text{If } B_{l_1, r_1} \subset B_{l_2, r_2} \text{ then } B_{l_1, r_1/s} \subset B_{l_2, r_2/s} . \tag{3.75}$$

(Here r_1 plays the role of ρ, r_2 plays the role of C, and the coordinate transformation from $(z_1^{l_1}, \ldots, z_n^{l_1})$ to $(z_1^{l_2}, \ldots, z_n^{l_2})$ plays the role of g in the previous statement.)

Using Formula (3.75) and Condition (i)$_l$ recursively, we get

$$B_{l,\rho} \supset B_{l-1,\exp(-1/l^2)\rho} \supset B_{l-2,\exp(-1/l^2-1/(l-1)^2)\rho}$$
$$\supset \cdots \supset B_{K,\exp(-\sum_{k=K}^{l} 1/k^2)\rho} .$$

Since $c = \exp(-\sum_{k=K}^{\infty} 1/k^2)$ is a positive number, we have $\bigcap_{l=K}^{\infty} B_{l,\rho} \supset B_{K,c\rho}$, which clearly contains an open neighborhood of 0. Lemma 3.5.3 is proved. $\qquad \square$

Proof of Lemma 3.5.4. Suppose that Condition (ii)$_{l+1}$ is satisfied. For simplicity of exposition, we will assume that the coordinate system $(z_1^l, \ldots, z_n^l)$ coincides with the coordinate system $(x_1^l, \ldots, y_{n-m}^l)$ (The more general case, when $(z_1^l, \ldots, z_n^l)$ is obtained from $(x_1^l, \ldots, y_{n-m}^l)$ by a constant linear transformation, is essentially the same.) Suppose that we have

$$|\psi_{l+1}|_{l,\exp(1/l-\varepsilon/l^2)\rho} < \rho .$$

Then it follows from Lemma 3.5.1 that, provided that l is large enough:

$$|d\psi_{l+1}|_{l,\exp(1/l-2\varepsilon/l^2)\rho} < \frac{1}{2n} .$$

(In order to define $|d\psi_{l+1}|_{l,\exp(1/l-2\varepsilon/l^2)\rho}$, consider $d\psi_{l+1}$ as an n^2-vector-valued function in variables $(z_1^l, \ldots, z_n^l)$.) Hence the map $\phi_{l+1} = \mathrm{Id} + \psi_{l+1}$ is injective in $B_{l,\exp(1/l-2\varepsilon/l^2)\rho}$: if $x, y \in B_{l,\rho_l}, x \neq y$, then $\|\phi_{l+1}(x) - \phi_{l+1}(y)\| \geq \|x - y\| - \|\psi_{l+1}(x) - \psi_{l+1}(y)\| \geq \|x-y\| - n|d\psi_{l+1}|_{\exp(1/l-2\varepsilon/l^2)\rho}\|x-y\| \geq (1-1/2)\|x-y\| > 0$. (Here $(x - y)$ means the vector $(z_1^l(x) - z_1^l(y), \ldots, z_n^l(x) - z_n^l(y))$, i.e., their difference is taken with respect to the coordinate system $(z_1^l, \ldots, z_n^l)$.)

It follows from Lemma 3.5.1 that

$$|\phi_{l+1}|_{l,\exp(1/l-2\varepsilon/l^2)\rho} = |\mathrm{Id} + \psi_{l+1}|_{l,\exp(1/l-2\varepsilon/l^2)\rho}$$
$$\leq |\mathrm{Id}|_{l,\exp(1/l-2\varepsilon/l^2)\rho} + |\psi_{l+1}|_{l,\exp(1/l-2\varepsilon/l^2)\rho}$$
$$< \exp(1/l - 2\varepsilon/l^2)\rho + \frac{\varepsilon}{4l^2} \exp(1/l - 2\varepsilon/l^2)\rho < \exp(1/l - \varepsilon/l^2)\rho .$$

In other words, we have

$$\phi_{l+1}(B_{l,\exp(1/l-2\varepsilon/l^2)\rho}) \subset B_{l,\exp(1/l-\varepsilon/l^2)\rho} . \tag{3.76}$$

Applying Formula (3.75) to the above relation, noticing that $1/l - 2\varepsilon/l^2 > 1/(l+1)$, and simplifying the obtained formula a little bit, we get

$$\phi_{l+1}(B_{l,\exp(1/(l+1)-2\varepsilon/(l+1)^2)\rho}) \subset B_{l,\exp(1/(l+1))\rho} \ . \tag{3.77}$$

We will show that ϕ_{l+1}^{-1} is well defined in $B_{l,\exp(1/(l+1))\rho}$, and

$$\phi_{l+1}^{-1}(B_{l,\exp(1/(l+1))\rho}) = B_{l+1,\exp(1/(l+1))\rho} \subset B_{l,\exp(1/(l+1)+2\varepsilon/(l+1)^2)\rho} \ . \tag{3.78}$$

Indeed, denote by $S_{l,\exp(1/l-2\varepsilon/l^2)\rho}$ the boundary of $B_{l,\exp(1/l-2\varepsilon/l^2)\rho}$. Then

$$\phi_{l+1}(S_{l,\exp(1/l-2\varepsilon/l^2)\rho}) \subset B_{l,\exp(1/l-\varepsilon/l^2)\rho}$$

and is homotopic to $S_{l,\exp(1/l-2\varepsilon/l^2)\rho}$ via a homotopy which does not intersect $B_{l,\exp(1/(l+1))\rho}$. It implies, via the classical Brower's fixed point theorem, that $\phi_{l+1}(B_{l,\exp(1/l-2\varepsilon/l^2)\rho})$ must contain $B_{l,\exp(1/(l+1))\rho}$. Because ϕ_{l+1} is injective in $(B_{l,\exp(1/l-2\varepsilon/l^2)\rho})$, it means that the inverse map is well defined in $B_{l,\exp(1/(l+1))\rho}$, with

$$\phi_{l+1}^{-1}(B_{l,\exp(1/(l+1))\rho}) \subset B_{l,\exp(1/l-2\varepsilon/l^2)\rho} \ .$$

In particular, $B_{l+1,\exp(1/(l+1))\rho} = \phi_{l+1}^{-1}(B_{l,\exp(1/(l+1))\rho})$ is well defined. Lemma 3.5.4 then follows from (3.77) and (3.78). $\qquad\square$

Proof of Lemma 3.5.5. Suppose that Condition (iii)$_l$ is satisfied. Then according to (3.28) we have:

$$\|f_{ij}^l\|_{l,\exp(1/l)\rho} \le |f_{ij}^l|_{l,\exp(1/l)\rho} = \left|\{x_i^l, x_j^l\} - \sum_k c_{ij}^k x_k^l\right|_{l,\exp(1/l)\rho}$$

$$\le C_1|\Pi|_{l,\exp(1/l)\rho} + \sum_k |c_{ij}^k| |x_k^l|_{l,\rho} \le C_1.C.\rho + C_2.\exp(1/l)\rho \sum_k |c_{ij}^k| < C_3\rho \ , \tag{3.79}$$

where C_3 is some positive constant (which does not depend on l).

We can apply the above inequality $\|f_{ij}^l\|_{l,\exp(1/l)\rho} < C_3\rho$ and Lemma 3.4.3 to find a positive constant C_4 (which does not depend on l) and a solution w_i^l of (3.37), such that

$$\|w_i^l\|_{l,\exp(1/l)\rho} < C_4\rho \ . \tag{3.80}$$

Together with Lemma 3.5.1, the above inequality yields

$$|dw_i^l|_{l,\exp(1/l-\varepsilon/2l^2)\rho} < C_4, \tag{3.81}$$

provided that l is large enough. Applying Lemma 3.5.1 and the assumption that $|\Pi|_{l,\exp(1/l)\rho} < C\rho$ to the above inequality, we get

$$|\{w_i^l, y_j^l\}|_{l,\exp(1/l-\varepsilon/2l^2)\rho} < C_5\rho$$

for some constant C_5 (which does not depend on l). Using this inequality, and inequalities similar to (3.79), we get that the norm $\|.\|_{l,\exp(1/l-\varepsilon/2l^2)\rho}$ of the 1-cocycle given in Formula (3.41) is bounded from above by $C_6\rho$, where C_6 is some constant which does not depend on L. Using Lemma 3.4.3, we find a solution v_i^L to Equation 3.43 such that

$$\left\|v_i^l\right\|_{l,\exp(1/l-\varepsilon/2l^2)\rho} < C_6\rho \, , \tag{3.82}$$

where C_6 is some constant which does not depend on l. Lemma 3.5.5 (fr l large enough compared to C_6) now follows directly from Inequalities (3.80), (3.82) and Lemma 3.5.1. $\qquad\square$

Proof of Lemma 3.5.6. Suppose that Condition (ii)$_{l+1}$ is satisfied. By Lemma 3.5.4, Condition (i)$_{l+1}$ is also satisfied. In particular,

$$B_{l+1,\exp(1/(l+1))\rho} \subset B_{l,\exp(1/(l+1)+2\varepsilon/(l+1)^2)\rho} \subset B_{l,\exp(1/l-2\varepsilon/l^2)\rho}$$

(for $\varepsilon < 1/4$ and l large enough). Thus we have

$$|\{z_i^{l+1}, z_j^{l+1}\}|_{l+1,\exp(1/(l+1))\rho} \leq |\{z_i^{l+1}, z_j^{l+1}\}|_{l,\exp(1/l-2\varepsilon/l^2)\rho} \leq T^1 + T^2 + T^3 + T^4$$

where

$$T^1 = |\{z_i^l, z_j^l\}|_{l,\exp(1/l-2\varepsilon/l^2)\rho} \, ,$$
$$T^2 = |\{z_i^{l+1} - z_i^l, z_j^{l+1}\}|_{l,\exp(1/l-2\varepsilon/l^2)\rho} \, ,$$
$$T^3 = |\{z_i^{l+1}, z_j^{l+1} - z_j^l\}|_{l,\exp(1/l-2\varepsilon/l^2)\rho} \, ,$$
$$T^4 = |\{z_i^{l+1} - z_i^l, z_j^{l+1} - z_j^l\}|_{l,\exp(1/l-2\varepsilon/l^2)\rho} \, .$$

For the first term, we have

$$T^1 \leq |\{z_i^l, z_j^l\}|_{l,\exp(1/l)\rho} \leq |\Pi|_{l,\exp(1/l)\rho} \leq C.\exp(-1/\sqrt{l})\rho \, .$$

Notice that $C\exp(-1/\sqrt{l+1})\rho - C\exp(-1/\sqrt{l})\rho > \frac{C}{l^2}\rho$ (for l large enough). So to verify Condition (iii)$_{l+1}$, it suffices to show that $T^2 + T^3 + T^4 < \frac{C}{l^2}\rho$. But this last inequality can be achieved easily (provided that l is large enough) by Conditions (ii)$_{l+1}$, (iii)$_l$ and Lemma 3.5.1. Lemma 3.5.6 is proved. $\qquad\square$

3.6   The smooth case

In this section we will give a sketch of the proof of Theorem 3.2.9, referring the reader to [263] for the details, which are quite long. Or rather, we will show what modifications need to be made to the proof of the analytic Levi decomposition theorem 3.2.6 in order to obtain the proof of Theorem 3.2.9.

In this section, $\mathfrak{g}$ will be a *compact* semisimple Lie algebra.

Denote by $(\xi_1, \ldots, \xi_m)$ a fixed basis of $\mathfrak{g}$, which is orthonormal with respect to a fixed positive definite invariant metric on $\mathfrak{g}$. Denote by c_{ij}^k the structural constants of $\mathfrak{g}$ with respect to this basis:

$$[\xi_i, \xi_j] = \sum_k c_{ij}^k \xi_k . \qquad (3.83)$$

Since $\mathfrak{g}$ is compact, we may extend $(\xi_1, \ldots, \xi_m)$ to a basis

$$(\xi_1, \ldots, \xi_m, y_1, \ldots, y_{n-m})$$

of $\mathfrak{l}$ such that the corresponding Euclidean metric is preserved by the adjoint action of $\mathfrak{g}$. The algebra $\mathfrak{g}$ acts on $\mathfrak{l}^* = \mathbb{R}^n$ via the coadjoint action of $\mathfrak{l}$ $\zeta(z) := \mathrm{ad}_\zeta^*(z)$ for $\zeta \in \mathfrak{g} \subset \mathfrak{l}$, $z \in \mathbb{R}^n = \mathfrak{l}^*$. The basis $(\xi_1, \ldots, \xi_m, y_1, \ldots, y_{n-m})$ of $\mathfrak{l}$ may be viewed as a coordinate system $(x_1, \ldots, x_m, y_1, \ldots, y_{n-m})$ on $\mathbb{R}^n$ (with $x_i = \xi_i$).

Denote by G the compact simply-connected Lie group whose Lie algebra is $\mathfrak{g}$. Then the above action of $\mathfrak{g}$ on $\mathbb{R}^n$ integrates into an action of G on $\mathbb{R}^n$ (the coadjoint action). The action of G on $\mathbb{R}^n$ preserves the Euclidean metric of $\mathbb{R}^n$ given by $\|z\|^2 = \sum |x_i(z)|^2 + \sum |y_j(z)|^2$.

For each positive number $r > 0$, denote by B_r the closed ball of radius r in $\mathbb{R}^n$ centered at 0. The group G (and hence the algebra $\mathfrak{g}$) acts linearly on the space of functions on B_r via its action on B_r: for each function F and element $g \in G$ we put

$$g(F)(z) := F(g^{-1}(z)) = F(\mathrm{Ad}_{(g^{-1})}z). \qquad (3.84)$$

In the smooth case, we will use C^k-norms and Sobolev norms. For each nonnegative integer $k \geq 0$ and each pair of real-valued functions F_1, F_2 on B_r, we will define the Sobolev inner product of F_1 with F_2 with respect to the Sobolev H_k-norm as follows:

$$\langle F_1, F_2 \rangle_{k,r}^H := \sum_{|\alpha| \leq k} \int_{B_r} \left(\frac{|\alpha|!}{\alpha!} \right) \left(\frac{\partial^{|\alpha|} F_1}{\partial z^\alpha}(z) \right) \left(\frac{\partial^{|\alpha|} F_2}{\partial z^\alpha}(z) \right) d\mu(z). \qquad (3.85)$$

The Sobolev H_k-norm of a function F on B_r is

$$\|F\|_{k,r}^H := \sqrt{\langle F, F \rangle_{k,r}^H} . \qquad (3.86)$$

We will denote by $\mathcal{C}_r$ the subspace of the space $\mathcal{C}^\infty(B_r)$ of smooth real-valued functions on B_r, which consists of functions vanishing at 0 whose first derivatives also vanish at 0. Then the action of G on $\mathcal{C}_r$ defined by (3.84) preserves the Sobolev inner products (3.85).

Denote by $\mathcal{Y}_r$ the space of smooth vector fields on B_r of the type

$$u = \sum_{i=1}^{n-m} u_i \partial/\partial y_i , \qquad (3.87)$$

such that u_i vanish at 0 and their first derivatives also vanish at 0. Then $\mathcal{Y}_r$ is a $\mathfrak{g}$-module under the following action:

$$\xi_i \cdot \sum_j u_j \partial/\partial y_j := \left[\sum_{jk} c_{ij}^k x_k \frac{\partial}{\partial x_j} + \sum_{jk} a_{ij}^k y_k \frac{\partial}{\partial y_j} \ , \ \sum_j u_j \partial/\partial y_j \right] , \qquad (3.88)$$

where $X_i = \sum_{jk} c_{ij}^k x_k \partial/\partial x_j + \sum_{jk} a_{ij}^k y_k \partial/\partial y_j$ are the linear vector fields which generate the linear orthogonal coadjoint action of $\mathfrak{g}$ on $\mathbb{R}^n$.

Equip $\mathcal{Y}_r$ with Sobolev inner products:

$$\langle u, v \rangle_{k,r}^H := \sum_{i=1}^{n-m} \langle u_i, v_i \rangle_{k,r} , \qquad (3.89)$$

and denote by $\mathcal{Y}_{k,r}^H$ the completion of $\mathcal{Y}_r$ with respect to the corresponding $H_{k,r}$-norm. Then $\mathcal{Y}_{k,r}^H$ is a separable real Hilbert space on which $\mathfrak{g}$ and G act orthogonally.

The C^k-norms can be defined as follows:

$$\|F\|_{k,r} := \sup_{|\alpha| \le k} \sup_{z \in B_r} |D^\alpha F(z)| \qquad (3.90)$$

for $F \in \mathcal{C}_r$, where the sup runs over all partial derivatives of degree $|\alpha|$ at most k. Similarly, for $u = \sum_{i=1}^{n-m} u_i \partial/\partial y_i \in \mathcal{Y}_r$ we put

$$\|u\|_{k,r} := \sup_i \sup_{|\alpha| \le k} \sup_{z \in B_r} |D^\alpha u_i(z)|. \qquad (3.91)$$

The C^k norms $\|.\|_{k,r}$ are related to the Sobolev norms $\|.\|_{k,r}^H$ as follows:

$$\|F\|_{k,r} \le C\|F\|_{k+s,r}^H \text{ and } \|F\|_{k,r}^H \le C(n+1)^k \|F\|_{k,r} \qquad (3.92)$$

for any F in $\mathcal{C}_r$ or $\mathcal{Y}_r$ and any $k \ge 0$, where $s = \left[\frac{n}{2}\right] + 1$ and C is a positive constant which does not depend on k. In other words, C^k norms and Sobolev norms are "tamely equivalent". A priori, the constant C depends on r, but later on we will always assume that $1 \le r \le 2$, and so may assume C to be independent of r. The above inequality is a version of the classical Sobolev's lemma for Sobolev spaces.

Similarly to the analytic case, we will need the following normed version of Whitehead's lemma (cf. Proposition 2.1 of [81]):

Lemma 3.6.1 (Conn). *For any given positive number r, and $W = \mathcal{C}_r$ or $\mathcal{Y}_r$ with the above action of $\mathfrak{g}$, consider the (truncated) Chevalley–Eilenberg complex*

$$W \xrightarrow{\delta_0} W \otimes \wedge^1 \mathfrak{g}^* \xrightarrow{\delta_1} W \otimes \wedge^2 \mathfrak{g}^* \xrightarrow{\delta_2} W \otimes \wedge^3 \mathfrak{g}^*.$$

Then there is a chain of operators

$$W \xleftarrow{h_0} W \otimes \wedge^1 \mathfrak{g}^* \xleftarrow{h_1} W \otimes \wedge^2 \mathfrak{g}^* \xleftarrow{h_2} W \otimes \wedge^3 \mathfrak{g}^*$$

such that

$$\delta_0 \circ h_0 + h_1 \circ \delta_1 = \mathrm{Id}_{W \otimes \wedge^1 \mathfrak{g}^*} \, ,$$
$$\delta_1 \circ h_1 + h_2 \circ \delta_2 = \mathrm{Id}_{W \otimes \wedge^2 \mathfrak{g}^*} \, . \tag{3.93}$$

Moreover, there exists a constant $C > 0$, which is independent of the radius r of B_r, such that

$$\|h_j(u)\|_{k,r}^H \le C \|u\|_{k,r}^H \quad j = 0, 1, 2 \tag{3.94}$$

for all $k \ge 0$ and $u \in W \otimes \wedge^{j+1} \mathfrak{g}^*$. If u vanishes to an order $l \ge 0$ at the origin, then so does $h_j(u)$.

Strictly speaking, Conn [81] proved the above lemma only in the case when $\mathfrak{g} = \mathfrak{l}$ and for the module $\mathcal{C}_r$, but his proof is quite general and works perfectly in our situation without any modification. In fact, in order to prove Lemma 3.6.1, it is sufficient to show that W is an infinite direct sum of finite-dimensional orthogonal modules, and then repeat the proof of Lemma 3.4.1. □

For simplicity, in the sequel we will denote the homotopy operators h_j in the above lemma simply by h. Homotopy relation (3.93) will be rewritten simply as

$$\mathrm{Id} - \delta \circ h = h \circ \delta \, . \tag{3.95}$$

The meaning of the last equality is as follows: if u is a 1-cocycle or 2-cocycle, then it is also a coboundary, and $h(u)$ is an explicit primitive of u: $\delta(h(u)) = u$. If u is a "near cocycle" then $h(u)$ is also a "near primitive" for u.

Combining Inequality (3.94) with Sobolev inequalities, we get the following estimate for the homotopy operators h with respect to C^k norms:

$$\|h(u)\|_{k,r} \le C(n+1)^{k+s} \|u\|_{k+s,r} \quad \forall \, j = 0, 1, 2 \tag{3.96}$$

for all $k \ge 0$ and $u \in W \otimes \wedge^{j+1} \mathfrak{g}^*$ where $W = \mathcal{C}_r$ or $\mathcal{Y}_r$. Here $s = [\frac{n}{2}] + 1$, C is a positive constant which does not depend on k (and r provided that $1 \le r \le 2$).

It is well known that the space $C^\infty(B_r)$ with C^k norms (3.90) is a *tame Fréchet space* (see, e.g., [167] for the theory of tame Fréchet spaces). Since $\mathcal{C}_r$ is a tame direct summand of $C^\infty(B_r)$, it is also a tame Fréchet space. Similarly, $\mathcal{Y}_r$ with norms (3.91) is a tame Fréchet space as well. What we will use here is the fact that tame Fréchet spaces admit smoothing operators and interpolation inequalities:

For each $t > 1$ there is a linear operator $S(t) = S_r(t)$ from $\mathcal{C}_r$ to itself, called a *smoothing operator*, with the following properties:

$$\|S(t)F\|_{p,r} \le C_{p,q} t^{(p-q)} \|F\|_{q,r} \tag{3.97}$$

and

$$\|(\mathrm{Id} - S(t))F\|_{q,r} \le C_{p,q} t^{(q-p)} \|F\|_{p,r} \tag{3.98}$$

for any $F \in \mathcal{C}_r$, where p, q are any nonnegative integers such that $p \ge q$, Id denotes the identity map, and $C_{p,q}$ denotes a constant which depends on p and q.

The second inequality means that $S(t)$ is close to identity and tends to identity when $t \to \infty$. The first inequality means that F becomes "smoother" when we apply $S(t)$ to it. That's why $S(t)$ is called a smoothing operator. A priori, the constants $C_{p,q}$ also depend on the radius r. But later on, we will always have $1 \le r \le 2$, and so we may choose $C_{p,q}$ to be independent of r.

There is a similar smoothing operator from $\mathcal{Y}_r$ to itself, which by abuse of language we will also denote by $S(t)$ or $S_r(t)$. We will assume that inequalities (3.97) and (3.98) are still satisfied when F is replaced by an element of $\mathcal{Y}_r$.

For any F in $\mathcal{C}_r$ or $\mathcal{Y}_r$, and nonnegative integers $p_1 \ge p_2 \ge p_3$, we have the following *interpolation inequality*:

$$(\|F\|_{p_2,r})^{p_3-p_1} \le C_{p_1,p_2,p_3}(\|F\|_{p_1,r})^{p_3-p_2}(\|F\|_{p_3,r})^{p_2-p_1}, \tag{3.99}$$

where C_{p_1,p_2,p_3} is a positive constant which may depend on p_1, p_2, p_3.

Similarly to the analytic case, in order to prove Theorem 3.2.9, we will construct by recurrence a sequence of local smooth coordinate systems $(x^d, y^d) := (x_1^d, \dots, x_m^d, y_1^d, \dots, y_{n-m}^d)$, which converges to a local coordinate system $(x^\infty, y^\infty) = (x_1^\infty, \dots, x_m^\infty, y_1^\infty, \dots, y_{n-m}^\infty)$, in which the Poisson structure Π has the desired form. Here $(x^0, y^0) = (x_1, \dots, x_m, y_1, \dots, y_{n-m})$ is the original linear coordinate system.

For simplicity of exposition, we will assume that Π is C^∞-smooth. However, in every step of the proof of Theorem 3.2.9, we will only use differentiability of Π up to some finite order, and that's why our proof will also work for finitely (sufficiently highly) differentiable Poisson structures.

We will denote by Θ_d the local diffeomorphisms of $(\mathbb{R}^n, 0)$ such that

$$(x^d, y^d)(z) = (x^0, y^0) \circ \Theta_d(z), \tag{3.100}$$

where z denotes a point of $(\mathbb{R}^n, 0)$.

Denote by Π^d the Poisson structure obtained from Π by the action of Θ_d:

$$\Pi^d = (\Theta_d)_* \Pi. \tag{3.101}$$

Of course, $\Pi^0 = \Pi$. Denote by $\{.,.\}_d$ the Poisson bracket with respect to the Poisson structure Π^d. Then we have

$$\{F_1, F_2\}_d(z) = \{F_1 \circ \Theta_d, F_2 \circ \Theta_d\}(\Theta_d^{-1}(z)). \tag{3.102}$$

Assume that we have constructed $(x^d, y^d) = (x, y) \circ \Theta_d$. Let us now construct $(x^{d+1}, y^{d+1}) = (x, y) \circ \Theta_{d+1}$. Similarly to the analytic case, this construction consists of two steps: 1) find an "almost Levi factor", i.e., coordinates x_i^{d+1} such that the error terms $\{x_i^{d+1}, x_j^{d+1}\} - \sum_k c_{ij}^k x_k^{d+1}$ are small, and 2) "almost linearize" it, i.e., find the remaining coordinates y^{d+1} such that in the coordinate system (x^{d+1}, y^{d+1}) the Hamiltonian vector fields of the functions x_i^{d+1} are very close to linear ones. In fact, we will define a local diffeomorphism θ_{d+1} of $(\mathbb{R}^n, 0)$ and then put $\Theta_{d+1} = \theta_{d+1} \circ \Theta_d$. In particular, we will have $\Pi^{d+1} = (\theta_{d+1})_* \Pi^d$ and $(x^{d+1}, y^{d+1}) = (x^d, y^d) \circ (\Theta_d)^{-1} \circ \theta_{d+1} \circ \Theta_d$.

Similarly to the analytic case, consider the 2-cochain

$$f^d = \sum_{ij} f^d_{ij} \otimes \xi^*_i \wedge \xi^*_j \qquad (3.103)$$

of the Chevalley–Eilenberg complex associated to the $\mathfrak{g}$-module $\mathcal{C}_r$, where now

$$f^d_{ij}(x,y) = \{x_i, x_j\}_d - \sum_{k=1}^m c^k_{ij} x_k, \qquad (3.104)$$

and $r = r_d$ depends on d and can be chosen as follows:

$$r_d = 1 + \frac{1}{d+1}. \qquad (3.105)$$

In particular, $r_0 = 2$, $r_d/r_{d+1} \sim 1 + \frac{1}{d^2}$, and $\lim_{d\to\infty} r_d = 1$ is positive. This choice of radii r_d means in particular that we will be able to arrange it so that the Poisson structure $\Pi^d = (\Theta_d)_*\Pi$ is defined in the closed ball of radius r_d. (For this to hold, we will have to assume that Π is defined in the closed ball of radius 2, and show by recurrence that $B_{r_d} \subset \theta_d(B_{r_{d-1}})$ for all $d \in \mathbb{N}$.)

Put

$$\varphi^{d+1} := \sum_i \varphi^{d+1}_i \otimes \xi^*_i = S(t_d)\big(h(f^d)\big), \qquad (3.106)$$

where h is the homotopy operator as given in Lemma 3.6.1, S is the smoothing operator and the parameter t_d is chosen as follows: take a real constant $t_0 > 1$ (which later on will be assumed to be large enough) and define the sequence $(t_d)_{d \geq 0}$ by $t_{d+1} = t_d^{3/2}$. In other words, we have

$$t_d = \exp\left(\left(\frac{3}{2}\right)^d \ln t_0\right), \quad \ln t_0 > 0. \qquad (3.107)$$

The above choice of smoothing parameter t_d is a standard one in problems involving the Nash–Moser method, see, e.g., [166, 167]. The number $\frac{3}{2}$ in the above formula is just a convenient choice. The main point is that this number is greater than 1 (so we have a very fast increasing sequence) and smaller than 2 (where 2 corresponds to the fact that we have a fast convergence algorithm which "quadratizes" the error term at each step, i.e., goes from an "ε-small" error term to an "ε^2-small" error term).

According to Inequality (3.96), in order to control the C^k-norm of $h(f^d)$ we need to control the C^{k+s}-norm of f^d, i.e., we face a "loss of differentiability". That's why in the above definition of φ^{d+1} we have to use the smoothing operator S, which will allow us to compensate for this loss of differentiability. This is a standard trick in the Nash–Moser method.

Next, consider the 1-cochain

$$\hat{g}^d = \sum_i \left(\sum_\alpha \{x_i - h(f^d)_i, y_\alpha\}_d - \sum_{\beta=1}^{n-m} a_{i\alpha}^\beta y_\beta \frac{\partial}{\partial y_\alpha} \right) \otimes \xi_i^* \qquad (3.108)$$

of the Chevalley–Eilenberg complex associated to the $\mathfrak{g}$-module $\mathcal{Y}_r$, where $r = r_d = 1 + \frac{1}{d+1}$, and put

$$\psi^{d+1} := \sum_\alpha \psi_\alpha^{d+1} \frac{\partial}{\partial y_\alpha} = S(t_d)\big(h(\hat{g}^d)\big), \qquad (3.109)$$

where h is the homotopy operator as given in Lemma 3.6.1, and $S(t_d)$ is the smoothing operator (with the same t_d as in the definition of φ^{d+1}).

Now define θ_{d+1} to be a local diffeomorphism of $\mathbb{R}^n$ given by

$$\theta_{d+1} := Id - (\varphi_1^{d+1}, \ldots, \varphi_m^{d+1}, \psi_1^{d+1}, \ldots, \psi_{n-m}^{d+1}) . \qquad (3.110)$$

This finishes our construction of $\Theta_{d+1} = \theta_{d+1} \circ \Theta_d$ and $(x^{d+1}, y^{d+1}) = (x, y) \circ \Theta_{d+1}$. This construction is very similar to the analytic case, except mainly for the use of the smoothing operator. Another difference is that, for technical reasons, in the smooth case we use the original coordinate system and the transformed Poisson structures Π^d for determining the error terms, while in the analytic case the original Poisson structure and the transformed coordinate systems are used. In particular, the closed balls used here are always balls with respect to the original coordinate system – this allows us to easily compare the Sobolev norms of functions on them, i.e., bigger balls correspond to bigger norms.

The technical part of the proof (see [263]) consists of a series of lemmas which show that the above construction actually yields a smooth Levi normalization in the limit, provided that Π is defined on the closed ball of radius 2 and is sufficiently close to its linear part there. If Π does not satisfy these conditions, then we may use the following homothety trick to make it satisfy: replace Π by $\Pi^t = \frac{1}{t} G(t)_* \Pi$ where $G(t) : z \mapsto tz$ is a homothety, $t > 0$. The limit $\lim_{t \to \infty} \Pi^t$ is equal to the linear part of Π. So by choosing t high enough, we may assume that Π^t is defined on the closed ball of radius 2 and is sufficiently close to its linear part there. If Θ is a local smooth Levi normalization for Π^t, then $G(1/t) \circ \Theta \circ G(t)$ will be a local smooth Levi normalization for Π.

Remark 3.6.2. In [263] there is also an abstract *Nash–Moser normal form theorem*, which can be applied to the problem of smooth Levi decomposition of Poisson structures, and hopefully to other smooth normal form problems as well.

Chapter 4

Linearization of Poisson Structures

4.1 Nondegenerate Lie algebras

Let Π be a Poisson structure which vanishes at a point z: $\Pi(z) = 0$. Denote by

$$\Pi = \Pi^{(1)} + \Pi^{(2)} + \cdots , \tag{4.1}$$

the Taylor expansion of Π in a local coordinate system centered at z, where $\Pi^{(k)}$ is a homogeneous 2-vector field of degree k. Recall that, the terms of degree k of the equation $[\Pi, \Pi] = 0$ give

$$\sum_{i=1}^{k} \left[\Pi^{(i)}, \Pi^{(k+1-i)} \right] = 0. \tag{4.2}$$

In particular, $[\Pi^{(1)}, \Pi^{(1)}] = 0$, i.e., the linear part $\Pi^{(1)}$ of Π is a linear Poisson structure. One says that Π is locally smoothly (resp. analytically, resp. formally) *linearizable* if there is a local smooth (resp. analytic, resp. formal) diffeomorphism ϕ (a coordinate transformation) such that $\phi_* \Pi = \Pi^{(1)}$.

Definition 4.1.1 ([346]). A finite-dimensional Lie algebra $\mathfrak{g}$ is called formally (resp. analytically, resp. smoothly) *nondegenerate* if any formal (resp. analytic, resp. smooth) Poisson structure Π which vanishes at a point and whose linear part at that point corresponds to $\mathfrak{g}$ is formally (resp. analytically, resp. smoothly) linearizable.

In other words, a Lie algebra is nondegenerate if any Poisson structure, whose linear part corresponds to this algebra, is completely determined by its linear part up to local isomorphisms.

The above definition begs the question: which Lie algebras are nondegenerate and which are degenerate? This question is the main topic of this chapter. One of the main tools for studying it is the Levi decomposition, treated in the previous chapter. The question is still largely open, though we now know of several series of nondegenerate Lie algebras, and many degenerate ones (it is much easier in general to find degenerate Lie algebras than to find nondegenerate ones).

As explained in Chapter 2, formal deformations of Poisson structures are governed by Poisson cohomology, and for linear Poisson structures Poisson cohomology is a special case of Chevalley–Eilenberg cohomology of the corresponding Lie algebras. In particular, if $\mathfrak{l}$ is a Lie algebra such that $H^2(\mathfrak{l}, \mathcal{S}^k \mathfrak{l}) = 0 \ \forall \ k \geq 2$, then it is formally nondegenerate (Theorem 2.3.8). Let us recall here a special case of this result:

Theorem 4.1.2 ([346]). *Any semisimple Lie algebra is formally nondegenerate.*

In general, it is much more difficult to study smooth or analytic nondegeneracy of Lie algebras than to study their formal nondegeneracy, because the former problem involves not only algebra (cohomology of Lie algebras) but also geometry and analysis (to show the analyticity or smoothness of coordinate transformations).

The first significant results about analytic and smooth nondegenerate Lie algebras are due to Conn [80, 81], and are already mentioned in Chapter 3 as special cases of Levi decomposition theorems. Let us recall here Conn's results.

Theorem 4.1.3 ([80]). *Any semisimple Lie algebra is analytically nondegenerate.*

Theorem 4.1.4 ([81]). *Any compact semisimple Lie algebra is smoothly nondegenerate.*

On the other hand, most non-compact real semisimple Lie algebras are smoothly degenerate (see Section 4.3).

Related to the notion of (formal) nondegeneracy is the notion of *rigidity* of Lie algebras, mentioned in Subsection 2.3.3, and also the notion of *strong rigidity* [39]. Recall that $H^2(\mathfrak{g}, \mathfrak{g})$ is the cohomology group which governs infinitesimal deformations of a Lie algebra $\mathfrak{g}$. This group is somehow related to the group $\oplus_{k \geq 2} H^2(\mathfrak{g}, S^k \mathfrak{g})$, but they are not the same. Not surprisingly, there are Lie algebras which are rigid but degenerate (e.g., $\mathfrak{saff}(2)$, see Example 4.1.5), Lie algebras which are non-rigid but nondegenerate (e.g., a three-dimensional solvable Lie algebra $\mathbb{K} \ltimes_A \mathbb{K}^2$, where A is a nonresonant 2×2 matrix, see Theorem 4.2.2), Lie algebras which are both rigid and nondegenerate (e.g., semisimple Lie algebras), and Lie algebras which are both non-rigid and degenerate (e.g., Abelian Lie algebras).

Example 4.1.5. Denote by $\mathfrak{saff}(2, \mathbb{K}) = sl(2, \mathbb{K}) \ltimes \mathbb{K}^2$ the Lie algebra of infinitesimal area-preserving affine transformations on $\mathbb{K}^2$. Then $\mathfrak{saff}(2, \mathbb{K})$ is rigid but degenerate. The linear Poisson structure corresponding to $\mathfrak{saff}(2)$ has the form $\Pi^{(1)} = 2e\partial h \wedge \partial e - 2f \partial h \wedge \partial f + h \partial e \wedge \partial f + y_1 \partial h \wedge \partial y_1 - y_2 \partial h \wedge \partial y_2 + y_1 \partial e \wedge \partial y_2 + y_2 \partial f \wedge \partial y_1$ in a natural system of coordinates. Now put $\Pi = \Pi^{(1)} + \tilde{\Pi}$ with $\tilde{\Pi} = (h^2 + 4ef) \partial y_1 \wedge \partial y_2$.

. Then Π is a Poisson structure, vanishing at the origin, with a linear part corresponding to $\mathfrak{saff}(2)$. For $\Pi^{(1)}$ the set where the rank is less than or equal to 2 is a codimension 2 linear subspace (given by the equations $y_1 = 0$ and $y_2 = 0$). For Π the set where the rank is less than or equal to 2 is a two-dimensional cone (the cone given by the equations $y_1 = 0$, $y_2 = 0$ and $h^2 + 4ef = 0$). So these two Poisson structures are not isomorphic, even formally. The rigidity of $\mathfrak{saff}(2)$ will be left to the reader as an exercise (see [300, 63]).

Example 4.1.6. The Lie algebra $\mathfrak{e}(3) = so(3) \ltimes \mathbb{R}^3$ of rigid motions of the Euclidean space $\mathbb{R}^3$ is degenerate and non-rigid. The linear Poisson structure corresponding to $\mathfrak{e}(3)$ has the form $\Pi^{(1)} = x_1\partial x_2 \wedge \partial x_3 + x_2\partial x_3 \wedge \partial x_1 + x_3\partial x_1 \wedge \partial x_2 + y_1\partial x_2 \wedge \partial y_3 + y_2\partial x_3 \wedge \partial y_1 + y_3\partial x_1 \wedge \partial y_2$ in a natural system of coordinates. Now put $\Pi = \Pi^{(1)} + \tilde{\Pi}$ with $\tilde{\Pi} = (x_1^2 + x_2^2 + x_3^2)(x_1\partial x_2 \wedge \partial x_3 + x_2\partial x_3 \wedge \partial x_1 + x_3\partial x_1 \wedge \partial x_2)$. For $\Pi^{(1)}$ the set where the rank is less than or equal to 2 is a dimension 3 subspace (given by the equation $y_1 = y_2 = y_3 = 0$), while for Π the set where the rank is less than or equal to 2 is the origin. A Lie algebra not isomorphic to $\mathfrak{e}(3)$ but adjacent to $\mathfrak{e}(3)$ is $so(3,1)$, the Lie algebra of infinitesimal linear automorphisms of the Minkowski space. Here the adjective *adjacent* means that, in the variety of all Lie algebra structures of dimension 6 (see Subsection 2.3.3), the $GL(6)$-orbit which corresponds to $\mathfrak{e}(3)$ lies in the closure (with respect to the Euclidean topology) of the $GL(6)$-orbit which corresponds to $so(3,1)$. One also says that $e(3)$ is a contraction of $so(3,1)$. If a Lie algebra is a contraction of another Lie algebra, then it is not rigid.

A *strongly rigid Lie algebra* is a Lie algebra $\mathfrak{g}$ whose universal enveloping algebra $\mathcal{U}(\mathfrak{g})$ is rigid as an associative algebra [39]. It is easy to see that if $\mathfrak{g}$ is strongly rigid then it is rigid. A sufficient condition for $\mathfrak{g}$ to be strongly rigid is $H^2(\mathfrak{g}, S^k\mathfrak{g}) = 0 \ \forall \ k \geq 0$, and if this condition is satisfied then $\mathfrak{g}$ is called *infinitesimally strongly rigid* [39]. Obviously, if $\mathfrak{g}$ is infinitesimally strongly rigid, then it is formally nondegenerate. In fact, we have the following result, due to Bordemann, Makhlouf and Petit:

Theorem 4.1.7 ([39]). *If $\mathfrak{g}$ is a strongly rigid Lie algebra, then it is formally nondegenerate.*

We refer to [39] for the proof of the above theorem, which is based on Kontsevich's theorem [195] on the existence of deformation quantization of Poisson structures (see Appendix A.9).

4.2 Linearization of low-dimensional Poisson structures

4.2.1 Two-dimensional case

Up to isomorphisms, there are only two Lie algebras of dimension 2: the trivial, i.e., Abelian one, and the solvable Lie algebra $\mathbb{K} \ltimes \mathbb{K}$, which has a basis (e_1, e_2)

with $[e_1, e_2] = e_1$. This Lie algebra is isomorphic to the Lie algebra of infinitesimal affine transformations of the line, so we will denote it by $\mathfrak{aff}(1)$.

The Abelian Lie algebra of dimension 2 is of course degenerate. For example, the quadratic Poisson structure $(x_1^2 + x_2^2)\frac{\partial}{\partial x_1} \wedge \frac{\partial}{\partial x_2}$ is nontrivial and is not locally isomorphic to its linear part, which is trivial.

On the other hand, we have:

Theorem 4.2.1 ([14]). *The Lie algebra $\mathfrak{aff}(1)$ is formally, analytically and smoothly nondegenerate.*

Proof. We begin with $\{x, y\} = x + \cdots$. Putting $x' = \{x, y\}, y' = y$, we obtain $\{x', y'\} = \frac{\partial x'}{\partial x}\{x, y\} = x'a(x', y')$, where a is a function such that $a(0) = 1$. We finish with a second change of coordinates $x'' = x', y'' = f(x', y')$, where f is a function such that $\frac{\partial f}{\partial y'} = 1/a$. $\square$

4.2.2   Three-dimensional case

Every Lie algebra of dimension 3 over $\mathbb{R}$ or $\mathbb{C}$ is of one of the following three types, where (e_1, e_2, e_3) denote a basis:

- $so(3)$ with brackets $[e_1, e_2] = e_3, [e_2, e_3] = e_1, [e_3, e_1] = e_2$.

- $sl(2)$ with brackets $[e_1, e_2] = e_3, [e_1, e_3] = e_1, [e_2, e_3] = -e_2$. (Recall that $so(3, \mathbb{R}) \not\cong sl(2, \mathbb{R})$, $so(3, \mathbb{C}) \cong sl(2, \mathbb{C})$.)

- semi-direct products $\mathbb{K} \ltimes_A \mathbb{K}^2$ where $\mathbb{K}$ acts linearly on $\mathbb{K}^2$ by a matrix A. In other words, we have brackets $[e_2, e_3] = 0, [e_1, e_2] = ae_2 + be_3, [e_1, e_3] = ce_2 + de_3$, and A is the 2×2-matrix with coefficients a, b, c and d. (Different matrices A may correspond to isomorphic Lie algebras.)

The Lie algebras $sl(2)$ and $so(3)$ are simple, so they are formally and analytically nondegenerate, according to Weinstein's and Conn's theorems.

The fact that the compact simple Lie algebra $so(3, \mathbb{R})$ is smoothly nondegenerate (it is a special case of Conn's Theorem 4.1.4) is due to Dazord [94]. In Chapter 6, we will extend this result of Dazord to the case of elliptic singularities of Nambu structures, using arguments similar to his.

On the other hand, $sl(2, \mathbb{R})$ is known to be smoothly degenerate (see [346]). A simple construction of a smooth nonlinearizable Poisson structure whose linear part corresponds to $sl(2, \mathbb{R})$ is as follows: In a linear coordinate system (y_1, y_2, y_3), write

$$\Pi^{(1)} = y_3 \frac{\partial}{\partial y_1} \wedge \frac{\partial}{\partial y_2} - y_2 \frac{\partial}{\partial y_1} \wedge \frac{\partial}{\partial y_3} - y_1 \frac{\partial}{\partial y_2} \wedge \frac{\partial}{\partial y_3} = X \wedge Y,$$

where $X = y_2 \frac{\partial}{\partial y_3} - y_3 \frac{\partial}{\partial y_2}$, and $Y = \frac{\partial}{\partial y_1} + \frac{y_1}{y_2^2 + y_3^2}\left(y_2 \frac{\partial}{\partial y_2} + y_3 \frac{\partial}{\partial y_3}\right)$. This linear Poisson structure corresponds to $sl(2, \mathbb{R})$ and has $C = y_2^2 + y_3^2 - y_1^2$ as a Casimir function. Denote by Z a vector field on $\mathbb{R}^3$ such that $Z = 0$ when $y_2^2 + y_3^2 - y_1^2 \geq 0$, and $Z = \frac{G(C)}{\sqrt{y_2^2 + y_3^2}}\left(y_2 \frac{\partial}{\partial y_2} + y_3 \frac{\partial}{\partial y_3}\right)$ when $y_2^2 + y_3^2 - y_1^2 > 0$, where G is a flat function such that $G(0) = 0$ and $G(C) > 0$ when $C > 0$. Then Z is a flat vector field such that $[Z, X] = [Z, Y] = 0$. Hence $\Pi = (X - Z) \wedge Y$ is a Poisson structure whose linear part is $\Pi^{(1)} = X \wedge Y$. While X is a periodic vector field, the integral curves of $X - Z$ in the region $\{y_2^2 + y_3^2 - y_1^2 > 0\}$ are spiraling towards the cone $\{y_2^2 + y_3^2 - y_1^2 = 0\}$. Thus, while almost all the symplectic leaves of $\Pi^{(1)}$ are closed, the symplectic leaves of Π in the region $\{y_2^2 + y_3^2 - y_1^2 > 0\}$ contain the cone $\{y_2^2 + y_3^2 - y_1^2 = 0\}$ in their closure (also locally in a neighborhood of 0). This implies that the symplectic foliation of Π is not locally homeomorphic to the symplectic foliation of $\Pi^{(1)}$. Hence Π can't be locally smoothly equivalent to $\Pi^{(1)}$.

For solvable Lie algebras $\mathbb{K} \ltimes_A \mathbb{K}^2$, we have the following result:

Theorem 4.2.2 ([110]). *The Lie algebra $\mathbb{R}^2 \times_A \mathbb{R}$ is smoothly (or formally) nondegenerate if and only if A is nonresonant in the sense that there are no relations of the type*

$$\lambda_i = n_1 \lambda_1 + n_2 \lambda_2 \quad (i = 1 \text{ or } 2), \tag{4.3}$$

where λ_1 and λ_2 are the eigenvalues of A, n_1 and n_2 are two nonnegative integers with $n_1 + n_2 > 1$.

Proof. Let Π be a Poisson structure on a three-dimensional manifold which vanishes at a point with a linear part corresponding to $\mathbb{R}^2 \times_A \mathbb{R}$ with a nonresonant A. In a system of local coordinates (x, y, z), centered at the considered point, we have

$$\{z, x\} = ax + by + O(2), \quad \{z, y\} = cx + dy + O(2), \quad \{x, y\} = O(2), \tag{4.4}$$

where a, b, c, d are the coefficients of A, and $O(2)$ means terms of degree at least 2. It follows that the curl vector field $D_\Omega \Pi$ (see Section 2.6), with respect to any volume form Ω, has the form $(a + c)\partial/\partial z + Y$, where Y is a vector field vanishing at the origin. But the nonresonance hypothesis imposes that the trace of A is not zero; so $D_\Omega \Pi$ doesn't vanish in a neighborhood of the origin. We can straighten it and suppose that the coordinates (x, y, z) are chosen such that $D_\Omega \Pi = \partial z$.

Now the three-dimensional hypothesis implies $\Pi \wedge \Pi = 0$ and, using formula

$$[\Pi, \Pi] = D_\Omega(\Pi \wedge \Pi) \pm D_\Omega(\Pi) \wedge \Pi$$

(see Section 2.6), we obtain

$$D_\Omega(\Pi) \wedge \Pi = 0.$$

In the above coordinates this gives $\partial/\partial z \wedge \Pi = 0$ and, so,

$$\Pi = \partial/\partial z \wedge X,$$

where X is a vector field. Now we recall the basic formula

$$[D_\Omega(\Pi), \Pi] = 0$$

which have the consequence that we can suppose that X depends only on the coordinates x and y.

Because of the form of the linear part of Π, X is a vector field which vanishes at the origin but with a nonresonant linear part. Hence, up to a smooth (or formal) change of coordinates x and y, we can linearize X (see Appendix A.5). This gives the smooth (or formal) nondegeneracy of $\mathbb{R}^2 \times_A \mathbb{R}$.

To prove the "only if" part, we start with a linear Poisson structure

$$\Pi^{(1)} = \partial/\partial z \wedge X^{(1)},$$

where $X^{(1)}$ is a linear resonant vector field. Every resonance relation permits the construction of a polynomial perturbation X of $X^{(1)}$ which is not smoothly isomorphic to $X^{(1)}$, even up to a product with a function (see Appendix A.5). Then it is not difficult to prove that $\partial/\partial z \wedge X$ is a polynomial perturbation of $\Pi^{(1)}$ which is not equivalent to it. $\square$

Remark 4.2.3. The same proof shows that algebras of type $\mathbb{K}^2 \times_A \mathbb{K}$ (where $\mathbb{K} = \mathbb{R}$ or $\mathbb{C}$) are analytic nondegenerate if we add to the nonresonance condition a Diophantine condition on the eigenvalues of A (see Appendix A.5).

4.2.3   Four-dimensional case

The results on (non)degeneracy of four-dimensional Lie algebras presented in this subsection are taken from Molinier's thesis [258].

According to [292], every four-dimensional Lie algebra over $\mathbb{K}$, where $\mathbb{K} = \mathbb{R}$ or $\mathbb{C}$, belongs to (at least) one of the following four types:

Type 1:   direct products $\mathbb{K} \times L_3$ where L_3 is a three-dimensional algebra (see the preceding paragraph for a classification).

Type 2:   semi-direct products $\mathbb{K} \ltimes_A \mathbb{K}^3$, where $\mathbb{K}^3$ is the commutative Lie algebra of dimension 3, and $\mathbb{K}$ acts on $\mathbb{K}^3$ by a matrix A.

Type 3:   semi-direct products $\mathbb{K} \ltimes_A H_3$, where H_3 is the three-dimensional Heisenberg Lie algebra: H_3 has a basis (x, y, z) such that $[x, y] = z, [x, z] = [y, z] = 0$.

Type 4:   semi-direct products $\mathbb{K}^2 \ltimes \mathbb{K}^2$.

In the first type we have $\mathbb{K} \times sl(2)$ and $\mathbb{K} \times so(3)$, which are the same when $\mathbb{K} = \mathbb{C}$, and the cases where L_3 is a semi-direct product $\mathbb{K} \ltimes \mathbb{K}^2$. The Levi decomposition theorems from Chapter 3 imply that $\mathbb{K} \times sl(2)$ and $\mathbb{K} \times so(3)$ are formally and analytically nondegenerate, and that $\mathbb{R} \times so(3, \mathbb{R})$ is smoothly nondegenerate. However, $\mathbb{R} \times sl(2, \mathbb{R})$ is smoothly degenerate: just repeat the proof, given in the previous subsection, of the fact that $sl(2)$ is smoothly degenerate. The case where L_3 is a semi-direct product is degenerate (formally, analytically and smoothly): if we choose coordinates (u, v, x, y) such that the corresponding linear Poisson tensor has the form

$$(ax + by)\frac{\partial}{\partial u} \wedge \frac{\partial}{\partial x} + (cx + dy)\frac{\partial}{\partial v} \wedge \frac{\partial}{\partial y}, \tag{4.5}$$

then we can add a quadratic term $v^2 \frac{\partial}{\partial u} \wedge \frac{\partial}{\partial v}$ to get a nonlinearizable Poisson tensor.

Every algebra of the second type is degenerate. To prove this we choose co-ordinates (u, x_1, x_2, x_3) such that in these coordinates the corresponding linear Poisson structure has brackets $\{u, x_i\} = \sum_j a_i^j x_j$, (the others brackets are trivial), where a_i^j are the coefficients of the matrix A. Up to isomorphisms, we can suppose that A is in Jordan form and we can also replace A by λA where λ is any nonvanishing constant. So we have the following list of cases:

$$\{u, x_1\} = 0, \qquad \{u, x_2\} = x_1, \qquad \{u, x_3\} = x_2; \tag{4.6}$$
$$\{u, x_1\} = ax_1, \quad \{u, x_2\} = x_2, \qquad \{u, x_3\} = x_2 + x_3; \tag{4.7}$$
$$\{u, x_1\} = x_1, \quad \{u, x_2\} = 0, \qquad \{u, x_3\} = x_2; \tag{4.8}$$
$$\{u, x_1\} = x_1, \quad \{u, x_2\} = x_1 + x_2, \quad \{u, x_3\} = x_2 + x_3; \tag{4.9}$$
$$\{u, x_1\} = x_1, \quad \{u, x_2\} = ax_2, \qquad \{u, x_3\} = bx_3; \tag{4.10}$$
$$\{u, x_1\} = ax_1, \quad \{u, x_2\} = bx_2 - x_3, \quad \{u, x_3\} = x_2 + bx_3. \tag{4.11}$$

Each of the above cases can be perturbed to a nonlinearizable Poisson structure by adding a quadratic term: $\partial x_3 \wedge (x_1^2 \partial x_1 + x_1 x_2 \partial x_2)$ for (4.6); $x_2^2 \partial x_3 \wedge \partial x_2$ for (4.7) and (4.8); $x_1^2 \partial x_3 \wedge \partial x_2$ for (4.9); $x_2 x_3 \partial x_3 \wedge \partial x_2$ for (4.10); and $(x_2^2 + x_3^2) \partial x_3 \wedge \partial x_2$ for (4.11).

Similarly, every algebra of the third type is also degenerate. To prove this we choose coordinates (u, x_1, x_2, x_3) such that the corresponding linear Poisson structure has brackets $\{x_3, x_2\} = x_1$, $\{u, x_i\} = \sum_j a_i^j x_j$, (the other brackets are zero), where a_i^j are the coefficients of the matrix A. Up to isomorphisms, we have the following cases:

$$\{u, x_1\} = 2x_1, \qquad \{u, x_2\} = x_2, \qquad \{u, x_3\} = x_2 + x_3; \tag{4.12}$$
$$\{u, x_1\} = (1 + b)x_1, \quad \{u, x_2\} = x_2, \qquad \{u, x_3\} = bx_3; \tag{4.13}$$
$$\{u, x_1\} = 2ax_1, \qquad \{u, x_2\} = ax_2 - x_3, \quad \{u, x_3\} = x_2 + ax_3. \tag{4.14}$$

Each case can be perturbed to a nonlinearizable Poisson structure by adding a quadratic term: $x_2^2 \partial x_3 \wedge \partial x_2$ for (4.12); $x_2 x_3 \partial x_3 \wedge \partial x_2$ for (4.13); and $(x_2^2 + x_3^2) \partial x_3 \wedge \partial x_2$ for (4.14).

Finally, if $\mathfrak{g}$ is a Lie algebra of the last type, which does not belong to the previous three types, then it admits a basis (u, v, x, y), with either the brackets

$$[u, x] = x, \quad [v, y] = y; \tag{4.15}$$

or the brackets

$$[u, x] = x, \quad [u, y] = y \ , [v, x] = -y, \quad [v, y] = x. \tag{4.16}$$

When $\mathbb{K} = \mathbb{C}$ then the above two cases are the same and are isomorphic to $\mathfrak{aff}(1, \mathbb{C}) \times \mathfrak{aff}(1, \mathbb{C})$. When $\mathbb{K} = \mathbb{R}$, these are the two real versions of $\mathfrak{aff}(1, \mathbb{C}) \times \mathfrak{aff}(1, \mathbb{C})$. We will see in Section 4.4 that (4.15) is formally, analytically and smoothly nondegenerate, and therefore (4.16) is also formally and analytically nondegenerate. We don't know whether (4.16) is smoothly nondegenerate.

4.3   Poisson geometry of real semisimple Lie algebras

We refer to [170, 194, 272] for the theory of real semisimple Lie algebras used in this section. Recall that, if $\mathfrak{g}$ is a real semisimple Lie algebra, then there is a (unique up to isomorphisms) *Cartan involution* $\theta : \mathfrak{g} \to \mathfrak{g}$, and we can write $\mathfrak{g} = \mathfrak{k} + \mathfrak{s}$, where $\mathfrak{k} = \{x \in \mathfrak{g}, \theta(x) = x\}$ and $\mathfrak{s} = \{x \in \mathfrak{g}, \theta(x) = -x\}$. Denote by $\mathfrak{g}_{\mathbb{C}}$ the complexification of $\mathfrak{g}$, and by $\mathfrak{g}_0 = \mathfrak{k} + \sqrt{-1}\mathfrak{s} \subset \mathfrak{g}_{\mathbb{C}}$. Then $\mathfrak{g}_0$ is a compact real semisimple Lie algebra (the compact part of $\mathfrak{g}_{\mathbb{C}}$; if $\mathfrak{g}$ is compact then $\mathfrak{s} = \{0\}$ and $\mathfrak{g} = \mathfrak{g}_0 = \mathfrak{k}$). The decomposition $\mathfrak{g} = \mathfrak{k} + \mathfrak{s}$ is called the *Cartan decomposition* of $\mathfrak{g}$, and $\mathfrak{k}$ is a subalgebra of $\mathfrak{g}$, called the *compact part* of $\mathfrak{g}$. We also have $[\mathfrak{k}, \mathfrak{s}] \subset \mathfrak{s}, [\mathfrak{s}, \mathfrak{s}] \subset \mathfrak{k}$. At the group level, K is the maximal compact subgroup of G, and $S = G/K$ is a *symmetric space*. The *real rank* of $\mathfrak{g}$ is the same as the rank of the symmetric space S, and is equal to the dimension of a maximal Abelian subalgebra of $\mathfrak{g}$ which is contained in $\mathfrak{s}$.

In this section, we will explain the following result of Weinstein [348] about the existence of smooth non-Hamiltonian Poisson vector fields for linear non-compact real semisimple Poisson structures.

Theorem 4.3.1 ([348]). *Let $\mathfrak{g}$ be a real semisimple Lie algebra of real rank $r \geq 1$. Then for the corresponding linear Poisson structure $\Pi^{(1)}$ on the dual space $\mathfrak{g}^*$, there are r smooth Poisson vector fields $X_1, \ldots, X_r$ which are flat at the origin and which commute pair-wise, and such that, in an open subset of $\mathfrak{g}^*$ whose closure contains the origin, these vector fields span an r-dimensional distribution whose intersection with the characteristic distribution (i.e., tangent spaces to coadjoint orbits) is trivial.*

A direct consequence of Theorem 4.3.1 is the following:

Theorem 4.3.2 (Weinstein [348]). *Any non-compact real semisimple Lie algebra $\mathfrak{g}$ of real rank at least 2 is smoothly degenerate.*

Proof of Theorem 4.3.2. Let $X_1, \ldots, X_2$ be smooth Poisson vector fields on $\mathfrak{g}^*$ given by Theorem 4.3.1, where $r \geq 2$ is the real rank of $\mathfrak{g}$. Then since $[X_1, X_2] = 0$, $[X_1, \Pi^{(1)}] = [X_2, \Pi^{(1)}] = 0$ and $[\Pi^{(1)}, \Pi^{(1)}] = 0$, where $\Pi^{(1)}$ denotes the linear Poisson structure, by the Leibniz rule we obtain that $[\Pi^{(1)} + X_1 \wedge X_2, \Pi^{(1)} + X_1 \wedge X_2] = 0$. In other words, $\Pi = \Pi^{(1)} + X_1 \wedge X_2$ is a smooth Poisson structure. The linear part of Π is $\Pi^{(1)}$ because X_1 and X_2 are flat. In an open region whose closure contains the origin, where X_1 and X_2 are linearly independent and are transversal to the symplectic leaves of $\Pi^{(1)}$, the characteristic distribution of Π is spanned by X_1, X_2 and the characteristic distribution of $\Pi^{(1)}$. Hence the rank of Π (in any neighborhood of the origin) is equal to the rank of $\Pi^{(1)}$ plus 2. Since Π and $\Pi^{(1)}$ don't have the same rank (locally near 0), they can't be locally isomorphic. In other words, Π is not locally smoothly linearizable.   □

Proof of Theorem 4.3.1. Denote by $G = KAN$ the *Iwasawa decomposition* of G, where G is the connected simply-connected Lie group whose Lie algebra is $\mathfrak{g}$. K

is the compact part of G, A is Abelian, N is nilpotent. Denote by $M \subset K$ the centralizer of A in K, and by $P = MAN$ the minimal parabolic subgroup. Denote by $\mathfrak{g} = \mathfrak{k} + \mathfrak{a} + \mathfrak{n}$ the corresponding Iwasawa decomposition of $\mathfrak{g}$ at the algebra level, by $\mathfrak{m}$ the Lie algebra of M, $\mathfrak{m} \subset \mathfrak{k}$, and by $\mathfrak{p} = \mathfrak{m} + \mathfrak{a} + \mathfrak{n}$ the parabolic subalgebra. The real rank of $\mathfrak{g}$ is $r = \dim \mathfrak{a} \geq 1$ by assumptions. The adjoint action of $\mathfrak{a}$ on $\mathfrak{g}$ is diagonalizable with real weights, and we have a linear decomposition of $\mathfrak{g}$ into a sum of eigenspaces of this adjoint action of $\mathfrak{a}$, called the real root decomposition. $\mathfrak{n}$ is spanned by the eigenspaces of positive roots, while $\mathfrak{m} + \mathfrak{a}$ is the eigenspace of weight 0. Denote by $\mathfrak{a}_+$ the *open* (without boundary) positive Weyl chamber of $\mathfrak{a}$ with respect to a root system of the real root decomposition of $\mathfrak{g}$. Denote

$$\mathfrak{g}_+ = \mathrm{Ad}_K(\mathfrak{m} + \mathfrak{a}_+ + \mathfrak{n}) = \mathrm{Ad}_K \mathfrak{p}_+, \quad \mathfrak{p}_+ = \mathfrak{m} + \mathfrak{a}_+ + \mathfrak{n} . \tag{4.17}$$

We will need the following lemma of Weinstein:

Lemma 4.3.3 ([348]). *If two elements of $\mathfrak{p}_+$ are conjugate by an element g of G, then g belongs to P.*

See ([348], Lemma (2.3)) for the proof of the above lemma, which uses standard results from the theory of real semisimple Lie algebras. □

For each $a \in \mathfrak{a}$, denote

$$\mathcal{P}_a = \mathrm{Ad}_K(\mathfrak{m} + a + \mathfrak{n}) . \tag{4.18}$$

Consider the following map $\mathcal{E}_{ab} : \mathcal{P}_a \to \mathcal{P}_b$, for each pair $a, b \in \mathfrak{a}$ with $a \neq 0$:

$$\mathcal{E}_{ab}(\mathrm{Ad}_h(m + a + n)) = \mathrm{Ad}_h(m + b + n) \ \forall \ h \in K, m \in \mathfrak{m}, n \in \mathfrak{n} . \tag{4.19}$$

Lemma 4.3.4. *With the above notations, we have:*
a) *$\mathfrak{g}_+ = \bigcup_{a \in \mathfrak{a}_+} \mathcal{P}_a$ is an open cone in $\mathfrak{g}$.*
b) *Each $\mathcal{P}_a$ is saturated by adjoint orbits of G: if $x \in \mathcal{P}_a$ then $\mathrm{Ad}_G(x) \subset \mathcal{P}_a$.*
c) *If $a, b \in \mathfrak{a}_+$, $a \neq b$, then $\mathcal{P}_a$ is disjoint from $\mathcal{P}_b$: $\mathcal{P}_a \cap \mathcal{P}_b = \emptyset$.*
d) *$\dim \mathcal{P}_a = \dim \mathfrak{g} - r \ \forall \ a \in \mathfrak{a}$, where $r = \dim \mathfrak{a} \geq 1$ is the real rank. If $a \in \mathfrak{a}_+$ then $\mathcal{P}_a$ is a smooth submanifold of $\mathfrak{g}$.*
e) *The map $\mathcal{E}_{ab} : \mathcal{P}_a \to \mathcal{P}_b$ is well defined for all $a \in \mathfrak{a}_+$ and $b \in \mathfrak{a}$. If $a, b \in \mathfrak{a}_+$ then $\mathcal{E}_{ab}$ is a diffeomorphism. When $b = 0$, $\mathcal{E}_{a0}$ is a local diffeomorphism almost everywhere.*

The proof of the above lemma follows directly from the root decomposition of $\mathfrak{g}$ with respect to $\mathfrak{a}$ and Lemma 4.3.3. It will be left to the reader as an exercise. □

Identify $\mathfrak{g}$ with $\mathfrak{g}^*$ via the Killing form. Then $\mathfrak{g}$ is a linear Poisson manifold whose symplectic leaves are the adjoint orbits. Each $\mathcal{P}_a$ is saturated by symplectic leaves of $\mathfrak{g}$ and has a natural induced Poisson structure.

Proposition 4.3.5. *$\mathcal{E}_{ab} : \mathcal{P}_a \to \mathcal{P}_b$ is a Poisson isomorphism for any $a, b \in \mathfrak{a}_+$.*

Proof of Proposition 4.3.5. Put $c = a - b$, and denote $\mathcal{E}_{ab}$ simply by $\mathcal{E}$. Since $\mathcal{E}$ is equivariant under the adjoint action of K, it is enough to show that $\mathcal{E}$ preserves the Poisson tensor at points of the type $x = m + a + n \in \mathfrak{m} + a + \mathfrak{n} \subset \mathcal{P}_a$.

For an element $g \in \mathfrak{g}$, we will write $g = g_{\mathfrak{k}} + g_a + g_{\mathfrak{n}}$, where $g_{\mathfrak{k}} \in \mathfrak{k}, g_a \in a, g_{\mathfrak{n}} \in \mathfrak{n}$. The map $\mathcal{E}$ maps x to $x_0 = m + b + n$. By direct computations, one verifies that

$$\mathcal{E}_*(x)([g, x]) = [g, x_0] + [g_{\mathfrak{n}}, c] = [g, x] - [g_{\mathfrak{k}}, c] . \tag{4.20}$$

In other words, the element $[g, x] \in \mathfrak{g}$, considered as a tangent vector at x to $\mathrm{Ad}_G(x) \subset \mathcal{P}_a$, will be mapped under $\mathcal{E}$ to $[g, x_0] + [g_{\mathfrak{n}}, c]$, considered as a tangent vector to $\mathcal{P}_b$ at x_0.

If $y^* \in \mathfrak{g}^*$ is an element of $\mathfrak{g}^*$, considered as a covector at x_0, then its pullback by $\mathcal{E}$ is an element $\widetilde{y}^* \in \mathfrak{g}^*$, considered as a covector at x and determined only up to elements which vanish on $T_x(\mathcal{P}_a)$, such that

$$\langle \mathrm{ad}_x^* \widetilde{y}^*, g \rangle = \langle \widetilde{y}^*, [g, x] \rangle = \langle y^*, [g, x] - [g_{\mathfrak{k}}, c] \rangle = \langle \mathrm{ad}_x^* y^* - (\mathrm{ad}_c^* y^*)_{\mathfrak{k}}, g \rangle, \tag{4.21}$$

where $(\mathrm{ad}_c^* y^*)_{\mathfrak{k}}$ denotes the $\mathfrak{k}^*$-component of $\mathrm{ad}_a^* y^*$ with respect to the decomposition $\mathfrak{g}^* = \mathfrak{k}^* + a^* + \mathfrak{n}^*$ which is dual to the decomposition $\mathfrak{g} = \mathfrak{k} + a + \mathfrak{n}$.

Denote by $y, \widetilde{y}$ the elements of $\mathfrak{g}$ which correspond to $y^*, \widetilde{y}^*$ under the identification of $\mathfrak{g}$ with $\mathfrak{g}^*$ via the Killing form, then we can rewrite the above equation as:

$$\mathrm{ad}_x \widetilde{y} = \mathrm{ad}_x y - (\mathrm{ad}_c y)^{\mathfrak{k}} , \tag{4.22}$$

where $(\mathrm{ad}_c y)^{\mathfrak{k}}$ is the image of $(\mathrm{ad}_c^* y^*)_{\mathfrak{k}}$ in $\mathfrak{g}$ under the identification of $\mathfrak{g}^*$ with $\mathfrak{g}$, and not the $\mathfrak{k}$-component of $\mathrm{ad}_c y$ (under this identification via the Killing form, $\mathfrak{k}^*$ is identified with $\mathfrak{m} + \mathfrak{n}$ rather than $\mathfrak{k}$).

The Hamiltonian vector (i.e., the image under the map $\sharp$ corresponding to the linear Poisson structure on $\mathfrak{g}$) of $\widetilde{y}^*$ at x is $\mathrm{ad}_x \widetilde{y}$ and the Hamiltonian vector of y^* at x_0 is $\mathrm{ad}_{x_0} y$. To show that $\mathcal{E}$ preserves the Poisson structure, we must show that

$$\mathcal{E}_*(x)(\mathrm{ad}_x \widetilde{y}) = \mathrm{ad}_{x_0} y. \tag{4.23}$$

According to Formula (4.22), we have

$$\mathcal{E}_*(x)(\mathrm{ad}_x \widetilde{y}) = \mathcal{E}_*(x)(\mathrm{ad}_x y) - \mathcal{E}_*(x)((\mathrm{ad}_c y)^{\mathfrak{k}}) .$$

Notice that $(\mathrm{ad}_c y)^{\mathfrak{k}} = [c, y_{\mathfrak{n}}] \in \mathfrak{n}$ (because $\langle \mathfrak{n}, a \rangle = \langle \mathfrak{k}, a \rangle = 0$). Hence $(\mathrm{ad}_c y)^{\mathfrak{k}}$ is invariant under $\mathcal{E}_*(x)$, because $\mathcal{E}_*(x)$ is identity on $\mathfrak{m} + \mathfrak{n}$. In other words, we have $\mathcal{E}_*(x)((\mathrm{ad}_c y)^{\mathfrak{k}}) = (\mathrm{ad}_c y)^{\mathfrak{k}} = [c, y_{\mathfrak{n}}]$. Now applying Formula (4.20), we get $\mathcal{E}_*(x)(\mathrm{ad}_x \widetilde{y}) = \mathcal{E}_*(x)(\mathrm{ad}_x y) - [c, y_{\mathfrak{n}}] = [x_0, y]$. Proposition 4.3.5 is proved.   □

An immediate consequence of Proposition 4.3.5 is that the open cone $\mathfrak{g}_+ = \bigcup_{a \in a_+} \mathcal{P}_a$, as a Poisson manifold, is isomorphic to the direct product $\mathcal{P}_a \times a_+$ (for any $a \in a_+$), where the Poisson structure on a_+ is trivial.

Since $\dim a_+ = r$, we can construct r smooth vector fields $Y_1, \ldots, Y_r$ on a_+ with the following properties: they commute pairwise and are linearly independent also everywhere on $\mathfrak{a}_+$, and they extend to smooth vector fields on $\mathfrak{a}$ which vanish on the complement of $\mathfrak{a}_+$. (We will leave the actual construction of these vector fields to the reader as an exercise.) Then lifting $Y_1, \ldots, Y_r$ to $\mathfrak{g}_+$ via the Poisson isomorphism $\mathfrak{g}_+ \cong \mathcal{P}_a \times \mathfrak{a}_+$, we get r smooth pairwise commuting Poisson vector fields on $\mathfrak{g}_+$, which can be extended to $\mathfrak{g}$ by making them vanish on the complement of $\mathfrak{g}_+$. These are the Poisson vector fields that we are looking for. Theorem 4.3.1 is proved.   □

Remark 4.3.6. The above proof of Theorem 4.3.1 is a simplified version of Weinstein's original proof in [348], where he interpreted $\mathfrak{g}_+$ as a classical Yang–Mills–Higgs phase space (see Example 1.7.12): $\mathfrak{g}_+$ is isomorphic to $\mathcal{Y}(P, G, \mathfrak{m}^* \times \mathfrak{a}_+^*)$, the classical Yang–Mills–Higgs phase space of configuration space G/P, gauge group P and internal phase space $\mathfrak{m}^*$, where $P = MAN$ (the real Borel subgroup of G), G is viewed as a principal P bundle over G/P, and the action of P on $\mathfrak{m}^* \times \mathfrak{a}_+^* \subset \mathfrak{p}^*$ is the coadjoint action. Weinstein [348] showed that classical Yang–Mills–Higgs setups form a category with natural morphisms, and deduced from this fact that $\mathfrak{g}_+ \cong \mathcal{Y}(P, G, \mathfrak{m}^* \times \mathfrak{a}_+^*)$ is isomorphic to $\mathcal{Y}(M, K, \mathfrak{m}^* \times \mathfrak{a}^*)$ where K is viewed as a principal bundle over K/M (basically because $K/M = G/P$). On the other hand, $\mathcal{Y}(M, K, \mathfrak{m}^* \times \mathfrak{a}_+^*)$ is clearly isomorphic to $\mathcal{Y}(M, K, \mathfrak{m}^*) \times \mathfrak{a}_+^*$, with the trivial Poisson structure on $\mathfrak{a}_+^*$. We refer to [348] for more details and interesting connections with representation theory.

When $\mathfrak{g}$ is a non-compact real semisimple Lie algebra with a Cartan decomposition $\mathfrak{g} = \mathfrak{k} + \mathfrak{s}$, then the semi-direct product

$$\mathfrak{g}_1 = \mathfrak{k} \ltimes \mathfrak{s}, \tag{4.24}$$

where $\mathfrak{s}$ is considered as a vector space with a trivial bracket on which $\mathfrak{k}$ acts, is called the *Cartan motion algebra* of $\mathfrak{g}$. Let us mention here another interesting result from [348]:

Theorem 4.3.7 (Weinstein [348]). *With the above notations, there is a K-equivariant Poisson isomorphism from $\mathfrak{g}_+$ to an open cone in the dual of the Cartan motion algebra $\mathfrak{g}_1$.*

As an exercise, the reader may try to find for himself a proof of Theorem 4.3.7 similar to the above proof of Proposition 4.3.5.   □

Weinstein's Theorem 4.3.2 leaves out the case of real semisimple Lie algebras of real rank 1. The rest of this section is devoted to the discussion of this case.

From the classification of real simple Lie algebras (see, e.g., [170, 194, 272]), we have the following complete list of real simple Lie algebras of real rank 1: $so(n, 1)$, $su(n, 1)$, $sp(n, 1)$, and a real form of F_2. (There is an isomorphism $so(2, 1) \cong su(1, 1) \cong sl(3, \mathbb{R})$; the other algebras in the list are distinct.) A general real semisimple Lie algebra of real rank 1 is a direct sum of one of the above algebras with a compact semisimple Lie algebra. The rank-1 symmetric spaces corresponding to $so(n, 1)$, $su(n, 1)$ and $sp(n, 1)$ are the hyperbolic space $\mathbb{H}^n$, the complex hyperbolic space, and the quaternionic hyperbolic space, respectively.

Conjecture 4.3.8. A real semisimple Lie algebra $\mathfrak{g}$ of real rank 1 is smoothly non-degenerate if and only if its compact part $\mathfrak{k}$ is semisimple.

Among the real simple Lie algebras of real rank 1, only the algebras $su(n, 1)$ have a non-semisimple compact part. (The compact part of $su(n, 1)$ is isomorphic to $su(n) \oplus \mathbb{R}$; at the group level, the compact part of $SU(n, 1)$ is $S(U(n) \times U(1)) \cong SU(n) \times \mathbb{T}^1$.) Hence, the "only if" part of Conjecture 4.3.8 is provided by the following theorem.

Theorem 4.3.9 (Monnier–Zung). *The Lie algebra $su(n, 1)$ is smoothly degenerate for any $n \in \mathbb{N}$.*

Proof. Denote the linear Poisson structure on $su^*(n, 1)$ by $\Pi^{(1)}$. Denote by K the maximal compact subgroup in $SU(n, 1)$, and $\mathfrak{k} \cong su(n) \oplus \mathbb{R}$ its Lie algebra. Denote by Y the non-Hamiltonian Poisson vector field on $su^*(n, 1)$ given by Theorem 4.3.1. By construction, Y is invariant under the coadjoint action of K. Denote by $X = X_c$ the Hamiltonian vector field generated by a nontrivial element c in the center of $\mathfrak{k}$ (i.e., c lies in $\mathbb{R}$ in the decomposition $\mathfrak{k} \cong su(n) \oplus \mathbb{R}$). Note that X is a periodic vector field. We will choose c so that the period of the flow of X on $su^*(n, 1)$ is exactly 1, and we have an analytic Hamiltonian $\mathbb{T}^1$-action on $su^*(n, 1)$ generated by X.

Choose a (singular) closed 1-form α such that α is K-invariant, Y-invariant, and

$$\langle \alpha, X \rangle = 1.$$

Denote by $Z = i_\alpha \Pi^{(1)}$ the singular locally Hamiltonian vector field generated by α.

The 1-form α is singular at points $x \in su^*(n, 1)$ such that $X(x) = 0$. A simple direct verification shows that the set $S = \{x \in su^*(n, 1), X(x) = 0\}$ is a linear subspace of $su^*(n, 1)$ which does not intersect the "positive" part $su^*(n, 1)_+$ of $su^*(n, 1)$ (using the notations of the proof of Theorem 4.3.1 and identifying $su(n, 1)$ with $su^*(n, 1)$), so we can choose α smooth on $su^*(n, 1)_+$. Since, by construction, the vector field Y is trivial outside of $su^*(n, 1)_+$, we can arrange it so that $Z \wedge Y$ is smooth. Since α is Y-invariant, Z is also Y-invariant, i.e., $[Z, Y] = 0$. Now put

$$\Pi = \Pi^{(1)} + Z \wedge Y.$$

Then Π is a Poisson structure (of the same rank as $\Pi^{(1)}$). Look at the Hamiltonian vector field of c with respect to Π: by construction, this vector field is equal to $X + Y$ instead of X. So instead of being periodic, its flow is now spiraling. As a consequence, there are symplectic leaves of Π in $su^*(n, 1)_+$ which are not closed (in an arbitrary small neighborhood of the origin), so Π can't be locally isomorphic to $\Pi^{(1)}$ for topological reasons. $\square$

Let us give here some arguments in support of the "if" part of Conjecture 4.3.8. Let Π be a C^∞-smooth Poisson structure which vanishes at a point, and whose linear part corresponds to a semisimple Lie algebra $\mathfrak{g} = \mathfrak{k} + \mathfrak{s}$ of real rank 1, such that its compact part $\mathfrak{k}$ is semisimple. Then one can apply the smooth Levi decomposition Theorem 3.2.9 to $\mathfrak{k}$. After that, it remains to linearize the "$\mathfrak{s}$-part" of Π. But this part has a "hyperbolic behavior" because $\mathfrak{g}$ is of real rank 1, so one can probably use some recent results and techniques on local linearization of hyperbolic dynamical systems due to Chaperon [74, 75] and other people. For example, consider the simplest case $\mathfrak{g} = so(3,1) = so(3) + \mathfrak{s}$, $\dim \mathfrak{g} = 6$, $\dim \mathfrak{s} = 3$. The rank of the corresponding linear Poisson structure is 4. By geometric arguments, one can probably show that the rank of a smooth nonlinear Poisson structure Π with this linear part is also 4 (it is at least 4, and can't be 6). The smooth Levi normal form gives a linear Hamiltonian $Spin(3)$-action. This action is locally free almost everywhere, and its orbits lie on the symplectic leaves of Π. Taking the quotient of $\mathbb{R}^6$ by this action, one gets a three-dimensional cone, on which the quotient of the symplectic leaves are curves, i.e., one gets a one-dimensional foliation in a three-dimensional cone. Due to the hyperbolic behavior, one can probably linearize this foliation. Lifting this linearization to $\mathbb{R}^6$, one gets a smooth linearization of the characteristic foliation of Π. The last step is to use Moser's path method to linearize Π.

4.4 Nondegeneracy of $\mathfrak{aff}(n)$

Theorem 4.4.1 ([120]). *For any natural number n, the Lie algebra $\mathfrak{aff}(n, \mathbb{K}) = \mathfrak{gl}(n, \mathbb{K}) \ltimes \mathbb{K}^n$ of affine transformations of $\mathbb{K}^n$, where $\mathbb{K} = \mathbb{R}$ or $\mathbb{C}$, is formally and analytically nondegenerate.*

Proof. We will prove the above theorem in the analytic case. The formal case is absolutely similar, if not simpler. Denote by $\mathfrak{l} = \mathfrak{g} \ltimes \mathfrak{r}$ a Levi decomposition for a (real or complex) Lie algebra $\mathfrak{l}$, where $\mathfrak{s}$ is semisimple and $\mathfrak{r}$ is the solvable radical of $\mathfrak{l}$. Let Π be an analytic Poisson structure vanishing at a point 0 in a manifold whose linear part at 0 corresponds to $\mathfrak{l}$. According to the analytic Levi decomposition Theorem 3.2.6, there exists a local analytic system of coordinates $(x_1, \ldots, x_m, y_1, \ldots, y_d)$ in a neighborhood of 0, where $m = \dim \mathfrak{g}$ and $d = \dim \mathfrak{r}$, such that in these coordinates we have

$$\{x_i, x_j\} = \sum c_{ij}^k x_k , \quad \{x_i, y_r\} = \sum a_{ir}^s y^s , \qquad (4.25)$$

where c_{ij}^k are structural constants of $\mathfrak{s}$ and a_{ir}^s are constants. This gives what we call a semilinearization for Π. Note that the remaining Poisson brackets $\{y_r, y_s\}$ are nonlinear in general.

We now restrict our attention to the case where $\mathfrak{l} = \mathfrak{aff}(n)$, $m = n^2 - 1$, $d = n + 1$, $\mathfrak{g} = \mathfrak{sl}(n)$, $\mathfrak{r} = \mathbb{R}(\mathrm{Id}) \ltimes \mathbb{K}^n$ where Id acts on $\mathbb{K}^n$ by the identity map. The following lemma says that we may have a semilinearization associated to the decomposition $\mathfrak{aff}(n) = \mathfrak{gl}(n) \ltimes \mathbb{K}^n$ (which is slightly better than the Levi decomposition).

Lemma 4.4.2. *There is a local analytic coordinate system*

$$(x_1, \ldots, x_{n^2-1}, y_0, y_1, \ldots, y_n)$$

which satisfy Relations (4.25), with the following extra properties: $\{y_0, y_r\} = y_r$
for $r = 1, \ldots, n$; $\{x_i, y_0\} = 0 \ \forall i$.

Proof. We can assume that the coordinates y_r are chosen so that Relations (4.25)
are already satisfied, and y_0 corresponds to Id in $\mathbb{R}(\mathrm{Id}) \ltimes \mathbb{K}^n$. Then the Hamiltonian
vector fields X_{x_i} are linear and form a linear action of $\mathfrak{sl}(n)$. Because of (4.25),
we have that $\{x_i, y_0\} = 0$, which implies that $[X_{x_i}, X_{y_0}] = 0$, i.e., X_{y_0} is invariant
under the $\mathfrak{sl}(n)$ action. Moreover we have $X_{y_0}(x_i) = 0$ (i.e., X_{y_0} does not contain
components $\partial/\partial x_i$), and $X_{y_0} = \sum_1^n y_i \partial/\partial y_i +$ nonlinear terms. Hence we can
use (the parametrized equivariant version of) Poincaré linearization theorem to
linearize X_{y_0} in a $\mathfrak{sl}(n)$-invariant way. After this linearization, we have that $X_{y_0} = \sum_1^n y_i \partial/\partial y_i$. In other words, Relations (4.25) are still satisfied, and moreover we
have $\{y_0, y_i\} = X_{y_0}(y_i) = y_i$. $\square$

Remark 4.4.3. Lemma 4.4.2 still holds if we replace $\mathfrak{aff}(n)$ by any Lie algebra of
the type $(\mathfrak{g} \oplus \mathbb{K}e_0) \ltimes \mathfrak{n}$ where $\mathfrak{g}$ is semisimple and e_0 acts on $\mathfrak{n}$ by the identity map
(or any matrix whose corresponding linear vector field is nonresonant and satisfies
a Diophantine condition).

We will redenote y_0 in Lemma 4.4.2 by x_{n^2}. Then Relations (4.25) are still
satisfied. We will work in a coordinate system $(x_1, \ldots, x_{n^2}, y_1, \ldots, y_n)$ provided
by this lemma . We will fix the variables $x_1, \ldots, x_{n^2}$, and consider them as linear
functions on $\mathfrak{gl}^*(n)$ (they give a Poisson projection from our $(n^2 + n)$-dimensional
space to $\mathfrak{gl}^*(n)$). Denote by $F_1, \ldots, F_n$ the n basic Casimir functions for $\mathfrak{gl}^*(n)$.
(If we identify $\mathfrak{gl}(n)$ with its dual via the Killing form, then $F_1, \ldots, F_n$ are basic symmetric functions of the eigenvalues of $n \times n$ matrices.) We will consider
$F_1(x), \ldots, F_n(x)$ as functions in our $(n^2 + n)$-dimensional space, which do not
depend on variables y_i. Denote by $X_1, \ldots, X_n$ the Hamiltonian vector fields of
$F_1, \ldots, F_n$.

Lemma 4.4.4. *The vector fields* $X_1, \ldots, X_n$ *do not contain components* $\partial/\partial x_i$.
They form a system of n *linear commuting vector fields on* $\mathbb{K}^n$ *(the space of* $y = (y_1, \ldots, y_n)$*) with coefficients which are polynomial in* $x = (x_1, \ldots, x_{n^2})$. *The set
of* x *such that they are linearly dependent everywhere in* $\mathbb{K}^n$ *is an analytic space
of complex codimension strictly greater than 1 (when* $\mathbb{K} = \mathbb{C}$*).*

Proof. The fact that the X_i are y-linear with x-polynomial coefficients follows
directly from Relations (4.25). Since F_i are Casimir functions for $\mathfrak{gl}(n)$, we have
$X_i(x_k) = \{F_i, x_k\} = 0$, and $[X_i, X_j] = X_{\{F_i, F_j\}} = 0$.
    One checks that, for a given x, $X_1 \wedge \cdots \wedge X_n = 0$ identically on $\mathbb{K}^n$ if and
only if x is a singular point for the map $(F_1, \ldots, F_n)$ from $\mathfrak{gl}^*(n)$ to $\mathbb{K}^n$. The set
of singular points of the map $(F_1, \ldots, F_n)$ in the complex case is of codimension
greater than 1 (in fact, it is of codimension 3). $\square$

Lemma 4.4.5. *Write the Poisson structure* Π *in the form* $\Pi = \Pi^{(1)} + \tilde{\Pi}$, *where* $\Pi^{(1)}$ *is the linear part and* $\tilde{\Pi}$ *denote the higher-order terms. Then* $\tilde{\Pi}$ *is a Poisson structure which can be written in the form*

$$\tilde{\Pi} = \sum_{i<j} f_{ij} X_i \wedge X_j \ , \tag{4.26}$$

where the functions f_{ij} *are analytic functions which depend only on the variables* x, *and they are Casimir functions for* $\mathfrak{gl}^*(n)$ *(if we consider the variables* x *as linear functions on* $\mathfrak{gl}^*(n)$*).*

Proof. We work first locally near a point (x, y) where the vector fields X_k are linearly independent point-wise. As $\tilde{\Pi}$ is a 2-vector field in $\mathbb{K}^n = \{y\}$ (with coefficient depending on x) we have a local formula $\tilde{\Pi} = \sum_{i<j} f_{ij} X_i \wedge X_j$ where f_{ij} are analytic functions in variables (x, y). Since X_k are Hamiltonian vector fields for Π and also for $\Pi^{(1)}$, we have $[X_k, \tilde{\Pi}] = [X_k, \Pi] - [X_k, \Pi^{(1)}] = 0$ for $k = 1, \ldots, n$. This leads to $X_k(f_{ij}) = 0 \ \forall \ k, i, j$. Hence, because the X_k generate $\mathbb{K}^n$, the functions f_{ij} are locally independent of y. Using analytic extension, Hartog's theorem and the fact that the set of x such that $X_1, \ldots, X_n$ are linearly dependent point-wise everywhere in $\mathbb{K}^n$ is of complex codimension greater than 1, we obtain that f_{ij} are local analytic functions in a neighborhood of 0 which depend only on the variables x. The fact that $\tilde{\Pi}$ is a Poisson structure, i.e., $[\tilde{\Pi}, \tilde{\Pi}] = 0$, is now evident, because $X_k(f_{ij}) = 0$ and $[X_i, X_j] = 0$.

Relations $[X_{x_k}, \tilde{\Pi}] = [X_{x_i}, \Pi] - [X_{x_i}, \Pi^{(1)}] = 0$ imply that $X_{x_k}(f_{ij}) = 0$, which means that f_{ij} are Casimir functions for $\mathfrak{gl}^*(n)$. $\square$

Remark 4.4.6. Lemma 4.4.5 is still valid in the formal case. In fact, every homogeneous component of $\tilde{\Pi}$ satisfies a relation of type (4.26).

Lemma 4.4.7. *There exists a vector field* Y *of the form* $Y = \sum_{i=1}^n \alpha_i X_i$, *where the analytic functions* α_i *depend only on the variables* x *and are Casimir functions for* $\mathfrak{gl}^*(n)$, *such that*

$$[Y, \Pi^{(1)}] = -\tilde{\Pi} \ , \quad [Y, \tilde{\Pi}] = 0. \tag{4.27}$$

Proof. Since the functions f_{ij} of Lemma 4.4.5 are analytic Casimir functions for $\mathfrak{gl}(n)$, we have $f_{ij} = \phi_{ij}(F_1, \ldots, F_n)$ where $\phi_{ij}(z_1, \ldots, z_n)$ are analytic functions of n variables. On the other hand, since $\Pi^{(1)}, \tilde{\Pi}$ and $\Pi = \Pi^{(1)} + \tilde{\Pi}$ are Poisson structures, they are compatible, i.e., we have $[\Pi^{(1)}, \tilde{\Pi}] = 0$. Decomposing this relation, we get $\frac{\partial \phi_{ij}}{\partial z_k} + \frac{\partial \phi_{jk}}{\partial z_i} + \frac{\partial \phi_{ki}}{\partial z_j} = 0 \ \forall \ i, j, k$. This is equivalent to the fact that the 2-form $\phi := \sum_{ij} \phi_{ij} dz_i \wedge dz_j$ is closed. By Poincaré's lemma we get $\phi = d\alpha$ with a 1-form $\alpha = \sum_i \alpha_i dz_i$. Then we put $Y := \sum_i \alpha_i(F_1, \ldots, F_n) X_i$. An elementary calculation proves that Y is the desired vector field. $\square$

Return now to the proof of Theorem 4.4.1. Consider a path of Poisson structures given by $\Pi_t := \Pi^{(1)} + t\tilde{\Pi}$. As we have $[Y, \Pi_t] = \tilde{\Pi} = \frac{d}{dt}\Pi_t$, the time-1 map of the vector field Y moves $\Pi^{(1)} = \Pi_0$ into $\Pi = \Pi_1$. This shows that Π is locally analytically linearizable, thus proving our theorem. $\square$

Remark 4.4.8. For any $n \in \mathbb{N}$, the algebra $\mathfrak{aff}(n)$ is a *Frobenius Lie algebra*, in the sense that its coadjoint representation has an open orbit. In other words, its corresponding linear Poisson structure has rank equal to the dimension of the algebra almost everywhere. One may think that there must be some links between the nondegeneracy and the property of being a Frobenius Lie algebra. Unfortunately, the search for new nondegenerate Lie algebras among Frobenius Lie algebras, carried out by Wade and Zung [343], has not brought up any new nondegenerate example so far, though some of the degenerate Frobenius Lie algebras turn out to be finitely determined (see Subsection 4.5.5).

In the above proof of Theorem 4.4.1, we implicitly showed that

$$H^2_{CE}(\mathfrak{aff}(n), \mathcal{S}^k(\mathfrak{aff}(n))) = 0 \ \ \forall \ k \geq 2, \qquad (4.28)$$

where $\mathcal{S}^k$ denotes the symmetric product of order k and H_{CE} denotes the Chevalley–Eilenberg cohomology (it is hidden in the last two lemmas). A purely algebraic proof of this fact was obtained independently by Bordemann, Makhlouf and Petit in [39], who showed that $\mathfrak{aff}(n)$ is infinitesimally strongly rigid, i.e.,

$$H^2_{CE}(\mathfrak{aff}(n), \mathcal{S}^k(\mathfrak{aff}(n))) = 0 \ \ \forall \ k \geq 0. \qquad (4.29)$$

They also verified that

$$H^1_{CE}(\mathfrak{aff}(n), \mathcal{S}^k(\mathfrak{aff}(n))) = 0 \ \forall \ k \geq 1 \qquad (4.30)$$

and

$$H^1_{CE}(\mathfrak{aff}(n), \mathbb{K}) = \mathbb{K}. \qquad (4.31)$$

Since $\mathfrak{aff}(n)$ is a Frobenius Lie algebra, we also have

$$H^0_{CE}(\mathfrak{aff}(n), \mathcal{S}^k(\mathfrak{aff}(n))) = 0 \ \forall \ k \geq 1 \qquad (4.32)$$

(geometrically, it means that $\mathfrak{aff}^*(n)$ can't admit a homogeneous Casimir function of degree $k \geq 1$, because it has an open symplectic leaf), and

$$H^0_{CE}(\mathfrak{aff}(n), \mathbb{K}) = \mathbb{K}. \qquad (4.33)$$

Theorem 4.4.9. *Any finite direct sum* $\mathfrak{l} = \bigoplus \mathfrak{l}_i$, *where each* $\mathfrak{l}_i$ *is either simple or isomorphic to* $\mathfrak{aff}(n_i)$ *for some* $n_i \in \mathbb{N}$, *is formally nondegenerate.*

Proof (sketch). Using the above formulas, Whitehead's lemmas, and Hochschild–Serre spectral sequence, one can show that $H^2_{CE}(\mathfrak{l}, S^k\mathfrak{l}) = 0$ for any $k \geq 1$.   □

Remark 4.4.10. The Lie algebra $\mathfrak{aff}(n) \oplus \mathfrak{g}$ ($n \in \mathbb{N}$, $\mathfrak{g}$ semisimple) is infinitesimally strongly rigid, but the Lie algebra $\mathfrak{aff}(n_1) \oplus \mathfrak{aff}(n_2)$ ($n_1, n_2 \in \mathbb{N}$) is not infinitesimally strongly rigid, because $H^2_{CE}(\mathfrak{aff}(n_1) \oplus \mathfrak{aff}(n_2), \mathbb{K}) = \mathbb{K}$ (see [39]).

Conjecture 4.4.11. Any finite direct sum $\mathfrak{l} = \bigoplus \mathfrak{l}_i$, where each $\mathfrak{l}_i$ is either simple or isomorphic to $\mathfrak{aff}(n_i)$ for some $n_i \in \mathbb{N}$, is analytically nondegenerate.

Theorem 4.4.1 shows that the above conjecture is true when $\mathfrak{l} = \mathfrak{aff}(n)$. It is also true when $\mathfrak{l} = \mathfrak{g} \oplus \mathfrak{aff}(n)$, $\mathfrak{g}$ being semisimple, with the same proof. Another case where we know that the conjecture is true is the following:

Theorem 4.4.12 (Dufour–Molinier [113]). *The direct product*

$$\mathfrak{aff}(1, \mathbb{K}) \times \cdots \times \mathfrak{aff}(1, \mathbb{K})$$

of n copies of $\mathfrak{aff}(1, \mathbb{K})$ is formally and analytically nondegenerate for any natural number n.

Proof (sketch). Denote $\mathfrak{l} = \mathfrak{aff}(1) \times \cdots \times \mathfrak{aff}(1)$ (n times). As discussed above, simple direct computations show that $H^2_{CE}(\mathfrak{l}, \mathcal{S}^k \mathfrak{l}) = 0 \ \forall \ k \geq 2$, so $\mathfrak{l}$ is formally nondegenerate.

Consider now an analytic Poisson structure Π whose linear part $\Pi^{(1)}$ corresponds to $\mathfrak{l}$. Consider the set

$$\Sigma = \{ x \in (\mathbb{K}^{2n}, 0) \mid \operatorname{rank} \Pi(x) < 2n \}$$

of singular points of Π. This set is given by the analytic equation

$$\det(\Pi_{ij}(x))^{2n}_{i,j=1} = 0 \ ,$$

where Π_{ij} are the coefficients of Π in a coordinate system.

When Π is linear, Σ is just a union of n hyperplanes in $\mathbb{K}^{2n}$ in generic position. Since Π is formally linearizable, there are n formal hyperplanes in generic position which are formal solutions of $\det(\Pi_{ij}(x))^{2n}_{i,j=1} = 0$. Applying Artin's theorem [18] about approximation of formal solutions of analytic equations by analytic solutions, we obtain that the equation $\det(\Pi_{ij}(x))^{2n}_{i,j=1} = 0$ admits n local hypersurfaces in generic position near 0 as its solutions. Thus, locally, Σ is a union of n analytic hypersurfaces. So there is a local analytic coordinate system $(x_1, y_1, \ldots, x_n, y_n)$ such that

$$\Sigma = \bigcup_{i=1}^{n} \{ x_i = 0 \} \ .$$

It is easy to see that in such a coordinate system we have

$$\{ x_i, x_j \} = x_i x_j \sigma_{ij}(x, y), \{ x_i, y_j \} = x_i \beta_{ij}(x, y) \ .$$

A first change of coordinates of the type $x'_i = x_i \nu_i(x, y)$, $y'_j = y_j$ leads to $\{ x_i, x_j \} = 0$. Then another change of coordinates of the type $x'_i = x_i$, $y'_j = y_j + b_j(x, y)$ gives $\{ x_i, y_j \} = \delta_{ij} x_i$. Finally, we obtain $\{ y_i, y_j \} = 0$ after a change of coordinates of the type $x'_i = x_i$, $y'_j = y_j + c_j(x)$. $\square$

Remark 4.4.13. It is also shown in [113] that $\mathfrak{aff}(1, \mathbb{R}) \times \mathfrak{aff}(1, \mathbb{R})$ is C^∞-nondegenerate. We don't know if other Lie algebras of the type $\bigoplus_i \mathfrak{aff}(n_i)$ are smoothly nondegenerate or not.

4.5   Some other linearization results

In this section we briefly discuss, mostly without proofs, some other known results about linearization of Poisson structures.

4.5.1   Equivariant linearization

Let Π be a Poisson structure on a manifold P which vanishes at a point $p \in P$, and G be a compact Lie group which acts on P in such a way that the action preserves the Poisson structure Π and fixes the point p. One is interested in G-equivariant linearization of Π, i.e., local coordinate transformations near p which linearize Π and the action of G at the same time. In this direction, we have:

Theorem 4.5.1 (Ginzburg [143]). *Assume that the Poisson structure Π is smoothly linearizable near p. Then it also admits a local smooth G-equivariant linearization near p.*

The above theorem is a consequence of the *rigidity principle* for structure-preserving actions of compact Lie groups. An example of this rigidity principle is the following:

Theorem 4.5.2 ([143]). *Let (P, Π) be a compact Poisson manifold, G a compact Lie group, and $\rho_t : G \times (P, \Pi) \to (P, \Pi)$ a smooth family of Π-preserving actions of G ($t \in [0, 1]$). Then there exists a smooth family of Poisson diffeomorphisms $\phi_t : P \to P$ which sends ρ_t to ρ_0, i.e., such that $\rho_t(g, x) = \phi_t(\rho_0(g, \phi_t^{-1}(x)))$ for all $x \in P$, $g \in G$, and $\phi_0 = Id$.*

The proof of Theorem 4.5.2 uses Moser's path method and is based on the vanishing of the first differentiable cohomology group $H^1(G, \mathcal{Z}_\Pi^1(P))$, where $\mathcal{Z}_\Pi^1(P)$ is the Fréchet space of Poisson vector fields on P. The G-module structure on $\mathcal{Z}_\Pi^1(P)$ is induced from the action of G on P and depends smoothly on the parameter t. See the last section of [143] for details.

4.5.2   Linearization of Poisson–Lie tensors

We refer to Sections 5.1, 5.2 and 5.3 for some basic facts about Poisson–Lie groups and Lie bialgebras used in this subsection.

A *Poisson–Lie tensor* is a Poisson tensor Π on a Lie group G such that the multiplication map $G \times G \to G$ is a Poisson morphism. If Π is a Poisson–Lie tensor on G and e denotes the neutral element of G then we automatically have $\Pi(e) = 0$, so one can talk about the linearization problem for Π et e. What makes the problem of linearization of Poisson–Lie tensors different from general Poisson structures is the fact that a Poisson–Lie tensor Π is completely determined by its linear part p at the neutral element e. More precisely, we have:

Lemma 4.5.3. *A Poisson–Lie tensor* Π *on a Lie group* G *is obtained from the corresponding cocycle* $p : \mathfrak{g} \to \mathfrak{g} \wedge \mathfrak{g}$ *by the following formula:*

$$\Pi(\exp x) = (R_{\exp x})_* \frac{\exp(\mathrm{ad}_x) - 1}{\mathrm{ad}_x} p(x) \quad \forall \ x \in \mathfrak{g} , \tag{4.34}$$

where $R_{\exp x}$ *denotes the right translation by* $\exp x$.

In the above lemma, (G, Π) denotes a Poisson–Lie group, $\mathfrak{g}$ is the Lie algebra of G, and $p : \mathfrak{g} \to \mathfrak{g} \wedge \mathfrak{g}$ is the associated 1-cocycle. The cocycle p may be viewed as a map which assigns to each element $x \in \mathfrak{g}$ the 2-vector $p(x) \in \Lambda^2 T_x \mathfrak{g} \cong \mathfrak{g} \wedge \mathfrak{g}$, and as such it is the linear part of Π at the neutral element e of G. The dual map $p^* : \mathfrak{g}^* \wedge \mathfrak{g}^* \to \mathfrak{g}^*$ is a Lie algebra structure on $\mathfrak{g}^*$, turning $(\mathfrak{g}, \mathfrak{g}^*)$ into a Lie bialgebra.

In fact, Formula (4.34) is true for any multiplicative tensor Π, i.e., it is a consequence of the multiplicativity condition

$$\Pi(gh) = (L_g)_* \Pi(h) + (R_h)_* \Pi(g) . \tag{4.35}$$

It is immediate from Lemma 4.5.3 that if $p = 0$ then Π also vanishes, and if $\mathfrak{g}$ is Abelian then Π is linear.

For Poisson–Lie tensors with a reductive linear part, we have:

Theorem 4.5.4 (Chloup [79]). *If* (G, Π) *is a Poisson–Lie group such that the corresponding Lie algebra structure on* $\mathfrak{g}^*$ *is reductive, then* Π *is locally analytically linearizable near the neutral element* e *of* G.

Chloup's proof of Theorem 4.5.4 follows from Lemma 4.5.3, Conn's Theorem 4.1.3, and the following relatively simple proposition.

Proposition 4.5.5 ([79]). *Consider a Lie bialgebra* $(\mathfrak{g}, \mathfrak{g}^*)$ *such that* $\mathfrak{g} = \mathfrak{g}_1 \oplus \mathfrak{g}_2$ *is the direct sum of a semisimple Lie algebra* $\mathfrak{g}_1$ *with a Lie algebra* $\mathfrak{g}_2$. *Write* $\mathfrak{g}^* = \mathfrak{g}_1^* \oplus \mathfrak{g}_2^*$ *the dual decomposition of* $\mathfrak{g}^*$ *as a vector space, i.e.,* $\langle \mathfrak{g}_1, \mathfrak{g}_2^* \rangle = 0$ *and* $\langle \mathfrak{g}_2, \mathfrak{g}_1^* \rangle = 0$. *Then* $\mathfrak{g}_1^*$ *and* $\mathfrak{g}_2^*$ *are Lie subalgebras of* $\mathfrak{g}^*$.

It is known that in the case when $\mathfrak{g}^*$ is semisimple compact, the linearization of Π can be made global:

Theorem 4.5.6 (Ginzburg–Weinstein [148]). *If* (G, Π) *is a simply connected Poisson–Lie group with a compact semisimple dual Lie algebra* $\mathfrak{g}^*$, *then as a Poisson manifold* (G, Π) *is globally smoothly Poisson-isomorphic to* $\mathfrak{g}$ *(with the linear Poisson structure corresponding to the semisimple Lie algebra structure of* $\mathfrak{g}^*$*).*

Of course, if we replace the word "globally" by the word "locally" in the above theorem, then it becomes a special case of Conn's Theorem 4.1.3. Ginzburg and Weinstein's proof of Theorem 4.5.6 is based on geometric arguments: they show that (G, Π) and $\mathfrak{g}$ have diffeomorphic symplectic foliations, and then use Moser's path method to show the existence of a Poisson isomorphism. Another proof of Theorem 4.5.6 was found by Alekseev in [5], where he showed a natural correspondence between Hamiltonian K-actions and Poisson K-actions which admit a momentum map in the sense of Lu, where K is a simply-connected compact semisimple Poisson–Lie group. (A *Poisson action* of a Poisson–Lie group K on a Poisson manifold M is an action of K on M such that the action map $K \times M \to M$ is a Poisson morphism. A *momentum map* in the sense of Lu for a Poisson K-action on M is a map from M to the dual group K^* which satisfies some natural properties, see Section 5.4.) An explicit Poisson isomorphism from $\mathfrak{g}^*$ to G^* in Ginzburg–Weinstein's theorem is given by Boalch [35] who interpreted $\mathfrak{g}$ as a moduli space of meromorphic connections with an irregular singularity and G^* as the space of monodromy data of these connections.

In particular, let K be a simply-connected compact semisimple Lie group. Denote by G the complexification of K. Then K admits a unique natural Poisson–Lie structure, called the *Iwasawa Poisson–Lie structure*, induced from the Iwasawa decomposition $G = KAN$ of G (see [224] and Section 5.3). The corresponding Poisson–Lie structure on $K^* = AN$ is globally linearizable according to Theorem 4.5.6. On the other hand, the Lie–Poisson structure on K is *not* locally linearizable in general:

Theorem 4.5.7 (Cahen–Gutt–Rawnsley [55]). *If K is a simply-connected compact simple Lie group different from $SU(2)$, then the Iwasawa Poisson–Lie structure on K is not locally linearizable at the neutral element.*

Cahen, Gutt, and Rawnsley proved the above theorem by showing that the symplectic foliation on K is not locally homeomorphic to the symplectic foliation of the corresponding linear Poisson structure.

Some other results about linearization of Poisson–Lie tensors, including examples of linearizable and nonlinearizable Poisson–Lie tensors on nilpotent Lie groups, can be found in [79].

4.5.3   Poisson structures with a hyperbolic $\mathbb{R}^k$-action

The following result on the smooth linearizability of a Poisson structure with a linear part of the type $\mathbb{R}^p \ltimes \mathbb{R}^n$ and a hyperbolic Hamiltonian $\mathbb{R}^p$-action may be viewed as a generalization of Theorem 4.2.2.

Theorem 4.5.8 ([111]). *Let Π be a smooth Poisson structure of maximal rank $2p$ in a neighborhood of 0 in $\mathbb{R}^{n+p}$, $n \geq p \geq 1$, $\Pi(0) = 0$. Suppose that there are p functions $f_1, \ldots, f_p$, $f_i(0) = 0$, such that $\{f_i, f_j\} = 0 \ \forall \ i, j = 1, \ldots, p$, $\mathrm{d}f_1 \wedge \cdots \wedge \mathrm{d}f_p(0) \neq 0$. Suppose moreover that the infinitesimal $\mathbb{R}^p$-action ρ generated by the Hamiltonian vector fields $X_i = X_{f_i}$ is hyperbolic on the submanifold $K = \{f_1 = \cdots = f_p = 0\}$ in the following sense: the weights $\lambda_1, \ldots, \lambda_n$ of the linear part $\rho^{(1)}$ of ρ (i.e., linear maps $\lambda_i : \mathbb{R}^p \longrightarrow \mathbb{C}$ which assign to each $a \in \mathbb{R}^p$ an eigenvalue of the linear automorphism $\rho^{(1)}(a)$ of $\mathbb{R}^n$), satisfy the following two properties:*

a) *The real parts of λ_i are p by p independent (we don't distinguish the real parts of λ_i and $\bar{\lambda}_i$).*
b) *There is no resonance relation of the type $\lambda_i = \sum_{j=1}^{n} k_j \lambda_j$, where k_j are non-negative integers with $k_1 + \cdots + k_n \geq 2$.*

Then Π is smoothly linearizable in a neighborhood of 0 in $\mathbb{R}^{n+p}$. Moreover, there are local smooth coordinates $x_1, \ldots, x_n, u_1, \ldots, u_p$ such that

$$\{x_i, x_j\} = \{u_r, u_s\} = 0, \quad \{x_i, u_r\} = \sum_{l=1}^{n} c_{ir}^l x_l \ \forall \ i, j, r, s \ ,$$

and the functions f_i depend only on the variables $u_1, \ldots, u_p$.

We refer the curious reader to [111] for a proof of the above theorem. A key ingredient in the proof is Chaperon's smooth linearization theorem [73] for hyperbolic actions of commutative groups. A generalization of Theorem 4.5.8 in the analytic case can be found in [324].

4.5.4   Transverse Poisson structures to coadjoint orbits

In [346], Weinstein erroneously claimed that the transverse Poisson structure to any coadjoint orbit in the dual of a finite-dimensional Lie algebra is linearizable. Molino [259] found out that Weinstein's proof needs an additional condition, namely the orbit must be *reductive*. More precisely, we have:

Theorem 4.5.9 ([259]). *Let $\mathfrak{g}$ be a Lie algebra, and $\mu \in \mathfrak{g}^*$ be a point in the dual of $\mathfrak{g}$ equipped with the corresponding linear Poisson structure. Denote by $\mathfrak{g}_\mu = \{x \in \mathfrak{g} \mid \mathrm{ad}_x^* z = 0\} \subset$ the isotropy algebra of μ. If there is a linear complement $\mathfrak{m}$ of $\mathfrak{g}_\mu$ in $\mathfrak{g}$, i.e., $\mathfrak{g} = \mathfrak{g}_\mu + \mathfrak{m}$ as a vector space, such that $[\mathfrak{g}_\mu, \mathfrak{m}] \subset \mathfrak{m}$, then the transverse Poisson structure at μ is linearizable.*

The above theorem follows directly from Dirac's formula (1.56): just take the affine space $N = \{\nu \in \mathfrak{g}^* \mid \langle x, \nu \rangle = \text{constant} \ \forall x \in \mathfrak{m}\}$. N is transverse to the coadjoint orbit at μ, and the transverse Poisson structure on N is linear with respect to the affine functions on N which vanish at μ.

If one drops the condition $[\mathfrak{g}_\mu, \mathfrak{m}] \subset \mathfrak{m}$ from Theorem 4.5.9 then it is false. The following simple example of a nonlinearizable transverse Poisson structure to a coadjoint orbit is provided by Givental (see the discussion in [347]): Identify $\mathfrak{g} = sl(3, \mathbb{R})$ with its dual via the Killing form, and consider the nilpotent element

$$\mu_0 = \begin{pmatrix} 0 & 0 & 1 \\ 0 & 0 & 0 \\ 0 & 0 & 0 \end{pmatrix}.$$

The adjoint orbit of μ_0 is four-dimensional. The transverse Poisson structure at μ_0 is a Poisson structure of rank 2 living in a four-dimensional space, with two

Casimir functions inherited from $sl(3, \mathbb{R})$. In an appropriate linear coordinate system (p, q, r, s) centered at μ_0, these Casimir functions have the form $3p^2 + r$ and $-2p^3 + qs + 2pr$; their common zero level set is a surface with an isolated singularity (of type A_2 in Arnold's classification). On the other hand, the linear part of this transverse structure corresponds to the centralizer $\mathfrak{g}_{\mu_0} = \{x \in \mathfrak{g} \mid [x, \mu_0] = 0\}$ of μ_0, which consists of the matrices

$$\begin{pmatrix} \alpha & \beta & \gamma \\ 0 & -2\alpha & \delta \\ 0 & 0 & \alpha \end{pmatrix}.$$

Thinking of $(\alpha, \beta, \gamma, \delta)$ as a basis of $\mathfrak{g}_{\mu_0}^*$ and denoting by (a, b, c, d) the dual basis of $\mathfrak{g}_{\mu_0}$ considered as linear functions on $\mathfrak{g}_{\mu_0}^*$, the common zero level set of the two Casimir functions c and $ac + 3bd$ on $\mathfrak{g}_{\mu_0}^*$ is a union of two 2-planes which intersect along a line. Just, for topological reasons, the transverse Poisson structure at μ_0 can't be isomorphic to its linear part.

An even simpler example is given by M. Duflo (see [347]): Consider the following four-dimensional linear Poisson structure:

$$\frac{\partial}{\partial x_1} \wedge \left(x_2 \frac{\partial}{\partial x_2} + x_3 \frac{\partial}{\partial x_3} + 2x_4 \frac{\partial}{\partial x_4} \right) + x_4 \frac{\partial}{\partial x_2} \wedge \frac{\partial}{\partial x_3}.$$

The corresponding Lie algebra is a Frobenius Lie algebra. The singular locus is $\{x_4 = 0\}$. The points with $x_4 = 0$ and $x_2^2 + x_3^2 > 0$ have Abelian isotropy. But of course the transverse Poisson structure is not trivial.

Remark 4.5.10. Damianou [92] proved in some special cases, and then Cushman and Roberts [89] proved in the general case, that the transverse Poisson structure to an arbitrary coadjoint orbit in the dual of a semisimple Lie algebra is *polynomial*, i.e., the Poisson tensor has polynomial coefficients in a linear system of coordinates. Probably, most of these polynomial Poisson structures (except the ones that fit into Theorem 4.5.9) are nonlinearizable.

Remark 4.5.11. Some results about transverse Poisson structures to coadjoint orbits of *infinite*-dimensional Lie algebras, e.g., the Virasoro algebra and the Adler–Gelfand–Dickey bracket, can be found in [289, 235, 236]

4.5.5 Finite determinacy of Poisson structures

Given a formal (resp. analytic, resp. smooth) Poisson structure $\Pi = \Pi^{(1)} + \Pi^{(2)} + \cdots$, we will say that it is formally (resp. analytically, resp. smoothly) *finitely determined* if there is a natural number k such that any other formal (resp. analytically, resp. smooth) Poisson structure $\Pi_1 = \Pi_1^{(1)} + \Pi_1^{(2)} + \cdots$ such that $\Pi_1^{(l)} = \Pi^{(l)} \; \forall \, l \leq k$ is formally (resp. analytically, resp. smoothly) locally isomorphic to Π.

In particular, if a Lie algebra $\mathfrak{g}$ is formally degenerate, one may still ask if it is *finitely determined*, in the sense that the space of formal Poisson structures which have $\mathfrak{g}$ as their linear part, modulo formal isomorphisms, is of finite dimension. If $\mathfrak{g}$ is finitely determined, then there is a natural number k (depending on $\mathfrak{g}$) such that any formal Poisson structure

$$\Pi = \Pi^{(1)} + \Pi^{(l)} + \Pi^{(l+1)} + \cdots$$

with $l > k$, where $\Pi^{(1)}$ corresponds to $\mathfrak{g}$, is formally linearizable. More generally, any two formal Poisson structures with the same linear part $\Pi^{(1)}$ and which coincide up to degree k are formally isomorphic.

It is clear that a sufficient condition for a Lie algebra $\mathfrak{g}$ to be finitely determined is the inequality

$$\dim H^2_{CE}(\mathfrak{g}, \mathcal{S}\mathfrak{g}) < \infty . \tag{4.36}$$

Example 4.5.12. The linear three-dimensional Poisson structure

$$\Lambda = \frac{\partial}{\partial x} \wedge (y\frac{\partial}{\partial y} + 2z\frac{\partial}{\partial z})$$

does not satisfy the conditions of Theorem 4.2.2 because of a resonance. In fact, it is a degenerate, but finitely determined Poisson structure: any formal (resp. analytic, resp. smooth) Poisson structure whose linear part is Λ is formally (resp. analytically, resp. smoothly) isomorphic to a Poisson structure of the type

$$\Pi = \frac{\partial}{\partial x} \wedge (y\frac{\partial}{\partial y} + 2z\frac{\partial}{\partial z} + cy^2\frac{\partial}{\partial z}),$$

where c is a constant.

Example 4.5.13. Consider a Poisson structure $\Pi = \Pi^{(1)} + \Pi^{(2)} + \cdots$, whose linear part is of the type

$$\Pi^{(1)} = \frac{\partial}{\partial x_0} \wedge \left(\sum_{i,j=1}^{n} a_{ij}x_i\frac{\partial}{\partial x_j} \right), \tag{4.37}$$

where the matrix $A = (a_{ij})$ is nonresonant in the sense that its eigenvalues $\gamma_1, \ldots, \gamma_n$ do not satisfy any nontrivial relation of the type

$$\lambda_i + \lambda_j = \sum_{k=1}^{n} \alpha_k\lambda_k \tag{4.38}$$

with $1 \leq i \leq j \leq n$, $\alpha_k \in \mathbb{N} \cup \{0\}$ (i.e., except the relations $\lambda_i + \lambda_j = \lambda_i + \lambda_j$). Such a Poisson structure Π is called nonresonant in [116]. Simple homological computations show that Π admits a formal nonhomogeneous quadratic normal form. See [116] for more details, and also a smooth nonhomogeneous quadratization result.

Example 4.5.14 ([343]). Consider the following six-dimensional linear Poisson structure:

$$\Pi_1 = \partial x_1 \wedge Y_1 + \partial x_2 \wedge Y_2 + \Lambda \qquad (4.39)$$

where

$$Y_1 = y_1 \partial y_1 + 2y_2 \partial y_2 + 3y_3 \partial y_3 + 4y_4 \partial y_4,$$
$$Y_2 = y_2 \partial y_2 + y_3 \partial y_3 + y_4 \partial y_4,$$
$$\Lambda = \partial y_1 \wedge (y_3 \partial y_2 + y_4 \partial y_3).$$

The corresponding solvable Lie algebra $\mathfrak{p} = \mathbb{K}^2 \ltimes L_4$, where L_4 is the nilpotent Lie algebra corresponding to Λ, is a six-dimensional Frobenius rigid solvable Lie algebra (see [151]) which is not strongly rigid. Lengthy direct computations, using the Hochschild–Serre formula and with the aid of MAPLE, show that $H^2(\mathfrak{p}, \mathcal{S}^k \mathfrak{p}) = \mathbb{K}$ for $k = 3, 4, 5$ and $H^2(\mathfrak{p}, \mathcal{S}^k \mathfrak{p}) = 0$ for all other values of $k \geq 1$. Thus $\mathfrak{p}$ is finitely determined. It is degenerate: the nonlinear Poisson tensor

$$\Pi = \Pi_1 + y_1^2 y_2 \frac{\partial}{\partial y_1} \wedge \frac{\partial}{\partial y_3} - y_1 y_2^2 \frac{\partial}{\partial y_2} \wedge \frac{\partial}{\partial y_3}$$

is not equivalent to its linear part, because the singular loci of Π and Π_1 are not locally isomorphic.

Chapter 5

Multiplicative and Quadratic Poisson Structures

After studying linearization and normal forms for Poisson structures with a non-trivial linear part in the previous two chapters, it is a logical next step to talk about Poisson structures which begin with a quadratic term. This leads us to the study of (homogeneous) quadratic Poisson structures, and then to another important subject of Poisson geometry, namely Poisson–Lie groups. The reason is that both Poisson–Lie groups and quadratic Poisson structures are in a sense multiplicative, and they both arise, most of the times, from classical r-matrices. For example, if we have a multiplicative Poisson structure on a finite-dimensional unital associative algebra (say the algebra $Mat_n(\mathbb{R})$ of $n \times n$-matrices), then it is a quadratic Poisson structure, and at the same time a Poisson–Lie structure on the group of reversible elements of the algebra. In this chapter, we will give a brief introduction to the theory of Poisson–Lie groups, and then discuss some results on quadratic Poisson structures and the quadratization problem.

5.1 Multiplicative tensors

Let G be a Lie group. A multi-vector field $\Lambda \in \mathcal{V}^\star(G)$, or more generally, a tensor field on G, is called a *multiplicative tensor field*, if Λ satisfies the following equation:

$$\Lambda(gh) = (L_g)_*\Lambda(h) + (R_h)_*\Lambda(g) \quad \forall \ g, h \in G, \tag{5.1}$$

where L_g denotes the left translation $h \mapsto gh$ and R_h denotes the right translation $g \mapsto gh$.

Exercise 5.1.1. Show that, if Λ is a multiplicative 2-vector field on a Lie group G, then the inversion map $g \mapsto g^{-1}$ of G sends Π to $-\Pi$.

If Λ is multiplicative, then it is clear that $\Lambda(e) = 0$, where e denotes the neutral element of G (put $g = h = e$ in Equation (5.1)). In particular, we can talk about the linear part $\Lambda^{(1)}$ of Λ at e.

Theorem 5.1.2 ([107, 346, 224]).

a) *A multi-vector field Λ on a connected Lie group G is multiplicative if and only if $\Lambda(e) = 0$ and $\mathcal{L}_X \Lambda$ is left-invariant for any left-invariant vector field X on G.*

b) *The Schouten bracket of two multiplicative multi-vector fields is a multiplicative multi-vector field.*

Proof. a) Let X be a left-invariant vector field. Then its flow is the one-dimensional group of right translations $R_{\exp(tX)}$. If Λ is multiplicative then we have

$$
\begin{aligned}
\mathcal{L}_X \Lambda(g) &= \left.\frac{\mathrm{d}}{\mathrm{d}t}\right|_{t=0} R_{\exp(-tX)}\Lambda(g.\exp(tX)) \\
&= \left.\frac{\mathrm{d}}{\mathrm{d}t}\right|_{t=0} R_{\exp(-tX)}[R_{\exp(tX)}\Lambda(g) + L_g\Lambda(\exp(tX))] \\
&= \left.\frac{\mathrm{d}}{\mathrm{d}t}\right|_{t=0} R_{\exp(-tX)}L_g\Lambda(\exp(tX)) \\
&= \left.\frac{\mathrm{d}}{\mathrm{d}t}\right|_{t=0} L_g R_{\exp(-tX)}\Lambda(\exp(tX)) \\
&= L_g\left(\left.\frac{\mathrm{d}}{\mathrm{d}t}\right|_{t=0} R_{\exp(-tX)}\Lambda(\exp(tX))\right) = L_g(\mathcal{L}_X\Lambda(e)).
\end{aligned}
$$

Thus $\mathcal{L}_X \Lambda$ is left-invariant if Λ is multiplicative and X is left-invariant. Conversely, if $\mathcal{L}(e) = 0$ and $\mathcal{L}_X\Lambda$ is left-invariant for any left-invariant X, then

$$
\frac{\mathrm{d}}{\mathrm{d}t}\left(R_{\exp(-tX)}\Lambda(g.\exp(sX+tX)) + L_{g.\exp(sX+tX)}\Lambda(\exp(-tX))\right) = 0,
$$

implying (by evaluating the above expression at $t = 0$ and $t = -s$) that

$$
\Lambda(g.\exp(sX)) = R_{\exp(sX)}\Lambda(g) + L_g\Lambda(\exp(sX)).
$$

Since G is connected, the last equation implies that Λ is multiplicative. Recall that left translations are generated by right-invariant vector fields. Hence Assertion a) of the lemma may be restated as follows: Λ is multiplicative if and only if

$$
\mathcal{L}_Y\mathcal{L}_X\Lambda = \mathcal{L}_X\mathcal{L}_Y\Lambda = 0 \tag{5.2}
$$

for any left-invariant vector field X and right-invariant vector field Y.

b) Let Λ_1 and Λ_2 be two multiplicative multi-vector fields. We have to show that

$$
\mathcal{L}_X\mathcal{L}_Y[\Lambda_1, \Lambda_2] = 0 \tag{5.3}
$$

for any left-invariant vector field X and right-invariant vector field Y.

Note that $\mathcal{L}_X \Lambda_1$ is left-invariant and $\mathcal{L}_Y \Lambda_2$ is right-invariant. Since right translations commute with left translations, left-invariant vector fields (which are generators of right translations) commute with right-invariant vector fields. By the Leibniz rule, it follows that the Schouten bracket of a left-invariant multi-vector field with a right-invariant multi-vector field vanishes. In particular, we have $[\mathcal{L}_X \Lambda_1, \mathcal{L}_Y \Lambda_2] = 0$. Similarly, $[\mathcal{L}_Y \Lambda_1, \mathcal{L}_X \Lambda_2] = 0$. Hence

$$\mathcal{L}_X \mathcal{L}_Y [\Lambda_1, \Lambda_2]$$
$$= [\mathcal{L}_X \mathcal{L}_Y \Lambda_1, \Lambda_2] + [\mathcal{L}_X \Lambda_1, \mathcal{L}_Y \Lambda_2] + [\mathcal{L}_Y \Lambda_1, \mathcal{L}_X \Lambda_2] + [\Lambda_1, \mathcal{L}_X \mathcal{L}_Y \Lambda_2]$$
$$= 0. \qquad \square$$

By left translations, we can identify $\wedge^\star TG$ with $G \times \wedge^\star \mathfrak{g}$, and if Λ is a k-vector field on G, we may view Λ as a map from G to $\wedge^k \mathfrak{g}$ via this identification. Then $\Lambda^{(1)}$ may be identified with the differential of Λ at e:

$$\Lambda^{(1)} = \mathrm{d}_e \Lambda : \mathfrak{g} \to \wedge^k \mathfrak{g}. \tag{5.4}$$

Theorem 5.1.3 ([224]). *If Λ is a multiplicative k-vector field on G, then $\mathrm{d}_e \Lambda$ is a 1-cocycle of the Chevalley–Eilenberg complex of the adjoint action of $\mathfrak{g}$ on $\wedge^k \mathfrak{g}$. Conversely, if G is a connected simply-connected Lie group, then for any 1-cocycle $p : \mathfrak{g} \to \wedge^k \mathfrak{g}$, there is a unique multiplicative k-vector field Λ on G such that $\mathrm{d}_e \Lambda = p$.*

Proof. Assume that Λ is a multiplicative k-vector field on G. Let $x, y \in \mathfrak{g}$ be two elements of $\mathfrak{g}$, considered as left-invariant vector fields on G. Denote $p = \mathrm{d}_e \Lambda$. We have to show that $p([x, y]) = \mathrm{ad}_x p(y) - \mathrm{ad}_y p(x)$.

By virtue of Theorem 5.1.2, we have:

$$
\begin{aligned}
\mathrm{ad}_x p(y) &= \mathrm{ad}_x \mathcal{L}_y \Lambda(e) = \left.\frac{\mathrm{d}}{\mathrm{d}t}\right|_{t=0} \mathrm{Ad}_{\exp(tx)}(\mathcal{L}_y \Lambda(e)) \\
&= \left.\frac{\mathrm{d}}{\mathrm{d}t}\right|_{t=0} R_{\exp(-tx)} L_{\exp(tx)}(\mathcal{L}_y \Lambda(e)) \\
&= \left.\frac{\mathrm{d}}{\mathrm{d}t}\right|_{t=0} R_{\exp(-tx)}(\mathcal{L}_y \Lambda(\exp(tx))) = (\mathcal{L}_x \mathcal{L}_y \Lambda)(e),
\end{aligned}
$$

hence $\mathrm{ad}_x p(y) - \mathrm{ad}_y p(x) = (\mathcal{L}_x \mathcal{L}_y \Lambda)(e) - (\mathcal{L}_y \mathcal{L}_x \Lambda)(e) = (\mathcal{L}_{[x,y]} \Lambda)(e) = p([x, y])$, which means that p is a 1-cocycle.

Conversely, assume that $p : \mathfrak{g} \to \wedge^k \mathfrak{g}$ is a 1-cocycle, and that G is connected simply-connected. The following construction of Λ from p is taken from [198]: Define the following 1-form $\widetilde{p}$ on G with values in $\wedge^k \mathfrak{g}$:

$$\widetilde{p}(X) = \mathrm{Ad}_g(p(L_{g^{-1}} X)) \quad \forall g \in G, X \in T_g G. \tag{5.5}$$

It is clear that $\widetilde{p}$ is an equivariant 1-form:

$$(L_g)^* \widetilde{p} = \mathrm{Ad}_g \circ p. \tag{5.6}$$

One verifies directly that, because p is a cocycle, $\widetilde{p}$ is a closed 1-form: $\mathrm{d}\widetilde{p} = 0$. Since G is simply-connected, $\widetilde{p}$ is exact, i.e., there is a unique $\wedge^k\mathfrak{g}$-valued function $P : G \to \wedge^k\mathfrak{g}$ such that $\mathrm{d}P = \widetilde{p}$ and $P(e) = 0$. Define the k-vector field Λ as

$$\Lambda(g) = (R_g)_* P(g). \tag{5.7}$$

It is clear that Λ is multiplicative if P satisfies the equation

$$P(gh) - P(g) = \mathrm{Ad}_g P(h) \quad \forall \; g, h \in G. \tag{5.8}$$

But this last equation follows from Equation (5.6), which is its infinitesimal version (for h near e), and the connectedness of G. $\qquad\square$

5.2   Poisson–Lie groups and r-matrices

Definition 5.2.1. A Lie group G equipped with a Poisson structure Π is called a *Poisson–Lie group* if the multiplication map $(G, \Pi) \times (G, \Pi) \to (G, \Pi)$, $(g, h) \mapsto gh$, is a Poisson morphism.

Equivalently, Π is a *Poisson–Lie structure* on a Lie group G if it is a multiplicative Poisson tensor on G, i.e., Π satisfies the equation

$$\Pi(gh) = (L_g)_* \Pi(h) + (R_h)_* \Pi(g) \quad \forall \; g, h \in G. \tag{5.9}$$

Poisson–Lie groups form a category: a morphism between two Poisson–Lie groups is a group homomorphism which is a Poisson morphism at the same time.

A Poisson–Lie group (G, Π) is called *exact* or *coboundary* if

$$\Pi = r^+ - r^- \tag{5.10}$$

for some $r \in \mathfrak{g} \wedge \mathfrak{g}$, where $\mathfrak{g}$ is the Lie algebra of G. Here, as usual, r^+ (resp. r^-) denotes the left-invariant (resp. right-invariant) tensor field whose value at the neutral element e of G is $r \in \mathfrak{g} \wedge \mathfrak{g} \cong \wedge^2 T_e G$. (See Notation 1.8.10.)

Lemma 5.2.2. *Any 2-vector field of the type $\Pi = r^+ - r^-$ with $r \in \mathfrak{g} \wedge \mathfrak{g}$ is multiplicative, i.e., satisfies Equation (5.9). Conversely, if G is a connected semisimple Lie group, then any multiplicative 2-vector field Π on G is of the type $\Pi = r^+ - r^-$ for some $r \in \mathfrak{g} \wedge \mathfrak{g}$.*

Proof. Notice that $R_{gh} = R_h R_g$ and $L_{gh} = L_g L_h$. Moreover, left translations commute with right translations, i.e., $R_g L_h = L_h R_g$. Hence we have: $\Pi(gh) = L_{gh} r - R_{gh} r = (R_h L_g r - R_{gh} r) + (L_{gh} r - L_g R_h r) = R_h(L_g r - R_g r) + L_g(L_h r - R_h r) = R_h \Pi(g) + L_g \Pi(h)$. Notice that the linear part of $r^+ - r^-$ at e is given by the 1-coboundary $\delta r : x \mapsto [x, r]$ ($x \in \mathfrak{g}$) of the Chevalley–Eilenberg complex of the $\mathfrak{g}$-module $\mathfrak{g} \wedge \mathfrak{g}$.

Conversely, if $\mathfrak{g}$ is semisimple, Π is multiplicative and $p = d_e \Pi : \mathfrak{g} \to \mathfrak{g} \wedge \mathfrak{g}$ its corresponding 1-cocycle, then by Whitehead's lemma, $H^1(\mathfrak{g}, \mathfrak{g} \wedge \mathfrak{g}) = 0$, and this 1-cocycle is a coboundary. In other words, $\exists \; r \in \mathfrak{g} \wedge \mathfrak{g}$ such that $p = \delta r$. By uniqueness (see Theorem 5.1.3), Π must be equal to $r^+ - r^-$. $\qquad\square$

Denote by $(\wedge^3 \mathfrak{g})^{\mathfrak{g}}$ the set of elements of $\wedge^3 \mathfrak{g}$ which are invariant under the adjoint action of $\mathfrak{g}$. In other words,

$$(\wedge^3 \mathfrak{g})^{\mathfrak{g}} = \{\alpha \in \wedge^3 \mathfrak{g} \mid [x, \alpha] = 0 \ \forall \ x \in \mathfrak{g}\},$$

where $[x, \alpha]$ is the Schouten bracket of x with α.

Theorem 5.2.3 ([107]). *A multiplicative 2-vector field $\Pi = r^+ - r^-$ with $r \in \mathfrak{g} \wedge \mathfrak{g}$ on a connected Lie group G with Lie algebra $\mathfrak{g}$ is a Poisson–Lie structure if and only if r satisfies the following condition:*

$$[r, r] \in (\wedge^3 \mathfrak{g})^{\mathfrak{g}}. \tag{5.11}$$

Proof. Recall that left-invariant multi-vector fields commute with right-invariant multi-vector fields. In particular, we have $[r^+, r^-] = 0$.

By Lemma 1.8.9, we have $[r^+, r^+] = [r, r]^+$. Similarly, $[r^-, r^-] = -[r, r]^-$. (The sign minus is due to the fact that $x \mapsto x^-$ is a Lie algebra *anti*-homomorphism from $\mathfrak{g}$ to $\mathcal{V}^1(G)$.) Hence we have

$$[\Pi, \Pi] = [r^+ - r^-, r^+ - r^-] = [r^+, r^+] + [r^-, r^-] - 2[r^+, r^-] = [r, r]^+ - [r, r]^-.$$

Thus, $[\Pi, \Pi] = 0$ if and only if $L_g[r, r] = R_g[r, r] \ \forall g \in G$. The last equation means that $[r, r]$ is invariant under the adjoint action of $\mathfrak{g}$. $\qquad \square$

In particular, if r satisfies the *CYBE* (*classical Yang–Baxter equation*)

$$[r, r] = 0, \tag{5.12}$$

then $\Pi = r^+ - r^-$ is an exact Poisson–Lie structure. For example, if $x, y \in \mathfrak{g}$ are two commuting elements, $[x, y] = 0$, then $r = x \wedge y \in \mathfrak{g} \wedge \mathfrak{g}$ satisfies the CYBE (5.12) and gives rise to a coboundary Poisson–Lie structure on G.

If an element $r \in \mathfrak{g} \wedge \mathfrak{g}$ satisfies Condition (5.11), then it is called a *classical r-matrix* of $\mathfrak{g}$. Condition (5.11) is called the *generalized classical Yang–Baxter equation* (*gCYBE*). If r satisfies the CYBE (5.12), then it is called a *triangular r-matrix*. The Poisson bracket of an exact Poisson–Lie structure $\Pi = r^+ - r^-$, where r is a classical r-matrix, is often called the *Sklyanin bracket*.

Remark that the CYBE (5.12) is preserved under homomorphisms: if r is a classical r-matrix of $\mathfrak{g}$, and $\rho : \mathfrak{g} \to \mathfrak{l}$ is a Lie algebra homomorphism, then $\rho(r)$ is a classical r-matrix of $\mathfrak{l}$.

The CYBE (5.12) is also often written as

$$[r^{12}, r^{13}] + [r^{12}, r^{23}] + [r^{13}, r^{23}] = 0. \tag{5.13}$$

In the above equation, $\mathfrak{g}$ may be viewed as a subalgebra of $End(V)$ where V is a vector space, and r is viewed as an anti-symmetric element of $End(V) \otimes End(V)$, which acts on $V \otimes V$. r^{12} is the endomorphism of $V_1 \otimes V_2 \otimes V_3$, where $V_1 = V_2 = V_3 = V$, which acts on $V_1 \otimes V_2$ by r and on V_3 by the identity map, i.e., $r^{12}(v_1 \otimes v_2 \otimes v_3) = r(v_1 \otimes v_2) \otimes v_3$. r^{13} and r^{23} are defined similarly. More intrinsically, one can replace $End(V)$ by the universal enveloping algebra $\mathcal{U}\mathfrak{g}$ of $\mathfrak{g}$. Equation (5.13) makes sense also for non-antisymmetric elements $r \in \mathfrak{g} \otimes \mathfrak{g}$, and a non-antisymmetric solution of (5.13) is called a *quasi-triangular r-matrix*.

Exercise 5.2.4. Show that, for an element $r \in \mathfrak{g} \wedge \mathfrak{g}$, Equation (5.13) is equivalent to Equation (5.12).

Suppose now that $\mathfrak{g}$ is a simple Lie algebra, and identify it with its dual $\mathfrak{g}^*$ via the Killing form $\langle x, y \rangle = trace(\mathrm{ad}_x \mathrm{ad}_y)$. On $\wedge^3 \mathfrak{g}$, we have the induced ad-invariant scalar product: if $(e_1, \ldots, e_m)$ is an orthonormal basis of $\mathfrak{g}$ then $(e_i \wedge e_j \wedge e_k, i < j < k)$ is an orthonormal basis of $\wedge^3 \mathfrak{g}$. The equation

$$\langle \eta, x \wedge y \wedge z \rangle = \langle [x, y], z \rangle \quad \forall\, x, y, z \in \mathfrak{g} \tag{5.14}$$

defines an element $\eta \in \wedge^3 \mathfrak{g}$, which is obviously ad-invariant, i.e., $\eta \in (\wedge^3 \mathfrak{g})^{\mathfrak{g}}$. In fact, $(\wedge^3 \mathfrak{g})^{\mathfrak{g}}$ is one-dimensional and is generated by η. Thus, an element $r \in \mathfrak{g} \wedge \mathfrak{g}$, where $\mathfrak{g}$ is simple, is a classical r-matrix if and only if it satisfies the following equation:

$$\langle [r, r], x \wedge y \wedge z \rangle = \lambda \langle [x, y], z \rangle \quad \forall\, x, y, z \in \mathfrak{g}, \tag{5.15}$$

where λ is a constant. Equation (5.15) is called the *modified classical Yang–Baxter equation (mCYBE)* of coefficient λ (when $\lambda \neq 0$; when $\lambda = 0$ it is the CYBE). If $\mathfrak{g}$ is simple and $r \in \mathfrak{g} \wedge \mathfrak{g}$ is a solution of the mCYBE, then there is a unique number c such that $r + c\Gamma$ is a quasi-triangular r-matrix, where $\Gamma \in \mathcal{U}\mathfrak{g}$ denotes the Casimir element of $\mathfrak{g}$, and vice versa.

A complete list of solutions of the mCYBE (5.15) for a complex simple Lie algebra was obtained by Belavin and Drinfeld [25].

Theorem 5.2.5 (Belavin–Drinfeld). *Let $\mathfrak{g}$ be a complex simple Lie algebra, and $r \in \mathfrak{g} \wedge \mathfrak{g}$ be a solution of the mCYBE (5.15), with $\lambda \neq 0$. Then r can be written as*

$$r = r_0 + \sqrt{-\lambda} \left(\sum_{\alpha \in \Delta^+} E_{-\alpha} \wedge E_\alpha + 2 \sum_{\alpha \in \hat{\Gamma}_+,\, \beta > \alpha} E_{-\beta} \wedge E_\alpha \right) \tag{5.16}$$

where:
(1) *Δ^+ is a system of positive roots of $\mathfrak{g}$ with respect to a Cartan subalgebra $\mathfrak{h}$ of $\mathfrak{g}$.*
(2) *(H_α, E_α) is a Weyl basis of $\mathfrak{g}$ with $\langle E_\alpha, E_{-\alpha} \rangle = 1$.*
(3) *Γ_+ is a subset of the set Φ of simple roots corresponding to Δ^+, $\hat{\Gamma}_+$ is the set of positive roots which can be written as a linear combination with integral coefficients of simple roots in Γ_+.*
(4) *There is another subset Γ_- of the set Φ of simple roots, and a map $\tau : \Gamma_+ \to \Gamma_-$ such that $\langle \tau(\alpha), \tau(\beta) \rangle = \langle \alpha, \beta \rangle \ \forall\, \alpha, \beta \in \Gamma_+$ where $\langle \alpha, \beta \rangle = \langle H_\alpha, H_\beta \rangle$. For any $\alpha \in \Gamma_+$ there is a positive integer k such that $r(\tau(\alpha), \beta) = r(\alpha, \beta) - k(\langle \alpha, \beta \rangle + \langle \tau(\alpha), \beta \rangle) \ \forall\, \alpha \in \Gamma_+, \beta \in \Phi$. The notation $\beta > \alpha$ means that $\beta = \tau^l(\alpha)$ for some $l \geq 1$.*
(5) *$r_0 \in \wedge^2 \mathfrak{h}$ is determined by $Q(\alpha, \beta)$, $\alpha, \beta \in \Phi$, such that*

$$Q(\tau(\alpha), \beta) - Q(\alpha, \beta) = -\sqrt{-\lambda}(\langle \alpha, \beta \rangle + \langle \tau(\alpha), \beta \rangle). \tag{5.17}$$

The proof of the above theorem is not very difficult, but lengthy. See, e.g., [56, 78], for a proof, where the case of real simple Lie algebras was also studied. □

Example 5.2.6. $r = \sum_{\alpha \in \Delta^+} E_{-\alpha} \wedge E_\alpha$ is a classical r-matrix, called the standard r-matrix.

There is another way to express the Yang–Baxter equations. To show it, we need the following lemma:

Lemma 5.2.7. *Let $\mathfrak{g}$ be an arbitrary Lie algebra, $r \in \mathfrak{g} \wedge \mathfrak{g}$ and $\alpha, \beta, \gamma \in \mathfrak{g}^*$. Then we have:*

$$\frac{1}{2}\langle [r, r], \alpha \wedge \beta \wedge \gamma \rangle = \langle [r\alpha, r\beta], \gamma \rangle + \langle [r\beta, r\gamma], \alpha \rangle + \langle [r\gamma, r\alpha], \beta \rangle, \qquad (5.18)$$

where $r\alpha \in \mathfrak{g}$ denotes the interior product of r with α: $\langle r\alpha, \zeta \rangle = \langle r, \alpha \wedge \zeta \rangle$.

Proof. Direct verification using a basis of $\mathfrak{g}$. □

Suppose that $\mathfrak{g}$ admits an ad-invariant nondegenerate scalar product. Then an element $r \in \mathfrak{g} \wedge \mathfrak{g}$ can be identified with an element $R \in End(\mathfrak{g})$ such that $\langle Rx, y \rangle = \langle r, x \wedge y \rangle \ \forall \ x, y \in \mathfrak{g}$. Note that R is antisymmetric: $\langle Rx, y \rangle = -\langle Ry, x \rangle$. Formula (5.18) can then be rewritten as follows:

$$\frac{1}{2}\langle [r, r], x \wedge y \wedge z \rangle = \langle [Rx, Ry], z \rangle + \langle [Ry, Rz], x \rangle + \langle [Rz, Rx], y \rangle$$
$$= \langle [Rx, Ry] - R([Rx, y] + [x, Ry]), z \rangle. \qquad (5.19)$$

Hence, in terms of $R \in End(\mathfrak{g})$, the CYBE (5.12) can be rewritten as

$$[Rx, Ry] - R([Rx, y] + [x, Ry]) = 0, \qquad (5.20)$$

while the mCYBE (5.15) can be rewritten as

$$[Rx, Ry] - R([Rx, y] + [x, Ry]) = \frac{\lambda}{2}[x, y]. \qquad (5.21)$$

Equation (5.21) was first studied by Semenov–Tian–Shansky [313] in relation with integrable systems, and its solutions (not necessarily anti-symmetric) are also called classical r-matrices. A remarkable property of these matrices is the following:

Theorem 5.2.8 ([313]). *Suppose that a matrix $R \in End(\mathfrak{g})$ satisfies the mCYBE*

$$[Rx, Ry] - R([Rx, y] + [x, Ry]) = -[x, y]. \qquad (5.22)$$

Put

$$[x, y]_R = \frac{1}{2}([Rx, y] + [x, Ry]). \qquad (5.23)$$

Then $[x, y]_R$ is a Lie bracket. Denote by $\mathfrak{g}_R$ the corresponding Lie algebra, and $R_\pm = \frac{1}{2}(R \pm Id)$. Then $R_\pm : \mathfrak{g}_R \to \mathfrak{g}$ are Lie algebra homomorphisms.

Proof. Direct verification. □

Example 5.2.9. If $\mathfrak{g} = \mathfrak{g}_1 + \mathfrak{g}_2$ as a vector space, where $\mathfrak{g}_1$ and $\mathfrak{g}_2$ are Lie subalgebras of $\mathfrak{g}$, and $p_1 : \mathfrak{g} \to \mathfrak{g}_1$ and $p_2 : \mathfrak{g} \to \mathfrak{g}_2$ are the corresponding projections, then $R = p_1 - p_2$ satisfies the mCYBE (5.22). The corresponding Lie bracket is

$$[x_1 + x_2, y_1 + y_2]_R = [x_1, x_2] - [y_1, y_2], \tag{5.24}$$

where $x_1, y_1 \in \mathfrak{g}_1, x_2, y_2 \in \mathfrak{g}_2$. There is a family of (integrable) Hamiltonian systems on $\mathfrak{g}_R^*$ (with respect to the new bracket) generated by Casimir functions of $\mathfrak{g}^*$ (with respect to the original bracket), called *Adler–Kostant–Symes systems*, see, e.g., [2].

5.3 The dual and the double of a Poisson-Lie group

Theorem 5.3.1 ([107]). *A multiplicative 2-vector field Π on a connected Lie group G is a Poisson–Lie structure if and only if its corresponding 1-cocycle $p = \mathrm{d}_e \Pi : \mathfrak{g} \to \wedge^2 \mathfrak{g}$ defines on the dual $\mathfrak{g}^*$ of $\mathfrak{g}$ a Lie algebra structure $[,]^*$ by the formula*

$$\langle [\alpha, \beta]^*, x \rangle = \langle p(x), \alpha \wedge \beta \rangle \quad \forall \ \alpha, \beta \in \mathfrak{g}^*, x \in \mathfrak{g}. \tag{5.25}$$

Proof. The necessity of the above condition is clear: If Π is Poisson, then its linear part $\Pi^{(1)}$ at e is a linear Poisson structure on $\mathfrak{g} = T_e G$, hence it corresponds to a Lie algebra structure on $\mathfrak{g}^*$, which is given exactly by Formula (5.25). Conversely, by Theorem 5.1.2 and Theorem 5.1.3, $[\Pi, \Pi]$ is a multiplicative 3-vector field and is determined uniquely by its linear part at e. The linear part of $[\Pi, \Pi]$ at e is $[\Pi^{(1)}, \Pi^{(1)}]$. If Formula (5.25) defines a Lie algebra structure on $\mathfrak{g}^*$, then $\Pi^{(1)}$ is the linear Poisson structure corresponding to this Lie algebra structure, hence $[\Pi^{(1)}, \Pi^{(1)}] = 0$, and therefore $[\Pi, \Pi] = 0$, i.e., Π is a Poisson structure, which is multiplicative by assumptions. $\square$

Definition 5.3.2. A *Poisson–Lie algebra* is a Lie algebra $\mathfrak{g}$ together with a 1-cocycle $p : \mathfrak{g} \to \mathfrak{g} \wedge \mathfrak{g}$ which defines on $\mathfrak{g}^*$ a Lie algebra structure by Formula (5.25).

It is clear from Theorem 5.1.3 and Theorem 5.3.1 that a Poisson–Lie algebra is the infinitesimal version of a Poisson–Lie group, and there is a natural one-to-one correspondence between Poisson–Lie algebras and connected simply-connected Poisson–Lie groups.

Lemma 5.3.3. *If $(\mathfrak{g}, [,])$ and $(\mathfrak{g}^*, [,]^*)$ are Lie algebras, and $p : \mathfrak{g} \to \mathfrak{g} \wedge \mathfrak{g}$ is defined by Formula (5.25), then p is a 1-cocycle if and only if the following equation is satisfied for any $x, y \in \mathfrak{g}$, $\alpha, \beta \in \mathfrak{g}^*$:*

$$\langle [\alpha, \beta]^*, [x, y] \rangle = -\langle \mathrm{ad}_x^* \alpha, \mathrm{ad}_\beta^* y \rangle + \langle \mathrm{ad}_x^* \beta, \mathrm{ad}_\alpha^* y \rangle + \langle \mathrm{ad}_y^* \alpha, \mathrm{ad}_\beta^* x \rangle - \langle \mathrm{ad}_y^* \beta, \mathrm{ad}_\alpha^* x \rangle. \tag{5.26}$$

The proof of the above lemma is a straightforward computation based on Formula (5.25). $\square$

Definition 5.3.4. A pair $(\mathfrak{g}, \mathfrak{g}^*)$ of Lie algebra, where $\mathfrak{g}^*$ is the dual vector space of $\mathfrak{g}$, which satisfies Equation (5.26), is called a *Lie bialgebra*.

Notice that Equation (5.26) is symmetric with respect to $\mathfrak{g}$ and $\mathfrak{g}^*$. Hence, by Lemma 5.3.3, if $(\mathfrak{g}, p)$ is a Poisson–Lie algebra, then $(\mathfrak{g}, \mathfrak{g}^*)$ is a Lie bialgebra, and $\mathfrak{g}^*$ is also a Poisson–Lie algebra, called the *dual Poisson–Lie algebra* of $(\mathfrak{g}, p)$, and vice versa. (The 1-cocycle on $\mathfrak{g}^*$ is the one which defines the Lie algebra structure on $\mathfrak{g}$.) If $(\mathfrak{g}, \mathfrak{g}^*)$ is a Lie bialgebra, then $(\mathfrak{g}^*, \mathfrak{g})$ is also a Lie bialgebra.

At the group level, each simply-connected Poisson–Lie group (G, Π) has a unique simply-connected *dual Poisson–Lie group*, denoted by (G^*, Π^*).

Theorem 5.3.5 ([108]). *If $(\mathfrak{g}_1, \mathfrak{g}_2, [,]_1, [,]_2)$ is a Lie bialgebra, $\mathfrak{g}_2 = \mathfrak{g}_1^*$, then the bracket*

$$[x + \alpha, y + \beta] = ([x,y]_1 + \mathrm{ad}_\alpha^* y - \mathrm{ad}_\beta^* x, [\alpha, \beta]_2 + \mathrm{ad}_x^* \beta - \mathrm{ad}_y^* \alpha), \qquad (5.27)$$

where $x, y \in \mathfrak{g}_1$, $\alpha, \beta \in \mathfrak{g}_2$, defines a Lie algebra structure on $\mathfrak{g}_1 + \mathfrak{g}_2$. Denote this Lie algebra by $\mathfrak{d} = \mathfrak{g}_1 \bowtie \mathfrak{g}_2$. Then

$$\langle x + \alpha, y + \beta \rangle := \langle x, \beta \rangle + \langle y, \alpha \rangle \qquad (5.28)$$

is a nondegenerate ad-invariant scalar product on $\mathfrak{d}$, with respect to which $\mathfrak{g}_1$ and $\mathfrak{g}_2$ are isotropic subspaces, i.e., $\langle \mathfrak{g}_1, \mathfrak{g}_1 \rangle = 0$ and $\langle \mathfrak{g}_2, \mathfrak{g}_2 \rangle = 0$.

Conversely, if a Lie algebra $\mathfrak{d}$ is a vector-space direct sum of its two subalgebras $\mathfrak{g}_1, \mathfrak{g}_2$, and $\mathfrak{d}$ admits a nondegenerate ad-invariant scalar product with respect to which $\mathfrak{g}_1$ and $\mathfrak{g}_2$ are isotropic subspaces, then $(\mathfrak{g}_1, \mathfrak{g}_2)$ admits a unique natural Lie bialgebra structure.

The proof of the above theorem is a straightforward verification. See, e.g., Section 10.5 of [333]. Let us verify, for example, the Jacobi identity

$$[\alpha, [x, y]] = [[\alpha, x], y] + [x, [\alpha, y]]$$

for $\alpha \in \mathfrak{g}_2, x, y \in \mathfrak{g}_1$. If $z \in \mathfrak{d} = \mathfrak{g}_1 \bowtie \mathfrak{g}_2$, we will denote by $z_1 \in \mathfrak{g}_1$ and $z_2 \in \mathfrak{g}_2$ its components in $\mathfrak{g}_1$ and $\mathfrak{g}_2$: $z = z_1 + z_2$. Then we have

$$([[\alpha, x], y] + [x, [\alpha, y]])_2 = ([-\mathrm{ad}_x^*\alpha, y])_2 + ([x, -\mathrm{ad}_y^*\alpha])_2$$
$$= \mathrm{ad}_y^*(\mathrm{ad}_x^*\alpha) - \mathrm{ad}_x^*(\mathrm{ad}_y^*\alpha) = -\mathrm{ad}_{[x,y]}^*\alpha = ([\alpha, [x,y]])_2.$$

On the other hand, for any $\beta \in \mathfrak{g}_2$ we have

$$\langle ([[\alpha, x], y] + [x, [\alpha, y]])_1, \beta \rangle$$
$$= \langle ([\mathrm{ad}_\alpha^* x - \mathrm{ad}_x^*\alpha, y])_1, \beta \rangle + \langle ([x, \mathrm{ad}_\alpha^* y - \mathrm{ad}_y^*\alpha])_1, \beta \rangle$$
$$= \langle [\mathrm{ad}_\alpha^* x, y], \beta \rangle - \langle \mathrm{ad}_{\mathrm{ad}_x^*\alpha}^* y, \beta \rangle + \langle [x, \mathrm{ad}_\alpha^* y], \beta \rangle + \langle \mathrm{ad}_{\mathrm{ad}_y^*\alpha}^* x, \beta \rangle$$
$$= \langle \mathrm{ad}_y^*\beta, \mathrm{ad}_\alpha^* x \rangle + \langle \mathrm{ad}_x^*\alpha, \mathrm{ad}_\alpha^* y \rangle - \langle \mathrm{ad}_x^*\beta, \mathrm{ad}_\alpha^* y \rangle - \langle \mathrm{ad}_y^*\alpha, \mathrm{ad}_\beta^* x \rangle$$
$$= -\langle [\alpha, \beta], [x, y] \rangle = \langle \mathrm{ad}_\alpha^*[x, y], \beta \rangle = \langle ([\alpha, [x, y]])_1, \beta \rangle.$$

The rest is similar. $\qquad \square$

A triple $(\mathfrak{d}, \mathfrak{g}_1, \mathfrak{g}_2)$, where $(\mathfrak{g}_1, \mathfrak{g}_2)$ is a Lie bialgebra and $\mathfrak{d} = \mathfrak{g}_1 \bowtie \mathfrak{g}_2$ as in the above theorem, is called a *Manin triple* [108], and $\mathfrak{d}$ is called the *double* of $\mathfrak{g}_1$ and $\mathfrak{g}_2$.

Example 5.3.6. Consider $\mathfrak{d} = sl(n, \mathbb{C})$ with its real invariant scalar product $\langle X, Y \rangle := \Im(\text{trace}(XY))$ (the imaginary part of $\text{trace}(XY)$). Put $\mathfrak{g}_1 = su(n)$, and $\mathfrak{g}_2 = $ the subalgebra of $gl(n)$ consisting of traceless upper-triangular matrices whose diagonal entries are real. Then $\mathfrak{d} = \mathfrak{g}_1 + \mathfrak{g}_2$ and $\mathfrak{g}_1$ and $\mathfrak{g}_2$ are isotropic subspaces of $\mathfrak{d}$, hence $(\mathfrak{d}, \mathfrak{g}_1, \mathfrak{g}_2)$ is a Manin triple. As a consequence, the Lie group $SU(n)$ of Lie algebra $su(n)$ has a natural Poisson–Lie structure corresponding to this Manin triple, called the Iwasawa Poisson–Lie structure of $SU(n)$. More generally, given a compact Lie group K of Lie algebra $\mathfrak{k}$, its complexification $G = K_{\mathbb{C}}$ has a natural Iwasawa decomposition $G = KAN$ where A is Abelian and N is nilpotent. At the algebra level, we have $\mathfrak{d} = \mathfrak{k} + \mathfrak{l}$ where $\mathfrak{l} = \mathfrak{a} + \mathfrak{n}$. The triple $(\mathfrak{d}, \mathfrak{k}, \mathfrak{l})$ is a Manin triple, and the corresponding Poisson–Lie structure on K is called the *Iwasawa Poisson–Lie structure* [224].

Theorem 5.3.7 ([220, 224]). *Given a Manin triple $(\mathfrak{d}, \mathfrak{g}_1, \mathfrak{g}_2 = \mathfrak{g}_1^*)$, denote by D the simply-connected Lie group of $\mathfrak{d}$, and by G_1 and G_2 the connected Lie subgroups of D generated by $\mathfrak{g}_1$ and $\mathfrak{g}_2$ respectively. Then the corresponding Poisson–Lie structures Π_1 on G_1 and Π_2 on G_2 can be given as follows:*

$$(R_{g^{-1}}\Pi_1(g))(\alpha, \beta) = -\langle p_1 \text{Ad}_{g^{-1}}\alpha, p_2 \text{Ad}_{g^{-1}}\beta \rangle \; \forall \; g \in G_1, \alpha, \beta \in \mathfrak{g}_2, \quad (5.29)$$

$$(R_{u^{-1}}\Pi_2(u))(X, Y) = \langle p_1 \text{Ad}_{u^{-1}}X, p_2 \text{Ad}_{u^{-1}}Y \rangle \; \forall \; u \in G_2, X, Y \in \mathfrak{g}_1, \quad (5.30)$$

where $p_1 : \mathfrak{d} \to \mathfrak{g}_1$ and $p_2 : \mathfrak{g} \to \mathfrak{g}_2$ are the two natural projections.

Proof. We will prove Formula (5.29). Formula (5.30) is absolutely similar. Define tensor field Λ on G_1 by

$$(R_{g^{-1}}\Lambda(g))(\alpha, \beta) = \langle p_1 \text{Ad}_{g^{-1}}\alpha, p_2 \text{Ad}_{g^{-1}}\beta \rangle. \quad (5.31)$$

We must show that Λ coincides with $-\Pi_1$, i.e., it is a multiplicative Poisson structure on G_1 whose linear part at the neutral element defines the minus of Lie bracket on $\mathfrak{g}_2 = \mathfrak{g}_1^*$. First notice that Λ is anti-symmetric, i.e., it is a 2-vector field, because

$$0 = \langle \alpha, \beta \rangle = \langle \text{Ad}_{g^{-1}}\alpha, \text{Ad}_{g^{-1}}\beta \rangle = (R_{g^{-1}}\Lambda(g))(\alpha, \beta) + (R_{g^{-1}}\Lambda(g))(\beta, \alpha).$$

By putting $g = \exp(tx)$ $(x \in \mathfrak{g}_1)$ in Formula (5.31) and deriving both sides with respect to t at $t = 0$, we obtain that the linear part of Λ at the neutral element defines the minus of the Lie bracket on $\mathfrak{g}_2$. It remains to show that Λ is multiplicative. Notice that $p_2 \text{Ad}_g \alpha = \text{Ad}_g^* \alpha$, which is a consequence of the formula $[x, \alpha] = \text{ad}_x^* \alpha - \text{ad}_\alpha^* x$ $(x \in \mathfrak{g}_1, \alpha \in \mathfrak{g}_2)$.

We have:

$$
\begin{aligned}
(R_{(gh)^{-1}}\Lambda(gh))(\alpha,\beta) &= \langle p_1\mathrm{Ad}_{h^{-1}}\mathrm{Ad}_{g^{-1}}\alpha, p_2\mathrm{Ad}_{h^{-1}}\mathrm{Ad}_{g^{-1}}\beta\rangle \\
&= \langle p_1\mathrm{Ad}_{h^{-1}}(p_1\mathrm{Ad}_{g^{-1}}\alpha + p_2\mathrm{Ad}_{g^{-1}}\alpha), p_2\mathrm{Ad}_{h^{-1}}p_2\mathrm{Ad}_{g^{-1}}\beta\rangle \\
&= \langle \mathrm{Ad}_{h^{-1}}p_1\mathrm{Ad}_{g^{-1}}\alpha, \mathrm{Ad}_{h^{-1}}p_2\mathrm{Ad}_{g^{-1}}\beta\rangle \\
&\quad + \langle p_1\mathrm{Ad}_{h^{-1}}p_2\mathrm{Ad}_{g^{-1}}\alpha, p_2\mathrm{Ad}_{h^{-1}}p_2\mathrm{Ad}_{g^{-1}}\beta\rangle \\
&= (R_{g^{-1}}\Lambda(g))(\alpha,\beta) + (R_{h^{-1}}\Lambda(h))(p_2\mathrm{Ad}_{g^{-1}}\alpha, p_2\mathrm{Ad}_{g^{-1}}\beta) \\
&= (R_{g^{-1}}\Lambda(g))(\alpha,\beta) + (R_{h^{-1}}\Lambda(h))(\mathrm{Ad}^*_{g^{-1}}\alpha, \mathrm{Ad}^*_{g^{-1}}\beta) \\
&= \left(R_{g^{-1}}\Lambda(g) + \mathrm{Ad}_g(R_{h^{-1}}\Lambda(h))\right)(\alpha,\beta),
\end{aligned}
$$

which means that $R_{(gh)^{-1}}\Lambda(gh) = R_{g^{-1}}\Lambda(g) + \mathrm{Ad}_g(R_{h^{-1}}\Lambda(h))$, i.e., Λ is multiplicative. $\square$

The double of a Poisson–Lie group also admits a Poisson–Lie structure. More precisely, we have:

Theorem 5.3.8 ([313, 220]). *With the notations of Theorem 5.3.7, put*

$$
\Pi(d) = R_d\pi - L_d\pi \quad \forall\, d \in D, \tag{5.32}
$$

where $\pi \in \mathfrak{d}^* \wedge \mathfrak{d}^*$ is defined by

$$
\pi(\alpha + x, \beta + y) = \langle x, \beta\rangle - \langle y, \alpha\rangle \tag{5.33}
$$

$\forall\, (\alpha, x), (\beta, y) \in \mathfrak{g}_2 + \mathfrak{g}_1 \cong \mathfrak{g}_1^* + \mathfrak{g}_2^* \cong \mathfrak{d}^*$. *Then (D, Π) is a Poisson–Lie group whose corresponding Lie algebra structure on $\mathfrak{d}^*$ is isomorphic to the direct sum $\mathfrak{g}_2 \oplus (-\mathfrak{g}_1)$, where the Lie bracket on the component $\mathfrak{g}_1$ is taken with the minus sign. Moreover, the inclusions $(G_1, \Pi_1) \to (D, \Pi)$ and $(G_2, \Pi_2) \to (D, \Pi)$ are Poisson and anti-Poisson maps respectively.*

The proof of the above theorem will be left as an exercise. $\square$

5.4 Actions of Poisson–Lie groups

5.4.1 Poisson actions of Poisson–Lie groups

When G is a Lie group equipped with a (nontrivial) Poisson–Lie structure Π, then it is interesting to consider actions of G on Poisson manifolds, which do not preserve Poisson structures, but rather twist them by the Poisson–Lie structure of G:

Definition 5.4.1. A left (or right) action of a Poisson–Lie group (G, Π) on a Poisson manifold (M, Λ) is called a *Poisson action* if the corresponding action map $G \times M \to M$ is a Poisson map.

In particular, when a Poisson action $(G, \Pi) \times (M, \Lambda) \to (M, \Lambda)$ is transitive, then (M, Λ) is called a *Poisson homogeneous space* [109]. See, e.g., [109, 222, 216, 191, 126], for results about these Poisson homogeneous spaces.

Remark 5.4.2. To turn a left Poisson action of G into a right Poisson action or vice versa, we have to change the sign of the Poisson–Lie structure on G.

Example 5.4.3. If (G, Π) is a Poisson–Lie group, then the actions of G on itself by left and right multiplications are Poisson actions.

The following result provides a family of examples of Poisson actions.

Example 5.4.4 (Semenov–Tian–Shansky [314]). Suppose that G is a Lie group with Lie algebra $\mathfrak{g}$, and $r_1, r_2 \in \wedge^2 \mathfrak{g}$ satisfy the mCYBE (5.15) with the same coefficient λ. Then

$$\Pi_{r_1, r_2} = r_2^+ - r_1^- \tag{5.34}$$

is a Poisson structure on G. If we denote by G_{r_1, r_2} the group G equipped with the Poisson structure $\Pi_{r_1, r_2} = r_1^+ - r_2^-$, then the multiplication map $G \times G \to G$ induces Poisson morphisms $G_{r_1, r_1} \times G_{r_1, r_2} \to G_{r_1, r_2}$ and $G_{r_1, r_2} \times G_{r_2, r_2} \to G_{r_1, r_2}$. The proof of these facts is similar to the proof of Theorem 5.2.3 and Lemma 5.2.2.

Lemma 5.4.5 ([224]). *Let $\rho : G \times M \to M$ be a left (right) action of a connected Poisson–Lie group (G, Π) on a Poisson manifold (M, Λ), whose Lie algebra action is $\xi : \mathfrak{g} \to \mathcal{V}^1(M)$, extended to a morphism $\xi : \wedge^* \mathfrak{g} \to \mathcal{V}^*(M)$. Then ρ is a Poisson action if and only if one of the following three equivalent conditions is satisfied:*

1) $\forall\ g \in G, z \in M$ *one has*

$$\Lambda(\rho(g, z)) = (\rho_g)_* (\Lambda(z)) + (\rho_z)_* (\Pi(g)), \tag{5.35}$$

   where $\rho_g(z) = \rho_z(g) = \rho(g, z)$.
2) $\forall\ X \in \mathfrak{g}$ *one has*

$$\mathcal{L}_{\xi(X)} \Lambda = \mp\ \xi((\mathrm{d}_e \Pi)(X)), \tag{5.36}$$

   where the sign on the right-hand side is minus if ρ is a left action, and plus if ρ is a right action.
3) $\forall\ z \in M, \alpha, \beta \in \Omega^1(M), X \in \mathfrak{g}$ *one has*

$$(\mathcal{L}_{\xi(X)} \Lambda)_z(\alpha, \beta) = \mp\ [(\rho_z)^* \alpha(z), (\rho_z)^* \beta(z)]^*(X). \tag{5.37}$$

Proof (sketch). Equation (5.35) is just a restatement of the condition that the map $\rho : G \times M \to M$ is a Poisson map. Equation (5.36) is the infinitesimal version of Equation (5.35). Indeed, putting $g = \exp(tX)$ in (5.35), we get

$$\frac{1}{t} \left(\Lambda(\rho(\exp(tX), z)) - (\rho_{\exp(tX)})_* (\Lambda(z)) \right) = \frac{1}{t} (\rho_z)_* (\Pi(\exp(tX))). \tag{5.38}$$

The limit of (5.38) when $t \to 0$ is (5.36), with the sign minus if $\rho_{\exp(tX)} = \exp(-t\xi(X))$, and the sign plus if $\rho_{\exp(tX)} = \exp(t\xi(X))$. Equation (5.37) is just the value of (5.36) on the two arguments α, β. $\qquad\square$

If we have an action $\xi : \mathfrak{g} \to \mathcal{V}^1(M)$ of a Poisson–Lie algebra $(\mathfrak{g}, p)$, where p is a 1-cocycle $p : \mathfrak{g} \to \mathfrak{g} \wedge \mathfrak{g}$, on a Poisson manifold (M, Λ), such that the equation

$$\mathcal{L}_{\xi(X)}\Lambda = \mp \xi(p(X)) \tag{5.39}$$

is satisfied $\forall\ X \in \mathfrak{g}$, then we say it is a left (resp. right) *Poisson action* if the sign on the right-hand side is minus (resp. plus). It is the infinitesimal version of Poisson actions of Poisson–Lie groups.

Theorem 5.4.6 (Lu [221]). *Let (M, Λ) be a Poisson manifold, $(\mathfrak{g}, p)$ a Poisson–Lie algebra, and $\xi : \mathfrak{g} \to \mathcal{V}^1(M)$ a Lie algebra homomorphism which can be lifted to a linear map $\hat{\xi} : \mathfrak{g} \to \Omega^1(M)$, i.e., $\xi(X) = \sharp(\hat{\xi}(X))\ \forall\ X \in \mathfrak{g}$. Define the $\mathfrak{g}^*$-valued 1-form Θ on M by*

$$\langle \Theta(z)(v), X \rangle = \langle v, \hat{\xi}(X)(z) \rangle \quad \forall\ z \in M, v \in T_z M, X \in \mathfrak{g}. \tag{5.40}$$

Then ξ is a left (resp. right) Poisson action if and only if $\mathrm{d}\Theta + [\Theta, \Theta] = 0$ (resp. $\mathrm{d}\Theta - [\Theta, \Theta] = 0$) on each symplectic leaf of M.

Remark 5.4.7. In the above theorem, $[\Theta, \Theta]$ is defined by

$$[\Theta, \Theta](X, Y) = [\Theta(X), \Theta(Y)] \quad \forall\ X, Y \in T_z M,$$

using the Lie bracket of $\mathfrak{g}^*$. The equation $\mathrm{d}\Theta + [\Theta, \Theta] = 0$ is called the (left) *Maurer–Cartan equation*. It is also written as $\mathrm{d}\Theta + \frac{1}{2}[\Theta \wedge \Theta] = 0$, and sometimes even as $\mathrm{d}\Theta + \frac{1}{2}[\Theta, \Theta] = 0$ (with a different convention on the definition of $[\Theta, \Theta]$).

Proof. We will make use of the following formula: If ζ_1, ζ_2 are two 1-forms on a Poisson manifold (M, Λ), then

$$[\sharp\zeta_1, \sharp\zeta_2] = \sharp\left(\mathcal{L}_{\sharp\zeta_1}\zeta_2 - \mathcal{L}_{\sharp\zeta_2}\zeta_1 - \mathrm{d}\left(\Lambda(\zeta_1, \zeta_2)\right)\right). \tag{5.41}$$

This formula corresponds to the fact that the bracket

$$\begin{aligned}\{\zeta_1, \zeta_2\} &= \mathcal{L}_{\sharp\zeta_1}\zeta_2 - \mathcal{L}_{\sharp\zeta_2}\zeta_1 - \mathrm{d}(\Lambda(\zeta_1, \zeta_2)) \\ &= \mathrm{d}(\Lambda(\zeta_1, \zeta_2)) + i_{\sharp\zeta_1}\mathrm{d}\zeta_2 - i_{\sharp\zeta_2}\mathrm{d}\zeta_1 \end{aligned} \tag{5.42}$$

is a Lie bracket on the space $\Omega^1(M)$ of 1-forms on M, and the anchor map $\sharp :$ $\Omega^1(M) \to \mathcal{V}^1(M)$ is a Lie algebra homomorphism with respect to this bracket and the usual Lie bracket on $\mathcal{V}^1(M)$. This bracket is actually the natural Lie algebroid bracket on T^*M associated to a Poisson structure on M, and it will be discussed in more detail in Chapter 8. Formula (5.41) can be proved by a simple direct verification in the symplectic case using Cartan's formula, and the general Poisson case can be reduced to the symplectic case by restriction to symplectic leaves.

Using (5.41) and (5.42), we get, for any $\zeta_1, \zeta_2 \in \Omega^1(M)$ and $X \in \mathfrak{g}$,

$$
\begin{aligned}
\mathcal{L}_{\xi(X)}\Lambda(\zeta_1, \zeta_2) &= \langle \xi(X), \{\zeta_1, \zeta_2\}\rangle - \sharp\zeta_1(\langle \xi(X), \zeta_2\rangle) + \sharp\zeta_2(\langle \xi(X), \zeta_1\rangle) \\
&= -\langle \Theta([\sharp\zeta_1, \sharp\zeta_2]), X\rangle + \sharp\zeta_1(\langle \Theta(\sharp\zeta_2), X\rangle) - \sharp\zeta_2(\langle \Theta(\sharp\zeta_1), X\rangle) \\
&= \langle \mathrm{d}\Theta(\sharp\zeta_1, \sharp\zeta_2), X\rangle.
\end{aligned}
$$

On the other hand,

$$
\xi(p(X))(\zeta_1, \zeta_2) = \langle [\Theta, \Theta](\sharp\zeta_1, \sharp\zeta_2), X\rangle. \tag{5.43}
$$

Therefore ξ is left (right) Poisson if and only if $(\mathrm{d}\Theta \pm [\Theta, \Theta])(\sharp\zeta_1, \sharp\zeta_2) = 0$ for arbitrary 1-forms ζ_1, ζ_2 on M, i.e., $\mathrm{d}\Theta \pm [\Theta, \Theta] = 0$ on each symplectic leaf of M. $\qquad \square$

5.4.2   Dressing transformations

Given a Poisson–Lie group (G, Π), there are natural Poisson actions of its dual Lie algebra $\mathfrak{g}^*$ on G, which generate (local) Poisson actions of G^* on G, called dressing transformations of G [314, 351, 224], which can be constructed as follows.

For each element $\alpha \in \mathfrak{g}^*$, we denote by α^+ (resp. α^-) the left-invariant (resp. right-invariant) 1-form on G whose value at the neutral element e is α. Denote by $\sharp = \sharp_\Pi : T^*G \to TG$ the usual anchor map of Π.

Proposition 5.4.8 ([351]). *The maps $\alpha \mapsto \sharp(\alpha^+)$ and $\alpha \mapsto \sharp(\alpha^-)$ are Lie algebra homomorphisms from $\mathfrak{g}^*$ to $\mathcal{V}^1(G)$.*

Proof. Using Formula (5.41), in order to prove Proposition 5.4.8, it is enough to check that

$$
\{\alpha^+, \beta^+\} = ([\alpha, \beta]^*)^+ \quad \text{and} \quad \{\alpha^-, \beta^-\} = ([\alpha, \beta]^*)^-. \tag{5.44}
$$

It is clear from the second line of (5.42) and the definition of the Lie bracket of $\mathfrak{g}^*$ that the above equalities hold at the neutral element e of G. The rest of the proof is now a direct consequence of the following characterization of Poisson-Lie groups:

Theorem 5.4.9 ([351, 97]). *Let G be a connected Lie group with a Poisson tensor Π. Then Π is Poisson–Lie if and only if $\Pi(e) = 0$ and for any two left-invariant (resp., right-invariant) 1-forms ζ_1, ζ_2 on G, their bracket $\{\zeta_1, \zeta_2\}$ defined by (5.42) is also left-invariant (resp., right-invariant).*

Proof. Denote by ζ_1 and ζ_2 two arbitrary left-invariant 1-forms. Denote by X a left-invariant vector field and by Y a right-invariant vector field. Then $\{\zeta_1, \zeta_2\}$ is left-invariant if and only if

$$
(\mathcal{L}_Y\{\zeta_1, \zeta_2\})(X) = 0 \quad \forall\, X, Y.
$$

Using Formula (5.42), Cartan's formula, and the fact that $X(\zeta_1)$ and $X(\zeta_2)$ are constant, we get

$$
\begin{aligned}
(\mathcal{L}_Y\{\zeta_1,\zeta_2\})(X) &= Y(\{\zeta_1,\zeta_2\}(X)) \\
&= Y[X(\Pi(\zeta_1,\zeta_2)) + (\mathrm{d}\zeta_2)(\sharp\zeta_1,X) - (\mathrm{d}\zeta_1)(\sharp\zeta_2,X)] \\
&= Y[(\mathcal{L}_X\Pi)(\zeta_1,\zeta_2)) + \Pi(\mathcal{L}_X\zeta_1,\zeta_2) + \Pi(\zeta_1,\mathcal{L}_X\zeta_2) \\
&\quad + (\mathrm{d}\zeta_2)(\sharp\zeta_1,X) - (\mathrm{d}\zeta_1)(\sharp\zeta_2,X)] \\
&= Y[(\mathcal{L}_X\Pi)(\zeta_1,\zeta_2)] = (\mathcal{L}_Y\mathcal{L}_X\Pi)(\zeta_1,\zeta_2).
\end{aligned}
$$

Thus, $(\mathcal{L}_Y\{\zeta_1,\zeta_2\})(X) = 0 \ \forall X,Y,\zeta_1,\zeta_2$ if and only if $\mathcal{L}_Y\mathcal{L}_X\Pi = 0$ for any left-invariant X and right-invariant Y, i.e., Π is multiplicative. The case of right-invariant 1-forms is completely similar. $\qquad\square$

In view of Proposition 5.4.8, we have the following definition:

Definition 5.4.10. The actions of $\mathfrak{g}^*$ on (G,Π) given by the maps $\alpha \mapsto \sharp(\alpha^+)$ and $\alpha \mapsto \sharp(\alpha^-)$, and their corresponding (local or global) actions of G^* on G, are called *dressing actions* or *dressing transformations*.

The word *local* in the above definition means that maybe the action map $G^* \times G \to G$ is not well defined globally, but is well defined for a neighborhood of $(e^*,e) \in G^* \times G$.

It is clear from the definition that the *symplectic leaves of (G,Π) are orbits of the dressing actions*. In particular, if $\Pi = 0$ then the dressing actions are also trivial.

Example 5.4.11. Consider the case when G is simply-connected Abelian, i.e., $G = \mathfrak{g}$ is a vector space with a trivial Lie bracket. Then the two dressing actions coincide, and they are just the coadjoint action of $\mathfrak{g}^*$ on $\mathfrak{g} = (\mathfrak{g}^*)^*$.

The action given by $\sharp(\alpha^+)$ is called the *left dressing action*. The corresponding left dressing action of G^* on G is defined by $\rho(\exp(\alpha),z) = \exp(-\sharp\alpha^+)(z)$ for $\alpha \in \mathfrak{g}^*, z \in G$. The action given by $\sharp(\alpha^-)$ is called the *right dressing action*. These two dressing actions are intertwined by the inversion map $S : g \mapsto g^{-1}$ of G. In other words, we have

$$
S_*\sharp(\alpha^+) = \sharp(\alpha^-). \tag{5.45}
$$

This equality follows directly from $S_*\Pi = -\Pi$ and $S_*\alpha^+ = -\alpha^-$.

Let $(\mathfrak{d} = \mathfrak{g}_1 \bowtie \mathfrak{g}_2, \mathfrak{g}_1, \mathfrak{g}_2)$ be a Manin triple. Suppose that there are connected Lie groups D, G_1, G_2 with Lie algebras $\mathfrak{d}, \mathfrak{g}_1, \mathfrak{g}_2$ such that $D = G_1G_2$, in the sense that $G_1, G_2 \subset D$, and any element $d \in D$ can be factorized in a unique way into a product $d = g_1g_2$ with $g_1 \in G_1, g_2 \in G_2$. Then the dressing actions have the following simple geometric interpretation: for any $g_1 \in G_1, g_2 \in G_2$, there is a unique element $g_1^{g_2} \in G_1$ and a unique element $g_2^{g_1} \in G_2$ such that

$$
g_2{\cdot}g_1 = g_1^{g_2}{\cdot}g_2^{g_1}. \tag{5.46}
$$

Then the map $(g_2, g_1) \mapsto g_1^{g_2}$ is a left action of G_2 on G_1, called the *left* dressing transformation. One can verify directly that this action is generated by the Lie algebra action $\mathfrak{g}_2 \to \mathcal{V}^1(G_1) : \alpha \mapsto \sharp(\alpha^+)$. Similarly, $(g_1, g_2) \mapsto g_2^{g_1}$ is the *right* dressing action of G_1 on G_2.

Theorem 5.4.12 ([314, 224]). *The left and right dressing actions of G^* on G are Poisson actions.*

Proof. We will prove that the left dressing action is Poisson. The case of right dressing action is similar. Denote by Θ the left-invariant *Maurer–Cartan form* on G, i.e., the $\mathfrak{g}$-valued 1-form on G defined by

$$\Theta(X) = L_{g^{-1}}X \in \mathfrak{g} \quad \forall\, X \in T_g G. \tag{5.47}$$

This form satisfies the *Maurer–Cartan equation*

$$\mathrm{d}\Theta + [\Theta, \Theta] = 0. \tag{5.48}$$

(This classical fact can be checked easily by evaluating $\mathrm{d}\Theta$ on left-invariant vector fields.) Having this equation in mind, Theorem 5.4.12 is now just a special case of Theorem 5.4.6. $\qquad \square$

5.4.3 Momentum maps

Momentum maps for Poisson actions of Poisson–Lie groups were introduced by Lu [221], in analogy with momentum maps for Hamiltonian actions of Lie groups.

Let $\varrho : G \times M \to M$ be a left (resp. right) Poisson action of a Poisson–Lie group G on a Poisson manifold M. Let G^* be the dual Poisson-Lie group. Denote by $\xi : \mathfrak{g} \to \mathcal{V}^1(M)$ the corresponding Lie algebra homomorphism. For each $X \in \mathfrak{g}$, denote by X^+ (resp. X^-) the left- (resp. right-) invariant 1-form on G^* with value X at the neutral element.

Definition 5.4.13. A smooth map $J : M \to G^*$ is called a *momentum map* for the left (resp. right) Poisson action $\varrho : G \times M \to M$ if for each $X \in \mathfrak{g}$,

$$\xi(X) = \sharp(J^*(X^+)) \quad (\text{resp. } \xi(X) = \sharp(J^*(X^-))). \tag{5.49}$$

If, moreover, J is a Poisson map, then it is called an *equivariant momentum map*.

Example 5.4.14. For both the left and the right dressing action of G^* on G, the identity map of G is an equivariant momentum map.

Remark 5.4.15. In general, it is not clear whether a given Poisson action of a Poisson–Lie group admits an equivariant momentum map or not, see [143]. One can view Lu's theory of momentum maps of Poisson–Lie group actions as a particular case of Xu's theory of momentum maps for quasi-symplectic groupoid actions [361].

5.5 r-matrices and quadratic Poisson structures

In this section, we will discuss some known constructions of quadratic Poisson structures from r-matrices.

Theorem 5.5.1 ([297]). *Let $r \in \mathfrak{g} \wedge \mathfrak{g}$ be a triangular r-matrix of a Lie algebra $\mathfrak{g}$, and $\rho : \mathfrak{g} \to End(V)$ be a linear action of $\mathfrak{g}$ on a vector space V. Identify $End(V)$ with the space of linear vector fields on V. Then the image r_V of r under the map $\wedge^2 \rho : \mathfrak{g} \wedge \mathfrak{g} \to End(V) \wedge End(V)$ is a quadratic Poisson structure on V.*

Proof. The fact that $\wedge^2 \rho(r)$ is quadratic is clear, because the wedge product of two linear vector fields is a quadratic 2-vector field. The fact that $\wedge^2 \rho(r)$ is Poisson is a particular case of the following more general result:

Theorem 5.5.2. *If $r \in \mathfrak{g} \wedge \mathfrak{g}$ is a triangular r-matrix and $\mathfrak{g}$ acts on a manifold M, then the image of r under the induced map $\mathfrak{g} \wedge \mathfrak{g} \to \Gamma(\wedge^2 TM) = \mathcal{V}^2(M)$ is a Poisson tensor on M, which we will denote by r_M. If Π is a Poisson structure on M which is invariant under the action of $\mathfrak{g}$, then r_M is compatible with Π : $[r_M, \Pi] = 0$.*

Proof. If $\rho : \mathfrak{g} \to \mathcal{V}^1(M)$ is a Lie algebra homomorphism, then $\wedge \rho : \wedge^\star \mathfrak{g} \to \mathcal{V}^\star(M)$ preserves the Schouten bracket. Hence if $[r, r] = 0$ then $[\wedge^2 \rho(r), \wedge^2 \rho(r)] = 0$, i.e., $r_M = \wedge^2 \rho(r)$ is a Poisson structure. Similarly, if Π is invariant under the action of $\mathfrak{g}$, then $[\rho(x), \Pi] = 0 \ \forall \, x \in \mathfrak{g}$, and by Leibniz's rule we have $[\wedge \rho(\alpha), \Pi] = 0 \ \forall \, \alpha \in \wedge^\star \mathfrak{g}$; in particular $[\wedge^2 \rho(r), \Pi] = 0$. $\qquad \Box$

Corollary 5.5.3 ([161]). *If $r \in \mathfrak{g} \wedge \mathfrak{g}$ is a triangular r-matrix, then its induced quadratic Poisson structure on $\mathfrak{g}^*$ via the coadjoint action is compatible with the linear Poisson structure on $\mathfrak{g}^*$.*

Example 5.5.4. When $\mathfrak{g} = gl(n)$ then by taking the standard linear action of $gl(n)$ on $\mathbb{R}^n$ or $\mathbb{C}^n$, we get an n-dimensional quadratic Poisson structure associated to each triangular r-matrix $r \in gl(n) \wedge gl(n)$. On the other hand, the adjoint action of $gl(n)$ on itself will give rise to n^2-dimensional quadratic Poisson structures.

When $\mathfrak{g}$ acts on V or M and $r \in \mathfrak{g} \wedge \mathfrak{g}$ is an r-matrix which is *not* triangular, i.e., $[r, r] \in (\wedge^3 \mathfrak{g})^{\mathfrak{g}}$ and $[r, r] \neq 0$, then sometimes r_V or r_M is still a Poisson structure. For example, Donin and Gurevich [104] showed that if M is a symmetric space of a semisimple Lie group G (i.e., $M = G/H$ such that $\mathfrak{g}$ has a linear decomposition $\mathfrak{g} = \mathfrak{h} + \mathfrak{m}$ with $[\mathfrak{h}, \mathfrak{m}] \subset \mathfrak{m}, [\mathfrak{m}, \mathfrak{m}] \subset \mathfrak{h}$), and r is an r-matrix of $\mathfrak{g}$, then r_M is a Poisson structure on M. The following result of Zakrzewski [365] gives several series of quadratic Poisson structures on $\mathbb{R}^n$ related to r-matrices.

Theorem 5.5.5 ([365]). *In the following three cases, for any classical r-matrix in $\mathfrak{g} \wedge \mathfrak{g}$ the induced quadratic 2-vector field r_V on V is Poisson:*

1. $\mathfrak{g} = so(n, \mathbb{R}), V = \mathbb{R}^n$ *(the action of $\mathfrak{g}$ on V is the natural one);*
2. $\mathfrak{g} = sl(n, \mathbb{R}), V = \mathbb{R}^n$;
3. $\mathfrak{g} = sp(2n, \mathbb{R}), V = \mathbb{R}^{2n}$.

For $\mathfrak{g} = su(n)$ and $V = \mathbb{C}^n = \mathbb{R}^{2n}$, the induced 2-vector field is Poisson if and only if r is triangular.

Proof (sketch). Recall that, if $\mathfrak{g}$ is a simple Lie algebra, then $(\wedge^3 \mathfrak{g})^{\mathfrak{g}}$ is one-dimensional and is generated by an element η whose Killing-transported version to $\wedge^3 \mathfrak{g}^*$ is $\eta^\dagger$ defined by $\eta^\dagger(X, Y, Z) = \langle [X, Y], Z \rangle \; \forall \; X, Y, Z \in \mathfrak{g}$. Let M be a manifold on which $\mathfrak{g}$ acts. If $[r, r] \in (\wedge^3 \mathfrak{g})^{\mathfrak{g}}$ and $[r, r] \neq 0$, then r_M is Poisson if and only if $\eta_M = 0$, where η_M is the image of η in $\mathcal{V}^3(M)$ under the extension of the homomorphism $\mathfrak{g} \to \mathcal{V}^1(M)$.

Now, $\eta_M(x) = 0$ for a point $x \in M$ if and only if the image of η under the projection $\wedge^3 \mathfrak{g} \to \wedge^3(\mathfrak{g}/\mathfrak{g}_x)$ is zero, where $\mathfrak{g}_x$ denotes the isotropy algebra of the action of $\mathfrak{g}$ on M at x. Equivalently, for any $X, Y, Z \in \mathfrak{g}_x^\perp$ we have $\langle [X, Y], Z \rangle = \langle \nu, X \wedge Y \wedge Z \rangle = 0$, where $\mathfrak{g}_x^\perp$ denotes the orthogonal complement of $\mathfrak{g}_x$ in $\mathfrak{g}$ with respect to the Killing form. It means that $[\mathfrak{g}_x^\perp, \mathfrak{g}_x^\perp] \subset \mathfrak{g}_x$. In other words, we have the following criterion:

$$[r_M, r_M](x) = 0 \iff [\mathfrak{g}_x^\perp, \mathfrak{g}_x^\perp] \subset \mathfrak{g}_x . \tag{5.50}$$

The rest of the proof is a straightforward case by case analysis, based on the above criterion. $\square$

Example 5.5.6 ([365]). Consider the standard r-matrix for $sl(n, \mathbb{R})$, which is given by

$$r = c \sum_{j<k} e_j^k \wedge e_k^j \quad (c \in \mathbb{R}), \tag{5.51}$$

where e_j^k denotes the matrix whose entry at line j column k is 1 and all the other entries are 0. Direct calculations show that r induces the following bracket on $\mathbb{R}^n$ via the natural action of $sl(n, \mathbb{R})$ on $\mathbb{R}^n$:

$$\{x_j, x_k\} = c x_j x_k \quad \forall \; 1 \le j < k \le n. \tag{5.52}$$

Let A be a (finite-dimensional) associative algebra. Then, similarly to the case of Lie groups, we can talk about multiplicative tensors on A. In particular, a *multiplicative Poisson structure* on A is a Poisson structure Π on A which satisfies Equation (5.9), i.e.,

$$\Pi(ab) = (L_a)_* \Pi(b) + (R_b)_* \Pi(a) \quad \forall \; a, b \in A, \tag{5.53}$$

where L_a denotes the left multiplication by a, $L_a(x) = a.x$, and R_b denotes the right multiplication by b. In other words, a Poisson structure Π on A is multiplicative if the product map $A \times A \to A$ o A is a Poisson morphism.

Example 5.5.7. If $\Pi = r^+ - r^-$ is an exact Poisson–Lie structure on $G = GL(n, \mathbb{K})$, where $r \in gl(n, \mathbb{K}) \wedge gl(n, \mathbb{K})$ is a classical r-matrix of $gl(n, \mathbb{K})$, then Π can naturally be extended to a multiplicative Poisson structure on the algebra $Mat_n(\mathbb{K})$ of $n \times n$ matrices by the same formula: $\Pi(M) = (L_M)_* r - (R_M)_* r \; \forall \; M \in Mat_n(\mathbb{K})$. By construction, it is a *quadratic* Poisson structure on $Mat_n(\mathbb{K})$. The corresponding Poisson bracket on $Mat_n(\mathbb{K})$ is called the *Sklyanin bracket* [318].

In fact, as was shown by Balinsky and Burman [21], multiplicative Poisson structures on associative algebras are usually automatically quadratic.

Theorem 5.5.8 ([21]). *If A is a finite-dimensional associative algebra with unit, then any smooth multiplicative Poisson structure on A is quadratic.*

Proof. Denote by I the unit element of A, by (e_i) a linear basis of A, and by (x_i) the dual basis of A^*, considered as a linear coordinate system of A. Let $\Pi = \sum_{i<j} \Pi_{ij} \frac{\partial}{\partial x_i} \wedge \frac{\partial}{\partial x_j}$ be a smooth multiplicative Poisson tensor on A.

Note that, for each $t \in \mathbb{R}$, the left and right multiplication maps $L_{t\mathrm{I}}$ and $R_{t\mathrm{I}}$ are the dilatation by a factor of t in A, considered as a vector space. Putting $a = b = t\mathrm{I}$ in Equation (5.53), we get

$$\Pi_{ij}(t^2\mathrm{I}) = 2t^2\Pi_{ij}(t\mathrm{I}).$$

The only smooth function $g(t)$ of one variable which satisfies the functional equation $g(t^2) = 2t^2 g(t)$ is the trivial function $g(t) = 0$. Thus we have $\Pi(t\mathrm{I}) = 0 \; \forall \, t \in \mathbb{R}$.

Now, putting $a = t\mathrm{I}$ in Equation (5.53) and taking into account the fact that $\Pi(t\mathrm{I}) = 0$, we get

$$\Pi_{ij}(tb) = t^2\Pi_{ij}(b) \;\; \forall \, b \in A, t \in \mathbb{R}.$$

In other words, Π_{ij} are homogeneous functions of degree 2 on A. Since they are smooth, they are homogeneous quadratic functions. □

Remark 5.5.9. Quadratic Poisson structures play an important role in integrable systems, and many other examples of them (all related to r-matrices) can be found in the literature. See, e.g., [208, 283, 326, 136, 282]. In [208, 283], natural *cubic* Poisson structures arising from *r*-matrices are also given.

5.6 Linear curl vector fields

Let

$$\Pi = \frac{1}{4} \sum_{i,j,k,l} \Pi_{ij}^{kl} x_k x_l \frac{\partial}{\partial x_i} \wedge \frac{\partial}{\partial x_j} \tag{5.54}$$

be a quadratic Poisson structure ($\Pi_{ij}^{kl} = \Pi_{ij}^{lk} = -\Pi_{ji}^{kl}$). Its curl vector field $X = D_\Omega \Pi$ with respect to the volume form $\Omega = dx_1 \wedge \cdots \wedge dx_n$ is a linear vector field which has the following expression:

$$X = \sum_{i,j,k} \Pi_{ij}^{kj} x_k \frac{\partial}{\partial x_i}. \tag{5.55}$$

Recall that, if we change the linear coordinate system linearly, the corresponding volume form will be multiplied by a constant function, and therefore the curl vector field will not be changed. So we may write the curl vector field as $X = D\Pi$, without reference to Ω. Recall from Lemma 2.6.9 that $X = D\Pi$ preserves Π and the volume form Ω: $\mathcal{L}_X \Pi = 0$, $\mathcal{L}_X \Omega = 0$.

Definition 5.6.1. If Π is a quadratic Poisson structure, then the eigenvalues of its linear curl vector field $X = D\Pi$ will be called the *eigenvalues* of Π.

Recall that an n-dimensional linear vector field X is *isochore*, i.e., it preserves the standard volume form, if and only if its eigenvalues $\lambda_1, \ldots, \lambda_n$ satisfy the resonance relation

$$\sum_{j=1}^{n} \lambda_j = 0. \tag{5.56}$$

In particular, the eigenvalues of a quadratic Poisson structure Π will satisfy the above relation.

Definition 5.6.2. A quadratic Poisson structure will be called *nonresonant* if its eigenvalues $\lambda_1, \ldots, \lambda_n$ do not satisfy any relation of resonance other than (5.56). In other words, if $\sum_{j=1}^{n} c_j \lambda_j = 0$ with $c_j \in \mathbb{Z}$ then $c_1 = \cdots = c_n$.

Definition 5.6.3. A quadratic Poisson structure of the type

$$\Pi = \sum_{1 \leq i < j \leq n} c_{ij} x_i x_j \frac{\partial}{\partial x_i} \wedge \frac{\partial}{\partial x_j}, \tag{5.57}$$

where c_{ij} are constants, is called a *diagonal* quadratic Poisson structure.

Remark 5.6.4. Any quadratic 2-vector field of the form (5.57) is automatically a Poisson structure, and can be given by a triangular r-matrix. But quadratic Poisson structures arising from r-matrices (via representations) need not be diagonal in general.

One may argue that a "generic" (in a disputable sense) quadratic Poisson structure is nonresonant. The following theorem says that "generic" quadratic Poisson structures are diagonalizable.

Theorem 5.6.5 (Dufour–Haraki [118]). *If the eigenvalues $\lambda_1, \ldots, \lambda_n$ of a quadratic Poisson structure Π does not verify any relation of the type*

$$\lambda_i + \lambda_j = \lambda_r + \lambda_s \tag{5.58}$$

with $i < j$ and $\{r, s\} \neq \{i, j\}$, then Π is diagonalizable, i.e., there is a linear coordinate system in which Π is diagonal. In particular, nonresonant quadratic Poisson structures are diagonalizable.

Proof. Notice that the condition of the above theorem implies that the eigenvalues $\lambda_1, \ldots, \lambda_n$ are pairwise different ($\lambda_1 = \lambda_2$ leads to $\lambda_1 + \lambda_2 = \lambda_1 + \lambda_1$). Thus the linear curl vector field $X = D\Pi$ is diagonalizable, i.e., there is a linear coordinate system in which X is diagonal:

$$X = \sum \lambda_i x_i \frac{\partial}{\partial x_i}. \tag{5.59}$$

Then the equation $[X, \Pi] = 0$ can be written as

$$0 = \left[\sum_i \lambda_i x_i \frac{\partial}{\partial x_i}, \sum_{rsuv} \Pi_{rs}^{uv} x_u x_v \frac{\partial}{\partial x_r} \wedge \frac{\partial}{\partial x_s} \right]$$

$$= \sum_{rsuv} \Pi_{rs}^{uv} (\lambda_u + \lambda_v - \lambda_r - \lambda_s) x_u x_v \frac{\partial}{\partial x_r} \wedge \frac{\partial}{\partial x_s},$$

whence $\Pi_{rs}^{uv} = 0$ for $\{r, s\} \neq \{u, v\}$, hence the result. $\square$

When the conditions of the above theorem are not satisfied, but the curl vector field is nontrivial, then the following decomposition theorem may be useful.

Theorem 5.6.6 (Liu–Xu [217]). *If Π is an n-dimensional quadratic Poisson structure, then it can be decomposed in a unique way as a sum of two terms:*

$$\Pi = \Pi_0 + \frac{1}{n} X \wedge I, \qquad (5.60)$$

where Π_0 is a quadratic Poisson structure with a trivial curl vector field, I denotes the radial linear vector field $\sum x_i \partial/\partial x_i$, and X is a linear vector field such that $[X, \Pi_0] = 0$.

Proof. If $\Pi = \Pi_0 + \frac{1}{n} X \wedge I$ with $D(\Pi_0) = 0$, then $D(\Pi) = D(\Pi_0) + \frac{1}{n} D(X \wedge I) = X$, and we must have $\Pi_0 = \Pi - \frac{1}{n} X \wedge I$ where X is the curl vector field of Π. Conversely, we have $[X, \Pi] = 0$, and $[I, \Pi] = 0$ (because P is a quadratic 2-vector field), therefore $[\Pi, X \wedge I] = 0$, i.e., the Poisson structure P is compatible with the Poisson structure $X \wedge I$. Hence $\Pi_0 = \Pi - \frac{1}{n} X \wedge I$ is also a Poisson structure. $\square$

Definition 5.6.7. A quadratic Poisson structure is called *exact* if its curl vector field is zero.

Example 5.6.8. A quadratic Poisson structure Π in $\mathbb{R}^3$ is exact if and only if

$$i_\Pi(dx_1 \wedge dx_2 \wedge dx_3) = df$$

where f is a homogeneous polynomial function of degree 3.

In a sense, Theorem 5.6.6 reduces the problem of classification of quadratic Poisson structures to the problem of classification of exact quadratic Poisson structures and linear vector fields which preserve them.

For example, in the case of three-dimensional quadratic Poisson structures, the curl vector field and Theorem 5.6.6 lead to the following classification:

Theorem 5.6.9 ([118, 217]). *Any quadratic Poisson structure in $\mathbb{R}^3$ can be written in a linear system of coordinates (x, y, z) as*

$$\left(\frac{\partial f}{\partial x} \frac{\partial}{\partial y} \wedge \frac{\partial}{\partial z} + \frac{\partial f}{\partial y} \frac{\partial}{\partial z} \wedge \frac{\partial}{\partial x} + \frac{\partial f}{\partial z} \frac{\partial}{\partial x} \wedge \frac{\partial}{\partial y} \right) + X \wedge \left(x \frac{\partial}{\partial x} + y \frac{\partial}{\partial y} + z \frac{\partial}{\partial z} \right) \quad (5.61)$$

where f is a homogeneous cubic polynomial and X is a linear vector field with trace 0 such that $X(f) = 0$, and the pair (X, f) belongs to one of the following nine cases:

(1) $X = 0$, f any cubic polynomial;

(2) $X = \alpha x \frac{\partial}{\partial x} + \beta y \frac{\partial}{\partial y} + \gamma z \frac{\partial}{\partial z}$ with $\alpha \neq \beta \neq \gamma \neq 0$, $f = axyz$;

(3) $X = \alpha x \frac{\partial}{\partial x} - \alpha y \frac{\partial}{\partial y}$ with $\alpha \neq 0$, $f = axyz + bz^3$;

(4) $X = \alpha x \frac{\partial}{\partial x} + \alpha y \frac{\partial}{\partial y} - 2\alpha z \frac{\partial}{\partial z}$ with $\alpha \neq 0$, $f = axyz + bx^2 z + cy^2 z$;

(5) $X = y \frac{\partial}{\partial x} + z \frac{\partial}{\partial y}$, $f = az^3 + 2bz^2 x - bzy^2$;

(6) $X = y \frac{\partial}{\partial x}$, $f = ay^3 + by^2 z + cyz^2 + dz^3$;

(7) $X = \alpha x \frac{\partial}{\partial x} + y \frac{\partial}{\partial y} \alpha y \frac{\partial}{\partial y} - 2\alpha z \frac{\partial}{\partial z}$ with $\alpha \neq 0$, $f = ay^2 z$;

(8) $X = \beta y \frac{\partial}{\partial x} - \beta x \frac{\partial}{\partial y}$ with $\beta \neq 0$, $f = az(x^2 + y^2) + bz^3$;

(9) $X = \alpha x \frac{\partial}{\partial x} + \beta y \frac{\partial}{\partial x} \alpha y \frac{\partial}{\partial y} - \beta x \frac{\partial}{\partial y} - 2\alpha z \frac{\partial}{\partial z}$ with $\alpha, \beta \neq 0$, $f = az(x^2 + y^2)$. □

The problem of classification of four-dimensional quadratic Poisson structures was studied by El Galiou [124].

5.7   Quadratization of Poisson structures

In this section, we will discuss the problem of quadratization, or more generally, normal forms for Poisson structures with zero 1-jet at a point. So let

$$\Pi = \Pi^{(2)} + \Pi^{(3)} + \cdots \qquad (5.62)$$

be the Taylor expansion of a Poisson structure Π in a local system of coordinates $(x_1, \ldots, x_n)$, which begins with terms of degree 2. Each term $\Pi^{(k)}$ is a homogeneous 2-vector field of degree k. Then the quadratic part $\Pi^{(2)}$ of Π is a quadratic Poisson structure: the equation $[\Pi, \Pi] = 0$ implies that $[\Pi^{(2)}, \Pi^{(2)}] = 0$.

Denote by

$$X = X^{(1)} + X^{(2)} + \cdots \qquad (5.63)$$

the Taylor expansion of the curl vector field of $X = D_\Omega \Pi$ of Π with respect to the volume form $\Omega = \mathrm{d}x_1 \wedge \cdots \wedge \mathrm{d}x_n$. Then $X^{(k)} = D_\Omega \Pi^{(k+1)}$. In particular, $X^{(1)}$ is the curl vector field of $\Pi^{(2)}$.

Definition 5.7.1. The eigenvalues of the linear vector field $X^{(1)}$ are called the *eigenvalues* of Π. If $\Pi^{(2)}$ is nonresonant in the sense of Definition 5.6.2, then we say that Π is *nonresonant*.

The following theorem provides a normal form à la Poincaré–Dulac for Poisson structures with zero 1-jet.

Theorem 5.7.2. *Let Π be a Poisson structure on a neighborhood of 0 in $\mathbb{C}^n$ with zero 1-jet at 0, and eigenvalues $\lambda_1, \ldots, \lambda_n$. Then there is a formal change of variables which puts Π in the following form:*

$$\Pi = \sum_{\substack{\lambda_i + \lambda_j = \langle \lambda, I \rangle \\ |I| \geq 2}} a_{ij}^I x^I \frac{\partial}{\partial x_i} \wedge \frac{\partial}{\partial x_j} \ , \tag{5.64}$$

where $I = (I_1, \ldots, I_n)$ denotes a multi-index, a_{ij}^I are constants, $\langle \lambda, I \rangle = \sum \lambda_j I_j$, $|I| = \sum I_j$, $x^I = \prod x_j^{I_j}$.

Proof. According to the isochore version of the classical Poincaré–Dulac theorem, there is a formal volume-preserving transformation of variables which puts X in Poincaré–Dulac normal form. In other words, there is a formal system of coordinates $(x_1, \ldots, x_n)$ in which the curl vector field X of Π with respect to the volume form $dx_1 \wedge \cdots \wedge dx_n$ has the following property:

$$[X, X^s] = 0. \tag{5.65}$$

where X^s denotes the semisimple part of the linear part $X^{(1)}$ of X (see Appendix A.5). The semisimple linear vector field X^s is diagonalizable, i.e., we may assume furthermore that

$$X^s = \sum_{j=1}^{n} \lambda_j x_j \frac{\partial}{\partial x_j}. \tag{5.66}$$

Then we have the following formula:

$$\left[X^s, \sum_{i,j,I} a_{ij}^I x^I \frac{\partial}{\partial x_i} \wedge \frac{\partial}{\partial x_j} \right] = \sum_{i,j,I} \left(\langle \lambda, I \rangle - \lambda_i - \lambda_j \right) a_{ij}^I x^I \frac{\partial}{\partial x_i} \wedge \frac{\partial}{\partial x_j}. \tag{5.67}$$

This formula implies that, if $\Pi = \sum_{i,j,I} a_{ij}^I x^I \frac{\partial}{\partial x_i} \wedge \frac{\partial}{\partial x_j}$ and $[X^s, \Pi] = 0$, then $a_{ij}^I = 0$ unless $\langle \lambda, I \rangle - \lambda_i - \lambda_j = 0$. Thus the theorem is proved modulo the following lemma. $\qquad\square$

Lemma 5.7.3. *With the above notations, if X is in Poincaré–Dulac normal form, then we have $[X^s, \Pi] = 0$.*

Proof. For each $q \in \mathbb{N}$ we denote by $\mathcal{A}_1^{[q]}$ (resp., $\mathcal{A}_2^{[q]}$) the space of polynomial vector fields (resp., polynomial 2-vector fields) of degree at most q in the variables $x_1, \ldots, x_n$ (which normalize X and puts X^s in diagonal form). Denote by $\mathcal{X}^{[q]}, \mathcal{S}^{[q]}$ and $\mathcal{N}^{[q]}$ the maps from $\mathcal{A}_2^{[q]}$ to itself, induced respectively from the maps $P \mapsto [X, P]$, $P \mapsto [X^s, P]$ and $P \mapsto [X - X^s, P]$ by truncating at degree q. It follows directly from the formula $[X, A \wedge B] = [X, A] \wedge B + A \wedge [X, B]$ and $[X, X^s] = 0$ that $\mathcal{N}^{[q]}$ and $\mathcal{S}^{[q]}$ commute. $\mathcal{N}^{[q]}$ is nilpotent because the map from $\mathcal{A}_1^{[q]}$ to itself defined by $Z \mapsto [X - X^s, Z]$ is nilpotent. Finally, Formula (5.67) shows that $\mathcal{S}^{[q]}$

is semisimple (with the eigenvalues $\lambda I - \lambda_i - \lambda_j$ and eigenvectors $x^I \dfrac{\partial}{\partial x_i} \wedge \dfrac{\partial}{\partial x_j}$ for $|I| \leq q$).

Return now to the relation $[X, \Pi] = 0$ (property of the curl vector field). It induces the relations

$$\mathcal{X}^{[q]}\big(\Pi^{[q]}\big) = 0,$$

where $\Pi^{[q]} = \Pi^{(2)} + \cdots + \Pi^{(q)}$ denotes the q-jet of Π. Thus $\Pi^{[q]}$ lies in the kernel of $\mathcal{X}^{[q]}$. Since $\mathcal{X}^{[q]}$ admits the Jordan decomposition $\mathcal{X}^{[q]} = \mathcal{S}^{[q]} + \mathcal{N}^{[q]}$, we also have

$$\mathcal{S}^{[q]}\big(\Pi^{[q]}\big) = 0$$

for any $q \in \mathbb{N}$. The result then follows by taking limit $q \to \infty$. $\qquad\square$

As a corollary of Theorem 5.7.2, we have:

Proposition 5.7.4. *If $\Pi = \Pi^{(2)} + \cdots$ is a Poisson structure with zero 1-jet and nonresonant quadratic part, then it admits a formal normal form of the type*

$$\Pi = \sum_{i<j} \alpha_{ij}(\rho) Y_i \wedge Y_j = \frac{1}{2} \sum_{i \neq j} \alpha_{ij}(\rho) Y_i \wedge Y_j \ , \tag{5.68}$$

where $Y_i = x_i \dfrac{\partial}{\partial x_i}$, $\rho = \prod_{k=1}^{n} x_k$ and $\alpha_{ij}(\rho) = -\alpha_{ji}(\rho)$. If, moreover, the eigenvalues of Π satisfies the ω-condition of Bruno, and Π is analytic, then there is a local analytic system of coordinates which puts Π in the above normal form.

The ω-condition of Bruno in the above theorem is the following small-divisor condition on the eigenvalues $\lambda_1, \ldots, \lambda_n$: denote

$$\omega_k = \min_{|I| \leq 2^k} \Big\{ \Big| \sum I_j \lambda_j \Big| \ ; \ I \in \mathbb{Z}^n \text{ admissible}, \sum I_j \lambda_j \neq 0 \Big\}, \tag{5.69}$$

where $|I| = \sum I_j$, $I = (I_j) \in \mathbb{Z}^n$ is called admissible if there is an index j such that $q_i \geq 0$ if $i \neq j$ and $I_j \geq -1$. We will say that the ω-condition of Bruno is satisfied if

$$\sum_{k=1}^{\infty} \frac{1}{2^k} \log \frac{1}{\omega_k} < \infty. \tag{5.70}$$

Proof. Since X is volume-preserving and its eigenvalues $\lambda_1, \ldots, \lambda_n$ do not admit any resonance relation other than $\sum \lambda_j = 0$, the isochore version of the classical Poincaré–Dulac theorem (see Appendix A.5) implies that X has a formal normal form of the type

$$X = b(\rho) \Big(\sum_{i=1}^{n} \lambda_i x_i \frac{\partial}{\partial x_i} \Big).$$

According to Bruno's theorem [48] (see Appendix A.5) that, under the ω-condition, there is a local complex analytic system of coordinates which puts X in the above normal form. (Because X preserves the volume form, the required coordinate transformation can be chosen to preserve the volume form too.) In this analytic coordinate system Π will automatically have the form $\Pi = \sum_{i<j} \alpha_{ij}(\rho) Y_i \wedge Y_j$, where $Y_i = x_i \partial/\partial x_i$. □

In fact, in the nonresonant case, we can improve Proposition 5.7.4 to obtain the following quadratization result.

Theorem 5.7.5. *If $\Pi = \Pi^{(2)} + \cdots$ is a Poisson structure with zero 1-jet at 0, whose quadratic part is nonresonant, then Π is formally quadratizable, i.e., there is a formal coordinate system in which Π is quadratic. If, moreover, the eigenvalues of Π satisfies the ω-condition of Bruno, and Π is analytic, then it admits a local analytic quadratization.*

Theorem 5.7.5 follows directly from Proposition 5.7.4 and the following two lemmas, which are valid in the formal, analytic, C^∞, as well as C^1, categories.

Lemma 5.7.6. *Consider a Poisson structure Π of the type*

$$\Pi = \frac{1}{2} \sum_{i,j} \alpha_{ij}(\rho) Y_i \wedge Y_j$$

as in Proposition 5.7.4, with $\alpha_{ij} = -\alpha_{ji}$. Put $\Gamma_i = \sum_{j=1}^{n} \alpha_{ij}$. Then we have the relation

$$\frac{d\alpha_{ij}}{d\rho}\Gamma_k + \frac{d\alpha_{jk}}{d\rho}\Gamma_i + \frac{d\alpha_{ki}}{d\rho}\Gamma_j = 0 \tag{5.71}$$

for every triplet (i, j, k). Moreover, if at least one of the eigenvalues of Π is nontrivial, then there is a function $\Gamma = \Gamma(\rho)$ such that $\Gamma(0) = 1$ and

$$\Gamma_i = \lambda_i \Gamma \quad \forall \quad i = 1, \ldots, n. \tag{5.72}$$

Proof. The first relation follows directly from the identities $[\Pi, \Pi] = 0$, $Y_i(\rho) = \rho$, $[Y_i, Y_j] = 0$, and the formula

$$[A \wedge B, C \wedge D] = [A, C] \wedge B \wedge D - [A, D] \wedge B \wedge C - [B, C] \wedge A \wedge D + [B, D] \wedge A \wedge C,$$

where A, B, C, D are arbitrary vector fields. In turn, it follows from the first relation that

$$\sum_{k=1}^{n} \left(\frac{d\alpha_{ij}}{d\rho}\Gamma_k + \frac{d\alpha_{jk}}{d\rho}\Gamma_i + \frac{d\alpha_{ki}}{d\rho}\Gamma_j \right) = 0,$$

or $\frac{d\Gamma_j}{d\rho}\Gamma_i - \frac{d\Gamma_i}{d\rho}\Gamma_j = 0$ (we have $\sum_k \Gamma_k = 0$ because the matrix (α_{ij}) is antisymmetric). It implies that the ratios Γ_i/Γ_j are constants. But we have $\Gamma_i(0) = \lambda_i$, which leads to the second relation. □

Lemma 5.7.7. *Under the assumptions and notations of the previous lemma, there is a change of variables of the form*

$$(x_1, \ldots, x_n) \mapsto (x_1 \theta_1(\rho), \ldots, x_n \theta_n(\rho)) \tag{5.73}$$

with $\theta_i(0) = 1$, which quadratizes the Poisson structure Π.

Proof. Put $x_i' = x_i \theta_i(\rho)$, where θ_i are to be determined. We have

$$
\begin{aligned}
\{x_i', x_j'\} &= \theta_i \theta_j \{x_i, x_j\} + \theta_i x_j \{x_i, \theta_j\} + \theta_j x_i \{\theta_i, x_j\} \\
&= \theta_i \theta_j x_i x_j \alpha_{ij} + \rho \theta_i \frac{\mathrm{d}\theta_j}{\mathrm{d}\rho} x_i x_j \sum_e \alpha_{ie} - \rho \theta_j \frac{\mathrm{d}\theta_i}{\mathrm{d}\rho} x_i x_j \sum_e \alpha_{je} \\
&= x_i x_j \left[\alpha_{ij} \theta_i \theta_j + \rho \left(\theta_i \frac{\mathrm{d}\theta_j}{\mathrm{d}\rho} \lambda_i - \theta_j \frac{\mathrm{d}\theta_i}{\mathrm{d}\rho} \lambda_j \right) \Gamma \right] \\
&= x_i' x_j' \alpha_{ij}(0) + x_i x_j \left[(\alpha_{ij} - \alpha_{ij}(0)) \theta_i \theta_j + \rho \left(\theta_i \frac{\mathrm{d}\theta_j}{\mathrm{d}\rho} \lambda_i - \theta_j \frac{\mathrm{d}\theta_i}{\mathrm{d}\rho} \lambda_j \right) \Gamma \right].
\end{aligned}
$$

We want to find the θ_i such that

$$\{x_i', x_j'\} = x_i' x_j' \alpha_{ij}(0) = x_i x_j \theta_i \theta_j \alpha_{ij}(0),$$

or equivalently,

$$\alpha_{ij} - \alpha_{ij}(0) = \rho \Gamma \left(\frac{\frac{\mathrm{d}\theta_i}{\mathrm{d}\rho}}{\theta_i} \lambda_j - \frac{\frac{\mathrm{d}\theta_j}{\mathrm{d}\rho}}{\theta_j} \lambda_i \right). \tag{5.74}$$

In view of Relation (5.72), Relation (5.71) becomes

$$\frac{\mathrm{d}\alpha_{ij}}{\mathrm{d}\rho} \lambda_k + \frac{\mathrm{d}\alpha_{jk}}{\mathrm{d}\rho} \lambda_i + \frac{\mathrm{d}\alpha_{ki}}{\mathrm{d}\rho} \lambda_j = 0,$$

therefore

$$(\alpha_{ij} - \alpha_{ij}(0)) \lambda_k + (\alpha_{jk} - \alpha_{jk}(0)) \lambda_i + (\alpha_{ki} - \alpha_{ki}(0)) \lambda_j = 0,$$

and in particular (supposing that $\lambda_1 \neq 0$)

$$\alpha_{ij} - \alpha_{ij}(0) = \mu_i \lambda_j - \mu_j \lambda_i$$

with

$$\mu_i = \frac{\alpha_{1i} - \alpha_{1i}(0)}{\lambda_1}.$$

We deduce from this that the system of equations (5.74) can be solved, proving the lemma. □

Let us mention here another result, about normal forms of Poisson structure with a diagonal quadratic part. We will use the notation $Y_i = x_i \frac{\partial}{\partial x_i}$.

Theorem 5.7.8 (Dufour–Wade [114]). *Let* $\Pi = \Pi^{(2)}$ *be a Poisson structure with zero 1-jet, whose quadratic part* $\Pi^{(2)} = \sum_{i<j} a_{ij} Y_i \wedge Y_j$ *is diagonal and satisfies the following hypothesis* (**H**):

(**H**) *If* $(\sum_{ij} a_{ij} I_j Y_i) \wedge Y_i \wedge Y_j = 0$ *for a multi-index* I *with* $I_j = I_j = -1$, $I_k \geq 0 \; \forall \, k \neq i, j$ *and* $\sum_{k=1}^{n} I_k \geq 1$, *then* $A.I = 0$ *where* $A = (a_{ij})$, *i.e.,* $\sum_j a_{ij} I_j = 0 \; \forall i$. *Then* Π *is formally isomorphic to a Poisson structure of the type*

$$\Pi = \sum_{\substack{i<j \\ A.I=0}} x^I \alpha_{ij}^I Y_i \wedge Y_j \tag{5.75}$$

Example 5.7.9. Let us interpret hypothesis (**H**) in the case $n = 3$. Consider a diagonal quadratic Poisson structure on a three-dimensional linear space given by

$$\Pi^{(2)} = cxy \frac{\partial}{\partial x} \wedge \frac{\partial}{\partial y} + ayz \frac{\partial}{\partial y} \wedge \frac{\partial}{\partial z} + bzx \frac{\partial}{\partial z} \wedge \frac{\partial}{\partial x},$$

where at least one of the three coefficients a, b, c is nonzero. Any multi-index I with $|I| \geq 1$ containing two negative components has one of the following forms:

1) $I = (I_1, -1, -1)$ with $I_1 > 2$;
2) $I = (-1, I_2, -1)$ with $I_2 > 2$;
3) $I = (-1, -1, I_3)$ with $I_3 > 2$.

So hypothesis (**H**) is equivalent to the combination of the following three conditions:

1) If $c - b = 0$ then $cI_1 + a = 0$ and $bI_1 + a = 0$ for any $I_1 > 2$;
2) If $c - a = 0$ then $cI_2 + b = 0$ and $aI_2 + b = 0$ for any $I_2 > 2$;
3) If $a - b = 0$ then $bI_3 + c = 0$ and $aI_3 + c = 0$, for any $I_3 > 2$.

Since at least one of the three numbers a, b, c is non-zero, it follows that hypothesis (**H**) is equivalent to the condition $a \neq b \neq c$.

We refer to [114] for a proof of Theorem 5.7.8, where the following smooth quadratization theorem for Poisson structures with a diagonal quadratic part is also obtained.

Let $\Pi^{(2)} = \sum_{i<j} a_{ij} x_i x_j \frac{\partial}{\partial x_i} \wedge \frac{\partial}{\partial x_j}$ be a diagonal quadratic Poisson structure on $\mathbb{R}^n$. Denote by $A_1 = (a_{1j}), \ldots, A_n = (a_{nj})$ the rows of $A = (a_{ij})$, and by $\lambda_1 = \sum_j a_{1j}, \ldots, \lambda_n = \sum_j a_{nj}$ the sums of the components of $A_1, \ldots, A_n$ respectively.

Theorem 5.7.10 ([114]). *Suppose that the following three conditions are satisfied:*

i) $\forall \, I = (I_1, \ldots, I_n) \in \mathbb{Z}^n$ *such that* $I_i \geq 0, \sum_{i=1}^{n} I_i > 1$, *we have*

$$A_i \neq \sum_{j=1}^{n} A_j I_j \quad \forall \, i.$$

ii) $\forall\ I = (I_1, \ldots, I_n) \in \mathbb{Z}^n$ such that $I_i = I_j = -1$, $I_k \geq 0$ if $k \neq i, j$, and $\sum_{i=1}^n I_i > 2$, there is an index $k \neq i, j$ such that

$$a_{ik} + a_{jk} \neq \sum_{l \neq i, j} a_{kl} I_l.$$

iii) *There is an index i such that the coefficients a_{ik}, for all $k \neq i$, are of same sign and such that $\lambda_j \lambda_i < 0$, $\forall j \neq i$.*

Then any C^∞-smooth Poisson structure $\Pi = \Pi^{(2)} + \Pi'$ on $\mathbb{R}^n$, where the 2-jet of Π' at 0 is zero, is locally C^∞-smoothly isomorphic to $\Pi^{(2)}$.

Theorem 5.7.10 follows from Theorem 5.7.8 and a parameterized version of Sternberg's smooth linearization theorem [321], due to Roussarie [304].

Remark 5.7.11. Some generalizations of improvements of Theorem 5.7.8 and Theorem 5.7.5 were obtained recently by Lohrmann [219]. See also [341, 169] for some other results on smooth normal forms and quadratization of Poisson structures in dimension 3 with a trivial 1-jet.

5.8   Nonhomogeneous quadratic Poisson structures

In this section, we will briefly discuss nonhomogeneous quadratic, or polynomial of degree 2, Poisson structures of the type

$$\Pi = \Pi^{(1)} + \Pi^{(2)}, \tag{5.76}$$

where $\Pi^{(1)}$ is linear and $\Pi^{(2)}$ is quadratic. Note that, in this case, the Jacobi identity $[\Pi, \Pi] = 0$ is equivalent to $[\Pi^{(1)}, \Pi^{(1)}] = [\Pi^{(2)}, \Pi^{(2)}] = [\Pi^{(1)}, \Pi^{(2)}] = 0$. Thus, if we have a nonhomogeneous quadratic Poisson structure then we have a quadratic Poisson structure which is compatible with a linear Poisson structure, and vice versa.

If Π is a homogeneous quadratic structure on a vector space $\mathbb{K}^n$, and $z \in \mathbb{K}^n$ is a point at which $\Pi(z) = 0$, then in the affine coordinate system centered at z, Π will become a nonhomogeneous quadratic Poisson structure (without the constant term). For example, the Sklyanin structure $\Pi = r^+ - r^-$ on $Mat_n(\mathbb{K})$, where r is a classical r-matrix of $gl(n, \mathbb{K})$, is homogeneous quadratic near 0, and is nonhomogeneous quadratic near the identity element $Id \in Mat_n(\mathbb{K})$. More generally, we have:

Proposition 5.8.1. *If A is an associative algebra with a unit element e and a smooth multiplicative Poisson structure Π, then in an affine coordinate system centered at e, Π is the sum of a quadratic Poisson structure Π^2 with a linear Poisson structure $\Pi^{(1)}$.*

Proof. It is a direct corollary of Theorem 5.5.8. $\square$

Remark 5.8.2. In the above theorem, if $\mathfrak{g}$ is the Lie algebra of A, $G \subset A$ is the Lie group of invertible elements of A, then (G, Π) is a Poisson–Lie group, and the linear part $\Pi^{(1)}$ of Π at the unit element e actually corresponds to the dual Lie algebra structure on $\mathfrak{g}^*$.

A large family of nonhomogeneous quadratic Poisson structures is given by the following result of Oh [284].

Theorem 5.8.3 ([284]). *Let $\mathfrak{g}$ be a Lie algebra, and μ a point in $\mathfrak{g}^*$. If the isotropy subalgebra $\mathfrak{g}_\mu = \{x \in \mathfrak{g} \mid \mathrm{ad}_x^* \mu = 0\}$ has a complementary Lie subalgebra $\mathfrak{k}$ in $\mathfrak{g}$ (i.e., $\mathfrak{g}$ is the vector-space direct sum of its subalgebras $\mathfrak{g}_\mu$ and $\mathfrak{k}$), then the transverse Poisson structure at μ of the linear Poisson structure of $\mathfrak{g}^*$ is (nonhomogeneous) quadratic.*

Proof. Denote by (x_i) a basis of $\mathfrak{g}_\mu$, and (y_k) a basis of $\mathfrak{k}$. Take $N = \mu + \mathfrak{k}^\perp$ to be the affine subspace which is transverse at μ to the coadjoint orbit of μ. According to Dirac's formula (1.56), the transverse Poisson structure on a neighborhood of μ in N is given as follows:

$$\{x_i|_N, x_j|_N\}_N(z) = \{x_i, x_j\}(z) - \sum_{k,h} \{x_i, y_k\}(z) r_{kl}(z) \{y_l, x_j\}(z), \qquad (5.77)$$

where $(r_{kl}(z))$ is the inverse of the matrix $(\{y_k, y_l\}(z))$.

In the case when $\mathfrak{g} = \mathfrak{g}_\mu + \mathfrak{k}$ and (y_k) is a basis of the Lie subalgebra $\mathfrak{k}$, the matrix $(\{y_k, y_h\})$ is a constant matrix on N, and therefore the functions r_{kh} are constant on N. On the other hand, the functions $\{x_i, x_j\}$, $\{x_i, y_k\}$ and $\{y_h, x_j\}$ are affine functions on N. Hence the Dirac formula shows that the Poisson structure on N is nonhomogeneous quadratic with respect to the coordinate system (x_i). $\square$

Remark 5.8.4. As observed by Bhaskara and Rama [29], the matrix (r_{kl}) in the above proof is actually a triangular r-matrix. More precisely, $r = \sum_{i<j} r_{ij} y_i \wedge y_j \in \mathfrak{k} \wedge \mathfrak{k} \subset \mathfrak{g} \wedge \mathfrak{g}$ satisfies the CYBE $[r, r] = 0$.

Another interesting result about nonhomogeneous quadratic Poisson structures, which seems to be strongly related to Proposition 5.8.1 and Theorem 5.8.3, is the following:

Theorem 5.8.5 (Diatta–Medina [101]). *Let (G, ω^+) be a Lie group with a left-invariant symplectic form ω^+. Denote by $r \in \mathfrak{g} \wedge \mathfrak{g}$ the triangular r-matrix associated to ω^+, and r^+ (resp. r^-) the corresponding left- (resp. right-) invariant Poisson tensor (r^+ is the Poisson tensor dual to ω^+). Then the Poisson–Lie tensor $\Pi = r^+ - r^-$ is polynomial of degree 2 with respect to the flat affine structure on G defined by ω^+.*

The flat affine structure ∇ on the symplectic Lie group (G, ω^+) is defined as follows [212]:

$$\omega^+(\nabla_{x^+} y^+, z^+) := -\omega^+(y^+, [x^+, z^+]), \qquad (5.78)$$

where x^+, y^+, z^+ denote left-invariant vector fields.

The proof of Theorem 5.8.5, given in [101], is based on direct algebraic calculations. It would be nice to find a more conceptual proof. $\square$

Chapter 6

Nambu Structures and Singular Foliations

6.1   Nambu brackets and Nambu tensors

Nambu structures were first introduced by Nambu [275] in a special case, and then generalized and formalized by Takhtajan [328].

Definition 6.1.1 ([328]). Let q be a natural number. A smooth *Nambu structure* of order q on a smooth manifold M is a skew-symmetric q-linear operator, called the *Nambu bracket* and denoted by $\{\ldots\}$, from $\mathcal{C}^\infty(M) \times \cdots \times \mathcal{C}^\infty(M)$ to $\mathcal{C}^\infty(M)$, which satisfies the following conditions:

a) Leibniz rule:

$$\{f_1, \ldots, f_{q-1}, g_1 g_2\} = \{f_1, \ldots, f_{q-1}, g_1\} g_2 + g_1 \{f_1, \ldots, f_{q-1}, g_2\}, \quad (6.1)$$

b) *fundamental identity*:

$$\{f_1,\ldots,f_{q-1},\{g_1,\ldots,g_q\}\} = \sum_{i=1}^{q} \{g_1,\ldots,g_{i-1},\{f_1,\ldots,f_{q-1},g_i\},g_{i+1},\ldots,g_q\} \quad (6.2)$$

$\forall \, f_i, g_i \in \mathcal{C}^\infty(M)$.

It is clear from the above definition that a Nambu structure of order 2 is nothing but a Poisson structure. In general, Nambu brackets may be viewed as q-ary generalizations of Poisson brackets.

Given a Nambu bracket of order q and a $(q-1)$-tuple of functions $(f_1, \ldots, f_{q-1})$, one defines the *Hamiltonian vector field* of $(f_1, \ldots, f_{q-1})$ by the following formula:

$$X_{f_1,\ldots,f_{q-1}}(g) = \{f_1, \ldots, f_{q-1}, g\}. \quad (6.3)$$

It follows from the Leibniz rule that the vector field $X_{f_1,\ldots,f_{q-1}}$ is well defined, and from the anti-symmetricity of the Nambu bracket that the functions $f_1,\ldots,f_{q-1}$ are first integrals of $X_{f_1,\ldots,f_{q-1}}$.

Exercise 6.1.2. Show that, given a Nambu bracket of order $q > 1$ and a smooth function g, the following formula will give a Nambu bracket of order $q - 1$:

$$\{f_1,\ldots,f_{q-1}\}_g := \{f_1,\ldots,f_{q-1},g\} . \tag{6.4}$$

Similarly to the case of Poisson manifolds, the Leibniz rule together with the anti-symmetricity means that Nambu brackets can be characterized by q-vector fields: for each Nambu bracket $\{\ldots\}$ of order q there is a unique q-vector field Λ, called its *Nambu tensor*, such that

$$\{f_1,\ldots,f_q\} = \Lambda(f_1,\ldots,f_q) := \langle \mathrm{d}f_1 \wedge \cdots \wedge \mathrm{d}f_q, \Lambda \rangle . \tag{6.5}$$

Exercise 6.1.3. Show that the fundamental identity (6.2) is equivalent to the condition that every Hamiltonian vector field preserves the Nambu tensor:

$$\mathcal{L}_{X_{f_1,\ldots,f_q}} \Lambda = 0 . \tag{6.6}$$

Exercise 6.1.4. Show that a q-vector field Λ on a manifold M is a Nambu tensor if and only if its restriction to the open subset $M_0 = \{x \in M \mid \Lambda(x) \neq 0\}$ is a Nambu tensor.

When $q = 1$, any vector field is a Nambu tensor. Nambu tensors of order $q = 2$ are the same as Poisson tensors, as mentioned earlier. Nambu tensors of order $q \geq 3$ are characterized by the following decomposition theorem.

Theorem 6.1.5 ([140, 9, 274]). *Suppose that $q \geq 3$. Then a q-vector field Λ on an m-dimensional manifold M is a Nambu tensor if and only if for every point $z \in M$ such that $\Lambda(z) \neq 0$, there is a local system of coordinates $(x_1,\ldots,x_m)$ in a neighborhood of z, in which Λ has the following form:*

$$\Lambda = \frac{\partial}{\partial x_1} \wedge \cdots \wedge \frac{\partial}{\partial x_q} . \tag{6.7}$$

In such a local coordinate system, the Nambu bracket has the form

$$\{f_1,\ldots,f_q\} = \frac{\partial(f_1,\ldots,f_q)}{\partial(x_1,\ldots,x_q)} . \tag{6.8}$$

Proof. Since $\Lambda(z) \neq 0$, there is a Hamiltonian vector field $X = X_{f_1,\ldots,f_{q-1}}$ such that $X(z) \neq 0$. Let f_q be a smooth function defined in a neighborhood of z such that $X(f_q) = 1$. For $i = 1,\ldots,q$, we denote by X_i the Hamiltonian vector field of

$$F_i = (f_1,\ldots,f_{i-1},f_{i+1},\ldots,f_q) .$$

The vector fields X_i are linearly independent, because $X_i(f_i) = \pm 1$ and $X_i(f_j) = 0 \: \forall \: i \neq j$. Moreover, they commute pairwise. Indeed, we have $\mathcal{L}_{X_i}(df_j) = d(X_i(f_j)) = 0 \: \forall \: i, j$, and $\mathcal{L}_{X_i}\Lambda = 0$. In other words, X_i preserves $df_1, \ldots, df_q$ and Λ. So it also preserves anything which can be created from these entities:

$$[X_i, X_j] = \mathcal{L}_{X_i}X_j = \mathcal{L}_{X_i}\langle df_1 \wedge \cdots \wedge df_{j-1} \wedge df_{j+1} \wedge \cdots \wedge f_q, \Lambda \rangle = 0 \: .$$

Applying Frobenius Theorem 1.5.3, we get a local regular q-dimensional foliation generated by $X_1, \ldots, X_q$, and a local coordinate system

$$(x_1, \ldots, x_m) := (f_1, \ldots, f_q, y_1, \ldots, y_{m-q}) \qquad (6.9)$$

such that $y_1, \ldots, y_{m-q}$ are first integrals of the foliation (i.e., $X_i(y_j) = 0 \: \forall i, j$). In this coordinate system we have

$$\{x_1, \ldots, x_q\} = 1 \qquad (6.10)$$

and

$$\{x_1, \ldots, x_{i-1}, x_{i+1}, \ldots, x_q, y_j\} = X_i(y_j) = 0 \: \forall \: i, j \: . \qquad (6.11)$$

To prove that $\Lambda = \frac{\partial}{\partial x_1} \wedge \cdots \wedge \frac{\partial}{\partial x_q}$, it remains to show that for any $s \geq 2$ and any indices $i_1 \neq \cdots \neq i_{q-s}, j_1 \neq \cdots \neq j_s$ we have

$$\{x_{i_1}, \ldots, x_{i_{q-s}}, y_{j_1}, \ldots, y_{j_s}\} = 0 \: .$$

Consider the case $q - s \geq 1$. By rearranging the indices if necessary, we may assume that $q \neq i_2, \ldots, i_{q-2}$. Then using (6.10) and (6.11) repetitively together with the Leibniz rule and the fundamental identity, we get:

$$\begin{aligned}
&\{x_{i_1}, \ldots, x_{i_{q-s}}, y_{j_1}, \ldots, y_{j_s}\} \\
&= \{x_{i_1}\{x_1, \ldots, x_q\}, x_{i_2}, \ldots, x_{i_{q-s}}, y_{j_1}, \ldots, y_{j_s}\} \\
&= \frac{1}{2}\{\{x_1, \ldots, x_{i_1-1}, x_{i_1}^2, x_{i_1+1}, \ldots, x_q\}, x_{i_2}, \ldots, x_{i_{q-s}}, y_{j_1}, \ldots, y_{j_s}\} \\
&= \frac{1}{2}\{x_1, \ldots, x_{i_1-1}, x_{i_1}^2, x_{i_1+1}, \ldots, x_{q-1}, \{x_q, x_{i_2}, \ldots, x_{i_{q-s}}, y_{j_1}, \ldots, y_{j_s}\}\} \\
&= x_{i_1}\{x_1, \ldots, x_{q-1}, \{x_q, x_{i_2}, \ldots, x_{i_{q-s}}, y_{j_1}, \ldots, y_{j_s}\}\} \\
&= x_{i_1}\{\{x_1, \ldots, x_q\}, x_{i_2}, \ldots, x_{i_{q-s}}, y_{j_1}, \ldots, y_{j_s}\} \\
&= x_{i_1}\{1, x_{i_2}, \ldots, x_{i_{q-s}}, y_{j_1}, \ldots, y_{j_s}\} = 0 \: .
\end{aligned}$$

Similarly, when $s = q$ we have

$$\begin{aligned}
\{y_{j_1}, \ldots, y_{j_q}\} &= \{y_{j_1}\{x_1, \ldots, x_q\}, y_{j_2}, \ldots, y_{j_q}\} \\
&= \{\{y_{j_1}x_1, x_2, \ldots, x_q\}, y_{j_2}, \ldots, y_{j_q}\} \: .
\end{aligned}$$

The fundamental identity (6.2) allows us to decompose the last bracket into the sum of brackets which contain at least one entry of type x_i and one entry of type y_j. So the previous equalities imply that this bracket is also equal to 0. $\square$

Exercise 6.1.6. Find the place(s) in the above proof where we use the fact that $q \geq 3$.

Exercise 6.1.7. Show that if Λ is a Nambu structure of order $q \geq 3$ and $x_1, \ldots, x_r$ are r functions $(r < q)$ such that $i_{dx_1 \wedge \cdots \wedge dx_r} \Lambda(z) \neq 0$ at a point z, then in a neighborhood of z these functions can be completed into a canonical coordinate system $(x_1, \ldots, x_m)$, i.e., a coordinate system in which $\Lambda = \frac{\partial}{\partial x_1} \wedge \cdots \wedge \frac{\partial}{\partial x_q}$.

A q-vector field Λ is called *decomposable* if it can be written as the wedge product of q vector fields: $\Lambda = X_1 \wedge \cdots \wedge X_q$. According to Theorem 6.1.5, Nambu tensors of order $q \geq 3$ are locally decomposable near non-vanishing points. This is one of the differences between Nambu and Poisson structures: a Poisson tensor is not decomposable at a point of rank ≥ 4.

Example 6.1.8. (Direct product). If Λ_i is a Nambu tensor of order q_i on M_i $(i = 1, 2)$ then $\Lambda_1 \wedge \Lambda_2$ is a Nambu tensor of order $q_1 + q_2$ on $M_1 \times M_2$.

Remark 6.1.9. The q-vector field $\frac{\partial}{\partial x_1} \wedge \cdots \wedge \frac{\partial}{\partial x_q} + \frac{\partial}{\partial y_1} \wedge \cdots \wedge \frac{\partial}{\partial y_q}$ is not a Nambu tensor when $q \geq 3$ because it is not decomposable. So the direct sum of two Nambu tensor is not a Nambu tensors in general, unlike the Poisson case.

Exercise 6.1.10. Show that if Λ is a Nambu tensor of order $q \geq 3$ and g is a smooth function, then $g\Lambda$ is a Nambu tensor. In particular, any m-vector field on an m-dimensional manifold is a Nambu tensor.

Exercise 6.1.11. Show that if Λ is a Nambu tensor then the Schouten bracket of Λ with itself vanishes: $[\Lambda, \Lambda] = 0$. More generally, if Λ is a Nambu tensor of order q and $f_1, \ldots, f_r$ are smooth functions, $0 \leq r \leq q - 1$, then

$$[i_{df_1 \wedge \cdots \wedge df_r} \Lambda, \Lambda] = 0 . \tag{6.12}$$

(Hint: use Exercise 6.1.7).

A consequence of Theorem 6.1.5 is that a nontrivial Nambu structure of order $q \geq 3$ is essentially just a *singular q-dimensional foliation with a leaf-wise volume form*. Let us explain this statement:

Consider the open subset $M_0 = \{x \in M \mid \Lambda(x) \neq 0\}$ of M consisting of the regular points of Λ. (Generically, M_0 is dense in M.) Locally near a point $x \in M_0$ we have $\Lambda = \frac{\partial}{\partial x_1} \wedge \cdots \wedge \frac{\partial}{\partial x_q}$. As a consequence, the *characteristic distribution* $\mathcal{C}_\Lambda$ of Λ, which by definition is the distribution on M generated by the Hamiltonian vector fields of Λ, is a regular distribution, which near x is simply spanned by $\frac{\partial}{\partial x_1}, \ldots, \frac{\partial}{\partial x_q}$. This distribution is clearly integrable and is the tangent distribution of a regular q-dimensional foliation in M_0. Locally, in a coordinate system $(x_1, \ldots, x_m)$ where $\Lambda = \frac{\partial}{\partial x_1} \wedge \cdots \wedge \frac{\partial}{\partial x_q}$, the leaves of this associated foliation is given simply by the equations $\{x_{q+1} = \text{constant}, \ldots, x_m = \text{constant}\}$. Λ can be restricted to the leaves of this foliation. On each leaf, Λ becomes a non-vanishing multi-vector field of top order, so it is dual to a volume form η on the leaf, i.e.,

$\langle \eta, \Lambda \rangle = 1$. Conversely, given a regular q-dimensional foliation and a leaf-wise volume form, the q-vector field dual to the volume form will be a Nambu tensor.

When considered over M, the characteristic distribution $\mathcal{C}_\Lambda$ of Λ is still integrable, and the *associated singular foliation* $\mathcal{F}_\Lambda$ has two types of leaves: singular zero-dimensional leaves (i.e., points), which are the points of the set $M \backslash M_0 = \{x \in M \mid \Lambda(x) = 0\}$, and regular q-dimensional leaves lying in M_0.

Remark 6.1.12. The discussion above shows that, although their definition presents them as generalizations of general Poisson structures, it is more realistic to consider Nambu structures as generalizations of *Poisson structures of maximal rank* 2.

At first sight, the class of singular foliations arising from Nambu structures looks rather poor: they are allowed to have only two types of leaves, namely regular leaves and zero-dimensional leaves. This is however a wrong impression. We will say that two singular foliations $\mathcal{F}_1$ and $\mathcal{F}_2$ on a manifold M are *essentially the same* (or essentially coincide) if for almost every point $x \in M$ we have $T_x \mathcal{F}_1 = T_x \mathcal{F}_2$. Then any singular foliation can be essentially given by a Nambu structure:

Proposition 6.1.13. *Let $\mathcal{F}$ be a smooth singular foliation of dimension q (i.e., $\dim T_x \mathcal{F} = q$ almost everywhere) on a smooth manifold M. Then $\mathcal{F}$ can be essentially generated by a Nambu tensor, i.e., there is a smooth Nambu tensor Λ of order q on M such that $\mathcal{F}_\Lambda$ essentially coincides with $\mathcal{F}$.*

Proof (sketch). We can construct q smooth vector fields $X_1, \ldots, X_q$ on M, which are tangent to $\mathcal{F}$, and which are linearly independent almost everywhere. Now put $\Lambda = X_1 \wedge \cdots \wedge X_q$. $\square$

A priori, the foliation $\mathcal{F}_\Lambda$ in the above proposition may contain many more singularities than $\mathcal{F}$. There is a process, which we call saturation, for removing "unnecessary" singularities:

Let $\mathcal{F}_1$ and $\mathcal{F}_2$ be two singular foliations on a manifold M. We will say that $\mathcal{F}_2$ *contains* $\mathcal{F}_1$ if $T_x \mathcal{F}_2 \supset T_x \mathcal{F}_1 \; \forall \; x \in M$.

Definition 6.1.14. The *saturation* of a singular foliation $\mathcal{F}$ is a singular foliation, denoted by $Sat(\mathcal{F})$, which essentially coincides with $\mathcal{F}$ almost everywhere, and which contains any singular foliation which essentially coincides with $\mathcal{F}$. When $\mathcal{F} = Sat(\mathcal{F})$ we will say that $\mathcal{F}$ is a *saturated singular foliation*.

The above definition does not say that the saturation of a singular foliation always exists. What is obvious is that if $Sat(\mathcal{F})$ exists then it is unique. If $\mathcal{F}$ is a regular foliation then it is clear that $Sat(\mathcal{F}) = \mathcal{F}$.

The process of saturation (when it works) makes a singular foliation "smoother", leaving out some "details" from it. Sometimes these "details" are just cumbersome and it's best to forget about them. Sometimes they are what make a singular foliation interesting.

Example 6.1.15. Consider the symplectic foliation $\mathcal{F}_\Pi$ of the linear Poisson structure $\Pi = x \frac{\partial}{\partial x} \wedge \frac{\partial}{\partial y}$ in $\mathbb{R}^2$. It has two regular two-dimensional leaves, and a line of singular 0-dimensional leaves (the line $\{x = 0\}$). $Sat(\mathcal{F}_\Pi)$ consists of just one leaf, i.e., the whole plane.

Example 6.1.16. Consider the following (rather pathological) singular foliation on $\mathbb{R}^2 = \{(x, y)\}$: the half-plane $\{x < 0\}$ is a two-dimensional leaf, and each point (x, y) with $x \geq 0$ is a 0-dimensional leaf. This singular foliation does not admit a saturation.

Excluding pathological situations such as in the above example, then the saturation of a singular foliation really exists. More precisely, we have:

Theorem 6.1.17. *If $\mathcal{F}$ is a singular foliation of dimension q such that $\dim T_x\mathcal{F} = q$ almost everywhere, then $Sat(\mathcal{F})$ exists.*

The proof of Theorem 6.1.17 is postponed to the next section, where we will show another way of defining $Sat(\mathcal{F})$.

Exercise 6.1.18. Show that, if a compact Lie group G acts on a manifold M, then the associated singular foliation on M given by the orbits of G is saturated. More generally, any orbit-like singular foliation is saturated. (An *orbit-like foliation* [261] is a singular foliation which near each point is locally isomorphic to a singular foliation given by the orbits of an action of a compact Lie group on a manifold).

Similarly to the case of vector fields, we will say that two Nambu structures Λ_1, Λ_2 (not necessarily of the same order) on a manifold *commute* if their Schouten bracket vanishes: $[\Lambda_1, \Lambda_2] = 0$. This leads to the notion of *commuting singular foliations*, i.e., foliations which can be essentially generated by commuting Nambu tensors. This seems to be an interesting geometric notion, which has been very little studied, to our knowledge. We will list here a few basic facts about commuting Nambu structures.

Lemma 6.1.19. *If a Nambu structure Λ_1 of degree q_1 on a manifold M commutes with a Nambu structure Λ_2 of degree q_2, and $\Lambda_1 \wedge \Lambda_2(m) \neq 0$ at a point m, then there is a local system of coordinates $(x_1, \ldots, x_n)$ such that in a neighborhood of m we have*

$$\Lambda_1 = \partial/\partial x_1 \wedge \cdots \wedge \partial/\partial x_{q_1}, \tag{6.13}$$
$$\Lambda_2 = \partial/\partial x_{q_1+1} \wedge \cdots \wedge \partial/\partial x_{q_1+q_2}. \tag{6.14}$$

Lemma 6.1.20. *If two Nambu structures Λ_1, Λ_2 commute, then their exterior product $\Lambda_1 \wedge \Lambda_2$ is again a Nambu structure.*

Lemma 6.1.21. *If two Nambu structures Λ_1, Λ_2 commute, and X is a vector field such that $\mathcal{L}_X\Lambda_1 = [X, \Lambda_1] = 0$ and $X \wedge \Lambda_1 = 0$, then $\mathcal{L}_X(\Lambda_1 \wedge \Lambda_2) = 0$.*

Lemma 6.1.22. *Let Λ_1, Λ_2 be two commuting Nambu structures, and let L be a leaf of the associated foliation of Λ_1. If $\Lambda_1 \wedge \Lambda_2(x_0) \neq 0$ for some point $x_0 \in L$, then $\Lambda_1 \wedge \Lambda_2(x) \neq 0$ for any point $x \in L$. If $\Lambda_1 \wedge \Lambda_2(x_0) = 0$ for some point $x_0 \in L$, then $\Lambda_1 \wedge \Lambda_2(x) = 0$ for any point $x \in L$.*

The proofs of the above lemmas will be left as exercises. $\square$

6.2   Integrable differential forms

Let Ω be a smooth volume form on a smooth m-dimensional manifold M. Recall that the map $\Lambda \mapsto i_\Lambda \Omega$ is a linear isomorphism from the space $\mathcal{V}^q(M)$ of smooth q-vector fields on M to the space Ω^{m-q} of smooth differential $(m-q)$-forms on M. The following proposition characterizes Nambu structures in terms of differential forms.

Proposition 6.2.1. *Suppose that Λ is a q-vector field, and $q \geq 3$, Ω is a volume form on an m-dimensional manifold M, $m - q = p \geq 0$. Then Λ is a Nambu tensor if and only if the p-form $\omega = i_\Lambda \Omega$ satisfies the following conditions:*

$$i_A \omega \wedge \omega = 0 \; \forall \; (p-1)\text{-vector field } A \qquad (6.15)$$

and

$$i_A \omega \wedge d\omega = 0 \; \forall \; (p-1)\text{-vector field } A. \qquad (6.16)$$

Proof. Suppose that Λ is a Nambu tensor, and let x be an arbitrary point of M. If $\Lambda(x) = 0$ then $\omega(x) = 0$ and Relations (6.15) and (6.16) are obviously satisfied at point x. Suppose now that $\Lambda(x) \neq 0$. Then there is a local coordinate system $(x_1, \ldots, x_m)$ near x in which we have $\Lambda = \frac{\partial}{\partial x_1} \wedge \cdots \wedge \frac{\partial}{\partial x_q}$. Write the volume form as $\Omega = f dx_1 \wedge \cdots \wedge dx_m$, where f is a smooth function. Then we have

$$\omega = f dx_{q+1} \wedge \cdots \wedge dx_m \quad \text{and} \quad d\omega = df \wedge dx_{q+1} \wedge \cdots \wedge dx_m \ .$$

These formulae imply (6.15) and (6.16) in a neighborhood of x.

Conversely, suppose that ω satisfies (6.15) and (6.16). We will assume that $p > 0$ (the case $p = 0$ is obvious: any m-vector field is a Nambu tensor, see Theorem 6.1.5 and Exercise 6.1.10). Let x be a point such that $\Lambda(x) \neq 0$. Then $\omega(x) \neq 0$. Condition (6.15) together with $\omega(x) \neq 0$ means that ω is locally *decomposable* in a neighborhood of x. In other words, there are differential 1-forms $\alpha_1, \ldots, \alpha_p$, which are linearly independent in a neighborhood of x, such that

$$\omega = \alpha_1 \wedge \cdots \wedge \alpha_p \ . \qquad (6.17)$$

(See, e.g., [43].) Choose p vector fields $X_1, \ldots, X_p$ in a neighborhood of x, which satisfy $\langle X_i, \alpha_j \rangle = \delta_{ij}$ (where δ_{ij} is the Kronecker symbol: $\delta_{ii} = 1$ and $\delta_{ij} = 0$ if $i \neq j$). Take $A_j = X_1 \wedge \cdots \wedge \widehat{X_j} \wedge \cdots \wedge X_p$. (The hat means that the term $\widehat{X_j}$ is missing in the product.) We have $i_{A_j} \omega = \pm \alpha_j$ for every $j = 1, \ldots, p$. Condition (6.16) gives $\alpha_j \wedge d\omega = 0$. As we have

$$d\omega = \sum_{k=1}^{p} \alpha_1 \wedge \cdots \wedge \alpha_{k-1} \wedge d\alpha_k \wedge \alpha_{k+1} \wedge \cdots \wedge \alpha_p \ ,$$

we obtain that

$$d\alpha_j \wedge \alpha_1 \wedge \cdots \wedge \alpha_p = 0 \; \forall \; j = 1, \ldots, p. \qquad (6.18)$$

The above condition (6.18) is known as *Frobenius integrability condition* for a p-tuple of 1-forms $(\alpha_1, \ldots, \alpha_p)$. This condition is equivalent to the condition that the annulator distribution of $\alpha_1, \ldots, \alpha_p$, i.e., the distribution consisting of tangent vectors whose pairings with $\alpha_1, \ldots, \alpha_p$ vanish, is a (regular) *involutive* distribution of rank $q = m-p$. (Exercise: show this equivalence.) Frobenius Theorem 1.5.3 then implies that there is a local system of coordinates $(x_1, \ldots, x_m)$ in a neighborhood of x (such that $x_1, \ldots, x_p$ are first integrals for the associated foliation of the above annulator distribution), such that

$$\alpha_i = \sum_{j=1}^{p} a_{ij} \mathrm{d}x_j \quad \forall\, i = 1, \ldots, p \,, \tag{6.19}$$

where a_{ij} are some smooth functions in a neighborhood of x. The above expression for α_i under the integrability condition (6.18) is also known as the *Frobenius theorem*.

So locally we have $\omega = \alpha_1 \wedge \cdots \wedge \alpha_p = g \mathrm{d}x_1 \wedge \cdots \wedge \mathrm{d}x_p$ for some smooth function g. If the local expression of Ω is $f \mathrm{d}x_1 \wedge \cdots \wedge \mathrm{d}x_m$ $(f \neq 0)$ then

$$\Lambda = \pm \frac{g}{f} \frac{\partial}{\partial x_{p+1}} \wedge \cdots \wedge \frac{\partial}{\partial x_m}.$$

Now we can use Theorem 6.1.5 (together with Exercises 6.1.4 and 6.1.10) to deduce that Λ is a Nambu tensor. $\qquad\qquad\qquad\qquad\qquad\qquad\qquad\qquad\qquad\qquad\qquad \square$

Exercise 6.2.2. Show that Proposition 6.2.1 is also true in the case $q = 2$ under the additional assumption that $\operatorname{rank} \Lambda(x) \leq 2 \; \forall\, x \in M$.

Definition 6.2.3. An *integrable differential form* of order p is a differential p-form ω which satisfies Conditions (6.15) and (6.16).

Exercise 6.2.4. Show that if ω is an integrable p-form and f is a smooth function, then $f\omega$ is also an integrable p-form, and $\mathrm{d}\omega$ is an integrable $(p+1)$-form.

In particular, if ω is a 1-form, then Condition (6.15) is empty, and Condition (6.16) is written as

$$\omega \wedge \mathrm{d}\omega = 0 \tag{6.20}$$

and is known as the *integrability condition* for 1-forms.

Proposition 6.2.1 means that Nambu tensors are dual to integrable forms in a natural way. Hence each integrable differential p-form (which is nontrivial almost everywhere) defines a singular foliation of codimension p, and that's why it is called integrable. In the regular region $M_0 = \{x \in M \mid \omega(x) \neq 0\}$, the *annulator distribution* of ω, generated by smooth vector fields X which annulate ω, i.e., $i_X \omega = 0$, is a regular involutive distribution of codimension p, so it is the tangent distribution to a regular foliation of codimension p. We can extend this foliation in a stupid way to the singular set of ω, by declaring that each singular point of ω is a zero-dimensional leaf. This is exactly what we get by considering a dual Nambu structure Λ of ω (i.e., $i_\Lambda \Omega = \omega$ where Ω is a volume form). We will show how to define the saturation of this singular foliation.

It is easy to see that if ω is integrable then its annulator distribution, defined as above, is still involutive when considered over the whole manifold M. Unfortunately, as we mentioned in Chapter 1, for singular foliations, involutivity does not guarantee integrability, though under some additional assumptions it does.

Example 6.2.5. Consider a 1-form α defined on $\mathbb{R}^3 = \{(x_1, x_2, x_3)\}$ by the following formula: $\alpha = \exp(\frac{-1}{x_1})dx_2$ when $x_1 > 0$, $\alpha = \exp(\frac{1}{x_1})dx_3$ when $x_1 < 0$, and $\alpha = 0$ when $x_1 = 0$. Then α is a smooth integrable 1-form. Its annulator distribution is involutive but not integrable.

To avoid pathologies such as in the above example, we will introduce another distribution:

Definition 6.2.6. Let ω be a differential p-form on a manifold M. Then the *maximal invariant distribution* of ω, denoted by $\mathcal{D}_\omega$, is the singular distribution on M which is generated by smooth vector fields X which satisfy the following properties:

a) X annulates ω: $i_X\omega = 0$.
b) X preserves ω *projectively* in the following weak sense: denote by ϕ_X^t the local flow of X, then for any point $x \in M$ and $\tau \in \mathbb{R}$ such that $\phi_X^\tau(x)$ is well defined and $\omega(x) \neq 0$, there is a number $f(\tau, x)$ such that $((\phi_X^\tau)^*\omega)(x) = f(\tau, x)\omega(x)$.

Exercise 6.2.7. Show that, for an arbitrary given smooth p-form ω, the set of vector fields which satisfy Conditions a) and b) in the above definition is a $C^\infty(M)$-module.

Exercise 6.2.8. Show that, if f is a function such that $f(x) \neq 0$ almost everywhere, then $\mathcal{D}_{f\omega} = \mathcal{D}_\omega$ for any differential form ω.

Exercise 6.2.9. Show that if ω is an integrable p-form and x is a regular point, i.e., $\omega(x) \neq 0$, then $(\mathcal{D}_\omega)_x = (Ann_\omega)_x$ and $\dim(\mathcal{D}_\omega)_x = m - p$ where m is the dimension of the manifold.

Proposition 6.2.10. *If ω is an arbitrary smooth p-form which is non-vanishing almost everywhere, then its maximal invariant distribution $\mathcal{D}_\omega$ is an integrable distribution.*

Proof. Let X be an arbitrary vector field which preserves ω projectively and such that $i_X\omega = 0$. We have to show that X preserves $\mathcal{D}_\omega$. Then we can apply Theorem 1.5.1 to conclude that $\mathcal{D}_\omega$ is integrable. Let $x \in M$ be an arbitrary point, Y_x be an arbitrary vector in $(\mathcal{D}_\omega)_x$. Suppose that $\varphi_X^\tau(x)$ is well defined for some $\tau > 0$, where φ_X^t denotes the local flow of X. We have to show that $(\varphi_X^\tau)_*(Y_x) \in (\mathcal{D}_\omega)_{\varphi_X^\tau(x)}$. Let Y be a vector field which annulates ω and preserves ω projectively, and such that the value of Y at x is Y_x. We may assume that Y vanishes outside a small neighborhood of x (see Exercise 6.2.8), so that $Y^\tau := (\varphi_X^\tau)_*Y$ is well defined and is supported in a small neighborhood of $\varphi_X^\tau(x)$. Since φ_X^τ preserves ω projectively, and since $i_Y\omega = 0$, we obtain that $i_{(Y^\tau)}\omega = 0$. (We use the fact that $\omega \neq 0$

almost everywhere to show that this equality holds almost everywhere, but then by continuity it holds everywhere.) Similarly, one can show that Y^τ preserves ω projectively, since both X and Y preserves ω projectively. Hence, by definition, Y^τ lies in the family of vector fields which generate $\mathcal{D}_\omega$. In particular, we have $(\varphi_X^\tau)_*(Y_x) = Y_{\varphi_X^\tau(x)}^\tau \in (\mathcal{D}_\omega)_{\varphi_X^\tau(x)}.$ □

Definition 6.2.11. If ω is an integrable differential form, then the associated singular foliation $\mathcal{F}_\omega$ of its maximal invariant distribution $\mathcal{D}_\omega$ will be called the *maximal invariant foliation* of ω, or simply the *associated foliation* of ω.

Theorem 6.1.17 is now a direct consequence of Proposition 6.1.13 and the following result:

Proposition 6.2.12. *If ω is an integrable differential form which is non-vanishing almost everywhere, then its maximal invariant foliation $\mathcal{F}_\omega$ is saturated (i.e., it is the saturation of itself).*

Proof. Let $\mathcal{F}$ be any singular foliation which essentially coincides with $\mathcal{F}_\omega$. We must show that $\mathcal{F} \subset \mathcal{F}_\omega$. Let X be any smooth vector field tangent to $\mathcal{F}$. We will show that X satisfies the conditions of Definition 6.2.6. Note that X is tangent to $\mathcal{F}_\omega$ at least almost everywhere, so by definition of $\mathcal{F}_\omega$ we have $i_X\omega = 0$ almost everywhere, which implies by continuity that $i_X\omega = 0$ everywhere. Let us verify that X preserves ω projectively in the weak sense. Assume that x, y are two points such that $y = \phi_X^\tau(x)$ for some τ and $\omega(x), \omega(y) \neq 0$. We have to show that $((\phi_X^\tau)^*\omega)(x) = c\omega(x)$ for some number c. By continuity, it is enough to prove this in the case x is a generic point. In particular, we may assume that x and y are two regular points of $\mathcal{F}_\omega$, and $\mathcal{F}$ coincides with $\mathcal{F}_\omega$ at these points. But then $\phi_*^\tau(T_x\mathcal{F}_\omega) = T_y\mathcal{F}_\omega$. Since ω is integrable, $\omega(x)$ (resp. $\omega(y)$) is determined uniquely by $T_x\mathcal{F}_\omega$ (resp. $T_y\mathcal{F}_\omega$) up to a multiplicative constant, and we also have that $(\phi_X^\tau)^*(\omega(y))$ is equal to $\omega(x)$ up to a multiplicative constant.   □

6.3   Frobenius with singularities

Let $(\alpha_1, \ldots, \alpha_p)$ be a p-tuple of germs at 0 of holomorphic 1-forms in $(\mathbb{C}^n, 0)$. We will say that $\alpha = (\alpha_1, \ldots, \alpha_p)$ is *completely integrable* (resp. *formally completely integrable*) if there are germs of holomorphic (resp. formal) functions f_i, g_{ij} ($1 \leq i, j \leq p$) such that $\det(g_{ij}(0)) \neq 0$ and $\alpha_i = \sum_{j=1}^p g_{ij}\mathrm{d}f_j \; \forall \, i$.

Clearly, the Frobenius integrability condition

$$\mathrm{d}\alpha_j \wedge \alpha_1 \wedge \cdots \wedge \alpha_p = 0 \;\; \forall \, j = 1, \ldots, p \tag{6.21}$$

is a necessary condition for complete integrability. The p-form $\omega = \alpha_1 \wedge \cdots \wedge \alpha_p$, under the Frobenius condition, is an integrable form. If, moreover, $\omega(0) \neq 0$, then $(\alpha_1, \ldots, \alpha_p)$ is also completely integrable: there is a local holomorphic coordinate system $(x_1, \ldots, x_n)$ in which ω has the form $\omega = h\mathrm{d}x_1 \wedge \cdots \wedge \mathrm{d}x_p$. Since $\alpha_i \wedge \omega = 0$, we also have $\alpha_i \wedge dx_1 \wedge \cdots \wedge dx_p = 0$, which implies that α_i can be written as $\alpha_i = \sum_{j=1}^p g_{ij} dx_j$.

Note that, if $\alpha_i = \sum_{j=1}^p g_{ij} df_j \ \forall \ i$, and $df_1 \wedge \cdots \wedge df_p \not\equiv 0$, $\det(g_{ij}) \not\equiv 0$, then

$$\omega = \alpha_1 \wedge \cdots \wedge \alpha_p = \det(g_{ij}) \, df_1 \wedge \cdots \wedge df_p \not\equiv 0,$$

and the functions $f_1, \ldots, f_p$ are functionally independent first integrals of the associated foliation of ω, which is of codimension p. In other words, we have a complete set of first integrals $(f_1, \ldots, f_p)$, and the associated foliation of ω essentially coincides with the singular fibration given by the level sets of the map $(f_1, \ldots, f_p) : (\mathbb{C}^n, 0) \to \mathbb{C}^p$.

The *Frobenius with singularity* problem may be formulated as follows: under which additional conditions a p-tuple $(\alpha_1, \ldots, \alpha_p)$ of germs at 0 of holomorphic 1-forms in $(\mathbb{C}^n, 0)$, which satisfies the Frobenius condition, but with $\alpha_1 \wedge \cdots \wedge \alpha_p(0) = 0$, is completely integrable. More generally, under which additional conditions a germ of singular foliation essentially coincides with a germ of singular fibration, or admits at least a nontrivial first integral.

Several conjectures concerning the Frobenius with singularities problem (for codimension 1 singular foliations) were formulated by René Thom in the beginning of the 1970s, and later proved or disproved by Malgrange [233, 234], Mattei–Moussu [242], Cerveau–Mattei [70] and other people (see, e.g., [269]). We will present here the result of Malgrange, which will be used later in this chapter, in particular in the problem of linearization of Nambu structures.

Theorem 6.3.1 (Malgrange [233, 234]). *Let $\alpha = (\alpha_1, \ldots, \alpha_p)$ be a p-tuple of germs at 0 of holomorphic 1-forms in $(\mathbb{C}^n, 0)$, which satisfies the Frobenius condition (6.21). Denote by $S = \{x \in \mathbb{C}^n \mid \alpha_1 \wedge \cdots \wedge \alpha_p(x) = 0\}$ the singular set of α. Suppose that one of the following two conditions is satisfied:*

   i) *codim $S \geq 3$;*
  ii) *codim $S = 2$ and α is formally completely integrable.*

Then α is completely integrable.

We will give here a sketch of the proof of the above theorem for the case $p = 1$, following Malgrange [233].

We will first show that $\alpha = \alpha_1$ is *formally* completely integrable, i.e., it can be written as $\alpha = f dg$ where f and g are formal functions, and then show that f, g can be chosen convergent. A trick is to add one dimension to the space, and find a (formal) integrable 1-form β in $(\mathbb{C}^{n+1}, 0)$ of the type

$$\beta = dt + \sum_{k=0}^{\infty} \frac{t^k}{k!} \omega_k, \tag{6.22}$$

where t is the additional variable, ω_k are germs of holomorphic 1-forms in $(\mathbb{C}^n, 0)$, and $\omega_0 = \alpha$. Since β is integrable and regular at 0, by the classical theorem of

Frobenius we have $\beta = F dG$ where F and G are functions in $(\mathbb{C}^{n+1}, 0)$, which implies that $\alpha = f dg$ where f and g are the restrictions of F and G to $\{t = 0\}$.

The integrability condition $d\beta \wedge \beta = 0$ is equivalent to an infinite sequence of equations in ω_k:

$$\alpha \wedge \omega_{k+1} = \psi_k, \quad k = 0, 1, 2, \ldots. \tag{6.23}$$

$$\text{where } \psi_k = d\omega_k - \sum_{l=1}^{k} \binom{k}{l} \omega_l \wedge \omega_{k-l+1}. \tag{6.24}$$

The above equations can be solved inductively. For each k, assuming that ω_k is already found (recall that $\omega_0 = \alpha$), it follows from the equation $d\alpha \wedge \alpha$ that we have

$$\psi_k \wedge \alpha = 0. \tag{6.25}$$

If the codimension of the singular set $S = \{x \in (\mathbb{C}^n, 0), \alpha(x) = 0\}$ is at least 3, then by de Rham's division theorem (see Appendix A.2), ψ_k is dividable by α, i.e., there exists ω_{k+1} which satisfies Equation (6.23). If $\mathrm{codim} S = 2$ and α is formally completely integrable, then one verifies directly that ψ_k is also dividable by α. The formal part of the proof is finished.

The next step is to show that $\sum_{k=0}^{\infty} \frac{t^k}{k!} \omega_k$ converges. Fix a local coordinate system $(x_1, \ldots, x_n)$ in $(\mathbb{C}^n, 0)$, and denote by $\mathcal{O} = \mathbb{C}\{x_1, \ldots, x_n\}$ the ring of germs of holomorphic functions in $(\mathbb{C}^n, 0)$. For each n-tuple $\rho = (\rho_1, \ldots, \rho_n)$ of small positive numbers (called a *polyradius*) and a function $f \in \mathcal{O}$, $f = \sum_I a_I x^I$ where $I = (I_1, \ldots, I_n)$ means a multi-index and x^I means the monomial $\prod_{j=1}^{n} x_j^{I_j}$, put

$$|f|_\rho = \sum_I |a_I| \rho^I. \tag{6.26}$$

Note that $|fg| \leq |f|_\rho \cdot |g|_\rho$ $\forall f, g \in \mathcal{O}$. We will extend this pseudo-norm to $\mathcal{O}^N$, $N \in \mathbb{N}$, by the formula $|F|_\rho = \sup_{1 \leq i \leq N} |F_i|_\rho$, $F = (F_1, \ldots, F_n) \in \mathcal{O}^N$. In particular, for a 1-form $\omega = \sum f_i dx_i$ we will write $|\omega|_\rho = \sum_i |f_i|_\rho$, and similarly for 2-forms. In order to make estimates on $|\omega_k|_\rho$, we will use the following two lemmas:

Lemma 6.3.2 ([233]). *For any $\mathcal{O}$-linear map $\Phi : \mathcal{O}^N \to \mathcal{O}^M$, where $N, M \in \mathbb{N}$, there is a $\mathbb{C}$-linear map $\Psi : \mathcal{O}^M \to \mathcal{O}^N$ such that $\Phi \circ \Psi \circ \Phi = \Phi$ (such a Ψ is called a splitting of Φ), with the following additional property: for any positive polyradius ρ' there is a polyradius $\rho \leq \rho'$ (i.e., $\rho_i \leq \rho'_i$ $\forall i$) and a constant $c = c(\rho) > 0$ (which may depend on ρ) such that*

$$|\Psi(F)|_{\tau\rho} \leq c |F|_{\tau\rho} \quad \forall \frac{1}{2} \leq \tau \leq 1, F \in \mathcal{O}^M. \tag{6.27}$$

Lemma 6.3.3. *For any polyradius ρ there is a positive constant $c_1 = c_1(\rho)$ such that for any $1/2 < s < \tau < 1$ and any germ of holomorphic 1-form ω on $(\mathbb{C}^n, 0)$ we have*

$$|d\omega|_{s\rho} \leq \frac{c_1}{\tau - s} |\omega|_{\tau\rho}. \tag{6.28}$$

Lemma 6.3.3 is elementary, while Lemma 6.3.2 is a fine version of the "privileged neighborhoods theorem" in complex analysis which goes back to Henri Cartan. See [233] for a nice proof of Lemma 6.3.2. Applying the above lemmas, we obtain from Equation (6.23) the following inequality:

$$|\omega_{k+1}|_{s\rho} \leq \frac{c'}{\tau - s}|\omega_k|_{\tau\rho} + c\sum_{1 \leq l \leq k}\binom{k}{l}|\omega_l|_{\tau\rho}.|\omega_{k-l+1}|_{\tau\rho}$$

$$\forall \ k = 0, 1, 2, \ldots, \frac{1}{2} \leq s < \tau \leq 1, \quad (6.29)$$

where ρ is a multiradius such that $|\alpha|_\rho < \infty$ and which satisfies Lemma 6.3.3 with respect to the linear map $\omega \mapsto \alpha \wedge \omega$; c and c' are some positive constants (ρ, c and c' don't depend on k).

Define a sequence of positive numbers v_k by the following recursive formula:

$$v_0 = |\alpha|_\rho, \ v_{k+1} = c'ev_k + c\sum_{l=1}^{k}v_lv_{k+1-l} \quad (e = \exp(1)). \quad (6.30)$$

Then the series $F(t) = \sum v_kt^k$ converges because it is a solution of the equation $F(t) = c'ev_0t + c'etF(t) + cF(t)^2$. By recurrence, one can verify directly that it follows from Equation (6.30) and Inequality (6.29) that we have

$$|\omega_k|_{s\rho} \leq \frac{k!v_k}{(1 - s)^k} \ \forall \ k \in \mathbb{N}, 1/2 \leq s < 1. \quad (6.31)$$

This last inequality, together with the convergence of $\sum v_kt^k$, implies that $\sum \frac{t^k}{k!}\omega_k$ converges. $\quad\square$

The case $p > 1$ of Malgrange's Theorem 6.3.1 is similar to the case $p = 1$, though technically much more involved, especially the part about convergence, see [234].

Remark 6.3.4. The method, used in the above proof, of dividing repetitively by a 1-form, is called the *Godbillon–Vey algorithm*, because it was first used by Godbillon and Vey in the definition of a characteristic class (the *Godbillon–Vey class*) of codimension 1 foliations [149].

6.4 Linear Nambu structures

A Nambu tensor $\Lambda = \sum_{i_1 < \cdots < i_q} \Lambda_{i_1\ldots i_q}\frac{\partial}{\partial x_{i_1}} \wedge \cdots \wedge \frac{\partial}{\partial x_{i_1}}$ in some linear coordinate system $(x_1, \ldots, x_m)$ is called a *linear Nambu structure*, if its coefficients $\Lambda_{i_1\ldots i_q}$ are linear functions. More generally, if $\Lambda_{i_1\ldots i_q}$ are homogeneous functions of a given degree then we say that Λ is a homogeneous Nambu tensor of that degree. Linear and homogeneous differential forms are defined in an obvious similar way. It is clear that if Λ is a linear Nambu tensor of order $q \neq 2$ (if $q = 2$, i.e., Λ is Poisson, then we assume in addition that rank $\Lambda \leq 2$), and Ω is a constant volume form, then $i_\Lambda \Omega$ is a linear integrable differential form, and vice versa.

In this section we will classify linear integrable differential forms and Nambu structures. We first state the result for differential forms:

Theorem 6.4.1 ([119, 245]). *Let ω be a linear integrable differential p-form on an m-dimensional linear space V (over $\mathbb{R}$ or $\mathbb{C}$ – when considered over $\mathbb{C}$, ω is holomorphic). Then there exists a linear coordinate system $(x_1, \ldots, x_m)$ such that ω belongs to (at least) one of the following two types:*

Type I: $\omega = \mathrm{d}x_1 \wedge \cdots \wedge \mathrm{d}x_{p-1} \wedge \alpha$, *where α is an exact 1-form of the type* $\alpha = \mathrm{d}(\sum_{j=p}^{p+r} \pm x_j^2/2 + \sum_{i=1}^s x_i x_{p+r+i})$, *with* $-1 \leq r \leq q = m - p$, $0 \leq s \leq q - r$.

Type II: $\omega = \sum_{i=1}^{p+1} a_i \mathrm{d}x_1 \wedge \cdots \wedge \mathrm{d}x_{i-1} \wedge \mathrm{d}x_{i+1} \wedge \cdots \wedge \mathrm{d}x_{p+1}$ *with* $a_i = \sum_{j=1}^{p+1} a_i^j x_j$, *where a_j^i are constant. The matrix (a_i^j) can be chosen to be in Jordan form.*

Proof. We will assume that $p > 0$ (the case $p = 0$ is obvious and belongs to Type II). Put $\omega = \sum_{j=1}^n x_j \omega_j$ where ω_j are constant p-forms. Then $\omega = \omega_j$ at points $(x_1 = 0, \ldots, x_j = \varepsilon, \ldots, x_n = 0)$. At any point ω is either decomposable (i.e., a wedge product of covectors) or zero, so is ω_j since it is constant. Denote by E_j the span of ω_j, i.e.,

$$E_j = \mathrm{Span}(\omega_j) \stackrel{\mathrm{def}}{=} \mathrm{Span}\{i_A \omega_j, \ A \text{ is a } (p-1)\text{-vector}\}$$
$$= \mathrm{Annulator}\{x \in V, \ i_x \omega_j = 0\} \subset V^*.$$

Then $\dim E_j = p$ if $\omega_j \neq 0$, because of decomposability. We have:

Lemma 6.4.2. *If $\omega_i \neq 0$ and $\omega_j \neq 0$ for some indices i and j, then $\dim(E_i \cap E_j) \geq p - 1$.*

Proof of Lemma 6.4.2. Putting $x_k = 0$ for every $k \neq i, j$, we obtain that $x_i \omega_i + x_j \omega_j = \omega$ is decomposable or null for any x_i, x_j. In particular, $\omega_i + \omega_j$ is decomposable. If $\dim(E_i \cap E_j) = d < p$, then there is a basis $(e_1, \ldots, e_d, f_1, \ldots, f_{p-d}, g_1, \ldots, g_{p-d})$ of $E_i + E_j$ such that $\omega_i = e_1 \wedge \cdots \wedge e_d \wedge f_1 \wedge \cdots \wedge f_{p-d}$, $\omega_j = e_1 \wedge \cdots \wedge e_d \wedge g_1 \wedge \cdots \wedge g_{p-d}$ and

$$\omega_i + \omega_j = e_1 \wedge \cdots \wedge e_d \wedge [f_1 \wedge \cdots \wedge f_{p-d} + g_1 \wedge \cdots \wedge g_{p-d}].$$

It follows easily that if $p - d \geq 2$, then $\mathrm{Span}(\omega_i + \omega_j) = E_i + E_j$, $\dim \mathrm{Span}(\omega_i + \omega_j) > p$ and $\omega_i + \omega_j$ is not decomposable. $\square$

Return now to Theorem 6.4.1. We can assume that $E_1, \ldots, E_h \neq 0$ and $E_{h+1}, \ldots, E_n = 0$ for some number h. Put $E = E_1 \cap E_2 \cap \cdots \cap E_h$. Then there are two alternative cases: $\dim E \geq p - 1$ and $\dim E < p - 1$.

<u>Case I</u>. $\dim E \geq p - 1$. Then denoting by $(x_1, \ldots, x_{p-1})$ a set of $p-1$ linearly independent covectors contained in E, and which are considered as linear functions on V, we have

$$\omega_i = \mathrm{d}x_1 \wedge \mathrm{d}x_2 \wedge \cdots \wedge \mathrm{d}x_{p-1} \wedge \alpha_i \ \forall \, i = 1, \ldots, h$$

for some constant 1-forms α_i, and hence

$$\omega = \mathrm{d}x_1 \wedge \mathrm{d}x_2 \wedge \cdots \wedge \mathrm{d}x_{p-1} \wedge \alpha \tag{6.32}$$

where $\alpha = \sum x_i \alpha_i$ is a linear 1-form.

<u>Case II</u>. In this case, without loss of generality, we can assume that $\dim(E_1 \cap E_2 \cap E_3) < p - 1$. Then Lemma 6.4.2 implies that $\dim(E_1 \cap E_2 \cap E_3) = p - 2$. For an arbitrary index i, $3 < i \leq h$, put $F_1 = E_1 \cap E_i, F_2 = E_2 \cap E_i, F_3 = E_3 \cap E_i$. Recall that $\dim F_1, \dim F_2, \dim F_3 \geq p - 1$ according to Lemma 6.4.2, but $\dim(F_1 \cap F_2 \cap F_3) = \dim(E_1 \cap E_2 \cap E_3 \cap E_i) < p - 1$, hence we cannot have $F_1 = F_2 = F_3$. Thus we can assume that $F_1 \neq F_2$. Then either F_1 and F_2 are two different hyperplanes in E_i, or one of them coincides with E_i. In any case we have $E_i = F_1 + F_2 \subset E_1 + E_2 + E_3$. It follows that

$$\sum_1^m E_i = \sum_1^h E_i = E_1 + E_2 + E_3 \ .$$

On the other hand, we have $\dim(E_1 + E_2 + E_3) = \dim E_1 + \dim E_2 + \dim E_3 - \dim(E_1 \cap E_2) - \dim(E_1 \cap E_3) - \dim(E_2 \cap E_3) + \dim(E_1 + E_2 + E_3) = 3p - 3(p-1) + (p-2) = p + 1$. Thus

$$\dim(E_1 + E_2 + \cdots + E_m) = p + 1 \ .$$

It follows that there is a system of linear coordinates $(x_1, \ldots, x_m)$ on V such that $(x_1, \ldots, x_{p+1})$ span $E_1 + \cdots + E_m$ and therefore

$$\omega_i = \sum_{j=1}^{p+1} \gamma_i^j \mathrm{d}x_1 \wedge \cdots \wedge \mathrm{d}x_{j-1} \wedge \mathrm{d}x_{j+1} \wedge \cdots \wedge \mathrm{d}x_{p+1} \ .$$

Hence we have

$$\omega = \sum x_i \omega_i = \sum_{j=1}^{p+1} a_j \mathrm{d}x_1 \wedge \cdots \wedge \mathrm{d}x_{j-1} \wedge \mathrm{d}x_{j+1} \wedge \cdots \wedge x_{p+1} \ , \tag{6.33}$$

where a_j are linear functions on V.

To finish the proof of Theorem 6.4.1, we still need to normalize further the obtained forms (6.32) and (6.33).

Return now to Case I and suppose that $\omega = \mathrm{d}x_1 \wedge \cdots \wedge \mathrm{d}x_{p-1} \wedge \alpha$ where $\alpha = \sum \alpha_j \mathrm{d}x_j$ with α_j being linear functions. We can put $\alpha_j = 0$ for $j = 1, \ldots, p-1$

since it will not affect ω. Then we have $\alpha = \sum_{j \geq p, i=1,\ldots,m} \alpha^i_j x_i \mathrm{d}x_j$. Equation (6.16) implies that $\alpha \wedge \mathrm{d}x_1 \wedge \cdots \wedge \mathrm{d}x_{p-1} \wedge \mathrm{d}\alpha = 0$. If we consider $(x_1, \ldots, x_{p-1})$ as parameters and denote by d' the exterior derivation with respect to the variables $(x_p, \ldots, x_m)$, then the last equation means $\alpha \wedge \mathrm{d}'\alpha = 0$. That is, α can be considered as an integrable 1-form in the space of variables $(x_p, \ldots, x_m)$, parametrized by $(x_1, \ldots, x_{p-1})$. We will distinguish two subcases: $\mathrm{d}'\alpha = 0$ and $\mathrm{d}'\alpha \neq 0$.

Subcase a). Suppose that $\mathrm{d}'\alpha = 0$. Then according to Poincaré's Lemma we have $\alpha_j = \sum_{i=1}^{p-1} \alpha^i_j x_i + \partial/\partial x_j q^{(2)}$, where $q^{(2)}$ is a quadratic function in the variables $(x_p, \ldots, x_m)$. By a linear change of coordinates on $(x_p, \ldots, x_m)$, we have $q^{(2)} = \sum_{j=p}^{p+r} \pm x_j^2/2$, for some number $r \geq -1$, and accordingly

$$\alpha = \sum_{j=p}^{p+r} \left(\pm x_j + \sum_{i=1}^{p-1} \alpha^i_j x_i \right) \mathrm{d}x_j + \sum_{\substack{i=1,\ldots,p-1 \\ j=p+r+1,\ldots,m}} \alpha^i_j x_i \mathrm{d}x_j \; .$$

By a linear change of coordinates on $(x_1, \ldots, x_{p-1})$ on the one hand, and on $(x_{p+r+1}, \ldots, x_m)$ on the other hand, we can normalize the second part of the above expression to obtain

$$\alpha = \sum_{j=p}^{p+r} \left(\pm x_j + \sum_{i=1}^{p-1} \tilde{\alpha}^i_j x_i \right) + \sum_{j=1}^{s} x_j \mathrm{d}x_{p+r+j}$$

for some number s $(0 \leq s \leq \min(p-1, m-p-r))$.

Replacing x_j $(j = p, \ldots, p+r)$ by new $x_j = x_j \mp \tilde{\alpha}^i_j x_i$ we get

$$\omega = \mathrm{d}x_1 \wedge \cdots \wedge \mathrm{d}x_{p-1} \wedge \alpha \; ,$$

where

$$\alpha = \mathrm{d}\Big(\sum_{j=p}^{p+r} \pm x_j^2/2 + \sum_{i=1}^{s} x_i x_{p+r+i} \Big)$$

(with $-1 \leq r \leq q = m - p, 0 \leq s \leq q - r$). These are the linear integrable forms of Type I in Theorem 6.4.1.

Subcase b). Suppose that $\mathrm{d}'\alpha \neq 0$. Then since $\mathrm{d}'\alpha$ is of constant coefficients, we can change the coordinates $(x_p, \ldots, x_n)$ linearly so that $\mathrm{d}'\alpha = \mathrm{d}x_p \wedge \mathrm{d}x_{p+1} + \cdots + \mathrm{d}x_{p+2r} \wedge \mathrm{d}x_{p+2r+1}$ in these new coordinates, for some $r \geq 0$.

If $r \geq 1$, then considering the coefficients of the term $\mathrm{d}x_p \wedge \mathrm{d}x_{p+1} \wedge \mathrm{d}x_i$ $(i > p+1)$, $\mathrm{d}x_p \wedge \mathrm{d}x_{p+2} \wedge \mathrm{d}x_{p+3}$ and $\mathrm{d}x_{p+1} \wedge \mathrm{d}x_{p+2} \wedge \mathrm{d}x_{p+3}$ in $0 = \alpha \wedge \mathrm{d}'\alpha$, we obtain that all the coefficients of α are zero, i.e., $\alpha = 0$, which is absurd. Thus $\mathrm{d}'\alpha = \mathrm{d}x_p \wedge \mathrm{d}x_{p+1}$, and the condition $\alpha \wedge \mathrm{d}'\alpha = 0$ implies $\alpha = \alpha_1 \mathrm{d}x_p + \alpha_2 \mathrm{d}x_{p+1}$ with linear functions α_1 and α_2 depending only on $x_1, \ldots, x_{p-1}, x_p, x_{p+1}$. In this Subcase b), $\omega = \mathrm{d}x_1 \wedge \cdots \wedge \mathrm{d}x_{p-1} \wedge \alpha$ also has the form (6.33), as in Case II.

Suppose now that ω has the form (6.33), as in Case II or Subcase b) of Case 1:

$$\omega = \sum x_i \omega_i = \sum_{j=1}^{p+1} a_j \mathrm{d}x_1 \wedge \cdots \wedge \mathrm{d}x_{j-1} \wedge \mathrm{d}x_{j+1} \wedge \cdots \wedge \mathrm{d}x_{p+1}.$$

There are also two subcases:

a) $\partial a_j / \partial x_i = 0$ for $j = 1, \ldots, p+1, i = p+2, \ldots, m$. In other words,

$$\omega = \sum_{i,j=1}^{p+1} a_j^i x_i \mathrm{d}x_1 \wedge \cdots \wedge \mathrm{d}x_{j-1} \wedge \mathrm{d}x_{j+1} \wedge \cdots \wedge \mathrm{d}x_{p+1}$$

with constant coefficients a_j^i.

To see that (a_j^i) can be put in Jordan form, notice that the linear Nambu tensor Λ dual to ω (i.e., $\omega = i_\Lambda \Omega$ where Ω is a constant volume form) is, up to multiplication by a constant:

$$\Lambda = \Big(\sum_{i,j=1}^{p+1} \pm a_j^i x_i \partial / \partial x_j \Big) \wedge \partial / \partial x_{p+2} \wedge \cdots \wedge \partial / \partial x_m \ .$$

The first term in Λ is a linear vector field, which is uniquely defined by a linear transformation $\mathbb{R}^{p+1} \to \mathbb{R}^{p+1}$ given by the matrix (a_j^i), so this matrix can be put in Jordan form.

b) There is $j \leq p+1$ and $i \geq p+2$ such that $\partial a_j / \partial x_i \neq 0$. We can assume that $\partial a_1 / \partial x_m \neq 0$. Putting $A = \partial / \partial x_3 \wedge \cdots \wedge \partial / \partial x_{p+1}$ in $0 = i_A \omega \wedge \mathrm{d}\omega$, we obtain

$$0 = (a_1 \mathrm{d}x_2 + a_2 \mathrm{d}x_1) \wedge \sum_{\substack{i=1,\ldots,m \\ j=1,\ldots,p+1}} \mathrm{d}x_i \wedge \frac{\partial a_j}{\partial x_i} \mathrm{d}x_1 \wedge \cdots \wedge \mathrm{d}x_{j-1} \wedge \mathrm{d}x_{j+1} \wedge \cdots \wedge \mathrm{d}x_{p+1}.$$

Considering the coefficient of $\mathrm{d}x_1 \wedge \cdots \wedge \mathrm{d}x_{p+1} \wedge \mathrm{d}x_m$ in the above equation, we get

$$a_1 \partial a_2 / \partial x_m - a_2 \partial a_1 / \partial x_n = 0 \ .$$

Since $\partial a_1 / \partial x_m \neq 0$, it follows that a_2 is linearly dependent on a_1. Similarly, a_j is linearly dependent on a_1 for any $j = 1, \ldots, p+1$. Thus $\omega = a_1 \omega_1$ where ω_1 is decomposable and constant: $\omega_1 = \mathrm{d}x_1 \wedge \cdots \wedge \mathrm{d}x_p$ in some new linear system of coordinates. If a_1 is linearly independent of $(x_1, \ldots, x_p)$, then we can also assume that $x_{p+1} = a_1$ in this new coordinate system. Thus we get back to Subcase a) of Case II by a linear change of coordinates. $\qquad\square$

Remark 6.4.3. Theorem 6.4.1 first appeared in [119], and later in an independent work of de Medeiros (see Theorem A of [245]). The proof given in [245] is very nice, though less detailed than the one given here.

Rewriting Theorem 6.4.1 in terms of Nambu tensors, we have:

Theorem 6.4.4 ([119]). *Every linear Nambu tensor Λ of order $q \geq 3$, or of order $q = 2$ but under the additional assumption that rank $\Lambda \leq 2$ if $q = 2$, on an m-dimensional linear space, belongs to (at least) one of the following two types:*

Type I: $\Lambda = \sum_{j=1}^{r+1} \pm x_j \partial/\partial x_1 \wedge \cdots \wedge \partial/\partial x_{j-1} \wedge \partial/\partial x_{j+1} \wedge \cdots \wedge \partial/\partial x_{q+1} + $
$\sum_{j=1}^{s} \pm x_{q+1+j} \partial/\partial x_1 \wedge \cdots \wedge \partial/\partial x_{r+j} \wedge \partial/\partial x_{r+j+2} \wedge \partial/\partial x_{q+1}$
(with $-1 \leq r \leq q, 0 \leq s \leq \min(p-1, q-r)$).

Type II: $\Lambda = \partial/\partial x_1 \wedge \cdots \wedge \partial/\partial x_{q-1} \wedge (\sum_{i,j=q}^{m} b_j^i x_i \partial/\partial x_j)$.

Remark 6.4.5. Strictly speaking, since linear Nambu tensors are determined by their dual linear integrable forms only up to a non-zero multiplicative constant, Theorem 6.4.1 implies Theorem 6.4.4 only up to a multiplicative constant. But that constant in the expression of Λ can be killed by a homothety when $q \neq 1$.

Remark 6.4.6. There is an obvious and very interesting duality between Type I and Type II in the above theorems: the expression for Type I linear integrable differential forms looks very similar to the expression for Type II linear Nambu structures, and vice versa.

Remark 6.4.7. The case $q = 2$ of Theorem 6.4.4 gives us a classification of linear Poisson structures of rank ≤ 2. They correspond to Lie algebras whose coadjoint orbits are at most two-dimensional. So we recover a classification, obtained by Arnal, Cahen and Ludwig [12], of these Lie algebras: Type II tensors give semi-direct products of $\mathbb{K}^{n-1}$ with $\mathbb{K}$ ($\mathbb{K} = \mathbb{R}$ or $\mathbb{C}$), where $\mathbb{K}$ acts on $\mathbb{K}^{n-1}$ linearly by the matrix (b_j^i). Type I tensors give, according to values of r and s, the following cases:

$$1) \quad \pm x_1 \frac{\partial}{\partial x_2} \wedge \frac{\partial}{\partial x_3} \pm x_2 \frac{\partial}{\partial x_3} \wedge \frac{\partial}{\partial x_1} \pm x_3 \frac{\partial}{\partial x_1} \wedge \frac{\partial}{\partial x_2},$$

$$2) \quad \pm x_1 \frac{\partial}{\partial x_2} \wedge \frac{\partial}{\partial x_3} \pm x_2 \frac{\partial}{\partial x_3} \wedge \frac{\partial}{\partial x_1} + \varepsilon_1 x_4 \frac{\partial}{\partial x_1} \wedge \frac{\partial}{\partial x_2},$$

$$3) \quad \pm x_1 \frac{\partial}{\partial x_2} \wedge \frac{\partial}{\partial x_3} + \varepsilon_1 x_4 \frac{\partial}{\partial x_3} \wedge \frac{\partial}{\partial x_1} + \varepsilon_2 x_5 \frac{\partial}{\partial x_1} \wedge \frac{\partial}{\partial x_2},$$

$$4) \quad \varepsilon_1 x_4 \frac{\partial}{\partial x_2} \wedge \frac{\partial}{\partial x_3} + \varepsilon_2 x_5 \frac{\partial}{\partial x_3} \wedge \frac{\partial}{\partial x_1} + \varepsilon_3 x_6 \frac{\partial}{\partial x_1} \wedge \frac{\partial}{\partial x_2},$$

where ε_i take values 0 or 1. These Type I tensors give us direct products of some Lie algebras of dimension ≤ 6 with Abelian Lie algebras.

Theorems 6.4.1 and 6.4.4 give us a clear picture about the singular foliations associated to linear Nambu structures and integrable forms. A Type I foliation has $p = m - q$ first integrals, namely $x_1, \ldots, x_{p-1}$ and $\sum_{j=p}^{p+r} \pm x_j^2 + \sum_{j=1}^{s} x_j x_{p+r+j}$. Hence a Type 1 foliation looks like a *cabbage pile*. A Type II foliation is a Cartesean product of a one-dimensional foliation given by a linear vector field in a linear space with (a 1-leaf foliation on) another linear space. Hence a Type II foliation looks like a *book*. So a cabbage pile is dual to a book (?!).

It is natural to call singular foliations given by linear Nambu structures *linear foliations*. There is, however, another kind of linear foliations: those given by linear vector fields. To distinguish these two different cases, we make the following definition:

Definition 6.4.8. A *Nambu-linear foliation* is a singular foliation $\mathcal{F}_\lambda$ associated to a linear Nambu tensor Λ of order $q \neq 3$, or of order $q = 2$ but with the additional condition that rank $\Lambda \leq 2$. A *Lie-linear foliation* is a singular foliation generated by a Lie algebra of linear vector fields on a linear space.

Exercise 6.4.9. Show that any Nambu-linear foliation is a Lie-linear foliation, but the inverse is not true.

An advantage of Nambu tensors and integrable forms is that we can use them to define homogeneous foliations on a linear space: a *homogeneous foliation* is a singular foliation which can be given, up to saturation, by a homogeneous integrable differential form. Note that we can't use Lie algebras of homogeneous vector fields to define homogeneous foliations: a Lie bracket of two vector fields of degree k will be a vector field of degree $2k - 1$, which is greater than k if $k > 1$.

Exercise 6.4.10. Show that a Lie-linear foliation is a homogeneous foliation.

The problem of classification of homogeneous foliations (or homogeneous Nambu structures) is, to our knowledge, an interesting and largely open problem. In Section 6.8 we will consider the case of quadratic foliations of codimension 1.

Remark 6.4.11. Similarly to the fact that linear Poisson structures correspond to Lie algebras, linear Nambu structures correspond to the so-called *n*-ary Lie algebras, also known as *Filippov algebras* because they were first studied by Filippov [134]. By definition, an *n-ary Lie algebra* structure of order n on a vector space V is a skew-symmetric n-linear mapping $[\ldots] : V \times \cdots \times V \to V$ which satisfies the relation

$$[u_1, \ldots, u_{n-1}, [v_1, \ldots, v_n]] = \sum_{i=1}^{n} [v_1, \ldots, v_{i-1}, [u_1, \ldots, u_{n-1}, v_i], v_{i+1}, \ldots, v_n]$$

(6.34)

$\forall\, u_i, v_j \in V$. A warning though: when $n \geq 3$, not every n-ary Lie algebra corresponds to a linear Nambu structure. For example, consider two nontrivial n-ary algebras of order $n \geq 3$ on two vector spaces V_1, V_2. Then the formula

$$[(a_1, b_1), \ldots, (a_n, b_n)] := ([a_1, \ldots, a_n]_1, [b_1, \ldots, b_n]_2)$$

defines an n-ary Lie algebra on the direct sum $V_1 \oplus V_2$. However, this n-ary Lie algebra structure on $V_1 \oplus V_2$ can never correspond to a linear Nambu structure on $V_1^* \oplus V_2^*$, because the corresponding linear n-vector field on $V_1^* \oplus V_2^*$ is not locally decomposable.

Exercise 6.4.12. Show that any n-ary algebra of dimension 3 or 4 comes from a linear Nambu structure. Find an n-ary algebra of dimension 5 and degree 3 which does not come from a linear Nambu structure.

Remark 6.4.13. Similarly to the case of Poisson structures, one can talk about Nambu–Lie structures (= multiplicative Nambu structures) on Lie groups, and view linear Nambu structures as a special case (when the Lie algebra is free Abelian). See, e.g., [152, 334].

6.5   Kupka's phenomenon

The so-called *Kupka's phenomenon*, first observed by Kupka [204], is the following: if ω is an integrable 1-form such that $d\omega(z) \neq 0$ at some point z, then there is a local coordinate system $(x_1, \ldots, x_m)$ in a neighborhood of z such that ω depends only on the first two coordinates, i.e., we have

$$\omega = a(x_1, x_2)dx_1 + b(x_1, x_2)dx_2 \ . \tag{6.35}$$

This phenomenon is important because it reduces the study of singularities of integrable 1-forms ω near points where $d\omega \neq 0$ to the study of 1-forms in dimension 2, and implies the local stability of integrable 1-forms under certain conditions [204]. (Singularities of 1-forms and vector fields in dimension 2 can already be highly complicated).

In this section, we will explain Kupka's phenomenon in a more general setting, and translate it into the language of Nambu tensors.

Recall from Section 2.6 that, given a volume form Ω, the *curl operator* D_Ω is defined by $D_\Omega = \Omega^\sharp \circ d \circ \Omega^\flat$, where $\Omega^\flat$ is the map $\Lambda \mapsto i_\Lambda \Omega$ and $\Omega^\sharp = (\Omega^\flat)^{-1}$.

Lemma 6.5.1. *If Λ is a Nambu tensor of order q, then its curl $D_\Omega \Lambda$ with respect to an arbitrary volume form Ω is a Nambu tensor of order $q - 1$. If $q = 3$, then $D_\Omega \Lambda$ is a Poisson structure of rank ≤ 2.*

Proof. Λ is Nambu of order $\neq 2 \Rightarrow \Omega^\flat(\Lambda)$ is integrable $\Rightarrow$ d $\circ \Omega^\flat(\Lambda)$ is integrable $\Rightarrow \Omega^\sharp \circ d \circ \Omega^\flat(\Lambda)$ is Nambu. $\square$

Near a point z where $\Lambda(z) \neq 0$, $\Lambda = \frac{\partial}{\partial x_1} \wedge \cdots \wedge \frac{\partial}{\partial x_p}$, $\Omega = f dx_1 \wedge \cdots \wedge dx_m$, we have (see Proposition 2.6.5):

$$D_\Omega \Lambda = [\Lambda, \ln |f|] = (-1)^{q+1} i_{d \ln |f|} \Lambda \ . \tag{6.36}$$

Lemma 6.5.2. *Let Ω be a volume form and Λ a Nambu tensor of order $q > 2$ on a manifold M. Then for every $r = 0, 1, \ldots, q - 2$, and smooth functions $g_1, \ldots, g_r$, we have:*

a) $i_{(dg_1 \wedge \cdots \wedge dg_r)} D_\Omega \Lambda \wedge \Lambda = 0$. *In particular, the exterior product of any Hamiltonian vector field of $D_\Omega \Lambda$ with Λ vanishes.*

b) $[i_{(dg_1 \wedge \cdots \wedge dg_r)} D_\Omega \Lambda, \Lambda] = 0$. *In particular, any Hamiltonian vector field of $D_\Omega \Lambda$ is also an infinitesimal automorphism of Λ.*

Proof. Assertion a) follows from Equation (6.36) and the local decomposability of Λ near a non-singular point. Assertion b) follows from Equation (6.36) and Equation (6.12). $\qquad\square$

Theorem 6.5.3. *Let Λ be a Nambu tensor of order q on a manifold M, and Ω be a volume form on M. When $q = 2$ we will assume in addition that rank $\Lambda \leq 2$. Suppose that $D_\Omega\Lambda(z) \neq 0$ at a point $z \in M$. Then in a neighborhood of z there is a local coordinate system $(x_1, \ldots, x_{q-1}, y_1, \ldots, y_{p+1})$, and a vector field X which depends only on $(y_1, \ldots, y_{p+1})$, such that*

$$\Lambda = \frac{\partial}{\partial x_1} \wedge \cdots \wedge \frac{\partial}{\partial x_{q-1}} \wedge X. \tag{6.37}$$

Proof. Consider the case $q > 2$. Since $D_\Omega\Lambda$ is a Nambu tensor of order $q - 1 \geq 2$ (of rank ≤ 2 when $q - 1 = 2$) and $D_\Omega\Lambda(z) \neq 0$, in a neighborhood of z there is a local coordinate system in which we have

$$D_\Omega\Lambda = \frac{\partial}{\partial x_1} \wedge \cdots \wedge \frac{\partial}{\partial x_{q-1}}.$$

Then $\frac{\partial}{\partial x_i}$ are local Hamiltonian vector fields for $D_\Omega\Lambda$. Assertion a) of Lemma 6.5.2 implies that $\frac{\partial}{\partial x_i} \wedge \Lambda = 0$ ($\forall\, i = 1, \ldots, q - 1$), which means that we can write Λ as

$$\Lambda = \frac{\partial}{\partial x_1} \wedge \cdots \wedge \frac{\partial}{\partial x_{q-1}} \wedge X ,$$

where X is a vector field which does not contain the terms $\frac{\partial}{\partial x_1}, \ldots, \frac{\partial}{\partial x_{q-1}}$. Assertion b) of Lemma 6.5.2 implies that $[\frac{\partial}{\partial x_i}, \Lambda] = 0 \,\forall\, i = 1, \ldots, q - 1$. Hence X is also invariant with respect to $\frac{\partial}{\partial x_1}, \ldots, \frac{\partial}{\partial x_{q-1}}$.

The case $q = 2$ is similar. $\qquad\square$

Translating the above theorem into the language of integrable differential forms by duality, we obtain the following result, due to de Medeiros [244], which may be called *Kupka's phenomenon* for integrable p-forms:

Theorem 6.5.4 ([244]). *Let ω be an integrable p-form on an m-dimensional manifold M, $p < m$, such that $d\omega(z) \neq 0$ at a point z. Then in a neighborhood of z there is a coordinate system $(x_1, \ldots, x_m)$ in which ω depends only on the $(p + 1)$ first coordinates (i.e., locally ω is the pull-back of a p-form on $\mathbb{K}^{p+1}$ by a projection).*

Proof. Fix a volume form Ω, and put $\Lambda = \Omega^\sharp(\omega)$. Then Theorem 6.5.3 implies that there is a coordinate system $(x_1, \ldots, x_{q-1}, y_1, \ldots, y_{q+1})$ in which we have $\Omega = f dx_1 \wedge \cdots \wedge dx_{q-1} \wedge dy_1 \wedge \cdots \wedge dy_{p+1}$ (for some non-zero function f), $d\omega = f dy_1 \wedge \cdots \wedge dy_{p+1}$, and $\omega = f i_X(dy_1 \wedge \cdots \wedge dy_{p+1})$ where X is a vector field which depends only on the variables $(y_1, \ldots, y_{p+1})$. Since $df \wedge dy_1 \wedge \cdots \wedge dy_{p+1} = d(d\omega) = 0$, f is a function which depends only on the variables $y_1, \ldots, y_{p+1}$. $\qquad\square$

We will now present some generalizations of Kupka's phenomenon to the case when $d\omega(z) = 0$ (or $D_\Omega\Lambda = 0$).

Generalizing the Type II linear Nambu structures, we will say that a q-vector field A is of *Type IIr* at a point z if in a neighborhood of z there is a local coordinate system $(x_1, \ldots, x_m)$ such that

$$A = \partial/\partial x_1 \wedge \cdots \wedge \partial/\partial x_r \wedge B ,$$

where B is a $(q-r)$-vector field which is independent of the coordinates $x_1, \ldots, x_r$. Note that if A is Nambu then B will be automatically Nambu.

Theorem 6.5.5. *Let Λ be a Nambu tensor and Ω a volume form. If the curl $D_\Omega\Lambda$ is non-vanishing almost everywhere and is of Type IIr at a point z, then Λ is also of Type IIr at z.*

Proof. Write
$$D_\Omega\Lambda = \partial/\partial x_1 \wedge \cdots \wedge \partial/\partial x_r \wedge Y ,$$

where Y is a $(q - 1 - r)$-tensor independent of $(x_1, \ldots, x_r)$. Then for each $i = 1, \ldots, r$, we have $\partial/\partial x_i \wedge D_\Omega\Lambda = 0$ and $\mathcal{L}_{\partial/\partial x_i}(D_\Omega\Lambda) = 0$. Since $D_\Omega\Lambda \neq 0$ almost everywhere, it implies that $\partial/\partial x_i$ is a locally Hamiltonian vector field for $D_\Omega\Lambda$ almost everywhere (near points where $D_\Omega\Lambda \neq 0$). Applying Assertion a) of Lemma 6.5.2, we obtain that $\partial/\partial x_i \wedge \Lambda = 0$ (almost everywhere, hence everywhere in a neighborhood of z). Since this fact is true for any $i = 1, \ldots, r$, we can write

$$\Lambda = \partial/\partial x_1 \wedge \cdots \wedge \partial/\partial x_r \wedge B$$

for some $(q - r)$-vector field B such that $i_{dx_i}B = 0 \; \forall \; i = 1, \ldots, r$. Again, since $\partial/\partial x_i$ is locally Hamiltonian for $D_\Omega\Lambda$ almost everywhere, applying Assertion b) of Lemma 6.5.2, we get that $\mathcal{L}_{\partial/\partial x_i}\Lambda = 0 \; \forall \; i = 1, \ldots, r$. It means that B does not depend on $x_1, \ldots, x_r$. $\square$

Theorem 6.5.6. *Let ω be an integrable differential form on $\mathbb{R}^n$ or $\mathbb{C}^n$, and s be a natural number, $s \leq n$. If $d\omega$ is non-zero almost everywhere and depends on at most s coordinates in a neighborhood of 0, then the same holds true for ω.*

Proof. Suppose that, in a local system of coordinates $(x_1, \ldots, x_n)$, $d\omega$ depends only on $x_1, \ldots, x_s$. We denote by Λ the Nambu tensor associated to ω with respect to the volume form $\Omega = dx_1 \wedge \cdots \wedge dx_n$. We have

$$D_\Omega\Lambda = Y \wedge \partial/\partial x_{s+1} \wedge \cdots \wedge \partial/\partial x_n$$

where Y is a multi-field independent of $x_{s+1}, \ldots, x_n$. By the proof of the preceding theorem we get
$$\Lambda = P \wedge \partial/\partial x_{s+1} \wedge \cdots \wedge \partial/\partial x_n$$

where P is independent of $x_{s+1}, \ldots, x_n$. Returning to ω, we get the desired result. $\square$

Theorem 6.5.7 (de Medeiros [245]). *Let ω be a holomorphic integrable p-form in $(\mathbb{C}^n, 0)$ with coordinates $(x_1, \ldots, x_n)$, $2 \leq p \leq n-1$.*

i) *If $\operatorname{codim}\{i_\Pi \omega = 0\} \geq 3$, where $\Pi = \frac{\partial}{\partial x_1} \wedge \cdots \wedge \frac{\partial}{\partial x_{p-1}}$ and $\{i_\Pi \omega = 0\}$ means the set of zeroes of $i_\Pi \omega$ in $\mathbb{C}^n$, then there is a local holomorphic coordinate system $(y_1, \ldots, y_n)$ in which we have*

$$\omega = f \mathrm{d}g \wedge \mathrm{d}y_1 \wedge \cdots \wedge \mathrm{d}y_{p-1}, \tag{6.38}$$

where f, g are holomorphic functions in $(\mathbb{C}^n, 0)$, $f(0) \neq 0$.

ii) *If $\operatorname{codim}_V\{\omega|_V = 0\} \geq 3$, where $V \in \mathbb{C}^n$ is the $p+1$-dimensional linear subspace $\{x_{p+2} = \cdots = x_n = 0\}$, then the dual Nambu tensor Λ ($\omega = i_\Lambda(\mathrm{d}x_1 \wedge \cdots \wedge \mathrm{d}x_n)$) is of Type II$(q-1)$ at 0, where $q = n - p$.*

Proof (sketch). i) Put $\alpha = i_\Pi \omega$ and $\beta_j = i_{\Pi_j} \omega$, where $\Pi_j = \frac{\partial}{\partial x_1} \wedge \cdots \wedge \widehat{\frac{\partial}{\partial x_j}} \wedge \cdots \wedge \frac{\partial}{\partial x_{p-1}}$. It follows from the integrability of ω that $\beta_i \wedge \alpha = 0$. By de Rham's division theorem [98] (see Appendix A.2), we can write $\beta_j = \gamma_j \wedge \alpha$, and we also have $\omega \wedge \gamma_j = 0$ by integrability of ω. Note that $\langle \gamma_j, \frac{\partial}{\partial x_i} \rangle = \pm \delta_{ij}$. It follows that $\omega = \pm \alpha \wedge \gamma_1 \wedge \cdots \wedge \gamma_{p-1}$. Now apply Malgrange's Theorem 6.3.1 to the p-tuple $(\alpha, \gamma_1, \ldots, \gamma_{p-1})$ to conclude.

ii) Similarly to the proof of the first part, we can write

$$\Lambda = Z \wedge Y_1 \wedge \cdots \wedge Y_{q-1},$$

where $Y_1, \ldots, Y_{i-1}$ are locally linearly independent everywhere. Let x_1 be a function such that $Y_1(x_1) = 1$. By fixing Y_2 on $\{x_1 = 0\}$ and changing it at other points by the flow of Y_1, we can arrange it so that $[Y_1, Y_2] = 0$. The above decomposition of Λ still holds with respect to the new Y_2 (with a new Z). Similarly, we can arrange it so that $[Y_i, Y_j] = 0$ for any $i, j \leq q-1$. So we can write $Y_i = \frac{\partial}{\partial y_i}$ and

$$\Lambda = Z \wedge \frac{\partial}{\partial y_1} \wedge \cdots \wedge \frac{\partial}{\partial y_{q-1}}$$

in some system of coordinates. We may assume that $Z = \sum_{j \geq p} f_i \frac{\partial}{\partial y_i}$. The involutivity of the characteristic distribution of Λ (plus the fact that the set where $Z = 0$ is of codimension ≥ 2) implies that there are functions h_i such that $[Y_i, Z] = h_i Z$. We have $\frac{\partial h_i}{\partial y_j} = \frac{\partial h_j}{\partial y_i}$, and by Poincaré's lemma we can write $h_i = \frac{\partial g}{\partial y_i}$ for some function g. Then

$$\Lambda = \exp(g)\widehat{Z} \wedge \frac{\partial}{\partial y_1} \wedge \cdots \wedge \frac{\partial}{\partial y_{q-1}},$$

where $\widehat{Z} = \exp(-g)Z$ is independent of $y_1, \ldots, y_p$. Finally, the multiplicative factor $\exp(g)$ can be killed, say, by changing the coordinate function y_1 (and leaving the other coordinate functions intact). $\square$

The second part of the above theorem together with Theorem 6.5.6 implies the following result:

Corollary 6.5.8 ([59, 245]). *If ω is a holomorphic integrable p-form in $(\mathbb{C}^n, 0)$, $1 \le p \le n-2$, such that $\operatorname{codim}_W \{d\omega|_W = 0\} \ge 3$, where $W = \{x_{p+3} = \cdots = x_n = 0\}$ is a linear subspace of dimension $p+2$, then there is a local holomorphic coordinate system in which ω depends on at most $p+2$ variables.*

The case $p = 1$ of the above corollary was known to Camacho and Lins Neto [59], the general case $p \ge 1$ is due to de Medeiros [245].

6.6   Linearization of Nambu structures

Consider a singular point of a Nambu tensor Λ of order q. In a local coordinate system centered at that point, we will write the Taylor expansion of Λ as follows:

$$\Lambda = \Lambda^{(1)} + \Lambda^{(2)} + \cdots \tag{6.39}$$

where $\Lambda^{(k)}$ is a homogeneous q-vector field of degree k. Then $\Lambda^{(1)}$ is also a linear Nambu tensor, called the *linear part* of Λ. (The easiest way to see that $\Lambda^{(1)}$ is Nambu is by using the dual picture: if a differential form is integrable then its linear part is also integrable.) Up to linear automorphisms, $\Lambda^{(1)}$ is well defined, i.e., it does not depend on the choice of local coordinates.

In this section we present some results about the linearization of Λ, i.e., the existence of a local diffeomorphism (smooth, analytic, or formal) which transforms Λ into its linear part $\Lambda^{(1)}$. As usual, in order to get positive results, we will have to impose some conditions on Λ. And we will consider two separate cases, depending on whether $\Lambda^{(1)}$ belongs to Type I or Type II. A singular point of a Nambu structure is called of Type I (respectively, Type II) if its linear part is of Type I (respectively, Type II).

First let us consider the Type II case.

Proposition 6.6.1. *Let z be a nondegenerate singular point of Type II of a Nambu tensor λ of order $q \ge 3$ (or of order $q = 2$ but under the additional condition that $\operatorname{rank} \Lambda \le 2$), on an m-dimensional manifold M, whose linear part has the form $\Lambda^{(1)} = \partial/\partial x_1 \wedge \cdots \wedge \partial/\partial x_{q-1} \wedge (\sum_{i,j=q}^m b_j^i x_i \partial/\partial x_j)$. If the matrix $(b_j^i)_{i,j=q}^m$ has a non-zero trace, then there is a local system of coordinates $(x_1, \ldots, x_m)$ in which Λ can be written as*

$$\Lambda = \frac{\partial}{\partial x_1} \wedge \cdots \wedge \frac{\partial}{\partial x_{q-1}} \wedge X .$$

where $X = \sum_{i=q}^m c_i(x_q, \ldots, x_m)\partial/\partial x_i$ is a vector field which does not depend on $(x_1, \ldots, x_{q-1})$.

Proof. One checks that, with respect to any volume form Ω, we have

$$(D_\Omega \Lambda)(z) = \Big(\sum_{i=q}^m b_i^i \Big) \partial/\partial x_1 \wedge \cdots \wedge \partial/\partial x_{q-1}(z) \ne 0 .$$

So the above proposition is a direct consequence of Theorem 6.5.3. $\square$

Remark 6.6.2. The above proposition remains true in the case when $\sum b_i^i = 0$ but $\det(b_j^i) \neq 0$ and $q \leq m - 2$. (See Theorem 6.1 of [119] – there is a multiplicative factor in that theorem, but one can kill it by another coordinate transformation which leaves $x_q, \ldots, x_m$ unchanged.)

The above proposition reduces the problem of linearization of Λ to the problem of linearization of a vector field X. Applying Sternberg's linearization theorem [321] for vector fields in the smooth case, and Bruno's theorem [48] for the analytic case (see Appendix A.5), we get the following theorem:

Theorem 6.6.3 ([119]). *Let z be a singular point of Type II of a Nambu tensor Λ of order $q \geq 3$ (or of order $q = 2$ under the additional assumption that $\operatorname{rank} \Lambda \leq 2$) on an m-dimensional manifold M, whose linear part at z has the form $\Lambda^{(1)} = \partial/\partial x_1 \wedge \cdots \wedge \partial/\partial x_{q-1} \wedge (\sum_{i,j=q}^{m} b_j^i x_i \partial/\partial x_j)$. If the matrix (b_j^i) is nonresonant, i.e., if its eigenvalues $(\lambda_1, \ldots, \lambda_{p+1})$ do not satisfy any relation of the form $\lambda_i = \sum_{j=1}^{p+1} m_j \lambda_j$ with m_j being nonnegative integers and $\sum m_i \geq 2$, then Λ is smoothly linearizable, i.e., there is a local smooth system of coordinates $(x_1, \ldots, x_n)$ in a neighborhood of O, in which Λ coincides with its linear part:*

$$\Lambda = \partial/\partial x_1 \wedge \cdots \wedge \partial/\partial x_{q-1} \wedge \Big(\sum_{i,j=q}^{n} b_j^i x_i \partial/\partial x_j \Big).$$

The above linearization can be made analytic (i.e., real analytic or holomorphic) if Λ is analytic and the eigenvalues $\lambda_1, \ldots, \lambda_{p+1}$ of (b_j^i) satisfy Bruno's ω-condition.

Remark 6.6.4. It is easy to show that two Nambu tensors of order $q \geq 2$ of the type

$$\Lambda = \frac{\partial}{\partial x_1} \wedge \cdots \wedge \frac{\partial}{\partial x_{q-1}} \wedge Z$$

(where Z is a vector field which does not depend on $x_1, \ldots, x_{q-1}$) are locally isomorphic if and only if corresponding vector fields Z are locally *orbitally equivalent*, i.e., they are equivalent up to multiplication by a non-vanishing function. (Hint: consider coordinate transformations which change only x_1 and leave other coordinates unchanged.) So the right theory to use in the study of Type II singularities of Nambu structures is the theory of *orbital classification* of vector fields (see, e.g., [17], for results on this theory).

We now turn to singularities of Type I. We will call a singular point z of Type I of a Nambu structure Λ of order $q \geq 2$ *nondegenerate* if the linear part $\Lambda^{(1)}$ of Λ at z has the form

$$\Lambda = \sum_{j=1}^{q+1} \pm x_j \partial/\partial x_1 \wedge \cdots \wedge \partial/\partial x_{j-1} \wedge \partial/\partial x_{j+1} \wedge \cdots \wedge \partial/\partial x_{q+1} . \tag{6.40}$$

Equivalently, z is nondegenerate if the dual linear integrable p-form $\omega^{(1)}$ ($p + q = m$) of $\Lambda^{(1)}$ has the form

$$\omega^{(1)} = \mathrm{d}x_1 \wedge \cdots \wedge \mathrm{d}x_{p-1} \wedge \mathrm{d}Q^{(2)} , \tag{6.41}$$

where $Q^{(2)} = \frac{1}{2} \sum_{i=p}^{m} \pm x_i^2$ is a nondegenerate quadratic function of $q+1$ variables. (Here we changed the ordering of indices.) When $Q^{(2)} = \pm \frac{1}{2} (\sum_{i=p}^{m} x_i^2)$ is a positive or negative definite function, then we say that z is a nondegenerate *elliptic singular point* of Type I.

Theorem 6.6.5 ([119]). *Let z be a nondegenerate singular point of Type II of a smooth Nambu tensor Λ of order $q \geq 3$ (or of order $q = 2$ under the additional assumption that rank $\Lambda \leq 2$) on an m-manifold M. Then Λ is formally linearizable at z. If Λ is analytic (real or complex) then it is, up to multiplication by a non-vanishing function, analytically linearizable. If z is elliptic, then Λ is smoothly linearizable.*

The rest of this section is devoted to the proof of the above theorem. In the formal case, we will first show how to linearize the associated singular foliation formally, i.e., linearize Λ up to multiplication by a non-vanishing function. Then we will kill that function by another change of variables. In the analytic case, we will use the formal linearization together with Malgrange's Theorem 6.3.1 to show the existence of analytic first integrals of the associated foliation. These analytic first integrals allow us to linearize the foliation analytically. In the smooth case, we will use blowing-up techniques to linearize the associated singular foliation smoothly, then use Moser's path method to linearize Λ. Most of the time, we will work with $\omega = \Omega^\flat(\Lambda)$. Λ is determined by ω up to multiplication by a non-vanishing function.

6.6.1   Decomposability of ω

In this subsection, we will show an auxiliary result which says that, in the neighborhood of a nondegenerate singular point of Type I, ω can be decomposed into a wedge product:

$$\omega = \gamma_1 \wedge \cdots \wedge \gamma_{p-1} \wedge \alpha \, , \tag{6.42}$$

where $\gamma_i = \mathrm{d}x_i + \theta_i$ with $\theta_i(z) = 0$. In particular, we deduce from this decomposition that the set of singular points of ω forms a submanifold Σ of dimension p.

According to the definition of nondegenerate singularities of Type I, we can suppose that ω has a Taylor expansion $\omega = \omega^{(1)} + \omega^{(2)} + \cdots$, with $\omega^{(1)} = \mathrm{d}x_1 \wedge \cdots \wedge \mathrm{d}x_{p-1} \wedge \mathrm{d}Q^{(2)}$, where $Q^{(2)} = \frac{1}{2} \sum_{j=p}^{m} \pm x_j^2$. Express ω as a polynomial in $\mathrm{d}x_1, \ldots, \mathrm{d}x_{p-1}$:

$$\omega = dx_1 \wedge \cdots \wedge dx_{p-1} \wedge \alpha + \sum_{j=1}^{p-1} dx_1 \wedge \cdots \widehat{dx_j} \cdots \wedge dx_{p-1} \wedge \beta_j$$

$$+ \sum_{1 \leq i < j \leq p-1} dx_1 \wedge \cdots \widehat{dx_i} \cdots \widehat{dx_j} \cdots \wedge dx_{p-1} \wedge \gamma_{ij} + \cdots .$$

Here $\alpha, \beta_i, \gamma_{ij}, \ldots$ are differential forms which, when written in coordinates $(x_1, \ldots, x_m)$, do not contain the terms $\mathrm{d}x_1, \ldots, \mathrm{d}x_{p-1}$. Applying the equation $i_A \omega \wedge \omega = 0$ to $A = \partial/\partial x_1 \wedge \cdots \wedge \partial/\partial x_{p-1}$, we have $\alpha \wedge \omega = 0$. It follows that $\alpha \wedge \beta_j = 0, \alpha \wedge \gamma_{ij} = 0$, etc. We can consider α and β_j as differential forms on the space of variables $(x_p, \ldots, x_m)$, parametrized by $x_1, \ldots, x_{p-1}$, and by our assumption of nondegeneracy, we can apply de Rham division theorem [98] (see Appendix A.2), which says that, since the number of variables is $q + 1 > 2$ and the order of β_j is 2, β_j is divisible by α: $\beta_j = \alpha \wedge \theta_j$ where θ_j are smooth 1-forms.

Applying the equation $i_A \omega \wedge \omega = 0$ to $A = \partial/\partial x_1 \wedge \cdots \wedge \partial/\partial x_{j-1} \wedge \partial/\partial x_{j+1} \wedge \cdots \wedge \partial/\partial x_{p-1} \wedge \partial/\partial x_p$, we get

$$0 = \omega \wedge \left(\langle \alpha, \partial/\partial x_p \rangle ((-1)^{p-j} \mathrm{d}x_j + \theta_j) - \langle \theta_j, \partial/\partial x_p \rangle \alpha \right).$$

Since $\langle \alpha, \partial/\partial x_p \rangle = \langle \alpha^{(1)}, \partial/\partial x_p \rangle + \cdots = \pm x_p + \cdots \neq 0$, and we already have $\omega \wedge \alpha = 0$, we get that $\omega \wedge \gamma_j = 0$ where $\gamma_j = \mathrm{d}x_j + (-1)^{p-j} \theta_j$. Since γ_j do not vanish and are linearly independent at z, it follows that ω is divisible by the product of γ_j:

$$\omega = \gamma_1 \wedge \cdots \wedge \gamma_{p-1} \wedge \alpha'$$

for some 1-form α'. By adding a combination of γ_j to α', we can assume that α' does not contain the terms $\mathrm{d}x_1, \ldots, \mathrm{d}x_{p-1}$ when written in the coordinates $(x_1, \ldots, x_n)$. Then considering the terms containing $\mathrm{d}x_1 \wedge \cdots \wedge \mathrm{d}x_{p-1}$ on the two sides of the equation $\omega = \gamma_1 \wedge \cdots \wedge \gamma_{p-1} \wedge \alpha'$, it follows that in fact we have $\alpha' = \alpha$.

6.6.2 Formal linearization of the associated foliation

We will apply the Godbillon–Vey algorithm twice to arrive at a formal linearization of the associated foliation.

A consequence of the expression $\omega = \gamma_1 \wedge \cdots \wedge \gamma_{p-1} \wedge \alpha$, with $\gamma_i = \mathrm{d}x_i + \cdots$, $\alpha = \sum_{i=p}^{m} \pm x_i \mathrm{d}x_i + \cdots$, and the implicit function theorem, is that the set of singular points of ω near z is a $(p-1)$-dimensional submanifold, which we will denote by Σ. By rectifying Σ, we may assume that Σ is given by

$$\Sigma = \{ x_p = \cdots = x_m = 0 \}. \tag{6.43}$$

To simplify the formulas which follow, we will change the name of the last $q + 1$ coordinates and denote them by y:

$$y := (y_1, \ldots, y_{q+1}) := (x_p, \ldots, x_m).$$

In this subsection, if F is a differential form, then we will denote by

$$F = F^{(0)} + F^{(1)} + F^{(2)} + \cdots$$

the Taylor expansion of F with respect to the variables y, i.e., $F^{(k)}$ is homogeneous of degree k in y. In particular, we have $\alpha^{(0)} = 0$ and

$$\alpha^{(1)} = \sum \alpha_{ij}(x) y_j \mathrm{d}y_i$$

where $\alpha_{ij}(x) = \pm \delta_{ij} +$ higher-order terms.

By a linear change of variables leaving y unchanged, we may assume that

$$\gamma_i = \mathrm{d}x_i + \gamma_i^{(1)} + \gamma_i^{(2)} + \cdots \; \forall \, i = 1, \ldots, p-1 \; .$$

We will use transformations of coordinates to kill the terms $\gamma_i^{(k)}$ in the above expression consecutively. Suppose that we have already killed $(k-1)$ first terms (where $k \geq 1$), so we can write

$$\gamma_i = \mathrm{d}x_i + \gamma_i^{(k)} + \gamma_i^{(k+1)} + \cdots \; \forall \, i = 1, \ldots, p-1 \; .$$

Let us show how to kill the term $\gamma_i^{(k)}$. The integrability of ω implies that

$$\gamma_1 \wedge \cdots \wedge \gamma_{p-1} \wedge \alpha \wedge \mathrm{d}\gamma_i = 0 \;\; \forall \, i = 1, \ldots, p-1 \; .$$

The terms of degree k (in y) in the above equation give

$$\mathrm{d}x_1 \wedge \cdots \wedge \mathrm{d}x_p \wedge \alpha^{(1)} \wedge \mathrm{d}_y \gamma_i^k = 0 \; ,$$

where d_y denotes the exterior derivation with respect to y. This gives $\alpha^{(1)} \wedge \mathrm{d}_y \gamma_i^k = 0$, and, by the de Rham division theorem, we can write

$$\mathrm{d}_y \gamma_i^k = \alpha^{(1)} \wedge \beta_i^{(k-2)}. \tag{6.44}$$

Similarly, $\gamma_1 \wedge \cdots \wedge \gamma_{p-1} \wedge \alpha \wedge \mathrm{d}\alpha = 0$, which leads to $\alpha^{(1)} \wedge \mathrm{d}_y \alpha^{(1)} = 0$, then to $\mathrm{d}_y \alpha^{(1)} = \alpha^{(1)} \wedge \beta$ for some β, hence $\mathrm{d}_y \alpha^{(1)} = 0$ (because $\mathrm{d}_y \alpha^{(1)}$ is of degree 0 in y).

So the equation $\mathrm{d}_y \circ \mathrm{d}_y = 0$ applied to Equation (6.44) leads to $\alpha^{(1)} \wedge \mathrm{d}_y \beta_i^{(k-2)} = 0$, hence we can write

$$\mathrm{d}_y \beta_i^{(k-2)} = \alpha^{(1)} \wedge \beta_i^{(k-4)} \; . \tag{6.45}$$

Iterating the above procedure, we obtain a finite sequence $\beta_i^{(k-2)}$, $\beta_i^{(k-4)}$, etc. of 1-forms which are homogeneous in y. So there is an r ($r = 0$ or $r = 1$) such that $\mathrm{d}_y \beta_i^{(r)} = 0$ and $\beta_i^{(r)} = \mathrm{d}_y \phi_i^{(r+1)}$, where $\phi_i^{(r+1)}$ is an $(r+1)$-homogeneous function in y. From the equation $\mathrm{d}_y \beta_i^{(r+2)} = \alpha^{(1)} \wedge \beta_i^{(r)}$, we get $\mathrm{d}_y \beta_i^{(r+2)} = -\mathrm{d}_y[\phi_i^{(r+1)} \alpha^{(1)}]$, hence $\beta_i^{(r+2)} = -\phi_i^{(r+1)} \alpha^{(1)} + \mathrm{d}_y \phi_i^{(r+3)}$ for some function $\phi_i^{(r+3)}$ which is homogeneous of degree $r+3$ in y. Similarly, from the equation $\mathrm{d}_y \beta_i^{(r+4)} = \alpha^{(1)} \wedge \beta_i^{(r+2)}$, we get $\mathrm{d}_y \beta_i^{(r+4)} = -\mathrm{d}_y \left(\phi_i^{(r+3)} \alpha^{(1)} \right)$, hence $\beta_i^{(r+4)} = -\phi_i^{(r+3)} \alpha^{(1)} + \mathrm{d}_y \phi_i^{(r+5)}$. Going back this way, we arrive at the following expression for $\gamma_i^{(k)}$:

$$\gamma_i^{(k)} = -\phi_i^{(k-1)} \alpha^{(1)} + \mathrm{d}_y \phi_i^{(k+1)} \; . \tag{6.46}$$

Replacing γ_i by $\gamma_i' = \gamma_i + \phi_i^{(k)} \alpha$ (it won't change ω), and x_i by $x_i' = x_i + \phi_i^{(k+1)}$, we get

$$\omega = \gamma_1' \wedge \cdots \wedge \gamma_{p-1}' \wedge \alpha$$

with

$$\gamma_i' = dx_i' + (\gamma')_i^{(k+1)} + \cdots .$$

In other words, we have killed the terms of degree k (in y) in the expression of γ. Repeating the above process for k going from 1 to ∞ and taking the (formal) limit, we arrive at a formal coordinate system $(x_1, \ldots, x_{p-1}, y_1, \ldots, y_{q+1})$ near z in which ω has the following form:

$$\omega = dx_1 \wedge \cdots \wedge dx_{p-1} \wedge \alpha ,$$

where we may assume that α does not contain the terms $dx_1, \ldots, dx_{p-1}$. Under this form, the integrability condition on ω can be written as

$$\alpha \wedge d_y \alpha = 0 .$$

So we may view α as a family, parametrized by $x_1, \ldots, x_{p-1}$, of formal integrable 1-forms in variables $(y_1, \ldots, y_{q+1})$. with a nondegenerate linear part. It is a well-known result that such an integrable 1-form α is formally linearizable up to multiplication by a non-vanishing function (see, e.g., [267]). Let us show how to do it, in a way similar to the above normalization of γ_i.

By a change of coordinates leaving $(x_1, \ldots, x_{p-1})$ unchanged, we may assume that $\alpha^{(1)} = f(\sum_1^{q+1} \pm y_i dy_i)$, where f is a non-vanishing function. Since we want to linearize α up to multiplication by a non-vanishing function, we will write α as

$$\alpha = (\zeta^{(1)} + \zeta^{(2)} + \zeta^{(3)} + \cdots)h$$

where $\zeta^{(k)}$ is homogeneous of degree k in y, $\zeta^{(1)} = dQ^{(2)} = \sum_1^{q+1} \pm y_i dy_i$, and h is a non-vanishing function. Now we will show how to kill the nonlinear terms $\zeta^{(k)}$ in the expression of α consecutively. Suppose that we already have

$$\zeta^{(2)} = \cdots = \zeta^{(k-1)} = 0$$

for some $k \geq 2$. Then the terms of degree k in the equation $\alpha \wedge d_y \alpha = 0$ gives $\zeta^{(1)} \wedge d_y \zeta^{(k)} = 0$. So by the de Rham division theorem we can write $d_y \zeta^{(k)} = \zeta^{(1)} \wedge \beta^{(k-2)}$. Then we have $0 = d_y(d_y \zeta^{(k)}) = \zeta^{(1)} \wedge d_y \beta^{(k-2)}$, which implies that $d_y \beta^{(k-2)} = \zeta^{(1)} \wedge \beta^{(k-4)}$, and so on. Repeat this process until we get $d_y \beta^{(k-2r)} = 0$, which implies that $\beta^{(k-2r)} = d_y \phi^{(k-2r+1)}$. Now go back (the same way that we did with γ_i): $d\beta^{(k-2r+2)} = \zeta^{(1)} \wedge d_y \phi^{(k-2r+1)}$, $\beta^{(k-2r+2)} = -\phi^{(k-2r+1)} \zeta^{(1)} + d_y \phi^{(k-2r+3)}$, and so on, until we get

$$\zeta^{(k)} = -\phi^{(k-1)} \zeta^{(1)} + d_y \phi^{(k+1)} .$$

This last equation gives

$$\zeta^{(1)} + \zeta^{(k)} + \cdots = d_h Q^{(2)} - \phi^{(k-1)} d_h Q^{(2)} + d_y \phi^{(k+1)} + \cdots$$
$$= (1 - \phi^{(k-1)}) d_h (Q^{(2)} + \phi^{(k+1)}) + \text{terms of degree } \geq (k+1) .$$

Putting $Q' = Q^{(2)} + \phi^{(k+1)}$ and $f' = (1 - \phi^{(k-1)})f$, we get

$$\alpha = (\mathrm{d}_h Q' + \text{terms of degree} \geq (k+1))\, f' \,.$$

Since Q' is a Morse function in y with the quadratic term $Q^{(2)}$, we can transform Q' into Q by a coordinate transformation of the type $y' = y +$ terms of degree $\geq (k+1)$. This is the coordinate transformation which kills the term $\zeta^{(k)}$ in the expression of α.

Repeating the above process, we arrive at a linearization of α up to multiplication by a non-vanishing function.

Summarizing, in this subsection we have shown that there is a formal coordinate system $(x_1, \ldots, x_{p-1}, y_1, \ldots, y_{q+1})$ in which we have

$$\omega = h\mathrm{d}x_1 \wedge \cdots \wedge \mathrm{d}x_{p-1} \wedge \mathrm{d}(\frac{1}{2}\sum_{i=1}^{q+1} \pm y_i^2)$$

where h is a non-vanishing function. In these coordinates, $\Lambda = \Omega^f lat(\omega)$ is also linearized up to multiplication by a non-vanishing function (which depends on h and the volume form). The associated singular foliation does not depend on h, and is formally linearized in these coordinates.

6.6.3   The analytic case

The above formal linearization implies that the associated singular foliation admits formal first integrals $x_1, \ldots, x_{p-1}$ and $Q = \frac{1}{2}\sum_{i=1}^{q+1} \pm y_i^2$. In the analytic (i.e., real analytic or holomorphic case), the decomposability of ω into the wedge product of $\gamma_1, \ldots, \gamma_{p-1}, \alpha$, together with the existence of the above formal first integrals, allows us to apply Malgrange's Theorem 6.3.1, which tells us that these formal first integrals $x_1, \ldots, x_{p-1}$ and $Q = \frac{1}{2}\sum_{i=1}^{q+1} \pm y_i^2$ can in fact be chosen locally analytic. Thus we obtain a local analytic linearization of the associated singular foliation, implying that Λ (and ω) can be linearized analytically up to multiplication by an analytic non-vanishing function.

6.6.4   Formal linearization of Λ

We may assume that

$$\Lambda = f\Lambda_1 = f\Big(\sum_{i=1}^{q+1} \pm x_i \frac{\partial}{\partial x_1} \wedge \cdots \wedge \frac{\partial}{\partial x_{i-1}} \wedge \frac{\partial}{\partial x_{i+1}} \wedge \cdots \wedge \frac{\partial}{\partial x_n}\Big)$$

where f is a non-vanishing function.

We want to change the coordinates $(x_1, \ldots, x_{q+1})$ (and leave $x_{q+2}, \ldots, x_m$ unchanged) so as to make $f = 1$. We will forget about the parameters $(x_{q+2}, \ldots, x_m)$, and may assume for simplicity that $m = q + 1$.

Write $f = \sum f^{(k)}$, where $f^{(k)}$ is homogeneous of degree k in $(x_1, \ldots, x_{q+1})$. By a change of coordinates of the type $x'_1 = gx_1, \ldots, x'_{q+1} = gx_{q+1}$, we can make $f^{(0)} = 1$. We assume now that we already have $f^{(1)} = \cdots = f^{(k-1)} = 0$ for some $k \geq 1$. We will show that there is a change of coordinates which changes x_i by terms of degree $\geq r$, and which kills $f^{(k)}$. It amounts to finding a vector field X such that

$$\mathcal{L}_X \Lambda_1 = f^{(k)} \Lambda_1$$

where $\mathcal{L}$ denotes the Lie derivative. Consider the volume form $\Omega = dx_1 \wedge \cdots \wedge dx_{q+1}$. Then it is easy to see, by contracting Λ_1 with Ω, that the equation $\mathcal{L}_X \Lambda_1 = f^{(k)} \Lambda_1$ is equivalent to the equation

$$dX(Q) = (f^{(k)} + \mathrm{div}_\Omega X) dQ,$$

where $Q = (1/2) \sum_i \varepsilon_i x_i^2$, $\varepsilon_i = \pm 1$. In turn, this equation is equivalent to the following system of equations:

$$\mathrm{div}_\Omega X + f^{(r)} = \frac{d(2QF(Q))}{dQ},$$
$$X(Q) = 2QF(Q),$$

where F is an unknown function. Write $X = A + Y$, where $A = F(Q) \sum x_i \frac{\partial}{\partial x_i}$, and Y is a vector field such that $Y(Q) = 0$. Then the above system of equations is equivalent to a system of the type

$$Y(Q) = 0,$$
$$\beta(Q) + \mathrm{div}_\Omega Y = f^{(k)},$$

where β is an unknown function. The equation $Y(Q) = 0$ is equivalent to the fact that $Y = \sum_{i<j} f_{ij} Y_{ij}$ where $Y_{ij} = \varepsilon_i x_j \partial/\partial x_i - \varepsilon_j x_i \partial/\partial x_j$. For such a Y, we have $\mathrm{div}_\Omega Y = \sum_{i<j} Y_{ij}(f_{ij})$. Denote by J the set of homogeneous polynomials of degree k. The solvability of the above system of equation follows from the following facts, which can be verified easily by choosing appropriate f_{ij}:

1. If a monomial $x^I = x_1^{I_1} \ldots x_{q+1}^{I_{q+1}}$ has one of I_i to be an odd number, then it belongs to J.
2. Q^s is equivalent to λx_1^{2s} modulo J for some non-zero number λ.
3. Any monomial $x^I = x_1^{I_1} \ldots x_{q+1}^{I_{q+1}}$, with all I_i even, is equivalent to $\lambda x_1^{\sum I_i}$ modulo J for some number λ.

Thus, the above system of equations can always be solved, and therefore the non-constant terms in f can be killed one by one.

Remark 6.6.6. We don't know if in the analytic case we can also linearize Λ analytically (without a multiplicative factor) or not. Our guess would be that the answer is yes.

6.6.5   The smooth elliptic case

When Λ is smooth and z is a nondegenerate elliptic singular point of Type I, then we can use geometric arguments instead of formal power series in order to linearize Λ smoothly. First let us show (in a sketchy way) how to linearize the associated foliation smoothly by the blowing-up method.

We may assume that

$$\omega = \mathrm{d}x_1 \wedge \cdots \wedge \mathrm{d}x_{p-1} \wedge \mathrm{d}Q + \cdots ,$$

in a local coordinate system $(x, y) = (x_1, \ldots, x_{p-1}, y_1, \ldots, y_{q+1})$, where $Q = \sum_1^{p+1} y_i^2$, and that the local singular set of ω is $\Sigma = \{y = 0\}$.

We will perform an *oriented blow-up* along the singular set Σ, i.e., we will consider the mapping

$$E : \mathbb{R}^{p-1} \times \mathbb{R} \times S^q \longrightarrow \mathbb{R}^{p-1} \times \mathbb{R}^{q+1} = \mathbb{R}^m ,$$

given by $E(x, \rho, u) = (x, \rho u)$, where S^q is the unit sphere in $\mathbb{R}^{q+1}$.

Lemma 6.6.7. *Under the above map E, the local associated singular foliation of ω on $\mathbb{R}^m$ lifts to a regular smooth q-dimensional foliation on a neighborhood of $\{0\} \times \{0\} \times S^q$ in $\mathbb{R}^{p-1} \times \mathbb{R} \times S^q$, which admits the spheres $\{x\} \times \{0\} \times S^q$ as leaves.*

We will leave the above lemma as an exercise. (Hint: lift the tangent vector fields.)

Since $q \geq 2$, S^q is compact simply-connected, and we can use Reeb's stability theorem [356] (see Appendix A.3) to conclude that the induced foliation on $\mathbb{R}^{p-1} \times \mathbb{R} \times S^q$ is a trivial fibration. In particular, it admits p smooth independent first integrals $f_1, \ldots, f_p$.

Notice that, since the map E is generically 2-to-1, it induces an involution $\sigma : (x, \rho, u) \mapsto (x, -\rho, -u)$. If f is a first integral, then $f^\sigma := f \circ \sigma$ is also a first integral. In order to get the first integrals for the associated singular foliation $\mathcal{F}_\omega$, we have to look for first integrals on $\mathbb{R}^{p-1} \times \mathbb{R} \times S^q$ which are σ-invariant, and which project to smooth functions on $\mathbb{R}^m$. To turn a first integral f into a σ-invariant first integral is easy: just take $f + f^\sigma$. But not every σ-invariant smooth function on $\mathbb{R}^{p-1} \times \mathbb{R} \times S^q$ projects to a smooth function on $\mathbb{R}^m$. (For example, the function $y_1 y_2 / (\sum y_i^2)$ is not smooth on $\mathbb{R}^m$ but its pull-back is smooth on $\mathbb{R}^{p-1} \times \mathbb{R} \times S^q$.) So we need the following lemma.

Lemma 6.6.8. *Every σ-invariant smooth first integral of the induced foliation on $\mathbb{R}^{p-1} \times \mathbb{R} \times S^q$ projects to a smooth first integral of the associated singular foliation $\mathcal{F}_\omega$ on $\mathbb{R}^m$.*

Proof. Using results from the previous subsections, and Borel's theorem about the existence of smooth functions with an arbitrary given Taylor expansion, we can assume that

$$\omega = h\mathrm{d}x_1 \wedge \cdots \wedge \mathrm{d}x_{p-1} \wedge \mathrm{d}Q + \mu$$

where h is a smooth non-vanishing function, and μ is a smooth differential form which is flat *along the local singular submanifold* Σ (and not just at one point z – by looking more carefully at the formal linearization process, the reader will see that we do it along Σ). In particular, the vector fields

$$Y_{ij} = \pm\Omega^\sharp(dy_1 \wedge \cdots \widehat{dy_i} \cdots \widehat{dy_j} \cdots \wedge dy_{q+1} \wedge \omega/h) ,$$

which are tangent to the associated foliation $\mathcal{F}_\omega$, has the form

$$Y_{ij} = y_i\frac{\partial}{\partial y_j} - y_j\frac{\partial}{\partial y_i} + Z_{ij} ,$$

where Z_{ij} are vector fields which are flat along Σ. Lifting the vector fields Y_{ij} by E, we get vector fields

$$\widetilde{Y_{ij}} = u_i\frac{\partial}{\partial u_j} - u_j\frac{\partial}{\partial u_i} + \widetilde{Z_{ij}}$$

on $\mathbb{R}^{p-1} \times \mathbb{R} \times S^q$, which are tangent to the induced foliation, where $\widetilde{Z_{ij}}$ is flat with respect to ρ. ($u_i\frac{\partial}{\partial u_j} - u_j\frac{\partial}{\partial u_i}$ are infinitesimal rotations of S^q.)

Let f be a σ-invariant first integral on $\mathbb{R}^{p-1} \times \mathbb{R} \times S^q$. Write the Taylor expansion of f with respect to the variable ρ as follows:

$$f = f_0(x, u) + f_1(x, u)\rho + f_2(x, u)\rho^2 + \cdots .$$

The idea is to show that, $\forall\ i = 0, 1, 2, \ldots, f_i(x, u)$ does not depend on u. Then $f_i(x, u) = 0$ if i is odd (because f is σ-invariant), and the projection of f on $\mathbb{R}^m$ will be smooth. It is clear that f_0 does not depend on u, because f is constant on the spheres $\{x\} \times \{0\} \times S^q$. For the other f_i, use the equations $\widetilde{Y_{ij}}(f) = 0$ and the fact that $\widetilde{Y_{ij}} = u_i\frac{\partial}{\partial u_j} - u_j\frac{\partial}{\partial u_i}$ up to a flat term in ρ. $\square$

Using the above lemma, we can find appropriate smooth first integrals for $\mathcal{F}_\omega$, which allow us to linearize $\mathcal{F}_\omega$ smoothly.

Let us now linearize the Nambu tensor Λ. Linearizing the associated singular foliation smoothly, we may assume that

$$\Lambda = f\left(\sum_{i=1}^{q+1}(-1)^i\frac{\partial}{\partial y_1} \wedge \cdots \widehat{\frac{\partial}{\partial y_i}} \cdots \wedge \frac{\partial}{\partial y_{q+1}}\right) ,$$

where f is a smooth non-vanishing function. Using the formal linearizability of Λ (see Subsection 6.6.4), we may assume that $f = 1 + g$, where g is a smooth flat function along Σ. As we said in the beginning of this chapter, a Nambu structure is essentially just a foliation plus a leaf-wise volume form. The volume of the leaf $S_{x,\rho} = \{(x, y) \mid \sum y_i^2 = \rho^2\}$ with respect to Λ is $C\rho^{q-1}(1+\varepsilon)$, where C is a positive constant, and ε is a smooth function which is flat along Σ (which depends on g). By a change of variables of the type $(x', y') = (x, h(\rho, x)y)$, where $h(\rho, x) - \rho$

is a flat function in ρ, we may kill the term ε in the above volume form, i.e., make the volume form of $S_{s,\rho}$ with respect to Λ equal to the volume form of the linear Nambu tensor $\Lambda_0 = \sum_{i=1}^{q+1}(-1)^i \frac{\partial}{\partial y_1} \wedge \cdots \widehat{\frac{\partial}{\partial y_i}} \cdots \wedge \frac{\partial}{\partial y_{q+1}}$. Now we can use Moser's path method, and even Moser's result about the equivalence of volume forms with the same volume on a compact manifold (see Appendix A.1), to find an isomorphism from Λ to Λ_0 which preserves each leaf of the associated singular foliation. In fact, we can choose this isomorphism in such a way that it is a local diffeomorphism whose ∞-jet at points on Σ is Identity. The smooth linearization of Λ is proved. $\qquad\square$

Remark 6.6.9. When the quadratic function Q in $\omega^{(1)} = \mathrm{d}x_1 \wedge \cdots \wedge \mathrm{d}x_{p-1} \wedge \mathrm{d}Q$ is of the type $Q = \pm(y_1^2 + y_2^2 - \sum_{i=3}^{q+1} y_i^2)$, then the associated singular foliation of $\omega^{(1)}$ has local leaves of type $S^1 \times D^{q-1}$. In this case, the associated singular foliation of ω and $\Lambda = \Omega^\sharp(\omega)$ near a nondegenerate singular point of Type I and with the above type of Q is *not* smoothly linearizable in general. The reason is similar to the reason why the Lie algebra $sl(2, \mathbb{R})$ is smoothly degenerate: the local leaves of type $S^1 \times D^{q-1}$ are unstable under a small perturbation and can be made to spiral. When $Q = \sum_1^r y_i^2 - \sum_{r+1}^{q+1} y_i^2$ with $r \neq 2, 0$ and $q+1-r \neq 2, 0$, we conjecture that the associated singular foliation is still smoothly linearizable. (The results of Moussu [267, 268] on the existence of smooth first integrals for integrable 1-forms imply that this conjecture is true in the case of integrable 1-forms and Nambu tensors of order $m-1$.)

6.7 Integrable 1-forms with a non-zero linear part

In this section we will give formal normal forms for Nambu tensors of order $m-1$ and integrable 1-forms near a singular point z where they have a nontrivial linear part.

We will distinguish two cases, depending on whether the curl of the Nambu structure (or the differential of the dual 1-form) vanishes at z or not. The easy case is when the curl does not vanish: in that case, Kupka's phenomenon (see Section 6.5) says that our Nambu tensor has the local form

$$\partial/\partial x_1 \wedge \cdots \wedge \partial/\partial x_{m-2} \wedge X$$

where X is a vector field which depends only on two variables x_{m-1}, x_m. The local classification of such Nambu tensors is then reduced to the classification of vector fields in dimension 2 up to orbital equivalence.

When the curl of Λ vanishes at z, we have the following formal normal form:

Theorem 6.7.1. *Let Λ be a* (*smooth or formal*) *Nambu tensor of order $m-1$ on an m-dimensional manifold with $m \geq 3$. Suppose that $\Lambda(z) = 0$ and $D_\Omega \Lambda(z) = 0$ where Ω is some volume form, and that the linear part $\lambda^{(1)}$ of Λ at z is nontrivial.*

Then there is a formal coordinate system $(x_1, \ldots, x_m)$ *at* z *in which we have:*

$$\Lambda = x_m \frac{\partial}{\partial x_1} \wedge \cdots \wedge \frac{\partial}{\partial x_{m-1}} + \sum_{i=1}^{m-1} (-1)^{m-i} \left(\frac{\partial f}{\partial x_i} + x_m \frac{\partial g}{\partial x_i} \right) \frac{\partial}{\partial x_1} \wedge \cdots \widehat{\frac{\partial}{\partial x_i}} \cdots \wedge \frac{\partial}{\partial x_m} ,$$

(6.47)

where f *and* g *are two formal functions which are independent of* x_m *and such that* $df \wedge dg = 0$.

A dual integrable 1-form of Λ in the above coordinates will have the form

$$\omega = h . \left(\sum_{i=1}^{m-1} \left(\frac{\partial f}{\partial x_i} + x_m \frac{\partial g}{\partial x_i} \right) dx_i + x_m dx_m \right) = h . (df + x_m dg + x_m dx_m) , \quad (6.48)$$

where h is a formal function such that $h(0) \neq 0$.

Proof. It follows from the assumptions of the above theorem and the classification of linear Nambu structures, that we can find a coordinate system in which we have

$$\{x_1, \ldots, x_{m-1}\}^{(1)} = x_n \qquad (6.49)$$

for the linear Nambu bracket corresponding to the linear part $\Lambda^{(1)}$ of Λ.

Remark 6.7.2. In fact, the above theorem remains true if we replace the condition that $D_\Omega \Lambda(z) \neq 0$ by Equation (6.49).

In this section, we will use following notations:

$$x := (x_1, \ldots, x_{m-1}), \quad y := x_m .$$

For each function $h = h(x, y)$, we will denote by

$$h^{(0)} + h^{(1)} + \cdots + h^{(k)} + \cdots$$

the Taylor expansion of h with respect to the variables $x_1, \ldots, x_{m-1}$ (i.e., consider y as a parameter in this expansion).

Lemma 6.7.3. *Suppose that there are coordinates* $x = (x_1, \ldots, x_{m-1})$ *and* y *such that*

$$\{x_1, \ldots, x_{m-1}\} = y + c^{(r+2)}(x, y) + c^{(r+3)}(x, y) + \cdots$$

and

$$\{x_1, \ldots, x_{i-1}, x_{i+1}, \ldots, x_{m-1}, y\} = (-1)^{m-i} (a_i^{(0)}(x, y) + a_i^{(1)}(x, y) + \cdots) ,$$

where $a_i^{(0)}, \ldots, a_i^{(r-1)}$ *are affine functions in* y, *for some number* $r \geq 0$. *Then there is a coordinate transformation of the form*

$$x_1' = x_1 + \mu^{(r+2)}(x, y)$$
$$x_2' = x_2, \ldots, x_{m-1}' = x_{m-1}$$
$$y' = y + \gamma^{(r+1)}(x, y) + \gamma^{(r+2)}(x, y) ,$$

which gives

$$\{x'_1, \ldots, x'_{n-1}\} = y' + C^{(r+3)}(x', y') + C^{(r+4)}(x', y') + \cdots$$

and

$$\{x'_1, \ldots, x'_{i-1}, x'_{i+1}, \ldots, x'_{n-1}, y'\}$$
$$= (-1)^{n-i}(a_i^{(0)}(x', y') + \cdots + a_i^{(r-1)}(x', y') + A_i^{(r)}(x', y') + \cdots),$$

where $A_i^{(r)}$ is an affine function in y'.

Proof of Lemma 6.7.3. Making a coordinate transformation of the form $\tilde{x} = x$, $\tilde{y} = y(1 + e^{(r+1)}(x, y))$, we get

$$\{\tilde{x}_1, \ldots, \tilde{x}_{n-1}\} = \tilde{y} + \tilde{y}(\tilde{c}^{(r+1)}(\tilde{x}, \tilde{y})) + \tilde{c}^{(r+2)}(\tilde{x}, \tilde{y}) + \cdots$$

and

$$\{\tilde{x}_1, \ldots, \tilde{x}_{i-1}, \tilde{x}_{i+1}, \ldots, \tilde{x}_{n-1}, \tilde{y}\}$$
$$= (-1)^{n-i}(a_i^{(0)}(\tilde{x}, \tilde{y}) + \cdots + a_i^{(r-1)}(\tilde{x}, \tilde{y}) + A_i^{(r)}(\tilde{x}, \tilde{y}) + \cdots),$$

with

$$A_i^{(r)} = a_i^{(r)} - \frac{\partial e^{r+1}}{\partial x_i} y^2. \tag{6.50}$$

Let us show how to choose $e^{(r+1)}$ in such a way that $A_i^{(r)}$ becomes an affine function in y. Putting $\Omega = dx_1 \wedge \cdots \wedge dx_{m-1} \wedge dy$, we have $\omega := i_\Lambda \Omega = \Gamma dy + \sum_i \Delta_i dx_i$ with

$$\Gamma = \{x_1, \ldots, x_{m-1}\}, \quad \Delta_i = (-1)^{m-i}\{x_1, \ldots, x_{i-1}, x_{i+1}, \ldots, x_m\}.$$

Recall that we have $\omega \wedge d\omega = 0$. The terms containing $dx_i \wedge dx_j \wedge dy$ in this equation give

$$\Gamma(\partial \Delta_i / \partial x_j - \partial \Delta_j / \partial x_i) + \Delta_i(\partial \Delta_j / \partial y - \partial \Gamma / \partial x_j) - \Delta_j(\partial \Delta_i / \partial y - \partial \Gamma / \partial x_i) = 0. \tag{6.51}$$

We write $\Delta_k = \alpha_k(x) + y\beta_k(x) + y^2\delta_k(x, y)$. The hypothesis of the lemma says that the Taylor expansion of δ_k in the variables x begins with terms of degree $\geq k$: $\delta_k = \delta_k^{(r)} + \delta_k^{(r+1)} + \cdots$. The terms in Equation (6.51) which are cubic in y and of degree $r-1$ in x give:

$$\partial \delta_i^{(r)} / \partial x_j - \partial \delta_j^{(r)} / \partial x_i = 0. \tag{6.52}$$

Using Poincaré's lemma, we can choose $e^{(r+1)}$ such that

$$\delta_i^{(r)} = \partial e^{r+1} / \partial x_i .$$

Equation (6.50) now gives

$$A_i^{(r)} = \alpha_i^{(r)} + \beta_i^{(r)} + y^2(\delta_i^{(r)} - \partial e^{r+1}/\partial x_i) = \alpha_i^{(r)} + \beta_i^{(r)} \,, \qquad (6.53)$$

which means that $A_i^{(r)}$ is an affine function in y.

After the above coordinate transformation turning $A_i^{(r)}$ into an affine function in y, we can suppose

$$\Gamma := \{x_1, \ldots, x_{m-1}\} = y + yc^{(r+1)}(x, y) + c^{(r+2)}(x, y) + \cdots$$

and

$$\Delta_i := (-1)^{m-i}\{x_1, \ldots \widehat{x}_i \ldots, x_{m-1}, y\} = a_i^{(0)}(x, y) + a_i^{(1)}(x, y) + \cdots$$

where $a_i^{(s)}$ are affine in y for $s = 0, \ldots, r$.

In the second step we use a coordinate transformation of the type $\tilde{x}_1 = x_1 + \theta^{(r+2)}(x, y)$, $\tilde{x}_2 = x_2, \ldots, \tilde{x}_{n-1} = x_{n-1}, \tilde{y} = y$ with $\partial\theta^{(r+2)}/\partial x_1 = -c^{(r+1)}$. Then we obtain

$$\{\tilde{x}_1, \ldots, \tilde{x}_{n-1}\} = \tilde{y} + \tilde{c}^{(r+2)}(\tilde{x}, \tilde{y}) + \cdots$$

and

$$\{\tilde{x}_1, \ldots, \tilde{x}_{i-1}, \tilde{x}_{i+1}, \ldots, \tilde{x}_{n-1}, \tilde{y}\}$$
$$= (-1)^{n-i}(a_i^{(0)}(\tilde{x}, \tilde{y}) + \cdots + a_i^{(r)}(\tilde{x}, \tilde{y}) + A_i^{(r+1)}(\tilde{x}, \tilde{y}) + \cdots) \,.$$

So after this second step we can suppose that $\Gamma = y + c^{(r+2)} + \cdots$ and $\Delta_i = a_i^{(0)} + \cdots$, where the functions $a_i^{(s)}$ are affine in y for $s = 0, \ldots, r$. Finally, change y by $\tilde{y} = y + c^{(r+2)}$ to finish the proof of Lemma 6.7.3.   □

Now we return to the proof of Theorem 6.7.1. Using Equation (6.49), we can take $\{x_1, \ldots, x_{m-1}\}$ as the new variable y to get $\{x_1, \ldots, x_{m-1}\} = y$. Then apply Lemma 6.7.3 consecutively for $r = 0, 1, 2, \ldots$. Taking the formal limit, we obtain a formal coordinate system $(x_1, \ldots, x_{m-1}, y)$ in which we have $\{x_1, \ldots, x_{m-1}\} = y$ and $\{x_1, \ldots \widehat{x}_i \ldots, x_{m-1}, y\} = (-1)^{m-i}(\alpha_i(x) + y\beta_i(x))$ for $i = 1, \ldots, m-1$.

In these coordinates, the dual integrable 1-form $\omega = \Omega^\flat(\Lambda)$ of Λ with respect to the volume form $\Omega = dx_1 \wedge \cdots \wedge dx_{m-1} \wedge dy$ is

$$\omega = \sum_{i=1}^{m-1} (\alpha_i(x) + y\beta_i(x))dx_i + ydy \,.$$

The terms containing $dx_i \wedge dx_j \wedge dy$ in the equation $\omega \wedge d\omega = 0$ give

$$(\alpha_i\beta_j - \alpha_j\beta_i) + y(\partial\alpha_i/\partial x_j - \partial\alpha_j/\partial x_i) + y^2(\partial\beta_i/\partial x_j - \partial\beta_j/\partial x_i) = 0.$$

So we get, for every i and j,

$$\partial\alpha_i/\partial x_j = \partial\alpha_j/\partial x_i, \ \partial\beta_i/\partial x_j = \partial\beta_j/\partial x_i, \ \alpha_i\beta_j = \alpha_j\beta_i \,.$$

Poincaré's lemma gives $\sum_i \alpha_i \mathrm{d}x_i = \mathrm{d}f$ and $\sum_i \beta_i \mathrm{d}x_i = \mathrm{d}g$ for some formal functions f and g which do not depend on y. Relations $\alpha_i \beta_j = \alpha_j \beta_i$ mean that $\mathrm{d}f \wedge \mathrm{d}g = 0$. Theorem 6.7.1 is proved.   □

Corollary 6.7.4. *Let ω be an integrable 1-form, which vanishes at a point z, with a nontrivial linear part at z. Then, up to multiplication by a non-vanishing function, ω is formally the pull-back of an integrable 1-form on a two-dimensional space.*

Proof. The case $\mathrm{d}\omega(z) \neq 0$ is just Kupka's phenomenon. Consider now the case $\mathrm{d}\omega(z) = 0$. Then we can apply Theorem 6.7.1 to write

$$\omega = h \left(\sum_{i=1}^{m-1} \left(\frac{\partial f}{\partial x_i} + x_m \frac{\partial g}{\partial x_i} \right) \mathrm{d}x_i + y\mathrm{d}y \right) .$$

So, up to multiplication by a non-vanishing function h, we may assume that $\omega = \mathrm{d}f + y\mathrm{d}g + y\mathrm{d}y$. Since $\mathrm{d}f \wedge \mathrm{d}g = 0$, i.e., f and g are functionally dependent, they can be written as $f(x) = a(\nu(x))$ and $g(x) = b(\nu(x))$, where a and b are formal functions of one variable, and ν is a formal function. Then we have $\omega = (a'(\nu) + yb'(\nu))\mathrm{d}\nu + y\mathrm{d}y = \phi^* \omega_2$, with $\omega_2 = (a'(u) + vb'(u))\mathrm{d}u + v\mathrm{d}v$ and $\phi : (x,y) \mapsto (u,v) = (\nu(x), y)$. This ends the proof of the corollary.   □

In fact, Corollary 6.7.4 can be proven directly, without using Nambu structures. The crucial point of the proof is that, up to multiplication by a non-vanishing function, an integrable 1-form of the type $y\mathrm{d}y + \sum_{i=1}^{m-1} A_i \mathrm{d}x_i$ is formally equivalent to a form of the type $y\mathrm{d}y + \alpha_0 + y\alpha_1$ where α_0 and α_1 are 1-forms which do not depend on y. This last result has the following generalization.

Theorem 6.7.5. *Let $\omega = y^r \mathrm{d}y + \sum_{i=1}^{m-1} A_i \mathrm{d}x_i$ be a formal integrable 1-form on $\mathbb{R}^m$ or $\mathbb{C}^m$, $r \geq 1$. Then, up to multiplication by a non-vanishing function, ω is formally equivalent to an integrable 1-form ω_0 of the type*

$$\omega_0 = y^r \mathrm{d}y + \sum_{i=0}^{r} y^i \alpha_i ,$$

where α_i are 1-forms which depend only on $x_1, \ldots, x_{m-1}$.

The proof of the above theorem, written in terms of a dual Nambu tensor, is absolutely similar to the proof of Lemma 6.7.3 and Theorem 6.7.1.   □

Remark 6.7.6. It is easy to show that a formal integrable 1-form of the type $y^2 \mathrm{d}y + \sum_{i=0}^{2} y^i \alpha_i$, where α_i ($i = 0, 1, 2$) do not depend on y, is a pull-back of a 1-form on $\mathbb{K}^2$ ($\mathbb{K} = \mathbb{R}$ or $\mathbb{C}$) by a formal map from $\mathbb{K}^m$ to $\mathbb{K}^2$. We don't know whether or not the same is true for $\omega_0 = y^r \mathrm{d}y + \sum_{i=0}^{r} y^i \alpha_i$ when $r > 2$.

Remark 6.7.7. One may view the results of this section as generalizations of Bogdanov–Takens normal forms (for two-dimensional vector fields with a nontrivial nilpotent linear part). Due to recent results on the existence of analytic Bogdanov–Takens normal forms [325], it is natural to ask if there is an analytic version for the above normal forms of integrable 1-forms and Nambu tensors of order $m - 1$ as well. It turns out that Frank Loray did just that in a recent paper [218], i.e., he obtained the holomorphic version of the results presented in this section, and some other results, by using some beautiful geometric arguments (extension of the foliation to a neighborhood of $\mathbb{CP}^1$ in $\mathbb{CP}^1 \times \mathbb{C}^{m-1}$). See [218] for details. We don't know whether there is a smooth version.

6.8 Quadratic integrable 1-forms

In this section we will give a classification of quadratic integrable 1-forms or, equivalently, a classification of quadratic Nambu tensors of order $m - 1$ in dimension m, up to multiplication by a constant. The study of quadratic integrable 1-forms and of their perturbations was initiated by Lins Neto [214], who obtained a partial classification of them (see Remark 6.8.1).

Let Λ be a quadratic Nambu tensor of order $m - 1$ in dimension m. Then its curl $D_\Omega \Lambda$ (with respect to a constant volume form Ω) is a linear Nambu tensor. According to the classification of linear Nambu tensors, we have the following two cases:

1) $D_\Omega \Lambda$ is of Type II: In this case we will also say that Λ and the dual quadratic integrable 1-form are of Type II. Then $D_\Omega \Lambda$ is of Type II$(n - 3)$ (see Section 6.5). According to the generalized Kupka's phenomenon (Theorem 6.5.5), Λ is also of Type II$(n - 3)$. Hence we have

$$\Lambda = \partial/\partial x_4 \wedge \cdots \wedge \partial/\partial x_n \wedge \Lambda_3 ,$$

where Λ_3 is a quadratic Poisson structure which depends only on the variables x_1, x_2 and x_3. In other words, Λ_3 is a quadratic Poisson structure on a three-dimensional space. The classification of these three-dimensional quadratic Poisson structures was done in [118] and [217] (see Section 5.6).

Remark 6.8.1. The quadratic integrable 1-forms studied by Lins Neto [214] are of Type II. More precisely, he showed that, if there is a three-dimensional subspace V^3 of $\mathbb{K}^n$, such that the linear curl vector field of the "restriction" of Λ to V^3 (= the dual quadratic 2-vector field of the restriction of $i_\Lambda \Omega$ to V^3) is hyperbolic, then Λ can be decomposed as above (this is a particular case of Corollary 6.5.8), and moreover its dual 1-form is stable under small perturbations. See [214] for details, and also [59, 72] for some generalizations, in particular to the case of homogeneous integrable 1-forms of higher degrees.

2) $D_\Omega \Lambda$ is of Type I: In this case it is more convenient to work with a dual quadratic integrable 1-form ω.

Theorem 6.8.2. *Let ω be a quadratic integrable 1-form such that $d\omega$ is of Type I. Then in an appropriate linear system of coordinates $(x_1, \ldots, x_m)$ we have*

$$\omega = (\beta x_1^2 + \theta Q) dx_1 + \gamma x_1 dQ ,$$

where β, θ, γ are constants and Q is a homogeneous quadratic function. In particular, ω can be written as a pull-back of a 1-form on a two-dimensional space by a nonhomogeneous quadratic map.

According to the definition of Type I, we have $\mathrm{d}\omega = \mathrm{d}x \wedge \mathrm{d}Q$ where Q is a homogeneous quadratic function. We will consider the case when Q is of the type $Q = \sum_{i=1}^{r} \pm y_i^2/2 + xz$ in a system of coordinates $x, y_1, \ldots, y_r, z, t_1, \ldots, t_s$ with $r \geq 2, r + s = m - 2$. (The other cases, i.e., $r = 0, 1$ or Q does not depend on x, are similar and simpler.)

As we have $\mathrm{d}\omega = -\mathrm{d}(Q\mathrm{d}x)$, we can put $\omega = -q\mathrm{d}x + \mathrm{d}f$ where f is a homogeneous function of degree 3. Denote $\overline{Q} = \sum_{i=1}^{r} \pm y_i^2/2$. We have

$$0 = \omega \wedge \mathrm{d}\omega = \mathrm{d}f \wedge \mathrm{d}x \wedge \mathrm{d}Q$$
$$= \Big(\sum_i \frac{\partial f}{\partial y_i}\mathrm{d}y_i + \frac{\partial f}{\partial z}\mathrm{d}z + \sum_j \frac{\partial f}{\partial t_j}\mathrm{d}t_j \Big) \wedge \mathrm{d}x \wedge (\mathrm{d}\overline{Q} + x\mathrm{d}z) \ .$$

The terms in $\mathrm{d}t_j \wedge \mathrm{d}x \wedge \mathrm{d}y_i$ in this formula give $\partial f/\partial t_j = 0$, so f is independent of the t_j. The terms in $\mathrm{d}y_j \wedge \mathrm{d}x \wedge \mathrm{d}y_i$ give

$$\Big(\sum_i \frac{\partial f}{\partial y_i}\mathrm{d}y_i \Big) \wedge \mathrm{d}\overline{Q} = 0 \ ,$$

and an elementary calculation leads to

$$f = (\lambda x + \mu z)\overline{Q} + b(x, z) \ ,$$

where λ and μ are constant and b is a homogeneous cubic function. Putting this last expression into the equation $\mathrm{d}f \wedge \mathrm{d}x \wedge \mathrm{d}Q = 0$, we get

$$0 = ((\mu\overline{Q} + \partial b/\partial z)\mathrm{d}z + (\lambda x + \mu z)\mathrm{d}\overline{Q}) \wedge \mathrm{d}x \wedge (\mathrm{d}\overline{q} + x\mathrm{d}z)$$
$$= ((\lambda x + \mu z)x - \mu\overline{Q} - \partial b/\partial z)\mathrm{d}\overline{Q} \wedge \mathrm{d}x \wedge \mathrm{d}z.$$

The terms in y_i in this last equation lead first to $\mu = 0$, then to $\partial b/\partial z = \lambda x^2$, then to $b = \lambda x^2 z + \alpha x^3$, and finally to $f = \lambda xQ + \alpha x^3$, where α is a constant. Putting this in the expression of ω, we get $\omega = \theta Q\mathrm{d}x + \gamma x\mathrm{d}Q + \beta x^2\mathrm{d}x$, where θ, β and γ are constants such that $\gamma - \theta = 1$. $\qquad\square$

Remark 6.8.3. The results in this section and the previous section lend support to the following conjecture: every integrable 1-form on $\mathbb{R}^n$ or $\mathbb{C}^n$ with a non-trivial 2-jet at 0 is, up to multiplication by a non-vanishing function, a pull-back of an integrable 1-form in dimension 3. More generally we can ask whether or not every integrable 1-form on $\mathbb{R}^n$ or $\mathbb{C}^n$ with a non-zero q-jet at 0 is, up to multiplication by a non-vanishing function, a pull-back of an integrable 1-form in dimension $q + 1$.

6.9 Poisson structures in dimension 3

In Chapter 4 we have given the list of real Lie algebras of dimension 3: besides the semisimple algebras $sl(2)$ and $so(3)$, we also have semi-direct products $\mathbb{R} \ltimes_A \mathbb{R}^2$ where A is a 2×2 matrix which can be put in Jordan form by an isomorphism.

In this section we will denote by Π a Poisson structure on a three-dimensional manifold. It is the same as a Nambu structure of order 2 in dimension 3, and is dual to an integrable 1-form in dimension 3 via the contraction with a volume form.

Definition 6.9.1. Assume that Π vanishes at a point p_0. We will say that Π has a singularity of *type V* (resp. *B*, resp. *N*) if the linear part of Π at p_0 corresponds to an algebra $\mathbb{R} \ltimes_A \mathbb{R}^2$ where the trace of A is non-zero (resp. the trace of A is zero but A is invertible, resp. A is nilpotent non-zero).

The following results concerning the above three types of singularities were obtained by Dufour and Zhitomirskii in [117] (see also [225]). We refer to [117] for details and proofs.

Theorem 6.9.2. *If Π admits a singularity of type V at a point p_0, then Π is locally equivalent to $\partial/\partial x \wedge Z$ where Z is a vector field which depends only on the variables y and z and which vanishes at the origin. Moreover, if Z has an algebraically isolated singularity at the origin, then we have for Z, up to local isomorphisms, the following list of models:*

$$Z = z\frac{\partial}{\partial y} + (\theta y + z)\frac{\partial}{\partial z} \ ;$$

$$Z = (ny + \delta z^n)\frac{\partial}{\partial y} + z\frac{\partial}{\partial z} \ ;$$

$$Z = (-\frac{m}{n}y + \gamma y^{n+1}z^m + \gamma a y^{2n+1}z^{2m})\frac{\partial}{\partial y} + z\frac{\partial}{\partial z} \ ;$$

$$Z = ((\pm 1)^{n+1} + ay^{2n+1})\frac{\partial}{\partial y} + z\frac{\partial}{\partial z}.$$

Here θ is an arbitrary number in $\mathbb{R} \setminus \{0\}$, n and m are positive integers, δ is equal to 0 or 1, and a is an arbitrary real number.

The hypothesis that the trace of A is non-zero is equivalent to the non-vanishing of the curl of Π at m. So the above theorem is just a consequence of Theorem 6.5.3 and Remark 6.6.4 in the case $q = 2$. The singularities of type V are non-isolated: they appear along the curve $(y = z = 0)$.

We will say that p_0 is an *algebraically isolated singularity* of Π if, in a local system of coordinates (x, y, z), the ideal generated by the germs of the functions $\{x, y\}, \{y, z\}$ and $\{z, x\}$ is of finite codimension in the algebra of germs of functions of three variables (x, y, z). This condition is a generic condition, and it assures that the singular point p_0 is topologically isolated (see [268]).

Theorem 6.9.3. *If* Π *admits an algebraically isolated singularity of type B, then* Π *is locally formally isomorphic to a model*

$$H(x, y, z) \left(z \frac{\partial}{\partial x} \wedge \frac{\partial}{\partial y} \pm x \frac{\partial}{\partial y} \wedge \frac{\partial}{\partial z} \pm y^k \frac{\partial}{\partial z} \wedge \frac{\partial}{\partial x} \right)$$

where k *is an integer greater than or equal to 2 and* H *is a function which does not vanish at* p_0.

Theorem 6.9.3 can be proved by studying the dual integrable 1-form and using the results of Moussu [268]. When the symplectic leaves are locally closed, then the above normal form is also a smooth normal form [117]. Using Malgrange's results (Theorem 6.3.1), it is also shown in [117] that in the analytic case the above normal form is an analytic one. We can also normalize further to a formal normal form in which H depends only on y.

Theorem 6.9.4. *If* Π *admits a singularity of type N, then* Π *is locally formally equivalent to a model*

$$z \frac{\partial}{\partial x} \wedge \frac{\partial}{\partial y} + \left(\frac{\partial f(x, y)}{\partial x} + z \frac{\partial g(x, y)}{\partial x} \right) \frac{\partial}{\partial y} \wedge \frac{\partial}{\partial z} + \left(\frac{\partial f(x, y)}{\partial y} + z \frac{\partial g(x, y)}{\partial y} \right) \frac{\partial}{\partial z} \wedge \frac{\partial}{\partial x}$$

where f *and* g *are two functionally dependent functions (i.e.,* $\mathrm{d}f \wedge \mathrm{d}g \equiv 0$ *) whose 1-jets at 0 are zero.*

Theorem 6.9.4 is in fact the three-dimensional case of Theorem 6.7.1. This last case of type N singularities is by far the most difficult to study.

The above normal forms can be extended to families of Poisson structures (or integrable 1-forms). This allows one to study the bifurcations of singularities of Poisson structures in dimension 3. Without entering into the details, we will mention that singularities of type V are stable (they do not disappear under a small perturbation). Singularities of type B are not stable, but they appear generically in 1-parameter families of Poisson structures. The surprising fact is that singularities of type N also appear generically in 1-parameter families (while algebras of the type $\mathbb{R} \ltimes_A \mathbb{R}^2$ with A nilpotent appear generically only in 2-parameter families of Lie algebras).

Remark 6.9.5. Some results on *global* topological stability of integrable 1-forms, and hence of Poisson structures, on 3-manifolds, were obtained by Camacho [58]. We will assume that M is an orientable 3-manifold with a given volume form. We say that an integrable 1-form o M ω (resp. a Poisson structure Π) is *topologically stable* if, for any other integrable 1-form ω' (resp. Poisson structure Π'), close enough to ω (resp. Π) in $\mathcal{C}^k$-topology for some k, there is a homeomorphism which exchanges the regular leaves of the associated foliation of ω (resp. Π) with those of ω' (resp. Π'). Camacho [58] obtains topological stability with respect to $\mathcal{C}^2$-topology by imposing a set of local and semi-local conditions (too technical to be completely reproduced here).

The local conditions are that the only permitted singularities must be (in Poisson structures' terms):

- those with a non-zero linear part which corresponds, up to isomorphisms, to $so(3)$, $sl(2)$ or $\mathbb{R} \ltimes_A \mathbb{R}^2$, such that A has eigenvalues with non-zero real parts,

- those with trivial linear part but with a quadratic part isomorphic, up to complexification, to a diagonal quadratic Poisson structure

$$ayz\partial y \wedge \partial z + bzx\partial z \wedge \partial x + cxy\partial x \wedge \partial y,$$

where $a \neq b \neq c$.

Chapter 7

Lie Groupoids

In this chapter we will give a quick introduction to the theory of Lie groupoids, and then present some normal form theorems for Lie groupoids and symplectic groupoids in the proper case. For a more comprehensive introduction, the reader may consult other references, in particular the book by Mackenzie [228], and also [60, 254].

The relations between Lie groupoids and Poisson structures will become clear near the end of this chapter when we discuss symplectic groupoids, and in Chapter 8 where it is shown that Lie algebroids, i.e., the infinitesimal versions of Lie groupoids, are nothing but fiber-wise linear Poisson manifolds.

7.1 Some basic notions on groupoids

7.1.1 Definitions and first examples

Formally speaking, a groupoid is a (small) category in which each morphism is invertible. Let us spell it out, and fix some notations:

Definition 7.1.1. A *groupoid* over a set M is a set Γ together with the following structure maps:

1) Two maps $s, t : \Gamma \to M$, called the *source map* and the *target map*.
2) A *product map* $m : \Gamma_{(2)} \to \Gamma, (g, h) \mapsto g.h$, where

$$\Gamma_{(2)} = \{(g, h) \in \Gamma \times \Gamma \mid s(g) = t(h)\}, \qquad (7.1)$$

which satisfies the following conditions:
   a) Compatibility with s and t:

$$s(g.h) = s(h), \quad t(g.h) = t(g) \ \forall \ (g, h) \in \Gamma_{(2)}. \qquad (7.2)$$

b) Associativity:

$$(g.h).k = g.(h.k) \quad \forall \, (g, h, k) \in \Gamma_{(3)} \tag{7.3}$$

where, for $k \geq 1$,

$$\Gamma_{(k)} = \{(g_1, \ldots, g_k) \mid g_i \in \Gamma, \ s(g_i) = t(g_{i+1}) \, \forall \, i\}. \tag{7.4}$$

3) An embedding $\varepsilon : M \to \Gamma$, called the *identity section*, such that

$$g.\varepsilon(s(g)) = g = \varepsilon(t(g)).g. \tag{7.5}$$

In particular $\alpha \circ \varepsilon = \beta \circ \varepsilon$ is the identity map on M.

4) An *inversion map* $\imath : \Gamma \to \Gamma$ such that $s(\imath(g)) = t(g), t(\imath(g)) = s(g)$, and

$$\imath(g).g = \varepsilon(s(g)), \quad g.\imath(g) = \varepsilon(t(g)). \tag{7.6}$$

The inversion $\imath(g)$ of an element $g \in \Gamma$ is denoted by g^{-1}.

If Γ is a groupoid over M then M is also denoted by $\Gamma_{(0)}$ and is called the set of *objects*, or base points, and is often identified with the set $\varepsilon(M)$ of identity elements of Γ. Γ, also denoted by $\Gamma_{(1)}$, is called the set of *arrows*. An arrow $g \in \Gamma$ from $x = s(g)$ to $y = t(g)$ will be denoted as $g : x \to y$. A groupoid Γ over M will often be denoted by $\Gamma \rightrightarrows M$.

A *groupoid morphism* from a groupoid $\Gamma_1 \rightrightarrows M_1$ to a groupoid $\Gamma_2 \rightrightarrows M_2$ is a map $\phi : \Gamma_1 \to \Gamma_2$ which preserves the structure maps. In other words, if we view (M_1, Γ_1) and (M_2, Γ_2) as categories, then ϕ is a functor. In particular, ϕ induces a map between the base spaces, which we will also denote by $\phi : M_1 \to M_2$. A groupoid morphism is also called a *homomorphism*.

Example 7.1.2. A group is a groupoid over a point.

Example 7.1.3. For any set M, put $\mathcal{G} = M \times M$, with the structure maps defined as follows: $s(x, y) = x, t(x, y) = y, (x, y).(y, z) = (x, z), \varepsilon(x) = (x, x), (x, y)^{-1} = (y, x)$. Then $\mathcal{G}$ is a groupoid over M, called the *pair groupoid*. If $\Gamma \rightrightarrows M$ is an arbitrary groupoid over M, then the map $(s, t) : \Gamma \to M \times M$, which is sometimes called the *anchor* of Γ, is a homomorphism from Γ to the pair groupoid of M.

Example 7.1.4. Associated to each left action $\rho : G \times M \to M$ of a group G on a set M, there is a *transformation groupoid* defined as follows: $\Gamma = G \times M, s(g, x) = x, t(g, x) = \rho(g, x), \varepsilon(x) = (e, x), (g, x)^{-1} = (g^{-1}, \rho(g, x))$, where $g \in G, x \in M$, and e is the neutral element of G. The transformation groupoid will be denoted by $G \times M \rightrightarrows M$, or $G \ltimes M$, and viewed as a semi-direct product of G with M. For right actions, there are similar transformation groupoids denoted by $M \rtimes G$.

If $\Gamma \rightrightarrows M$ is a groupoid and x is a point of M, then

$$\Gamma_x = \{g \in \Gamma \mid s(g) = t(g) = x\} = s^{-1}(x) \cap t^{-1}(x) \tag{7.7}$$

is a group: if $g, h \in \Gamma_x$ then $g.h \in \Gamma_x$ and $g^{-1} \in \Gamma_x$. This group is called the *isotropy group* of x. The set

$$\mathcal{O}(x) = \{t(g) \mid g \in \Gamma, s(g) = x\} = \{s(g) \mid g \in \Gamma, t(g) = x\} \qquad (7.8)$$

is called the *orbit* of x, or of Γ through x. For example, when $\Gamma = G \ltimes M$ is a transformation groupoid, then the orbit of Γ through x is the same as the orbit of x under the action of G. If $\mathcal{O}(x) = \{x\}$, or equivalently, $s^{-1}(x) = t^{-1}(x) = \Gamma_x$, then x is called a *fixed point*. The *orbit space* of Γ is, as its name indicates, the space of orbits of Γ on M, i.e., the quotient space of M by the equivalence relation induced by Γ: two points of M are equivalent iff they lie on the same orbit.

In practice, $M = \Gamma_{(0)}$ is often a space of structures of some kind (for example, the space of flat connections on a given vector bundle), and the set of arrows $\Gamma_{(1)}$ consists of isomorphisms among the structures. Then the orbit space is called a *moduli space*.

A subset B of M is called *invariant* if it is saturated by the orbits of Γ; in other words, if $x \in B$ then $\mathcal{O}(x) \subset B$.

A groupoid $\Gamma \rightrightarrows M$ is called *transitive* if it has only one orbit, i.e., $\mathcal{O}(x) = M$ for some (hence any) $x \in M$. In other words, for any $x_1, x_2 \in M$ there is $g \in \Gamma$ such that $s(g) = x_1, t(g) = x_2$. A *totally intransitive groupoid* is a groupoid, each orbit of which consists of just one point. A totally intransitive groupoid is just a family of groups.

If $\Gamma \rightrightarrows M$ is a groupoid and U is a subset of M, then the restriction Γ_U of Γ to U,

$$\Gamma_U = \{g \in \Gamma \mid s(g), t(g) \in U\}, \qquad (7.9)$$

is a subgroupoid of Γ: the inclusion map $\Gamma_U \to \Gamma$ is a groupoid morphism.

More generally, if U, V are subsets of M, we will denote by

$$\Gamma_{UV} = t^{-1}(U) \cap s^{-1}(V) \qquad (7.10)$$

the set of arrows from V to U. If $U = \{x\}$ is a point then we will write Γ_{xV} for $\Gamma_{\{x\}V}$ and so on. Note that $\Gamma_{DD} = \Gamma_D$, and $\Gamma_{xx} = \Gamma_x$.

One may impose various topological and geometrical structures on a groupoid, depending on what one wants. We will be mainly interested in Lie groupoids, also known as differentiable groupoids, to be defined in the next subsection.

Example 7.1.5. Let Γ^q be the groupoid (over $\mathbb{R}^q$) of germs of local diffeomorphisms of $\mathbb{R}^q$ ($q \in \mathbb{N}$). If $\phi : U \to V$, where U, V are open subsets of $\mathbb{R}^q$, is a diffeomorphism, then denote by $\mathcal{U}_\phi$ the set of germs of ϕ at points $x \in U$. These sets $\mathcal{U}_\phi$ form a basis of open subsets of a topology of Γ^q. With respect to this topology, Γ^q is a *topological groupoid* in a natural sense, and it plays an important role in the theory of Γ-structures (generalized foliations) of Haefliger [164]. A Γ-*structure* on a topological space Y is represented by a 1-cocycle on Y with values in Γ: a family of continuous maps $\phi_{ij} : U_{ij} = U_i \cap U_j \to \Gamma$, where (U_i) is an open covering of Y, is called a 1-cocycle if $\forall\, i, j, k, \forall\, x \in U_i \cap U_j \cap U_k$ we have $\phi_{ik}(x) = \phi_{ij}(x).\phi_{jk}(x)$. Two 1-cocycles on Y with values in Γ define the same Γ-structure if they differ by a coboundary, and the set of Γ-structures on Y is $H^1(Y, \Gamma)$. A regular codimension q foliation on a manifold Y may be viewed as a Γ^q-structure, though not every Γ^q-structure on Y arises from a regular foliation.

7.1.2   Lie groupoids

Definition 7.1.6 ([123]). A *Lie groupoid* is a groupoid $\Gamma \rightrightarrows M$ such that Γ is a smooth not necessarily separated manifold, M is a smooth separated (i.e., Hausdorff) closed submanifold of Γ, the structure maps are smooth, and the maps s, t are submersions. A *Lie groupoid morphism* is a groupoid morphism between Lie groupoids, which is smooth.

Remark 7.1.7. A non-separated smooth manifold is a non-separated topological space, which admits a smooth atlas. For example, take two copies $\mathbb{R}_1$ and $\mathbb{R}_2$ of $\mathbb{R}$, and glue their negative parts together: $x \in \mathbb{R}_1$ is glued to $x \in \mathbb{R}_2$ if $x < 0$. The resulting space $(\mathbb{R}_1 \cup \mathbb{R}_2)/\sim$ is a non-separated one-dimensional smooth manifold, which is the leaf space of the foliation of $\mathbb{R}^2 \setminus \{(x, 0) \mid x \geq 0\}$ by vertical lines. The reason why one allows Γ to be non-separated is that there are many natural examples which are so. Except total spaces of groupoids, all other manifolds in this book are assumed to be Hausdorff paracompact.

Remark 7.1.8. In the definition of a Lie groupoid, the maps are usually assumed to be C^∞-smooth. When they are only C^k-smooth ($k \in \mathbb{N}$), we will say that it is a C^k-smooth groupoid.

Example 7.1.9. The transformation groupoid of a smooth Lie group action is a Lie groupoid.

Example 7.1.10. Let $\mathcal{F}$ be a regular smooth foliation on a manifold M. We can define its *fundamental groupoid* as follows: Γ is the space of triples (x, y, γ), where $x, y \in M$ lie in a same leaf of $\mathcal{F}$, and γ is a path in M lying on $\mathcal{F}$ connecting x to y, considered only up to homotopy (among such paths). The source and target maps are given by the origin and the end of γ, and the product is given by concatenation. The *holonomy groupoid* of a foliation $\mathcal{F}$ is defined similarly by the same triples (x, y, γ) as in the case of the fundamental groupoid, except that two paths γ_1 and γ_2 are considered to be equivalent if they are not necessarily homotopic but have the same holonomy, i.e., the holonomy of the foliation along the loop $\gamma_1 - \gamma_2$ is trivial (see Appendix A.3 for a definition of holonomy). These are Lie groupoids whose total spaces are not necessarily separated. For example, the fundamental groupoid and holonomy groupoid of the Reeb foliation of the three-dimensional sphere S^3 [356] are non-separated.

Example 7.1.11. In the previous example, if $\mathcal{F}$ consists of just one leaf, i.e., the connected manifold M itself, then its fundamental groupoid is called the *fundamental groupoid* of M. It is clear that the fundamental groupoid of M is transitive (provided that M is connected), and its isotropy groups are isomorphic to the fundamental group of M.

Lemma 7.1.12. *If $\Gamma \rightrightarrows M$ is a Lie groupoid, then the partition of M into orbits of Γ is a smooth singular foliation.*

Proof. The local foliation property follows directly from the fact that the source and target maps are submersions. $\square$

Lemma 7.1.13. *If $\Gamma \rightrightarrows M$ is a Lie groupoid, then for any $x \in M$, the isotropy group $\Gamma_x = s^{-1}(x) \cap t^{-1}(x)$ is a Lie group.*

The preimages of the source map $s : \Gamma \to M$ of a Lie groupoid are called *s-fibers*. Those of the target map $t : \Gamma \to M$ are called *t-fibers*. They are submanifolds of Γ.

Lemma 7.1.14. *Each s-fiber (t-fiber) of a Lie groupoid $\Gamma \rightrightarrows M$ is a Hausdorff closed submanifold of Γ.*

A submanifold $S \subset \Gamma$ is called a (smooth) *bisection* of a Lie groupoid $\Gamma \rightrightarrows M$ if $s|_S : S \to M$ and $t|_S : S \to M$ are diffeomorphisms.

Lemma 7.1.15. *The space of bisections of a groupoid is a group.*

Proof. The product $S_1.S_2$ of two bisections S_1 and S_2 is defined as follows:

$$S_1.S_2 = \{g.h \mid (g,h) \in (S_1 \times S_2) \cap \Gamma_{(2)}\}. \tag{7.11}$$

One verifies directly that the right-hand side of the above formula is a bisection, and that the usual axioms of a group are satisfied. $\square$

The group of smooth bisections of a Lie groupoid admits a natural topology which makes it into a Fréchet Lie group, but we know almost nothing about the topological structure of such groups.

Example 7.1.16. The group of bisections of a pair groupoid $M \times M \rightrightarrows M$ is nothing but the group $Diff(M)$ of diffeomorphisms of M.

Example 7.1.17. Let $P \to M$ be a principal bundle over a manifold M with structural group G, i.e., G acts freely on P on the right and $P/G = M$. Then the quotient $(P \times P)/G$ of the pair groupoid $P \times P$ by the diagonal action of G has a unique natural Lie groupoid structure for which the projection map $P \times P \to (P \times P)/G$ is a Lie groupoid morphism. $(P \times P)/G$ is a transitive groupoid over M. Its bisections can be identified with graphs of gauge transformations of P, i.e., diffeomorphisms from P to itself which preserve the action of G. For this reason, $(P \times P)/G \rightrightarrows M$ is called the *gauge groupoid* of P.

Exercise 7.1.18. Show that any transitive Lie groupoid is the gauge groupoid of some principal bundle.

7.1.3   Germs and slices of Lie groupoids

Let $\Gamma \rightrightarrows M$ be a Lie groupoid, and U an open subset of M. Then the restriction Γ_U of Γ to U is again a Lie groupoid. In particular, we can talk about the germ of Γ over a point $x \in M$.

On the other hand, let D be a submanifold of M, which is transversal to the orbital foliation (= singular foliation of M by the orbits of Γ), i.e., $T_x D + T_x \mathcal{O}(x) = T_x M \ \forall x \in D$, where $\mathcal{O}(x) = t(s^{-1}(x))$ is the orbit through x. Then Γ_D is also a Lie groupoid. In particular, if D intersects an orbit $\mathcal{O}$ transversally at a point x, i.e., $T_x D + T_x \mathcal{O}(x) = T_x M$ and $T_x D \cap T_x \mathcal{O}(x) = 0$, then D is called a *slice* to $\mathcal{O}$ at x, Γ_D is called a *slice groupoid* (or *slice* for short) of Γ at x, and the germ of Γ_D at x is called the *transverse groupoid structure* of Γ at x.

Exercise 7.1.19. Show that, up to isomorphisms, the transverse groupoid of a Lie groupoid at a point x on an orbit $\mathcal{O}$ does not depend on the choice of x on $\mathcal{O}$ nor on the slice.

7.1.4   Actions of groupoids

Definition 7.1.20. A left *groupoid action* of a groupoid $\Gamma \rightrightarrows M$ on a space N consists of a map $\mu : N \to M$ called the *momentum map*, or also *moment map*, and a map

$$\varrho : \Gamma * N \to N, \varrho(g, x) = g.x \qquad (7.12)$$

called the *action map*, where

$$\Gamma * N = \Gamma *_\mu N = \{(g, x) \in \Gamma \times N \mid s(g) = \mu(x)\}, \qquad (7.13)$$

such that the following compatibility conditions are satisfied:

$$\mu(g.x) = t(g) \quad \forall \ (g, x) \in \Gamma * N, \qquad (7.14)$$

$$(g.(h.x)) = (g.h).x \quad \forall \ (h, x) \in \Gamma * N, (g, h) \in \Gamma_{(2)}, \qquad (7.15)$$

$$(\varepsilon \mu(x)).x = x \quad \forall \ x \in N. \qquad (7.16)$$

The action is called *free* if $g.x \neq x \ \forall \ (g, x) \in \Gamma * N$ such that $g \notin M$.

When Γ acts on N, we will say that N is a Γ *module*, or Γ *space*. In the case when $\mu : N \to \mu(N) = M$ is a vector bundle over M and the action of Γ on N is fiber-wise linear, then we say that N is a *linear* Γ *module*, or a *linear representation* of Γ.

Exercise 7.1.21. Let $\Gamma \rightrightarrows M$ be a Lie groupoid, and $E \to M$ a vector bundle over M. Denote by $GL(E)$ the Lie groupoid whose arrows are linear isomorphisms among the fibers of E. Show that a linear representation of Γ on E is the same as a Lie groupoid morphism from Γ to $GL(E)$ which projects to the identity map on M.

Right groupoid actions are defined similarly. When Γ is a group, i.e., $M = \Gamma_{(0)}$ is a point, we recover the usual definition of a group action. In the case of a Lie groupoid action, one requires the moment map and the action map to be smooth.

Similarly to the case of group actions, if $\varrho : \Gamma * N \to N$ is a left groupoid action, then $\Gamma * N$ itself has a groupoid structure: its target map is ϱ, and its source map is the projection to N. The multiplication of $(g, x) \in \Gamma * N$ with $(h, y) \in \Gamma * N$ is defined when $x = h.y$ and the result is

$$(g, x).(h, y) = (g.h, y). \tag{7.17}$$

The inversion map is

$$(g, x)^{-1} = (g^{-1}, g.x), \tag{7.18}$$

and the identity section is

$$x \mapsto (\varepsilon\mu(x), x). \tag{7.19}$$

This groupoid is called the semi-direct product of Γ with N and can be denoted by $\Gamma \ltimes N \rightrightarrows N$.

Example 7.1.22. Γ acts on itself on the left and on the right. The moment map for the left (resp. action) of Γ on itself is the target map (resp. the source map). Γ also acts on its base space $\Gamma_{(0)}$ on the left and on the right, with the identity map of $\Gamma_{(0)}$ as the moment map.

Example 7.1.23. Consider a left action of a transformation groupoid $G \ltimes M$ on N with moment map $\mu : N \to M$. $(g, x) \in G \ltimes M$ acts on $y \in N$ if $\mu(y) = x$. We can forget about x and define $g.y = (g, \mu(y)).y$. Then it is a left action of G on N, which projects to the action of G on M via the map $\mu : N \to M$. One can show that the groupoid $(G \ltimes M) \ltimes N$ is isomorphic to $G \ltimes N$ (exercise).

7.1.5 Haar systems

A smooth *Haar system* on a Lie groupoid $\Gamma \rightrightarrows M$ is a smooth family of measures on the t-fibers of Γ, which are invariant under left translations. In other words, for each t-fiber $t^{-1}(x)$ of Γ, we have a smooth measure μ_x on $t^{-1}(x)$, which is given by a smooth nowhere vanishing density form on $t^{-1}(x)$. This density form also depends smoothly on x. The invariance under left translations means that, for each $g \in \Gamma$, the corresponding left translation map $L_g : t^{-1}(s(g)) \to t^{-1}(t(g))$, $L_g(h) = g.h$, is a measure-preserving diffeomorphism from $t^{-1}(s(g))$ to $t^{-1}(t(g))$.

Haar systems on groupoids are analogs of left-invariant measures on groups. To define a Haar system on $\Gamma \rightrightarrows M$, it suffices to define it at points of M, then extend it to other points of Γ by left translations.

Given a Haar system (μ_x) on $\Gamma \rightrightarrows M$, one can turn the space $C_c(\Gamma)$ of compactly supported continuous functions on Γ into an associative algebra with the following *convolution product*:

$$(\phi_1 * \phi_2)(g) = \int_{k \in t^{-1}(t(g))} \phi_1(k)\phi_2(k^{-1}.g)\mathrm{d}\mu_{t(g)}. \tag{7.20}$$

The above convolution product is associative, and the reason is precisely because (μ_x) is invariant under left translations. The closure of $C_c(\Gamma)$ in a suitable
norm is a C^*-algebra. These C^*-algebras play an important role in noncommutative geometry [82]. We refer to [299, 82, 207, 60] for the theory of these groupoid
algebras.

7.2   Morita equivalence

Let $\phi : \Gamma \to \mathcal{G}$ and $\psi : \Gamma' \to \mathcal{G}$ be Lie groupoid morphisms. Then the *fibered product*
$\Gamma \times_{\mathcal{G}} \Gamma'$ is the groupoid whose objects are triples (y, g, y') where $y \in \Gamma_{(0)}$ (the space
of objects of Γ), $y' \in \Gamma'_{(0)}$, and $g : \phi(y) \to \psi(y')$ in $\mathcal{G}$. Arrows $(y, g, y') \to (z, g', z')$
in $\Gamma \times_{\mathcal{G}} \Gamma'$ are pairs of arrows (h, k), $h : y \to z$ in Γ and $k : y' \to z'$ in Γ', with the
property that $g'.\phi(h) = \psi(k).g$:

$$
\begin{array}{ccc}
\phi(y) & \xrightarrow{\ g\ } & \psi(y') \\
{\scriptstyle \phi(h)}\downarrow & & \downarrow{\scriptstyle \psi(k)} \\
\phi(z) & \xrightarrow{\ g'\ } & \psi(z')
\end{array}
\tag{7.21}
$$

The product map in $\Gamma \times_{\mathcal{G}} \Gamma'$ is defined in an obvious way. The fibered product
$\Gamma \times_{\mathcal{G}} \Gamma'$ is a Lie groupoid as soon as its space of objects $(\Gamma \times_{\mathcal{G}} \Gamma')_{(0)} = \Gamma_{(0)} \times_{\mathcal{G}_{(0)}}$
$\mathcal{G} \times_{\mathcal{G}_{(0)}} \Gamma'_{(0)}$ is a manifold.

A homomorphism ϕ from a Lie groupoid Γ_1 over M_1 to a Lie groupoid Γ_2
over M_2 is called a *(weak) equivalence* if it satisfies the following two conditions:

i) ϕ induces a surjective submersion

$$t \circ \pi_1 : \Gamma_2 \times_{M_2} M_1 \to M_2, \tag{7.22}$$

where $\Gamma_2 \times_{M_2} M_1 = \{(g, x) \in \Gamma_2 \times M_1 \mid s(g) = \phi(x)\}$, and π_1 is the projection
to the first component (i.e., to Γ_2). In particular, every orbit of Γ_2 contains
a point in the image $\phi(M_1)$ of M_1.

ii) The square

$$
\begin{array}{ccc}
\Gamma_1 & \xrightarrow{\ \phi\ } & \Gamma_2 \\
{\scriptstyle (s,t)}\downarrow & & \downarrow{\scriptstyle (s,t)} \\
M_1 \times M_1 & \xrightarrow{\ \phi \times \phi\ } & M_2 \times M_2
\end{array}
\tag{7.23}
$$

is a fibered product. (Here $M_1 \times M_1$ and $M_2 \times M_2$ are viewed as pair groupoids.) In particular, for any $x, y \in M_1$, the set $\Gamma_{1,xy} = t^{-1}(x) \cap s^{-1}(y)$ of arrows from y to x is diffeomorphic to the set $\Gamma_{2,\phi(x)\phi(y)} = t^{-1}(\phi(x)) \cap s^{-1}(\phi(y))$ of arrows from $\phi(y)$ to $\phi(x)$.

Example 7.2.1. The identity map is clearly a weak equivalence from a groupoid to itself.

Example 7.2.2. Let $\mathcal{F}$ be a smooth foliation on a manifold M and let T be a *complete transversal* of $\mathcal{F}$ in M. In other words, T is an open submanifold of M (i.e., T is the interior of a submanifold with boundary), which may be disconnected, whose dimension is equal to the codimension of $\mathcal{F}$, and which is transversal to $\mathcal{F}$ and intersects each leaf of $\mathcal{F}$ at least one point. Let Γ be the holonomy groupoid of $\mathcal{F}$ (see Example 7.1.10). Then the restriction Γ_T of Γ to T is a Lie groupoid, called the *transversal holonomy groupoid* of $\mathcal{F}$ with respect to T. The inclusion map $\Gamma_T \to \Gamma$ is a weak equivalence.

Example 7.2.3. Let $\Gamma \rightrightarrows M$ be a transitive Lie groupoid, i.e., the gauge groupoid of a principal bundle over a connected manifold M, and x be a point of M. Then the inclusion map $\Gamma_x \to \Gamma$ is a weak equivalence from the isotropy group Γ_x to Γ. On the other hand, there is a weak equivalence from Γ to Γ_x if and only if the corresponding principal bundle over M is trivial.

Lemma 7.2.4. *If $\phi : \Gamma_1 \to \Gamma_2$ and $\psi : \Gamma_2 \to \Gamma_3$ are weak equivalences of Lie groupoids, then the composition $\psi \circ \phi$ is also a weak equivalence.*

Proof. Direct verification. $\square$

Lemma 7.2.5. *If $\phi : \mathcal{G} \to \Gamma$ is a homomorphism and $\psi : \mathcal{H} \to \Gamma$ is a weak equivalence of Lie groupoids, then $\mathcal{G} \times_\Gamma \mathcal{H}$ is a Lie groupoid and the projection to the first component $\pi_1 : \mathcal{G} \times_\Gamma \mathcal{H} \to \mathcal{G}$ is a weak equivalence.*

Proof. Direct verification. $\square$

Lemma 7.2.6. *If $\mathcal{G} \to \Gamma_1$, $\mathcal{G} \to \Gamma_2$, $\mathcal{H} \to \Gamma_2$, $\mathcal{H} \to \Gamma_3$ are weak equivalences of Lie groupoids, then there exists a Lie groupoid $\mathcal{K}$ and weak equivalences $\mathcal{K} \to \Gamma_1$, $\mathcal{K} \to \Gamma_3$.*

Proof. Put $\mathcal{K} = \mathcal{G} \times_{\Gamma_2} \mathcal{H}$. According to the previous lemmas, $\pi_1 : \mathcal{K} \to \mathcal{G}$ is a weak equivalence, hence $\phi \circ \pi_1 : \mathcal{K} \to \Gamma_1$ is a weak equivalence. Similarly, $\psi \circ \pi_2 : \mathcal{K} \to \Gamma_3$ is also a weak equivalence. $\square$

The above lemmas and examples show that the weak equivalence is not an equivalence relation (it is transitive but not reflexive), but it can be turned into an equivalence relation, called Morita equivalence, as follows:

Definition 7.2.7. Two Lie groupoids Γ_1 and Γ_2 are called *Morita equivalent* if there exists a third Lie groupoid $\mathcal{G}$ and weak equivalences from $\mathcal{G}$ to Γ_1 and Γ_2.

Roughly speaking, the Morita equivalence class of a Lie groupoid reflects its transverse structure. In particular, we have:

Proposition 7.2.8. *If two Lie groupoids Γ_1 and Γ_2 are Morita equivalent, then there is a homeomorphism from the orbit space of Γ_1 to the orbit space of Γ_2 (with induced topology), such that if $\mathcal{O}_1$ is an orbit of Γ_1 and $\mathcal{O}_2$ is the corresponding orbit of Γ_2, then the transverse groupoid structure of Γ_1 at $\mathcal{O}_1$ is isomorphic to the transverse groupoid structure of Γ_2 at $\mathcal{O}_2$.*

The proof of the above proposition is another direct verification, based on the definition of weak equivalence. We will leave it as an exercise.

Exercise 7.2.9. Show that if $\mathcal{G} \rightrightarrows \mathcal{G}_{(0)}$ and $\mathcal{H} \rightrightarrows \mathcal{H}_{(0)}$ are two Lie groups which are Morita equivalent, then $\dim \mathcal{G} - 2 \dim \mathcal{G}_{(0)} = \dim \mathcal{H} - 2 \dim \mathcal{H}_{(0)}$.

Let $\phi : \Gamma \to \mathcal{G}$ and $\psi : \Gamma \to \mathcal{H}$ be weak equivalences of Lie groupoids. Let $P_{\mathcal{G}\mathcal{H}} = \mathcal{P}_{\mathcal{G}\mathcal{H}} / \sim$ be the quotient space of the space

$$\mathcal{P}_{\mathcal{G}\mathcal{H}} = \{(g, \gamma, h) \in \mathcal{G} \times \Gamma \times \mathcal{H} \mid s(g) = t(\phi(\gamma)), t(h) = s(\psi(\gamma))\} \qquad (7.24)$$

by the equivalence relation

$$(g, \gamma_1.\gamma_2.\gamma_3, h) \sim (g.\phi(\gamma_1), \gamma_2, \psi(\gamma_3).h). \qquad (7.25)$$

One verifies directly that $P_{\mathcal{G}\mathcal{H}}$ is a manifold of dimension $\dim P_{\mathcal{G}\mathcal{H}} = \frac{1}{2}(\dim \mathcal{G} + \dim \mathcal{H})$, and the projections $p_1 : (g, \gamma, h) \mapsto t(g)$ and $p_2 : (g, \gamma, h) \mapsto s(h)$ are submersions from $P_{\mathcal{G}\mathcal{H}}$ to $\mathcal{G}_{(0)}$ and $\mathcal{H}_{(0)}$ respectively. $\mathcal{G}$ acts freely on $P_{\mathcal{G}\mathcal{H}}$ from the left by the formula $g'.(g, \gamma, h) = (g'.g, \gamma, h)$ with the moment map p_1. Similarly, $\mathcal{H}$ acts freely on $P_{\mathcal{G}\mathcal{H}}$ from the right by the formula $(g, \gamma, h).h' = (g, \gamma, h.h')$. These two actions obviously commute, i.e., $P_{\mathcal{G}\mathcal{H}}$ is a *free $(\mathcal{G}, \mathcal{H})$-bimodule*[1]. Moreover, the orbits of $\mathcal{G}$ (resp., $\mathcal{H}$) on $P_{\mathcal{G}\mathcal{H}}$ are precisely the fibers of the submersion $p_2 : P \to \mathcal{H}_{(0)}$ (resp., $p_1 : P_{\mathcal{G}\mathcal{H}} \to \mathcal{G}_{(0)}$), and we have $P_{\mathcal{G}\mathcal{H}}/\mathcal{G} = \mathcal{H}_{(0)}$ (resp. $P_{\mathcal{G}\mathcal{H}}/\mathcal{H} = \mathcal{G}_{(0)}$). Conversely, we have:

Proposition 7.2.10. *Let $\mathcal{G}$ and $\mathcal{H}$ be two Lie groupoids. Suppose that there is a free $(\mathcal{G}, \mathcal{H})$-bimodule $P_{\mathcal{G}\mathcal{H}}$ such that the moment maps $p_1 : P_{\mathcal{G}\mathcal{H}} \to \mathcal{G}_{(0)}$ and $p_2 : P_{\mathcal{G}\mathcal{H}} \to \mathcal{H}_{(0)}$ of the actions of $\mathcal{G}$ and $\mathcal{H}$ on $P_{\mathcal{G}\mathcal{H}}$ are submersions, and the orbits of $\mathcal{G}$ (resp. $\mathcal{H}$) on $P_{\mathcal{G}\mathcal{H}}$ are precisely the fibers of p_2 (resp. p_1). Then $\mathcal{G}$ is Morita equivalent to $\mathcal{H}$.*

Proof. Put $\Gamma = \mathcal{G} \ltimes P_{\mathcal{G}\mathcal{H}} \rtimes \mathcal{H}$. In other words, $\Gamma = \{(g, x, h) \in \mathcal{G} \times P_{\mathcal{G}\mathcal{H}} \times \mathcal{H} \mid s(g) = p_1(x), t(h) = p_2(x)\}$. The source map is $(g, x, h) \mapsto x$, the target map is $(g, x, h) \mapsto g.x.h$, and the product map is $(g', g.x.h, h').(g.x.h) = (g'.g, x, h.h')$. One verifies directly that the projections from Γ to $\mathcal{G}$ and $\mathcal{H}$ are weak Morita equivalences. $\qquad \square$

---

[1] A $(\mathcal{G}, \mathcal{H})$-bimodule is a manifold together with a left action of $\mathcal{G}$ and a right action of $\mathcal{H}$, such that the two actions commute.

An important fact about Morita equivalent groupoids is that they have equivalent "categories of modules": if N is a left $\mathcal{H}$ space then $P_{\mathcal{GH}} \times_{\mathcal{H}} N$ is a left $\mathcal{G}$ space and $P_{\mathcal{HG}} \times_{\mathcal{G}} P_{\mathcal{GH}} \times_{\mathcal{H}} N \cong N$.

The existence of a free bimodule which satisfies the conditions of Proposition 7.2.10 is often used as the definition of Morita equivalence. The preceding discussion shows that this definition is equivalent to Definition 7.2.7. A free bimodule which satisfies the conditions of Proposition 7.2.10 is called a *generalized isomorphism* from $\mathcal{H}$ to $\mathcal{G}$.

More generally, a $(\mathcal{G}, \mathcal{H})$-bimodule P is called a *generalized morphism* from $\mathcal{H}$ to $\mathcal{G}$ if the moment map $p_2 : P \to \mathcal{H}_{(0)}$ is a principal $\mathcal{G}$-bundle, i.e., the action of $\mathcal{G}$ is free and its orbits are precisely the fibers of p_2. The reason is that, if $\phi : \mathcal{H} \to \mathcal{G}$ is a morphism, then its "graph", namely the quotient space of $\{(g, h) \in \mathcal{G} \times \mathcal{H} \mid s(g) = t(\phi(h))\}$ by the equivalence relation $(g, h_1.h_2) \sim (g.\phi(h_1), h_2)$, together with the natural actions of $\mathcal{G}$ and $\mathcal{H}$, satisfies the conditions of a generalized morphism.

Exercise 7.2.11. Show that the graph of a Lie groupoid morphism defined above is a manifold which can also be written as $\{(g, q) \in \mathcal{G} \times \mathcal{H}_{(0)} \mid s(g) = t(\phi(1_q))\}$. Define the composition of two generalized morphisms, and show that Lie groupoids together with generalized morphisms form a category.

Remark 7.2.12. One can define Morita equivalence for topological groupoids in a similar way (see, e.g., [271]). Morita equivalence of groupoids was probably first used by Haefliger in his theory of Γ-structures (see [165]). As stressed by Haefliger [165], many important invariants and properties of foliations are preserved under Morita equivalence (of their holonomy groupoids). Probably, the same may be said about general Lie (or topological) groupoids. The name Morita comes from the analogy with a natural equivalence relation in algebra, first studied by Morita [264]: two unital rings A and B are called Morita-equivalent if the category of A-modules is isomorphic to the category of B-modules. Since then, the notion of Morita equivalence has been generalized to and studied in many other situations in algebra and geometry. For example, Muhly, Renault and Williams [271] proved that if two locally compact topological groupoids are Morita equivalent then their corresponding C^*-algebras (with respect to given Haar systems) are "strongly Morita equivalent". For Morita equivalence of Poisson manifolds see, e.g., [54, 85, 87, 145, 357, 358].

7.3 Proper Lie groupoids

7.3.1 Definition and elementary properties

Recall that a continuous map $\phi : X \to Y$ between two Hausdorff topological spaces is called **proper** if the preimages of compact subsets of Y under ϕ are compact subsets of X. A (left) action of a Lie group G on a manifold M is called a *proper action* if the map $G \times M \to M \times M, (g, x) \mapsto (g.x, x)$ is a proper map. If G is compact, then any smooth action of G is automatically proper. But non-compact groups can also act properly. For example, the action of any Lie group on itself by multiplication on the left is proper.

The following definition generalizes the notion of proper actions of Lie groups to the case of Lie groupoids.

Definition 7.3.1. A Lie groupoid $\Gamma \rightrightarrows M$ is called a *proper Lie groupoid* if Γ is Hausdorff and the map $(s, t) : \Gamma \to M \times M$ is proper.

Example 7.3.2. A smooth action of a Lie group G on a manifold M is proper if and only if the corresponding transformation groupoid $G \ltimes M$ is proper.

The above properness condition has some immediate topological consequences, which we put together into a proposition:

Proposition 7.3.3 ([254, 354]). *Let $\mathcal{G} \rightrightarrows M$ be a proper Lie groupoid. Then we have:*

i) *The isotropy group $\mathcal{G}_m = \{p \in \mathcal{G} \mid s(p) = t(p) = m\}$ of any point $m \in M$ is a compact Lie group.*

ii) *Each orbit $\mathcal{O}$ of $\mathcal{G}$ on M is a closed submanifold of M.*

iii) *The orbit space $M/\mathcal{G}$ together with the induced topology is a Hausdorff space.*

iv) *If $\mathcal{H}$ is a Hausdorff Lie groupoid which is Morita-equivalent to $\mathcal{G}$, then $\mathcal{H}$ is also proper.*

v) *If N is a submanifold of M which intersects an orbit $\mathcal{O}$ transversally at a point $m \in M$, and B is a sufficiently small open neighborhood of m in N, then the slice $\mathcal{G}_B = s^{-1}(B) \cap t^{-1}(B)$ is a proper Lie groupoid over B which has m as a fixed point.*

Proof. Points i) and v) follow directly from the definition. A sketchy proof of point iv) can be found in [254], the chapter on Lie groupoids, and will be left as an exercise. The proof of point ii) can be found in [354] and is similar to the following proof of point iii). Let us give here a proof of point iii): Let $x, y \in M$ such that their orbits are different: $\mathcal{O}(x) \cap \mathcal{O}(y) = \emptyset$, or equivalently, $s^{-1}(y) \cap t^{-1}(x) = \emptyset$. Denote by $D_1^z \supset D_2^z \supset \ldots \ni z$ a series of compact neighborhoods (i.e., compact sets which contain open neighborhoods) of z in M, where $z = x$ or y, such that $\bigcap_{n \in \mathbb{N}} D_n^z = \{z\}$. We have $\bigcap_{n \in \mathbb{N}} t^{-1}(D_n^x) \cap s^{-1}(D_n^y) = t^{-1}(x) \cap s^{-1}(y) = \emptyset$. Since Γ is proper, the sets $t^{-1}(D_n^x) \cap s^{-1}(D_n^y)$ are compact. It follows that there is $n \in \mathbb{N}$ such that $t^{-1}(D_n^x) \cap s^{-1}(D_n^y) = \emptyset$, or equivalently, $\mathcal{O}(D_n^x) \cap \mathcal{O}(D_n^y) = \emptyset$, where $\mathcal{O}(D_n^x)$ is the union of orbits through D_n^x. But the orbit space of $\mathcal{O}(D_n^x)$ (resp. $\mathcal{O}(D_n^y)$) is a (compact) neighborhood of x (resp., y) in the orbit space of M. Thus the orbit space of M is Hausdorff. $\square$

Definition 7.3.4. A smooth action of a Lie groupoid Γ on a manifold N is called a *proper action* if the corresponding semi-direct product $\Gamma \ltimes N$ is a proper Lie groupoid.

Exercise 7.3.5. Show that if Γ is a proper Lie groupoid, then its smooth actions on manifolds are automatically proper.

Exercise 7.3.6. Show that the adjoint action of a non-compact simple Lie group G on its Lie algebra $\mathfrak{g}$ is not proper.

7.3.2   Source-local triviality

A Lie groupoid $\Gamma \rightrightarrows M$ is called *source-locally trivial*, if the source map $s : \Gamma \to M$ is a locally trivial fibration.

For example, the transformation groupoid of a Lie group action is source-locally trivial.

Recall that the inversion map exchanges s-fibers with t-fibers, hence a source-locally trivial groupoid is also target-locally trivial and vice versa.

While the condition of properness is preserved under Morita equivalence of groupoids, source-local triviality is not, as the following simple example shows.

Example 7.3.7 ([354]). Let X_1 be the plane $\mathbb{R}^2$ with the origin removed. Let the groupoid $\mathcal{G}_1$ be the equivalence relation on X_1, with quotient space $\mathbb{R}$, consisting of all the pairs of points lying on the same vertical line. (In other words, the space of objects is X_1, and there is a unique arrow from a point $x \in X_1$ to a point $y \in X_1$ if and only if they lie on the same vertical line.) The anchor of $\mathcal{G}_1$ is proper, but the source map is not locally trivial over any point lying on the vertical line passing through the origin. Let $\mathcal{G}_2$ be the restriction of $\mathcal{G}_1$ to a horizontal line *not* passing through the origin. $\mathcal{G}_1$ is just a trivial groupoid over X_2, so it is proper and source-locally trivial. The inclusion map from $\mathcal{G}_1$ to $\mathcal{G}_2$ is a weak equivalence, so these two groupoids are Morita-equivalent.

See [354] for more examples of groupoids which are not source-locally trivial.

Proposition 7.3.8. *If $\Gamma \rightrightarrows B$ is a proper Lie groupoid with a fixed point $m \in B$, then any neighborhood of m in B contains an open neighborhood U such that Γ_U is a source-locally trivial proper Lie groupoid.*

In particular, a slice of a proper Lie groupoid is a proper Lie groupoid which can be chosen source-locally trivial.

Proof. Since the isotropy group $\Gamma_m = s^{-1}(m) \subset \Gamma$ is compact and s is a submersion, there is a neighborhood U of Γ_m in Γ such that $s|_U : U \to s(U)$ is a trivial fibration, and $s(U)$ is a neighborhood of m in B which can be chosen arbitrarily small. Fix a local coordinate system on B near x, and for each $n \in \mathbb{N}$ denote by B_n the open ball in B of radius $1/n$ centered at x with respect to this coordinate system. We can assume that $B_1 \subset s(U)$ is relatively compact. By properness of Γ, the sets $V_n = s^{-1}(B_n) \cap t^{-1}(B_n)$ are relatively compact, and $\bigcap_{n=1}^{\infty} \overline{V_n} = \Gamma_m$, where $\overline{V_n}$ means the closure of V_n. Since $\overline{V_n} \supset \overline{V_{n+1}}$ $\forall n$ and U is a neighborhood of Γ_m, there is $n \in \mathbb{N}$ such that $V_n \subset U$. We may assume that $V_1 \subset U$.

Define $U_n = t^{-1}(B_n) \cap U$ and $D_n = s(U_n)$. Then for n large enough, $D_n \subset B_1$ is a small open neighborhood of m in B, and $\Gamma_{D_n} = s^{-1}(D_n) \cap t^{-1}(D_n) = s^{-1}(D_n) \cap U \subset U$. Indeed, since the target map t is a submersion, U_n is a small neighborhood of Γ_m, and since $s(\Gamma_m) = m$, $D_n = s(U_n)$ is a small neighborhood of m, $\cap_{n=1}^{\infty} D_n = \{m\}$, so $D_n \subset B_1$ provided that n is large enough. Since $V_1 = s^{-1}(B_1) \cap t^{-1}(B_1) \subset U$, we also have $s^{-1}(D_n) \cap t^{-1}(D_n) \subset U$, hence $s^{-1}(D_n) \cap t^{-1}(D_n) \subset s^{-1}(D_n) \cap U$. Conversely, let g be an arbitrary element of $s^{-1}(D_n) \cap U$, we will show that $g \in t^{-1}(D_n)$, i.e., $t(g) \in D_n$ (provided that n is large enough). By definition, $g \in U$ and $s(g) \in D_n = s(t^{-1}(B_n) \cap U)$, and there is an element $h \in U$ such that $t(h) \in B_n$ and $s(h) = s(g) \in D_n$. Since D_n is small, g and h lie in a small neighborhood of Γ_m in Γ, and by continuity of the product map and the inversion map, we can assume that $h.g^{-1} \in U$. Then $t(h.g^{-1}) = t(h) \in B_n$, so $h.g^{-1} \in t^{-1}(B_n) \cap U$, and $t(g) = s(h.g^{-1}) \in s(t^{-1}(B_n) \cap U) = D_n$.

Since $\Gamma_{D_n} = s^{-1}(D_n) \cap U$, we have that $s : \Gamma_{D_n} \to D_n$ is a trivial bundle, i.e., Γ_{D_n} is source-locally trivial. The fact that Γ_{D_n} is proper is automatic. Note that $t : \Gamma_{D_n} \to D_n$ is also a trivial fibration for n large enough. $\square$

Proposition 7.3.9 ([354]). *If $\Gamma \rightrightarrows B$ is a source-locally trivial proper Lie groupoid with a fixed point $m \in B$, then any neighborhood of m in B contains an invariant open neighborhood of m.*

Proof. Use the neighborhoods D_n constructed in the proof of the previous proposition. In the source-locally trivial case, they are invariant. $\square$

7.3.3   Orbifold groupoids

Orbifold groupoids form an interesting class of proper groupoids. They were introduced by Moerdijk and Pronk [255, 256] as a convenient setting in which to study structures on orbifolds.

Definition 7.3.10. An *orbifold groupoid* is a proper Lie groupoid whose isotropy groups are finite.

Recall that a smooth *orbifold* is a space V which is locally "diffeomorphic" to the quotient of a smooth manifold by a finite group action. More precisely, there is an open covering (U_i) of V, open subsets $\widetilde{U}_i$ of $\mathbb{R}^n$, finite groups G_i which act on $\widetilde{U}_i$, and projections $\pi_i : \widetilde{U}_i \to U_i = \widetilde{U}_i/G_i$. If $\widetilde{x}_i$ and $\widetilde{x}_j$ are points of $\widetilde{U}_i$ and $\widetilde{U}_j$ such that $\pi_i(\widetilde{x}_i) = \pi_j(\widetilde{x}_j)$, then there is an open neighborhood $\widetilde{W}_i$ of $\widetilde{x}_i$, an open neighborhood $\widetilde{W}_j$ of $\widetilde{x}_j$ and an (automatically unique) smooth diffeomorphism $\phi_{ij} : \widetilde{W}_i \to \widetilde{W}_j$ such that $\pi_i|_{\widetilde{W}_i} = \pi_j|_{\widetilde{W}_j} \circ \phi_{ij}$. The above open covering $(U_i = \widetilde{U}_i/G_i)$ is called an *atlas* of V, and each open set $U_i = \widetilde{U}_i/G_i$ is called a defining *chart*. Orbifolds are natural generalizations of manifolds. They were introduced by Satake[2], who proved an analogue of the Gauss–Bonnet formula for them [309].

---

[2]Satake used the term V-manifold. The term *orbifold* was probably coined by Thurston, see [329].

Given an orbifold V with an atlas $(U_i = \widetilde{U}_i/G_i)$ as above, the disjoint union $P = \coprod_i \widetilde{U}_i$ is the base space of a Lie groupoid Γ^V, which consists of germs of diffeomorphisms generated by the elements of G_i, and of germs of the above local diffeomorphisms ϕ_{ij}. The orbit space of Γ^V is precisely V. Notice that the isotropy groups of Γ^V are finite (they are subgroups of the groups G_i). It is also clear that Γ^V is a proper Lie groupoid. Thus, any smooth orbifold can be represented as the orbit space of an orbifold groupoid [165]. The converse is also true: if Γ is an orbifold groupoid then its orbit space is an orbifold [255]. It is a consequence of the étale case of the local linearization theorem for proper Lie groupoids (Theorem 7.4.7).

7.4 Linearization of Lie groupoids

7.4.1 Linearization of Lie group actions

Probably the most well-known result about local linearization of Lie group actions near a fixed point is the following theorem of Bochner:

Theorem 7.4.1 (Bochner [36]). *Any C^n-action $(n = 1, 2, \ldots, \infty)$ of a compact Lie group G on a manifold V with a fixed point x is locally C^n-isomorphic in a neighborhood of x to a linear action.*

Proof. The proof of Bochner's theorem, based on the averaging method, is very simple [36]: Denote by $\rho : G \times M \to M$ an action of G on M with a fixed point $z \in M$. The linear part $\rho^{(1)}$ of ρ, i.e., the differential of ρ at z, is a linear action of G on $T_z M$. By a local coordinate system $(x_1, \ldots, x_m)$ centered at z, we will identify a neighborhood of z in M with a neighborhood of z in $T_z M$, and understand both ρ and $\rho^{(1)}$ as actions of G on (a neighborhood of 0 in) $V = T_z M$. Denote by μ the Haar probability measure on G. Then the following map ϕ,

$$\phi(x) = \int_G \rho^{(1)}(g^{-1}, \rho(g, x)) \mathrm{d}\mu, \qquad (7.26)$$

is a local diffeomorphism of V (whose differential at 0 is the identity) which intertwines ρ with $\rho^{(1)}$:

$$\rho^{(1)} \circ \phi = \phi \circ \rho. \qquad (7.27)$$

In other words, ϕ^{-1} is a linearization of the action ρ, and it has the same smoothness class as ρ. $\qquad \square$

Remark 7.4.2. Of course, the above theorem also holds for analytic actions of compact Lie groups (then we will have a local analytic linearization), with the same proof. Using the so-called unitary trick to turn the action of a non-compact semisimple Lie group (or Lie algebra) to that of a compact Lie group (in the complexified space), Guillemin and Sternberg showed that an analytic action of a non-compact semisimple Lie group (or Lie algebra) is also locally linearizable [160], a result which was also obtained independently by Kushnirenko [205].

Remark 7.4.3. Smooth actions of non-compact Lie groups are often locally non-linearizable. See, e.g., [57] for some results in the non-compact semisimple case.

Example 7.4.4. Consider an action of $\mathbb{R}$ on the plane generated by the vector field $X = x\partial/\partial y - y\partial/\partial x - (\sin^2 x)(x\partial/\partial x - y\partial/\partial y)$. Then the linearized action has closed orbits, while the orbits of X are spiralling towards 0. So for topological reasons, this action of $\mathbb{R}$ is locally nonlinearizable.

Consider now an orbit $\mathcal{O}(z) = G.z$ of an action of a Lie group G on a manifold M, where $z \in M$ is not a fixed point of the action. Then the isotropy group $G_z = \{g \in G, g.z = z\}$ acts naturally on the normal vector space $W = T_z M / T_z \mathcal{O}(z)$ (via the derivation of the action of G_z in the neighborhood of z). Similarly, G acts on the normal bundle $N\mathcal{O}$ of $\mathcal{O}$ in M in a fiber-wise linear fashion, and we can identify $N\mathcal{O}$ with $G \times_{G_z} W$. A natural question arises: does there exist a tubular neighborhood of $\mathcal{O}$ in M which is diffeomorphic, by a G-equivariant diffeomorphism, to a neighborhood of the zero section in $N\mathcal{O} \cong G \times_{G_z} W$? A positive answer to this linearization question was obtained by Koszul [200] in the case when G is compact, and then by Palais [291] in the case when G is not compact but its action on M is proper. More precisely, we have the following theorem, called the *slice theorem*. (Imagine the fibers of a neighborhood of $\mathcal{O}$, after the identification of this neighborhood with a neighborhood of the zero section of the normal bundle $N\mathcal{O}$, as slices.)

Theorem 7.4.5 (Slice theorem [200, 291]). *Let a Lie group G act properly on a manifold M, and z be a point of M. Then there is a G-equivariant diffeomorphism from a neighborhood of the orbit $\mathcal{O}(z) = G.z$ of z in M to a neighborhood of the zero section of $G \times_{G_z} W$, which sends $O(z)$ to the zero section, where $G_z = \{g \in G, g.z = z\}$ is the isotropy group at z, and $W = T_z M / T_z \mathcal{O}(z)$ is the normal space to $\mathcal{O}(z)$ at z (on which G_z acts linearly).*

The proof of Theorem 7.4.5 follows from Bochner's Theorem 7.4.1 (applied to the action of the compact isotropy group) and some relatively simple topological arguments. See, e.g., Chapter 2 of [121] and Appendix B of [158] for details and some applications.

7.4.2   Local linearization of Lie groupoids

Let $\Gamma \rightrightarrows B$ be a Lie groupoid with a fixed point $x \in B$. Then the isotropy group $G = \Gamma_x = s^{-1}(x) = t^{-1}(x)$ acts naturally on $T_x B$ as follows. For $g \in G, v \in T_x M$, denote by $\gamma(v), r \in [0,1]$ a smooth path in Γ such that $\gamma(0) = g$ and $\frac{d}{dr}|_{r=0} s(\gamma(r)) = v$, and put

$$g.v = \frac{d}{dr}\Big|_{r=0} t(\gamma(r)). \qquad (7.28)$$

One can check that this definition is independent of the choice of γ, and gives a linear action of G on $T_x M$. The corresponding transformation groupoid $G \ltimes M$

is called the *linear part* of Γ at x. The general local linearization problem for Lie groupoids is: given a Lie groupoid with a fixed point, is it locally isomorphic to its linear part? The infinitesimal version of this linearization problem for Lie groupoids is the linearization problem for Lie algebroids, which will be discussed in Chapter 8. To make the problem more precise, we will need the following definition:

Definition 7.4.6. A Lie groupoid $\Gamma \rightrightarrows B$ with a fixed point $x \in B$ is said to be *locally isomorphic* to a Lie groupoid $\mathcal{G} \rightrightarrows D$ with a fixed point $y \in D$ if there is an open neighborhood U (resp., V) of x (resp., y) in B (resp., D) such that Γ_U is isomorphic to $\mathcal{G}_V$.

The following local linearization theorem generalizes Bochner's Theorem 7.4.1 to the case of proper Lie groupoids.

Theorem 7.4.7 ([370]). *Any C^n-smooth ($n \in \mathbb{N} \cup \{\infty\}$) proper Lie groupoid $\Gamma \rightrightarrows B$ with a fixed point $x_0 \in B$ is C^n-smoothly locally linearizable, i.e., it is C^n-smoothly locally isomorphic to the transformation groupoid $G \ltimes V$ of a linear action of G on V, where $G = G_{x_0}$ is the isotropy group of x_0 and $V = T_{x_0}B$.*

Remark 7.4.8. The étale case of Theorem 7.4.7 (i.e., the case when the isotropy group G_{x_0} is finite) is relatively simple and was obtained by Moerdijk and Pronk [255] (see also [354]). The general case was conjectured by Weinstein [353, 354].

We will give here a sketch of the proof of Theorem 7.4.7, referring the reader to [370] for the details.

By Proposition 7.3.8, we can assume that the proper groupoid $\Gamma \rightrightarrows B$ is locally-source trivial, and by Proposition 7.3.9, we may shrink B to an arbitrarily small neighborhood of x_0 in B.

Note that Theorem 7.4.7 is essentially equivalent to the existence of a smooth *surjective* homomorphism ϕ from Γ to G (after shrinking B to a sufficiently small invariant neighborhood of x_0), i.e., a smooth map $\phi : \Gamma \to G$ which satisfies

$$\phi(p.q) = \phi(p).\phi(q) \ \forall \ (p, q) \in \Gamma_{(2)} := \{(p, q) \in \Gamma \times \Gamma, s(p) = t(q)\} , \qquad (7.29)$$

and such that the restriction of ϕ to $G = s^{-1}(x_0) \subset \Gamma$ is an automorphism of G. We may assume that this automorphism is identity.

Indeed, if there is an isomorphism from $\Gamma \rightrightarrows B$ to a transformation groupoid $G \ltimes U$, then the composition of the isomorphism map $\Gamma \to G \times U$ with the projection $G \times U \to G$ is such a homomorphism. Conversely, if we have a homomorphism $\phi : \Gamma \to G$, whose restriction to $G = s^{-1}(x_0) \subset \Gamma$ is the identity map of G, then shrinking B to a sufficiently small invariant neighborhood of z in B if necessary, we get a diffeomorphism

$$(\phi, s) : \Gamma \to G \times B. \qquad (7.30)$$

Denote by θ the inverse map of (ϕ, s). Then there is an action of G on B defined by $g.x = t(\theta(g, x))$, and the map (ϕ, s) will be an isomorphism from $\Gamma \rightrightarrows B$ to the transformation groupoid $G \ltimes B$. This transformation groupoid is linearizable by Bochner's Theorem 7.4.1, implying that the groupoid $\Gamma \rightrightarrows B$ is linearizable.

In order to find such a homomorphism from Γ to G, we will use the averaging method. The idea is to start from an arbitrary smooth map $\phi : \Gamma \to G$ such that $\phi|_G = \mathrm{Id}$. (Recall that $G = \Gamma_{x_0} = s^{-1}(x_0) = t^{-1}(x_0)$.) Then Equality (7.29) is not satisfied in general, but it is satisfied for $p, q \in G$. Hence it is "nearly satisfied" in a small neighborhood of $G = s^{-1}(x_0)$ in Γ. In other words, if the base B is small enough, then $\phi(p.q)\phi(q)^{-1}$ is near $\phi(p)$ for any $(p, q) \in \Gamma_{(2)}$. We will replace $\phi(p)$ by the average value of $\phi(p.q)\phi(q)^{-1}$ for q running on $t^{-1}(s(p))$ (it is to be made precise how to define this average value). This way we obtain a new map $\widehat{\phi} : \Gamma \to G$, which will be shown to be "closer" to a homomorphism than the original map ϕ. By iterating the process and taking the limit, we will obtain a true homomorphism ϕ_∞ from Γ to G.

Notice that the t-fibers of $\Gamma \rightrightarrows B$ are compact and diffeomorphic to $G = t^{-1}(x_0)$ by assumptions. As a consequence, there exists a smooth Haar probability system (μ_x) on Γ, i.e., a smooth Haar system such that for each $x \in B$, the volume of $t^{-1}(x)$ with respect to μ_x is 1. Such a Haar probability system (μ_x) can be constructed as follows: begin with an arbitrary Haar system (μ'_x) on Γ, then define $\mu = \mu'/I$ where I is the left-invariant function $I(g) = \int_{t^{-1}(t(g))} \mathrm{d}\mu'_{t(g)}$. We will fix a Haar probability system $\mu = (\mu_x)$ on Γ.

We fix a bi-invariant metric on the Lie algebra $\mathfrak{g}$ of G and the induced bi-invariant metric d on G itself. Denote by 1_G the neutral element of G. For each number $\rho > 0$, denote by $B_{\mathfrak{g}}(\rho)$ (resp., $B_G(\rho)$) the closed ball of radius ρ in $\mathfrak{g}$ (resp., G) centered at 0 (resp., 1_G). By resizing the metric if necessary, we will assume that the exponential map

$$\exp : B_{\mathfrak{g}}(1) \to B_G(1) \tag{7.31}$$

is a diffeomorphism. Denote by

$$\log : B_G(1) \to B_{\mathfrak{g}}(1) \tag{7.32}$$

the inverse of exp. Define the distance $\Delta(\phi)$ of $\phi : \Gamma \to G$ from being a homomorphism as follows:

$$\Delta(\phi) = \sup_{(p,q)\in\Gamma_{(2)}} d\left(\phi(p.q).\phi(q)^{-1}.\phi(p)^{-1}, 1_G\right). \tag{7.33}$$

Let $\phi : \Gamma \to G$ be a smooth map such that $\phi|_G$ is identity. We will assume that $\Delta(\phi) \leq 1$, so that the following map $\widehat{\phi} : \Gamma \to G$ is clearly well defined:

$$\widehat{\phi}(p) = \exp\left(\int_{q\in t^{-1}(s(p))} \log(\phi(p.q).\phi(q)^{-1}.\phi(p)^{-1})\mathrm{d}\mu_{s(p)}\right).\phi(p) . \tag{7.34}$$

It is clear that $\widehat{\phi}$ is a smooth map from Γ to G, and its restriction to $G = s^{-1}(x_0) \subset \Gamma$ is also identity. The proof of the following lemma, which says that when G is Abelian we are done, is straightforward:

Lemma 7.4.9. *With the above notations, if G is essentially commutative (i.e., the connected component of the neutral element of G is commutative) then $\widehat{\phi}$ is a homomorphism.*

In particular, if G is finite then $\widehat{\phi}$ is a homomorphism, and we obtain a proof of Theorem 7.4.7 in the étale case.

In general, due to the noncommutativity of G, $\widehat{\phi}$ is not necessarily a homomorphism, but $\Delta(\widehat{\phi})$ (the distance of $\widehat{\phi}$ from being a homomorphism) is of the order of $\Delta(\phi)^2$. More precisely, we have:

Lemma 7.4.10. *There is a positive constant $C_0 > 0, C_0 \leq 1$ (which depends only on G and on the choice of the bi-invariant metric on it) such that if $\Delta(\phi) \leq C_0$ then $\widehat{\phi}$ is well defined and*

$$\Delta(\widehat{\phi}) \leq (\Delta(\phi))^2/C_0 \leq \Delta(\phi) \ . \tag{7.35}$$

The above lemma implies that we have the following fast convergent iterative process: starting from an arbitrary given smooth map $\phi : \Gamma \to G$, such that $\phi|_G = \mathrm{Id}$, construct a sequence of maps $\phi_n : \Gamma \to G$ by the recurrence formula $\phi_1 = \phi$, $\phi_{n+1} = \widehat{\phi_n}$. Then this sequence is well defined (after shrinking B once to a smaller invariant neighborhood of x_0 if necessary), and

$$\phi_\infty = \lim_{n \to \infty} \phi_n \tag{7.36}$$

exists (in C^0-topology), and is a continuous homomorphism from Γ to G.

More elaborate estimates using C^k-norms (where k does not exceed the smoothness class of Γ) show that in fact we have $\phi_\infty = \lim_{n \to \infty} \phi_n$ in C^k-topology, and hence ϕ_∞ is a C^k-smooth homomorphism. This concludes the proof of Theorem 7.4.7. See [370] for the details.

Remark 7.4.11. The above iterative averaging process is inspired by a similar process which was deployed by Grove, Karcher and Ruh in [155] to prove the following theorem about approximation of near-homomorphisms between compact Lie groups by homomorphisms. The idea of using Grove–Karcher–Ruh's iterative averaging method was proposed by Weinstein [352, 353].

Theorem 7.4.12 (Grove–Karcher–Ruh [155]). *If G and K are two given compact Lie groups, then any map $\phi : G \to K$ which is a near-homomorphism (i.e., the map $G \times G \to K : (g, h) \mapsto \phi(g).\phi(h).\phi(h^{-1}g^{-1})$ is sufficiently close to the constant map $(g, h) \mapsto e_K$ in C^0-topology) can be approximated by a homomorphism $\phi_0 : G \to K$ (i.e., the map $G \to K : g \mapsto \phi(g)^{-1}.\phi_0(g)$ is close to the constant map $g \mapsto e_K$).*

An immediate consequence of Theorem 7.4.7 is the following:

Corollary 7.4.13. *The characteristic foliation of a proper Lie groupoid is an orbit-like foliation in the sense of Molino, with closed leaves. In particular, it is a singular Riemannian foliation with closed leaves.*

Recall that a *Riemannian foliation* [298] is a foliation which admits a transverse Riemannian metric, i.e., a Riemannian metric on a complete transversal which is invariant under holonomy. We refer to [260] for an introduction to Riemannian foliations, including singular Riemannian foliations. An orbit-like foliation [261] is a singular Riemannian foliation which is locally linearizable.

Another immediate consequence of Theorem 7.4.7 is that, if $\mathcal{G} \rightrightarrows M$ is a proper groupoid, then the orbit space $M/\mathcal{G}$ (together with the induced topology and smooth structure from M) locally looks like the quotient of a vector space by a linear action of a compact Lie group. (Locally, the orbit space $M/\mathcal{G}$ is the same as the orbit space of a slice $B/\mathcal{G}_B$.) In analogy with the fact that orbifolds are orbit spaces of étale proper groupoids (see Subsection 7.3.3), it would be natural to call the orbit space of a proper Lie groupoid a (smooth) *orbispace*.

7.4.3   Slice theorem for Lie groupoids

Consider a Lie groupoid $\mathcal{G} \rightrightarrows M$, and an orbit $\mathcal{O}$ of $\mathcal{G}$ on M. Then the restriction $\mathcal{G}_\mathcal{O} := \{p \in G \mid s(p), t(p) \in \mathcal{O}\}$ of $\mathcal{G}$ to $\mathcal{O}$ is a transitive Lie groupoid over $\mathcal{O}$. Similarly to the case of Lie group actions and the case of Lie groupoids with a fixed point, the structure of $\mathcal{G}$ induces a linear action of $\mathcal{G}_\mathcal{O}$ on the normal vector bundle $N_\mathcal{O}$ of $\mathcal{O}$ in M. The corresponding semi-direct product $\mathcal{G}_\mathcal{O} \ltimes N_\mathcal{O}$ is the linear model for $\mathcal{G}$ in the neighborhood of $\mathcal{O}$.

The following theorem, which generalizes Koszul–Palais' slice Theorem 7.4.5 to the case of Lie groupoids, and which can also be called a *slice theorem*, describes the structure of a proper Lie groupoid in the neighborhood of an orbit, under some mild conditions.

Theorem 7.4.14 (Slice theorem). *Let $\mathcal{G} \rightrightarrows M$ be a source-locally trivial proper Lie groupoid, and let $\mathcal{O}$ be an orbit of $\mathcal{G}$ which is a manifold of finite type. Then there is a neighborhood U of $\mathcal{O}$ in M such that the restriction of $\mathcal{G}$ to U is isomorphic to the restriction of the transformation groupoid $\mathcal{G}_\mathcal{O} \times_\mathcal{O} N\mathcal{O} \rightrightarrows N\mathcal{O}$ to a neighborhood of the zero section of $N\mathcal{O}$.*

In the above theorem, the condition that $\mathcal{O}$ is of finite type means that there is a proper map from $\mathcal{O}$ to $\mathbb{R}$ with a finite number of critical points. Theorem 7.4.14 was obtained by Weinstein [354] modulo Theorem 7.4.7. The original result of Weinstein [354] may be formulated as follows:

Theorem 7.4.15 ([354]). *Let $\Gamma \rightrightarrows M$ be a proper groupoid, and let $\mathcal{O}$ be an orbit of Γ which is a manifold of finite type. Suppose that the restriction Γ_D of Γ to a slice D through $x \in \mathcal{O}$ is isomorphic to the restriction of the transformation groupoid $\Gamma_x \times N_x\mathcal{O} \rightrightarrows N_x\mathcal{O}$ to a neighborhood of zero. Then there is a neighborhood $\mathcal{U}$ of $\mathcal{O}$ in M such that the restriction of Γ to $\mathcal{U}$ is isomorphic to the restriction of the transformation groupoid $G_\mathcal{O} \times_\mathcal{O} N\mathcal{O} \rightrightarrows N\mathcal{O}$ to a neighborhood of the zero section.*

We refer to [354] for the proof of Theorem 7.4.15. It makes use of the following auxiliary topological result to treat the case when the orbit is not compact:

Proposition 7.4.16 ([354]). *Let $f : X \to Y$ be a submersion. For any $y \in Y$, if $O = f^{-1}(y)$ is a manifold of finite type, then there is a neighborhood U of O in X such that $f|_U : U \to f(U)$ is a trivial fibration. In other words, there is a retraction $\rho : U \to O$ such that $(\rho, f) : U \to O \times f(U)$ is a diffeomorphism.*

If f is equivariant with respect to actions of a compact group K on X and Y, with y a fixed point, and $f^{-1}(y)$ is of finite type as a K-manifold, then U can be chosen to be K-invariant and ρ to be K-equivariant.

7.5 Symplectic groupoids

7.5.1 Definition and basic properties

Symplectic groupoids were introduced independently by Karasev [189], Weinstein [349], and Zakrzewski [364], in connection with symplectic realization and quantization of Poisson manifolds. One of the main motivations is the following question: Global objects which integrate Lie algebras are Lie groups. General Poisson manifolds may be viewed as non-linear analogs of Lie algebras. So what are the global objects which are analogous to Lie groups and which have Poisson manifolds as their infinitesimal objects? The answer is, they are symplectic groupoids.

Definition 7.5.1. A *symplectic groupoid* is a Lie groupoid $\Gamma \rightrightarrows P$, equipped with a symplectic form σ on Γ, such that the graph of the multiplication map

$$\Delta = \{(g, h, g.h) \in \Gamma \times \Gamma \times \overline{\Gamma} \mid (g, h) \in \Gamma_{(2)}\} \tag{7.37}$$

is a Lagrangian submanifold of $\Gamma \times \Gamma \times \overline{\Gamma}$, where $\overline{\Gamma}$ means the manifold Γ with the opposite symplectic form $-\sigma$.

The condition that Δ is a Lagrangian submanifold of $\Gamma \times \Gamma \times \overline{\Gamma}$ means that $\dim \Delta = \frac{3}{2} \dim \Gamma$ and Δ is isotropic. The condition that Δ is isotropic may be expressed more visually as follows: If $\phi_1, \phi_2, \phi_3 : D^2 \to \Gamma$ are three maps from a two-dimensional disk D^2 to Γ such that $\phi_1(x).\phi_2(x) = \phi_3(x)$ for any $x \in D^2$, then

$$\int_{D^2} \phi_1^* \omega + \int_{D^2} \phi_2^* \omega = \int_{D^2} \phi_3^* \omega. \tag{7.38}$$

In other words, roughly speaking, we have

$$\text{Area}(D_1) + \text{Area}(D_2) = \text{Area}(D_3), \tag{7.39}$$

where Area means the symplectic area, and $D_i = \phi_i(D^2)$.

Example 7.5.2. If (M, ω) is a symplectic manifold, then the pair groupoid $M \times \overline{M} \rightrightarrows M$ is a symplectic groupoid, where $\overline{M}$ means M with the opposite symplectic form $-\omega$.

Example 7.5.3. The cotangent bundle T^*N of a manifold N has a natural symplectic groupoid structure over N: the source map coincides with the target map and is the projection map from T^*N to N, the symplectic structure on T^*N is the standard one, and the multiplication map is given by the usual sum of covectors with the same base points.

Example 7.5.4. Consider the transformation groupoid $G \times \mathfrak{g}^* \rightrightarrows \mathfrak{g}^*$ of the coadjoint action of a Lie group G. Identify $G \times \mathfrak{g}^*$ with T^*G via left translations, and equip it with the standard symplectic form. Then it becomes a symplectic groupoid, which we will call a *standard symplectic groupoid* and denote by $T^*G \rightrightarrows \mathfrak{g}^*$. The corresponding Poisson structure on $\mathfrak{g}^*$ is the standard linear (Lie-) Poisson structure.

We will put together some basic facts about symplectic groupoids into the following theorem:

Theorem 7.5.5 ([189, 349, 83, 3]). *If $(\Gamma, \sigma) \rightrightarrows P$ is a symplectic groupoid with symplectic form σ, source map s, target map t, and inversion map $\imath$, then we have:*

a) $\dim P = \frac{1}{2} \dim \Gamma$.

b) *P is a Lagrangian submanifold of Γ (P is identified with the image $\varepsilon(P)$ of the identity section $\varepsilon : P \to \Gamma$).*

c) *The foliation by s-fibers is symplectically dual to the foliation by t-fibers. In other words, for any $g \in \Gamma$, $T_g s^{-1}(s(g)) = (T_g t^{-1}(t(g)))^\perp$.*

d) *The inversion map $\imath : \Gamma \to \Gamma$ is an anti-symplectomorphism:*

$$\imath^* \sigma = -\sigma. \tag{7.40}$$

e) *For any two functions ϕ, ψ on P,*

$$\{s^* \phi, t^* \psi\} = 0. \tag{7.41}$$

f) *For any function f on P, the Hamiltonian vector field X_{s^*f} is tangent to t-fibers and is invariant under left translations in Γ. Similarly, X_{t^*f} is tangent to s-fibers and is invariant under right translations.*

g) *There is a unique Poisson structure Π on P such that the source map s is a Poisson map, and the target map t is anti-Poisson.*

h) *If B is a sufficiently small open neighborhood of a point x in a submanifold in P which intersects the symplectic leaf $\mathcal{O} = \mathcal{O}(x)$ of x transversally at x, then the slice $(\Gamma_B, \omega|_{\Gamma_B}) \rightrightarrows B$ is a symplectic groupoid, and the corresponding Poisson structure on B is the transverse Poisson structure of P at x.*

i) *For any point $x \in P$, the Lie algebra of the isotropy group Γ_x corresponds to the linear part of the transverse Poisson structure of (P, Π) at x.*

j) *If $x \in P$ is a regular point (with respect to the characteristic foliation on P), then the isotropy group Γ_x is Abelian.*

k) *For each $g \in \Gamma$, the set $K_g = s^{-1}(s(g)) \cap t^{-1}(t(g)) = g.G_{s(g)} = G_{t(g)}.g$ is an isotropic submanifold of Γ, and if $h \in K_g$ then $K_h = K_g$.*

l) *Denote by P_{reg} the set of regular elements of P, and by Γ_{reg} the set of $g \in \Gamma$ such that $s(g) \in P_{\mathrm{reg}}$. Then the pull-back of the characteristic foliation on P by s or t is a coisotropic foliation on Γ_{reg} whose dual isotropic foliation has as leaves connected components of the submanifolds $K_g = s^{-1}(s(g)) \cap t^{-1}(t(g))$.*

m) *Conversely, suppose that $(\Gamma, \sigma) \rightrightarrows P$ is a Lie groupoid equipped with a symplectic structure σ and a Poisson structure Π such that the inversion map $\imath$ is anti-symplectic, and the foliation by s-fibers is symplectically dual to the foliation by t-fibers; then it is a symplectic groupoid.*

Proof. a) The graph Δ given by (7.37) has dimension $\dim \Delta = 2 \dim \Gamma - \dim P$. If it is a Lagrangian submanifold in $\Gamma \times \Gamma \times \overline{\Gamma}$, then its dimension is half the dimension of $\Gamma \times \Gamma \times \overline{\Gamma}$, i.e., $\dim \Delta = \frac{3}{2} \dim \Gamma$, implying that $\dim P = \frac{1}{2} \dim \Gamma$. Remark that, as a consequence, the dimension of each s-fiber and each t-fiber is equal to the dimension of P, i.e., half the dimension of Γ.

b) If X, Y are two tangent vectors to P and a point $x \in P$, then (X, X, X) and (Y, Y, Y) are two tangent vector fields to Δ at $(x, x, x) \in \Delta$. Since Δ is Lagrangian, we have $0 = (\sigma \oplus \sigma \oplus -\sigma)((X, X, X), (Y, Y, Y)) = \sigma(X, Y) + \sigma(X, Y) - \sigma(X, Y) = \sigma(X, Y)$. It means that P is an isotropic submanifold of Γ. Since $\dim P = \frac{1}{2} \dim \Gamma$, it is a Lagrangian submanifold.

c) If $X \in T_g s^{-1}(s(g))$ and $Y \in T_g t^{-1}(t(g))$ where $g \in \Gamma$, then $(X, \imath_* X, 0)$ and $(Y, 0, (L_{g^{-1}})_* Y)$ are two tangent vector fields to Δ at $(g, g^{-1}, g.g^{-1}) \in \Delta$, where $L_{g^{-1}} : t^{-1}(t(g)) \to t^{-1}(s(g))$ is the left translation by g^{-1}. Again, since Δ is Lagrangian, we have $0 = \sigma(X, Y) + \sigma(\imath_* X, 0) - \sigma(0, (L_{g^{-1}})_* Y) = \sigma(X, Y)$. Since $\dim T_g s^{-1}(s(g)) = \dim T_g t^{-1}(t(g)) = \frac{1}{2} \dim \sigma$, they are symplectically dual to each other.

d) If $X, Y \in T_g \Gamma$ then $(X, \imath_* X, X')$ and $(Y, \imath_*, Y')$ are tangent to Δ at $(g, g^{-1}, s(g))$, where $X' = s_* X, Y' = s_* Y \in T_{s(g)} P$. Using Assertion b) and the fact that Δ is Lagrangian, we get $\sigma(X, Y) + \sigma(\imath_* X, \imath_* Y) = 0$.

e) It is a direct consequence of c) that $X_{s^* \phi}(g) \in T_g t^{-1}(t(g))$ and $X_{t^* \psi}(g) \in T_g s^{-1}(s(g))$, therefore $\{s^* \phi, t^* \psi\}(g) = \omega(X_{s^* \phi}, X_{t^* \psi})(g) = 0$.

f) It follows directly from e).

g) If ϕ, ψ are two smooth functions on P, then it follows from d) and the Jacobi identity that $\{\{s^* \phi, s^* \psi\}, t^* \xi\} = 0$ for any function ξ on P. It implies that $\{s^* \phi, s^* \psi\}$ is invariant on s-fibers, so there is a unique function on P, which we will denote by $\{\phi, \psi\}$, such that $\{s^* \phi, s^* \psi\} = s^* \{\phi, \psi\}$. This defines a unique Poisson structure on P, which we will denote by Π, such that the map $s : \Gamma \to P$ is a Poisson map. Since $t = s \circ \imath$ and $\imath$ is anti-symplectic, t is anti-Poisson.

h) Since the graph Δ_B of the multiplication map in Γ_B is a subset of Δ, it is isotropic, and since $\dim \Gamma_B = \dim \Gamma - 2 \dim \mathcal{O}$ and $\dim B = \dim P - \dim \mathcal{O}$, we still have that $\dim \Delta_B = (3/2) \dim \Gamma_B$. In order to show that Γ_B is a symplectic groupoid, it remains to verify that the restriction of ω to Γ_B is nondegenerate. Let $f_1, \ldots, f_{2s}$ be a family of independent local functions on P such that $f_i|_B = 0$, where $2s = \dim \mathcal{O}$. Let g be an arbitrary point of Γ_B. We have $s(g), t(g) \in B$. Consider $4s$ tangent vectors to Γ at g, $X_i = X_{s^* f_i}(g)$ and $Y_i = X_{t^* f_i}(g)$, $i = 1, \ldots, 2s$. It follows from the previous points that we have $\omega(X_i, Y_j) = 0 \; \forall \; i, j$, while the matrix $\omega(X_i, X_j) = \{f_i, f_j\}(s(g))$ and the matrix $\omega(Y_i, Y_j) = -\{f_i, f_j\}(t(g))$ are nondegenerate provided that B is small enough. Thus the space W spanned by $X_1, \ldots, X_{2s}, Y_1, \ldots, Y_{2s}$ is a symplectic $4s$-dimensional subspace of $T_g\Gamma$. It also follows from the previous points of the theorem that W is symplectically orthogonal to $T_g\Gamma_B$, and $\dim T_g\Gamma_B + \dim W = \dim \Gamma$. Hence $T_g\Gamma_B$ is a symplectic subspace of $T_g\Gamma$. In other words, the restriction of ω to Γ_B is nondegenerate.

Consider now a local coordinate system $\psi_1, \ldots, \psi_m$ on B, where $m = \dim B$, and extend them to functions $\tilde{\psi}_1, \ldots, \tilde{\psi}_m$ in a neighborhood of x in P such that $\{f_i, \psi_j\}(y) = 0 \; \forall \; y \in B$, $\forall \; i, j$. Then one verifies directly that for any $g \in \Gamma_B$ we have $X_i(s^*\tilde{\psi}_j) = Y_i(\tilde{\psi}_j) = 0 \; \forall \; i, j$, where X_i, Y_i are defined as above, which implies that $\{s^*\psi_i, s^*\psi_j\}_{\omega|_{\Gamma_B}}(g) = \{s^*\tilde{\psi}_i, s^*\tilde{\psi}_j\}(g)$. Since the source map is a Poisson map, we have $\{\psi_i, \psi_j\}_B(s(g)) = \{\tilde{\psi}_i, \tilde{\psi}_j\}(s(g))$, where the Poisson structure on B is the one induced from the symplectic structure on Γ_B. But this last formula is also a special case of Dirac's formula for the restriction of the Poisson structure from P to B.

i) In view of point h), it is enough to prove point i) in the case when x is a fixed point of the symplectic groupoid $\Gamma \rightrightarrows P$. Consider a coordinate system $\psi_1, \ldots, \psi_m$ in a neighborhood of x in P. Then the vector fields $X_{s^*\psi_1}, \ldots, X_{s^*\psi_m}$ are tangent to the isotropy group $\Gamma_x = s^{-1}(x) = t^{-1}(x)$, and form a basis of left-invariant vector fields there. So we have

$$X_{\{s^*\psi_i, s^*\psi_j\}}|_{\Gamma_x} = \{X_{s^*\psi_i}, X_{s^*\psi_j}\}|_{\Gamma_x} = \sum_k c_{ij}^k X_{s^*\psi_k}|_{\Gamma_x}, \qquad (7.42)$$

where c_{ij}^k are structural constants of the Lie algebra $\mathfrak{g}$ of Γ_x. Since the source map is a Poisson map, it follows that

$$\{\psi_i, \psi_j\} = \sum_k c_{ij}^k \psi_k + O(2) \qquad (7.43)$$

on P, i.e., c_{ij}^k are structural constants are also structural constants for the linear part of the Poisson structure on P at x.

j) If x is a regular point of P then the transverse Poisson structure at x is trivial. Now apply point i).

k) It follows directly from point c).

l) It follows directly from point c) and point k).

m) The proof of the converse part will be left as an exercise. $\qquad \square$

We will also denote a symplectic groupoid by $(\Gamma, \sigma) \rightrightarrows (P, \Pi)$ to emphasize the fact that Γ is a symplectic manifold and P is a Poisson manifold. In particular, Assertions b) and g) of the above theorem say that $s : (\Gamma, \sigma, \varepsilon(P)) \to (P, \Pi)$ is a marked symplectic realization of P with the marked Lagrangian submanifold $\varepsilon(P)$.

Exercise 7.5.6. Show that any Poisson vector field on the base space (P, Π) of a symplectic groupoid $(\Gamma, \sigma) \rightrightarrows (P, \Pi)$ with connected s-fibers can be naturally lifted to a symplectic vector field on (Γ, σ) which preserves the groupoid structure of Γ. Find an example when this symplectic vector field is not (globally) Hamiltonian.

A Poisson manifold is called *integrable* [95] if it can be realized by a symplectic groupoid. For example, any symplectic manifold is an integrable Poisson manifold (see Example 7.5.2). The integrability and non-integrability of Poisson manifolds will be discussed in Section 8.8.

Example 7.5.7. Let G be a Lie group with a free proper Hamiltonian action on a symplectic manifold (M, ω) with an equivariant momentum map $\mu : M \to \mathfrak{g}^*$. Then the quotient M/G, together with the reduced Poisson structure from M, is an integrable Poisson manifold. A symplectic groupoid integrating M/G is

$$M * \overline{M}/G \rightrightarrows M/G, \tag{7.44}$$

where

$$M * \overline{M} = \{(x, y) \in M \times \overline{M}, \mu(x) = \mu(y)\}, \tag{7.45}$$

i.e., $M * \overline{M}/G$ is the Marsden–Weinstein reduction of the symplectic manifold $(M, \omega) \times (M, -\omega)$ with respect to the diagonal action of G. The structure maps of this groupoid are induced from the structure maps of the pair groupoid $M \times M \rightrightarrows M$ in an obvious way. This example is a symplectic groupoid version of reduction by Lie group actions: while the symplectic manifold M is reduced to the Poisson manifold M/G, the corresponding symplectic groupoid $M \times \overline{M}$ is reduced to $M * \overline{M}/G$.

Exercise 7.5.8. What are the isotropy groups of the above symplectic groupoid $M * \overline{M}/G \rightrightarrows M/G$?

7.5.2 Proper symplectic groupoids

Definition 7.5.9. A symplectic groupoid $(\Gamma, \sigma) \rightrightarrows (P, \Pi)$ is called *proper* if it is proper as a Lie groupoid.

Exercise 7.5.10. Find necessary and sufficient conditions for the symplectic groupoid $M * \overline{M}/G \rightrightarrows M/G$ given in Example 7.5.7 to be proper.

It follows point h) of Theorem 7.5.5 and point v) of Proposition 7.3.3 that a slice of a proper symplectic groupoid is a proper symplectic groupoid with a fixed point. The local structure of proper symplectic groupoids is given by the following theorem:

Theorem 7.5.11 ([370]). *If* $\Gamma \rightrightarrows P$ *is a proper symplectic groupoid with a fixed point* $x \in P$, *then it is locally symplectically isomorphic to the standard symplectic groupoid* $T^*G \rightrightarrows \mathfrak{g}^*$, *where* $G = \Gamma_x$ *is the isotropy group of* x. *In other words, there is an open neighborhood* U *of* x *in* P *and an open neighborhood* V *of* 0 *in* $\mathfrak{g}^*$

such that the restriction of Γ *to* U *is isomorphic to the restriction of* $T^*G \rightrightarrows \mathfrak{g}^*$
to V *by a symplectic Lie groupoid isomorphism.*

Proof. Recall from point i) of Theorem 7.5.5 that the linear part of the Poisson structure Π at x is isomorphic to the Lie–Poisson structure on $\mathfrak{g}^*$. Theorem 7.4.7 allows us to linearize Γ near x without the symplectic structure. The corresponding linear action of G must be (isomorphic to) the coadjoint action, so without losing generality we may assume that P is a neighborhood of $x = 0$ in $\mathfrak{g}^*$, and the orbits on P near 0 are nothing but the coadjoint orbits (though the symplectic form on each orbit may be different from the standard one). But then, as was shown by Ginzburg and Weinstein [148] using a standard Moser's path argument (see Appendix A.1), since G is compact, the Poisson structure on P is actually locally isomorphic to the Lie–Poisson structure of $\mathfrak{g}^*$. We can now apply the following proposition to finish the proof of Theorem 7.5.11:

Proposition 7.5.12. *If G is a (not necessarily connected) compact Lie group and $\mathfrak{g}$ is its Lie algebra, then any proper symplectic groupoid $(\Gamma, \omega) \rightrightarrows U$ with a fixed point 0 whose base Poisson manifold is a neighborhood U of 0 in $\mathfrak{g}^*$ with the Lie–Poisson structure and whose isotropy group at 0 is G is locally isomorphic to $T^*G \rightrightarrows \mathfrak{g}^*$.*

Proof of Proposition 7.5.12. Without loss of generality, we can assume that Γ is source-locally trivial.

We will first prove the above proposition for the case when G is connected. The Lie algebra $\mathfrak{g}$ can be written as a direct sum $\mathfrak{g} = \mathfrak{s} \oplus \mathfrak{l}$, where $\mathfrak{s}$ is semisimple and $\mathfrak{l}$ is Abelian. Denote by $(f_1, \ldots, f_n, h_1, \ldots, h_m)$ a basis of linear functions on $\mathfrak{g}^*$, where $f_1, \ldots, f_n$ correspond to $\mathfrak{s}$ and $h_1, \ldots, h_m$ correspond to $\mathfrak{l}$. Then the vector fields $X_{s^*f_i}, X_{s^*h_j}$ generate a Hamiltonian action of $\mathfrak{g}$ on (Γ, ω). When restricted to the isotropy group $G = \Gamma_0$ over the origin of $\mathfrak{g}^*$, the vector fields $X_{s^*f_i}, X_{s^*h_j}$ become left-invariant vector fields on G, and the action of $\mathfrak{g}$ integrates to the right action of G on itself by multiplication on the right. Assume that the above Hamiltonian action of $\mathfrak{g}$ integrates to a right action of G on Γ. Then we are done. Indeed, since the action is free on Γ_0, we may assume, by shrinking the base space U, that the action is free on Γ. Then one can verify directly that the map $(g, y) \mapsto \varepsilon(\mathrm{Ad}_g^* y) \circ g$, $g \in G$, $y \in U$, where $\varepsilon : U \to \Gamma$ denotes the identity section and $\circ g$ denotes the right action by g, is a symplectic isomorphism between the restriction of the standard symplectic groupoid $G \times \mathfrak{g}^* \cong T^*G \rightrightarrows \mathfrak{g}^*$ to $U \subset \mathfrak{g}^*$ and Γ.

In general, the action of $\mathfrak{g}$ on Γ integrates to an action of the universal covering of G on Γ, which does not factor to an action of G on Γ if the Abelian part of G is nontrivial, i.e., $\mathfrak{l} \neq 0$. So we may have to change the generators of this $\mathfrak{g}$ action, by changing $h_1, \ldots, h_m$ to new functions h_i' which are still Casimir functions of $\mathfrak{g}^*$. Such a change of variables (leaving f_i intact) will be a local Poisson isomorphism of $\mathfrak{g}^*$. We want to choose h_i' so that the Hamiltonian vector field $X_{s^*h_i'}$ are periodic, i.e., they generate Hamiltonian $\mathbb{T}^1$-actions.

Note that for each $y \in U \subset \mathfrak{g}^*$, the isotropy group $\Gamma_y = s^{-1}(y) \cap t^{-1}(y)$ of Γ at y admits a canonical injective homomorphism to G (via an a priori non-symplectic local linearization of Γ using Theorem 7.4.7). Denote by $\mathbb{T}_0^m$ the Abelian torus of dimension m in the center of G (the Lie algebra of $\mathbb{T}_0^m$ is $\mathfrak{l}$). The coadjoint action of $\mathbb{T}_0^m$ on $\mathfrak{g}^*$ is trivial. It follows that each isotropy group Γ_y contains a torus $\mathbb{T}_y^m$ whose image under the canonical injection to G is $\mathbb{T}_0^m$. For each $q \in \Gamma$, denote $\mathbb{T}_q^m = q.\mathbb{T}_{s(q)}^m = \mathbb{T}_{t(q)}^m.q$. Note that if $r \in \mathbb{T}_q^m$ then $\mathbb{T}_r^m = \mathbb{T}_q^m$.

Choose a basis $\gamma_1, \ldots, \gamma_m$ of one-dimensional sub-tori $\mathbb{T}^m$. Translate them to each point $q \in \Gamma$ as above, we get m curves $\gamma_{1,q}, \ldots, \gamma_{m,q} \subset \mathbb{T}_q^m \; \forall \; q \in \Gamma$. Recall that, due to the fact that $\mathbb{T}_0^m$ lies in the center of G, these curves are well defined and depend continuously on q.

Since $G = s^{-1}(0) = t^{-1}(0)$ is a Lagrangian submanifold of Γ, the symplectic form ω of Γ is exact (near G) and we can write $\omega = \mathrm{d}\alpha$. Define m functions H_i, $i = 1, \ldots, m$ on Γ via the following integral formula, known as Arnold–Mineur's formula for action functions of integrable Hamiltonian systems (see Appendix A.4):

$$H_i(q) = \int_{\gamma_{i,q}} \alpha. \tag{7.46}$$

Recall from Theorem 7.5.5 that the "regular" part Γ_{reg} of Γ admits a natural symplectically complete foliation by isotropic submanifolds $K_q = q.G_{s(q)}$, and since Γ is proper, these submanifolds are compact. So this foliation is a proper non-commutatively integrable Hamiltonian system (see Appendix A.4). Since $\gamma_{i,q} \subset K_q \; \forall i$, it follows from the classical Arnold–Liouville–Mineur theorem on action-angle variables of integrable Hamiltonian systems (Theorem A.4.5) that H_i are action functions, i.e., the Hamiltonian vector fields X_{H_i} are periodic (of period 1) and generate $\mathbb{T}^1$-actions, and they are tangent to the isotropic submanifolds $K_q, q \in \Gamma$. This fact is true in Γ_{reg}, which is dense in Γ, so by continuity it's true in Γ.

By construction, the action functions H_i are invariant on the leaves of the dual coisotropic foliation of the foliation by K_g, $g \in \Gamma_{\mathrm{reg}}$, so they project to (independent) Casimir functions on P. In other words, we have m independent Casimir functions $h'_1, \ldots, h'_m$ such that $s^* h'_i = t^* h'_i = H_i$.

The infinitesimal action of $\mathfrak{g}$ on Γ generated by Hamiltonian vector fields $X_{s^* f_i}, X_{s^* h'_j}$, where the functions $f_1, \ldots, f_n$ are as before, now integrates into an action of $S \times \mathbb{T}^m$ on Γ, where S is the connected simply-connected semisimple Lie group with Lie algebra $\mathfrak{s}$. The group $S \times \mathbb{T}^m$ is a finite covering of G, i.e., we have an exact sequence $0 \to \mathcal{G} \to S \times \mathbb{T}^m \to G \to 0$, where $\mathcal{G}$ is a finite group. Indeed, by construction, for every element $g \in \mathcal{G} \subset S \times \mathbb{T}^m$, the action $\phi(g)$ of g on Γ is identity on the isotropy group G, and its differential at the neutral element $e \in G \subset \Gamma$ is also the identity map of $T_e\Gamma$. Since a finite power of $\phi(g)$ is the identity map on Γ, it follows that $\phi(g)$ itself is the identity map. Hence the action of $\mathcal{G}$ on Γ is trivial, and the action of $S \times \mathbb{T}^m$ on Γ factors to a Hamiltonian action of G on Γ. The proposition is proved for the case when G is connected.

Consider now the case G is disconnected. Denote by G^0 the connected component of G which contains the neutral element, and by Γ^0 the corresponding connected component of Γ^0 (we assume that the base U is connected and sufficiently small). Then Γ_0 is a proper symplectic groupoid over U whose isotropy group at 0 is G^0. According to the above discussion, Γ^0 can be locally symplectically linearized, i.e., we may assume that Γ^0 is symplectically isomorphic to $(T^*G^0 \rightrightarrows \mathfrak{g}^*)_U$ with the standard symplectic structure. Consider a map $\phi : \Gamma \to G$, whose restriction to the isotropy group $G = s^{-1}(0) \cap t^{-1}(0)$ is identity, and whose restriction to Γ^0 is given by the projection $T^*G^0 \cong G^0 \times \mathfrak{g}^* \to G^0$ after the above symplectic isomorphism from Γ^0 to $(G^0 \times \mathfrak{g}^* \rightrightarrows \mathfrak{g}^*)_U$. We can arrange it so that $\phi(p) = \phi(p^{-1})^{-1}$ for any $p \in \Gamma$, and also $\phi(p).\phi(q) = \phi(p.q)$ for any $p \in \Gamma^0, q \in \Gamma$. (This is possible because $\phi|_{\Gamma_0} : \Gamma_0 \to G^0$ is a homomorphism.) Then the averaging process used in the proof of Theorem 7.4.7 does not change the value of ϕ on Γ. By repeating the proof of Theorem 7.4.7, we get a homomorphism $\phi_\infty : \Gamma \to G$, which coincides with ϕ on Γ^0.

Identifying Γ with $G \times U$ via the isomorphism $p \mapsto (\phi_\infty(p), s(p))$ as in the proof of Theorem 7.4.7, and then with $(T^*G \rightrightarrows \mathfrak{g}^*)_U$, we will assume that Γ, as a Lie groupoid, is nothing but the restriction $(T^*G \rightrightarrows \mathfrak{g}^*)_U$ of the standard symplectic groupoid $T^*G \rightrightarrows \mathfrak{g}^*$ to $U \subset \mathfrak{g}^*$, and the symplectic structure ω on $(T^*G)_U \cong G \times U$ coincides with the standard symplectic structure ω_0 on the connected component $(T^*G^0)_U \cong G^0 \times U$. For each $\theta \in G/G^0$, we will denote the corresponding connected component of G by G^θ and the corresponding connected component of Γ by Γ^θ. We will use Moser's path method to find a groupoid isomorphism of Γ which moves ω to ω_0.

Let $f : U \to \mathbb{R}$ be a function on U. Then according to Theorem 7.5.5, the Hamiltonian vector fields $X^\omega_{s^*f}$ and $X^{\omega_0}_{s^*f}$ of s^*f with respect to ω and ω_0 are both invariant under left translations in Γ, and since they coincide in Γ^0 they must coincide in Γ, because any element in Γ can be left-translated from an element in Γ^0. So we have a common Hamiltonian vector field X_{s^*f} for both ω and ω_0. Similarly, we have a common Hamiltonian vector field X_{t^*f} for both ω and ω_0. It means that $i_X(\omega - \omega_0) = 0$ for any $X \in T_p s^{-1}(s(p)) + T_p t^{-1}(t(p))$, which implies that $\omega - \omega_0$ is a basic closed 2-form with respect to the coisotropic singular foliation whose leaves are connected components of the sets $s^{-1}(s(t^{-1}(t(p))))$, $p \in \Gamma$. In particular, for any connected component Γ^θ of Γ, where $\theta \in G/G^0$, there is a unique closed 2-form β_θ on U, which is basic with respect to the foliation by the orbits of the coadjoint action of G^0 on U, such that

$$\omega - \omega_0 = s^*\beta_\theta \quad \text{on } \Gamma^\theta. \tag{7.47}$$

The coadjoint action of G on U induces an action ρ of G/G^0 on the space of connected coadjoint orbits (orbits of G^0) on U: if $\mathcal{O}$ is a connected coadjoint orbit on U, then $\rho(\theta)(\mathcal{O})$ is the orbit $\mathrm{Ad}^*_{G^\theta}\mathcal{O}$. Since Γ is a symplectic groupoid with respect to both ω and ω_0, the closed 2-form $\omega - \omega_0$ is also compatible with the product map in Γ, i.e., Equation (7.38) is still satisfied if we replace ω by $\omega - \omega_0$.

By projecting this compatibility condition to U, we get the following equality:

$$\beta_{\theta_1\theta_2} = \beta_{\theta_2} + \rho(\theta_2)^*\beta_{\theta_1} \quad \forall \ \theta_1, \theta_2 \in G/G^0. \tag{7.48}$$

Since the 2-forms β_θ are closed on U which are basic with respect to the foliation by connected coadjoint orbits (i.e., orbits of the coadjoint action of G^0), we can write

$$\beta_\theta = \mathrm{d}\alpha_\theta, \tag{7.49}$$

where α_θ are 1-forms on U which are also basic with respect to the foliation by connected coadjoint orbits. Indeed, write $\beta_\theta = \mathrm{d}\hat{\alpha}_\theta$, then define α_θ by the averaging formula

$$\alpha_\theta = \int_{G^0} (\mathrm{Ad}_g^*)^* \hat{\alpha}_\theta \mathrm{d}\mu_{G^0}, \tag{7.50}$$

where μ_{G^0} is the Haar measure on G^0. Then α_θ is invariant with respect to the coadjoint action of G^0, and $\mathrm{d}\alpha_\theta = \beta_\theta$. One verifies easily that α_θ must automatically vanish on vector fields tangent to the coadjoint orbits, or otherwise β_θ would not be a basic 2-form.

Moreover, by averaging α_θ with respect to the action of G/G^0 via the formula

$$\alpha_\theta^{new} = \frac{1}{|G/G^0|} \sum_{\theta' \in G/G^0} (\alpha_{\theta'\theta} - \rho(\theta)^*\alpha_\theta), \tag{7.51}$$

we may assume that the 1-forms α_θ satisfy the equation

$$\alpha_{\theta_1\theta_2} = \alpha_{\theta_2} + \rho(\theta_2)^*\alpha_{\theta_1} \quad \forall \ \theta_1, \theta_2 \in G/G^0. \tag{7.52}$$

Consider the vector field Z on Γ defined by

$$s^*\alpha_\theta = i_Z\omega = i_Z\omega_0 \quad \text{on} \ \ \Gamma^\theta. \tag{7.53}$$

One verifies directly that the flow ϕ_Z^t of Z preserves the groupoid structure of Γ, and ϕ_Z^1 moves ω to ω_0. $\Box$

Corollary 7.5.13 ([370]). *If $(\Gamma \rightrightarrows P, \omega + \Omega)$ is a proper symplectic groupoid whose isotropy groups are connected, then the orbit space P/Γ is a manifold with locally polyhedral boundary. Moreover, this orbit space admits a natural integral affine structure, and near each point is locally affinely isomorphic to a Weyl chamber (with the standard affine structure) of a compact Lie group (namely the corresponding isotropy group).*

The reason is that the quotient of the dual of the Lie algebra of a *connected* compact Lie group by its coadjoint action can be naturally identified to a Weyl chamber. The affine structure is provided by action functions defined by Formula 7.46. In fact, the foliation of the "regular" part Γ_{reg} of Γ by tori $s^{-1}(x) \cap t^{-1}(y)$ turns it to a proper noncommutatively integrable system, the interior of the orbit space P/Γ is the reduced base space of this integrable system, and the action functions provide an integral affine structure on this reduced base space (see Appendix A.4 and [370]). The fact that this intrinsically defined affine structure coincides with the affine structures coming from the Weyl chambers via local linearizations of Γ will be left as an exercise.

When the isotropy groups are not connected, then locally P/Γ is isomorphic to a quotient of the Weyl chamber by a finite group action. The reason is that, if G is a disconnected group, then its coadjoint action on $\mathfrak{g}^*$ may mix the connected coadjoint orbits (orbits of the connected part G^0 of G) by an action of G/G^0.

Example 7.5.14. Consider the following disconnected double covering $G = \mathbb{T}^2 \sqcup \theta.\mathbb{T}^2$ of $\mathbb{T}^2$, where θ is an element such that $\theta.g.\theta^{-1} = g^{-1} \ \forall \ g \in \mathbb{T}^2$. Then the coadjoint action of $G^0 = \mathbb{T}^2$ on $\mathbb{R}^2 = Lie(G)^*$ is trivial, but the coadjoint action of θ on $\mathbb{R}^2$ is given by the map $(x, y) \mapsto (-x, -y)$. The quotient space of $\mathbb{R}^2$ by the coadjoint action of G is the orbifold $\mathbb{R}^2/\mathbb{Z}_2$.

7.5.3   Hamiltonian actions of symplectic groupoids

Definition 7.5.15. A *Hamiltonian action* of a symplectic groupoid $(\Gamma, \sigma) \rightrightarrows (P, \Pi)$ on a Poisson manifold (M, Λ) is a Lie groupoid action such that its graph $\{(g, y, z) \mid (g, y) \in \Gamma * M, z = g.y\}$ is a coisotropic submanifold of $\Gamma \times M \times \overline{M}$, where $\overline{M}$ denotes M with the opposite Poisson structure $-\Lambda$.

Remark that the dimension of the graph $\{(g, y, z) \mid (g, y) \in \Gamma * M, z = g.y\}$ is half the dimension of $\Gamma \times M \times \overline{M}$. In the case when M is symplectic, to say that this graph is coisotropic is the same as to say that it is isotropic (or Lagrangian). The coisotropic condition is a compatibility condition which may be expressed by a formula similar to Equation (7.38).

Example 7.5.16. The action of a symplectic groupoid (Γ, σ) on itself by multiplication is Hamiltonian. The natural action of Γ on its Poisson base space is also Hamiltonian.

Exercise 7.5.17. Show that if a symplectic groupoid $(\Gamma, \sigma) \rightrightarrows (P, \Pi)$ acts Hamiltonianly on a Poisson manifold (M, Λ), then the corresponding momentum map $\mu : (M, \Lambda) \to (P, \Pi)$ is a Poisson map.

Theorem 7.5.18 ([248]). *Let G be a connected Lie group. Then there is a natural one-to-one correspondence between Hamiltonian actions of the standard symplectic groupoid $T^*G \rightrightarrows \mathfrak{g}^*$ and Hamiltonian actions of G.*

Proof. The identification between Hamiltonian G-actions and Hamiltonian $(T^*G \rightrightarrows \mathfrak{g}^*)$-actions on a Poisson manifold (M, Λ) is given by the formula

$$(L_g\alpha).x = g.x, \tag{7.54}$$

where $g \in G, x \in (M, \Lambda)$ such that $\mu(x) = \alpha \in \mathfrak{g}^*$. The rest of the proof is a direct verification. $\qquad\square$

The above theorem means that one may view the theory of Hamiltonian actions of Lie groups as a particular case of the theory of Hamiltonian actions of symplectic groupoids.

Exercise 7.5.19. Describe Hamiltonian actions of a symplectic pair groupoid $(M, \sigma) \times (M, -\sigma) \rightrightarrows (M, \sigma)$ on Poisson manifolds.

Remark 7.5.20. Lu's theory of equivariant momentum maps for Poisson actions of a compact Poisson–Lie group can also be embedded into the theory of Hamiltonian actions of symplectic groupoids, see [223]. In fact, Lu's theory is in a sense equivalent to the usual theory of equivariant momentum maps of Hamiltonian actions of a compact Lie group, see [5].

7.5.4   Some generalizations

Motivated in part by the theories of *quasi-Hamiltonian spaces* and *twisted Poisson structures* (see, e.g., [8, 6, 7, 315, 197]), Xu [361] and Bursztyn–Crainic–Weinstein–Zhu [53] introduced the notion of *quasi-symplectic groupoids*, also called *twisted presymplectic groupoids*. A *quasi-symplectic groupoid* is a Lie groupoid $\Gamma \rightrightarrows P$, equipped with a 2-form ω on Γ and a 3-form Ω on P, which satisfy the following four conditions:

i) $d\omega = t^*\Omega - s^*\Omega$.

ii) $d\Omega = 0$.

iii) The graph $\Delta = \{(p, q, p.q) \mid p, q \in \Gamma, s(p) = t(q)\}$ of the product operation of Γ is isotropic with respect to the 2-form $\omega \oplus \omega \oplus (-\omega)$ on $\Gamma \times \Gamma \times \Gamma$.

iv) Identify P with its unity section $\varepsilon(P)$ in Γ. Due to condition iii), for each point $m \in P$, the differential t_* of the target map t can be restricted to a map

$$t_* : \ker \omega_m \cap T_m s^{-1}(m) \to \ker \omega_m \cap T_m P , \qquad (7.55)$$

where $\ker \omega_m$ denotes the kernel of ω at m, and the condition is that this restricted map is bijective.

The first three conditions mean that ω is a *twisted presymplectic form* and Ω is the twisting term, and the last condition is a weak nondegeneracy condition on ω. If ω is nondegenerate and $\Omega = 0$ then one gets back to the notion of symplectic groupoids. The base space P of a quasi-symplectic groupoid is a *twisted Dirac manifold*, and if the 2-form ω is nondegenerate then P is a *twisted Poisson manifold*, see [53, 65, 52]. For Dirac structures, see Appendix A.8.

It is easy to check that a sufficiently small slice of a (proper) quasi-symplectic groupoid is again a (proper) quasi-symplectic groupoid. A result of Xu (Proposition 4.8 of [361]) says that if $(\Gamma \rightrightarrows P, \omega + \Omega)$ is a quasi-symplectic groupoid, and β is an arbitrary 2-form on P, then $(\Gamma \rightrightarrows P, \omega' + \Omega')$, where $\omega' = \omega + t^*\beta - s^*\beta$ and $\Omega' = \Omega + d\beta$, is again a quasi-symplectic groupoid, and moreover it is Morita-equivalent in a natural sense to $(\Gamma \rightrightarrows P, \omega + \Omega)$. This result together with Theorem 7.5.11 lead to the following proposition.

Proposition 7.5.21 ([370]). *If $(\Gamma \rightrightarrows P, \omega + \Omega)$ is a proper quasi-symplectic groupoid with a fixed point m, then it is locally isomorphic to a quasi-symplectic groupoid of the type $(T^*G \rightrightarrows \mathfrak{g}^*, \omega_0 + t^*\beta - s^*\beta + \mathrm{d}\beta)$, where $(T^*G \rightrightarrows \mathfrak{g}^*, \omega_0)$ is the standard symplectic groupoid of the isotropy group $G = \Gamma_m$ of m, and β is a 2-form on $\mathfrak{g}^*$. If, moreover, the isotropy groups of Γ are connected, then the orbit space P/Γ is an integral affine manifold with locally convex polyhedral boundary which locally looks like a Weyl chamber.*

The (local) convexity of orbit spaces P/Γ of symplectic and quasi-symplectic groupoids is very closely related to convexity properties of momentum maps in symplectic geometry, see [370].

Example 7.5.22. Consider the AMM (Alekseev–Malkin–Meinrenken) groupoid [361]: it is the transformation groupoid $G \times G \rightrightarrows G$ of the conjugation action of a compact Lie group G, equipped with a natural quasi-symplectic structure arising from Alekseev–Malkin–Meinrenken's theory of group-valued momentum maps [8]. Xu [361] developed a theory of quasi-Hamiltonian actions of quasi-symplectic groupoids, and showed a natural equivalence between quasi-Hamiltonian spaces with G-valued momentum maps and quasi-Hamiltonian spaces of the AMM groupoid. The orbit space of the AMM groupoid is naturally affine-equivalent to a Weyl alcove of G. In particular it is a convex affine polytope.

There is another interesting class of groupoids, called *Poisson groupoids* , introduced by Weinstein [350]. The definition of a Poisson groupoid is similar to the definition of a symplectic groupoid, except for the fact that the arrow space Γ is now equipped with a Poisson instead of a symplectic structure, and the graph of the multiplication map is now required to be coisotropic instead of Lagrangian. For example, the pair groupoid of a Poisson manifold is a Poisson groupoid. Poisson groupoids generalize at the same time symplectic groupoids and Poisson–Lie groups, and play an important role in Poisson geometry. They are, unfortunately, out of the scope of this book. See, e.g., [229, 230, 231, 228] and references therein. Just as Poisson–Lie groups are related to r-matrices, Poisson groupoids are very closely related to so-called *dynamical r-matrices*, see, e.g., [125] and references therein.

Chapter 8

Lie Algebroids

8.1 Some basic definitions and properties

8.1.1 Definition and some examples

Lie algebroids were introduced by Pradines [294] as infinitesimal versions of Lie groupoids.

Definition 8.1.1. A *Lie algebroid* $(A \to M, [,], \sharp)$ is a (finite-dimensional) vector bundle $A \to M$ over a manifold M, equipped with a linear bundle map $\sharp : A \to TM$ called the *anchor map*, and a Lie bracket $[,]$ on the space $\Gamma(A)$ of sections of A, such that the following Leibniz rule is satisfied:

$$[\alpha, f\beta] = (\sharp\alpha(f))\beta + f[\alpha, \beta] \tag{8.1}$$

for any sections α, β of A and function f on M.

We will often denote a Lie algebroid simply by a letter A, or by $(A, [,], \sharp)$.

Remark 8.1.2. The above definition makes sense in many categories: formal, smooth, real analytic, holomorphic, etc. (In the real analytic and holomorphic cases, $\Gamma(A)$ should be replaced by the sheaf of local analytic sections of A.) It also makes sense when M is a manifold with boundary.

Remark 8.1.3. Due to the Leibniz rule, the Lie bracket is a bi-differential operator of first order in each variable. In other words, the value of $[\alpha, \beta]$ at a point x depends only on the value of α, β and their first derivatives at x. As a consequence, the restriction of a Lie algebroid over M to an open subset of M is again a Lie algebroid. We can also talk about a germ of a Lie algebroid at a point $x \in M$.

Lemma 8.1.4. *If $(A, [,], \sharp)$ is a Lie algebroid, then the anchor map $\sharp$ is a Lie algebra homomorphism:*

$$\sharp[\alpha, \beta] = [\sharp\alpha, \sharp\beta] \quad \forall \; \alpha, \beta \in \Gamma(A). \tag{8.2}$$

Proof. By the Jacobi identity and the Leibniz rule, we have

$$
\begin{aligned}
0 &= [[\alpha, \beta], f\gamma] + [[\beta, f\gamma], \alpha] + [[f\gamma, \alpha], \beta] \\
&= f[[\alpha, \beta], \gamma] + (\sharp[\alpha, \beta](f))\gamma \\
&\quad + f[[\beta, \gamma], \alpha] - (\sharp\alpha(f))[\beta, \gamma] + (\sharp\beta(f))[\gamma, \alpha] - (\sharp\alpha(\sharp\beta(f)))\gamma \\
&\quad + f[[\gamma, \alpha], \beta] - (\sharp\beta(f))[\gamma, \alpha] - (\sharp\alpha(f))[\gamma, \beta] + (\sharp\beta(\sharp\alpha(f)))\gamma \\
&= ((\sharp[\alpha, \beta] - [\sharp\alpha, \sharp\beta])(f))\gamma.
\end{aligned}
$$

Since this is true for any $\alpha, \beta, \gamma \in \Gamma(A)$ and function f, one concludes that $\sharp[\alpha, \beta] = [\sharp\alpha, \sharp\beta]$. $\qquad\square$

Remark 8.1.5. Condition (8.2) is a fundamental property of Lie algebroids, and is often considered as a part of the definition of a Lie algebroid, though it is a consequence of the other conditions. Lemma 8.1.4 is implicit in [173]. See, e.g., [154] for a discussion on the axioms of Lie algebroids. If, instead of the Lie bracket, one imposes other algebraic structures on $\Gamma(A)$ and other compatibility conditions, then one arrives at other kinds of algebroids.

Remark 8.1.6. *Isomorphisms* of Lie algebroids can be defined in an obvious way. It is more tricky to define *morphisms* of Lie algebroids, which will be considered in Section 8.3

Example 8.1.7. A Lie algebra can be thought of as a Lie algebroid over a point.

Example 8.1.8. If M is a manifold then TM is a Lie algebroid: the anchor map is the identity map, and the Lie bracket is the usual Lie bracket of vector fields. This is called the *tangent algebroid* of M. More generally, if $\mathcal{F}$ is a regular foliation in M, then the *tangent algebroid* of $\mathcal{F}$ is the vector sub-bundle of TM consisting of tangent spaces to $\mathcal{F}$, with the usual Lie bracket, and the inclusion map as the anchor. Any Lie algebroid whose anchor map is injective is isomorphic to the tangent algebroid of some regular foliation.

Example 8.1.9. If Y is a hypersurface in a manifold M, then according to Melrose [246], there is a vector bundle $T(M, Y)$ over M, called the Y-*tangent bundle of* M, such that the space $\mathcal{V}_Y^1(M)$ of smooth vector fields on M which are tangent to Y is isomorphic to the space of smooth sections of $T(M, Y)$. The Lie bracket on $\mathcal{V}_Y^1(M)$ turns $T(M, Y)$ into a Lie algebroid.

Example 8.1.10. If $\xi : \mathfrak{g} \to \mathcal{V}^1(M)$ is an action of a Lie algebra $\mathfrak{g}$ on a manifold M, then we can associate to it the following *transformation algebroid*: the vector bundle is the trivial bundle $\mathfrak{g} \times M \to M$, the anchor map is $\sharp(X, z) = \xi(X)(z) \; \forall \, X \in \mathfrak{g}, z \in M$, and the Lie bracket on the sections of $\mathfrak{g} \times M \to M$, considered as maps from M to $\mathfrak{g}$, is defined as

$$
[\alpha, \beta](z) = [\alpha(z), \beta(z)] + (\xi(\alpha(z)))_z(\beta) - (\xi(\beta(z)))_z(\alpha). \tag{8.3}
$$

In particular, if α, β are two constant sections then their bracket is a constant section given by the Lie bracket of $\mathfrak{g}$. The last two terms in the above formula are due to the Leibniz rule. We will denote the transformation algebroid of an action of $\mathfrak{g}$ on M as $\mathfrak{g} \ltimes M$ (it is a kind of semi-direct product in the category of Lie algebroids). For a linear action of $\mathfrak{g}$ on a vector space V, to avoid confusion with the Lie algebra $\mathfrak{g} \ltimes V$, we will denote the corresponding transformation algebroid as $\mathfrak{g} \dot{\ltimes} V$.

Example 8.1.11. If the anchor map is identically zero, then the Lie bracket on $\Gamma(A)$ is a point-wise Lie bracket: due to the equality $[\alpha, f\beta] = f[\alpha, \beta]$, the value of $[\alpha, \beta]$ over a point $z \in M$ depends only on the value of α and β over z. This is the case of a *bundle of Lie algebras* over M.

Example 8.1.12. Attached to each Poisson manifold (M, Π), there is a natural Lie algebroid structure $(T^*M, [,], \sharp)$ on the cotangent bundle of M, whose anchor map $\sharp : T^*M \to TM$ is the usual anchor map of Π, $\langle \sharp\alpha, \beta \rangle = \Pi(\alpha, \beta) \; \forall \; \alpha, \beta \in \Omega^1(M)$, and whose Lie bracket is

$$
\begin{aligned}
[\alpha, \beta] &= \mathcal{L}_{\sharp\alpha}\beta - \mathcal{L}_{\sharp\beta}\alpha - \mathrm{d}\Pi(\alpha, \beta) \\
&= \mathrm{d}(\Pi(\alpha, \beta)) + i_{\sharp\alpha}\mathrm{d}\beta - i_{\sharp\beta}\mathrm{d}\alpha.
\end{aligned}
\tag{8.4}
$$

The above Lie bracket on differential 1-forms of (M, Π) probably first appeared in the work of Fuchssteiner [139]. It is immediate that this bracket satisfies the Leibniz rule:

$$
\begin{aligned}
[\alpha, f\beta] &= \mathrm{d}(f(\Pi(\alpha, \beta))) + i_{\sharp\alpha}\mathrm{d}(f\beta) - f i_{\sharp\beta}\mathrm{d}\alpha \\
&= f[\alpha, \beta] + \Pi(\alpha, \beta)\mathrm{d}f + i_{\sharp\alpha}(\mathrm{d}f \wedge \beta) \\
&= f[\alpha, \beta] + ((\sharp\alpha)(f))\beta.
\end{aligned}
$$

Let us verify that this bracket satisfies the Jacobi identity. If $\alpha = \mathrm{d}f, \beta = \mathrm{d}g, \gamma = \mathrm{d}h$ are exact 1-forms, then by definition $[\mathrm{d}f, \mathrm{d}g] = \mathrm{d}\{f, g\}$ and so on, and the Jacobi identity for the triple $(\mathrm{d}f, \mathrm{d}g, \mathrm{d}h)$ follows from the Jacobi identity for the triple (f, g, h) with respect to the Poisson bracket. For more general 1-forms, one can use the Leibniz rule to reduce to the case of exact 1-forms. $(T^*M, [,], \sharp)$ is called the *cotangent algebroid* of (M, Π).

Exercise 8.1.13. Show that a Lie algebroid structure $([,], \sharp)$ on T^*M comes from a Poisson structure on M if and only if $\sharp$ is anti-symmetric (i.e., $\langle \sharp\alpha, \beta \rangle = -\langle \sharp\beta, \alpha \rangle$ for any two 1-forms α, β on M) and the bracket of two arbitrary closed 1-forms is again a closed 1-form.

8.1.2   The Lie algebroid of a Lie groupoid

Let $\Gamma \rightrightarrows M$ be a Lie groupoid over a manifold M, with source map s and target map t. A tangent vector field X on Γ is called a *left-invariant vector field* if it satisfies the following two properties: i) X is tangent to t-fibers $t^{-1}(x)$ ($x \in M$). ii) X is invariant under the left action of Γ on itself: for each $g \in \Gamma$, the left translation $L_g : h \mapsto g.h$, which maps $t^{-1}(s(g))$ to $t^{-1}(t(g))$, preserves X.

Denote the space of smooth left-invariant vector fields on Γ by $\mathcal{V}_L(\Gamma)$. It is clear the Lie bracket of two left-invariant vector fields is again a left-invariant vector field:

$$[\mathcal{V}_L(\Gamma), \mathcal{V}_L(\Gamma)] \subset \mathcal{V}_L(\Gamma). \tag{8.5}$$

Denote by A the vector bundle over M consisting of tangent spaces to t-fibers at M:

$$A_x = T_x(t^{-1}(x)). \tag{8.6}$$

By left translations, each section of A gives rise to a unique left-invariant vector field on G. In other words, the space of smooth sections $\Gamma(A)$ of A may be identified with $\mathcal{V}_L(\Gamma)$, and therefore it inherits the Lie bracket from $\mathcal{V}_L(\Gamma)$. The anchor map $\sharp : A \to TM$ is defined to be the (restriction to A of the) differential of the source map $s : \Gamma \to M$. This anchor map can also be defined as follows: identify $C^\infty(M)$ with the space $C_L^\infty(\Gamma)$ of left-invariant functions on Γ. For each $X \in \mathcal{V}_L(\Gamma)$ and $f \in C_L^\infty(\Gamma)$, one has $X(f) \in C_L^\infty(\Gamma)$. This way sections of A can be mapped to derivations of $C^\infty(M)$, i.e., vector fields on M. One verifies directly that, equipped with the above Lie bracket and anchor map, A becomes a Lie algebroid, called the *Lie algebroid of the Lie groupoid* $\Gamma \rightrightarrows M$, and sometimes denoted by $Lie(\Gamma)$.

Remark 8.1.14. Some authors use right-invariant vector fields tangent to s-fibers to define the Lie algebroid of a Lie groupoid. The resulting Lie algebroid is the same as the one given by left-invariant vector fields, up to an isomorphism.

Example 8.1.15. If $G \ltimes M$ is the transformation groupoid of a smooth action of a Lie group G on a manifold M, then its Lie algebroid is the transformation algebroid $\mathfrak{g} \ltimes M$ of the corresponding Lie algebra action $\mathfrak{g} \to \mathcal{V}^1(M)$.

Exercise 8.1.16. Show that the Lie algebroid of a smooth pair groupoid $M \times M \rightrightarrows M$ is isomorphic to the tangent algebroid of M.

Exercise 8.1.17. Show that the Lie algebroid of a symplectic groupoid $(\Gamma, \sigma) \rightrightarrows (P, \Pi)$ is isomorphic to the cotangent algebroid of (P, Π).

8.1.3 Isotropy algebras

Let $(A, [,], \sharp)$ be a Lie algebroid over a manifold M. For a point $z \in M$, we will denote by A_z the fiber of A over z, and by $\ker \sharp(z)$ the kernel of the anchor map

$$\sharp_z : A_z \to T_z M. \tag{8.7}$$

The kernel $\ker \sharp(z)$ has a natural Lie algebra structure, defined as follows. For any $\alpha_z, \beta_z \in \ker \sharp(z)$, denote by α, β arbitrary sections of A whose value at z is α_z and β_z respectively, and put

$$[\alpha_z, \beta_z] = [\alpha, \beta](z). \tag{8.8}$$

The above bracket on $\ker \sharp(z)$ is well defined. Indeed, if $\widetilde{\alpha}$ is another section of A with $\widetilde{\alpha}(z) = \alpha_z$, then locally there are functions $f_1, \ldots, f_n$ on M which vanish at z and a basis $\alpha_1, \ldots, \alpha_n$ of A such that $\widetilde{\alpha} - \alpha = \sum f_i \alpha_i$. By the Leibniz rule, we have $[\widetilde{\alpha}, \beta](z) - [\alpha, \beta](z) = \sum f_i(z)[\alpha_i, \beta](z) - \sum (\sharp \beta)(f_i)(z)\alpha_i(z) = 0$, because $f_i(z) = 0$ and $\sharp \beta(z) = 0$.

The kernel $\ker \sharp(z)$ together with its natural Lie bracket is called the *isotropy algebra* of A at z.

8.1.4   Characteristic foliation of a Lie algebroid

The distribution $\mathcal{C} : x \in M \mapsto \mathcal{C}_x = \operatorname{Im} \sharp_x$, where $\sharp$ is the anchor map of a Lie algebroid $(A, [,], \sharp)$, is called the *characteristic distribution* of A. The dimension of $\operatorname{Im} \sharp_x$ is called the rank of $\sharp$ (or of A) at x. If A is smooth then its characteristic distribution is a smooth distribution generated by vector fields of the type $\sharp \alpha$, where α is a smooth section of A. Similarly to the case of Poisson manifolds (see Remark 8.5.6), it is an integrable distribution. The corresponding singular foliation is called the *characteristic foliation* of A. Its leaves are called *leaves* or *orbits* of A.

Exercise 8.1.18. Show that, if A is the Lie algebroid of a Lie groupoid $\Gamma \rightrightarrows M$, then the orbits of A are the same as the orbits of Γ, and the isotropy algebra of A at a point $x \in M$ is the Lie algebra of the isotropy group of Γ at x.

If O is an orbit of a Lie algebroid A, then the restriction A_O of A to O also is a Lie algebroid: If α, β are two sections of A over O, and $\widetilde{\alpha}, \widetilde{\beta}$ are their extensions to sections of A over M, then we can define the bracket of α with β as the restriction of $[\widetilde{\alpha}, \widetilde{\beta}]$ to O. The Leibniz rule implies that this bracket does not depend on the choice of the extensions $\widetilde{\alpha}, \widetilde{\beta}$. Note that A_O is a *transitive Lie algebroid*, i.e., its anchor map is surjective.

Remark 8.1.19. There is a natural question: can *every* singular foliation be realized, at least locally, as the characteristic foliation of a Lie algebroid? We don't know the answer to this question.

8.1.5   Lie pseudoalgebras

The purely algebraic version of a Lie algebroid is often called a Lie pseudoalgebra. By definition, a *Lie pseudoalgebra* is a pair (L, C), where C is a commutative algebra over a commutative ring R, and L is a Lie algebra over R, which is a C-module and which acts on C by derivations, i.e., $\forall \alpha, \beta \in L, f, g \in C$,

$$\alpha.(fg) = f(\alpha.g) + g(\alpha.f) \tag{8.9}$$

and

$$[\alpha, \beta].f = \alpha.(\beta.f) - \beta.(\alpha.f), \tag{8.10}$$

such that the following compatibility condition (Leibniz rule) is satisfied

$$[\alpha, f\beta] = f[\alpha, \beta] + (\alpha.f)\beta. \tag{8.11}$$

Clearly, if A is a smooth Lie algebroid over M then $(\Gamma(A), C^\infty(M))$ is a Lie pseudoalgebra. Many algebraic constructions involving Lie algebroids can in fact be made for Lie pseudoalgebras.

Remark 8.1.20. Lie pseudoalgebras are also called by many other names by different authors (see [227]): *Lie d-ring* [290], *Lie–Cartan pair* [192], *Lie–Rinehart algebra* [180], *differential Lie algebra* [199], etc. They were probably first studied by Herz [173] and Rinehart [301], before the appearance of Lie algebroids.

8.2   Fiber-wise linear Poisson structures

Let $\pi : E \to M$ be a vector bundle. A *basic function* on E is a function of the type $g \circ \pi$ where g is a function on M. A *fiber-wise linear function* on E is a function whose restriction to each fiber of E is linear.

Definition 8.2.1. Let $\pi : E \to M$ be a vector bundle. A Poisson structure Π on E is called a *fiber-wise linear Poisson structure* if it satisfies the following three conditions:

(i)   The Poisson bracket of any two basic functions is zero.

(ii) The Poisson bracket of a basic function and a fiber-wise linear function is a basic function.

(iii) The Poisson bracket of two fiber-wise linear functions is a fiber-wise linear function.

Recall that there is a natural one-to-one correspondence between finite-dimensional Lie algebras and linear Poisson structures. In this section, we will show a similar natural one-to-one correspondence between Lie algebroids and fiber-wise linear Poisson structures.

Let $(A, [,], \sharp)$ be a Lie algebroid over a manifold M. We will construct a fiber-wise linear bracket on the total space of the dual bundle A^* of A as follows:

$$\{f, g\} = 0, \tag{8.12}$$

$$\{\alpha, f\} = (\sharp\alpha)(f), \tag{8.13}$$

$$\{\alpha, \beta\} = [\alpha, \beta]. \tag{8.14}$$

Here f, g are functions on M, considered as basic functions on A^*, α, β are sections of A, considered as fiber-wise linear functions on A^*. On the right-hand side, $(\sharp\alpha)(f)$ is considered as a basic function on A^*, and $[\alpha, \beta]$ is considered as a fiber-wise linear function. This bracket can be extended to other functions on A^* by the Leibniz rule. In a local coordinate system $(\alpha_1, \ldots, \alpha_n, x_1, \ldots, x_m)$ on A^*, where $(x_1, \ldots, x_m)$ is a local coordinate system on M and $(\alpha_1, \ldots, \alpha_n)$ is a basis of local sections of A, the corresponding 2-vector field Π on A^* can be written as follows:

$$\Pi = \frac{1}{2}\sum_{i,j}[\alpha_i, \alpha_j]\frac{\partial}{\partial\alpha_i} \wedge \frac{\partial}{\partial\alpha_j} + \sum_{i,j}(\sharp\alpha_i)(x_j)\frac{\partial}{\partial\alpha_i} \wedge \frac{\partial}{\partial x_j}. \tag{8.15}$$

Let us show that the above bracket satisfies the Jacobi identity, i.e., it is a Poisson bracket. By the Leibniz rule, the verification of the Jacobi identity is reduced to a verification in the following four cases: 1) all the three functions are fiber-wise linear; 2) two of them are fiber-wise linear, the third one is basic; 3) two of them are basic, the third one is fiber-wise linear; 4) all three functions are basic. In the first case, the Jacobi identity is nothing else but the Jacobi identity for the Lie bracket $[,]$ on $\Gamma(A)$. In the second case, it is a consequence of the fact that the anchor map is a Lie algebra homomorphism:

$$\{\{\alpha, \beta\}, f\} = \{[\alpha, \beta], f\} = [\sharp\alpha, \sharp\beta](f) = \{\alpha, \{\beta, f\}\} - \{\beta, \{\alpha, f\}\}.$$

The third and the forth cases are trivial. Thus, we have shown that the above bracket is a Poisson bracket. And by construction, this Poisson bracket is fiber-wise linear on A^*.

Conversely, given a fiber-wise linear Poisson structure Π on A^*, Formulas (8.13) and (8.14) will define a Lie algebroid structure on A: the Jacobi identity for $\Gamma(A)$ and the Leibniz rule are just particular cases of the Jacobi identity for Π. In other words, we have proved the following result.

Theorem 8.2.2. *There is a natural one-to-one correspondence between Lie algebroids and fiber-wise linear Poisson structures, given by Formulas (8.12)–(8.14).*

Example 8.2.3. The linear Poisson structure on the dual of a Lie algebra is a particular case of the preceding construction.

When studying isomorphisms and infinitesimal automorphisms of Lie algebroids, it may be convenient to view them as fiber-wise linear Poisson structures. A Poisson isomorphism $\phi: E_1 \to E_2$ between a fiber-wise linear Poisson structure Π_1 on a vector bundle E_1 and a fiber-wise linear Poisson structure Π_2 on a vector bundle E_1 is called *fiber-wise linear* if it is a vector bundle isomorphism. Similarly, a Poisson vector field on a vector bundle E with a fiber-wise linear Poisson structure Π is called *fiber-wise linear* if it is an infinitesimal vector bundle isomorphism of E.

A vector field on (the total space of) a Lie algebroid A is called an *infinitesimal automorphism* of A, if its local flow maps fibers to fibers linearly (i.e., it is an infinitesimal vector bundle isomorphism), and preserves the Lie bracket on $\Gamma(A)$ and the anchor map.

Exercise 8.2.4. Show that there is a natural one-to-one correspondence between Lie algebroid isomorphisms and fiber-wise linear Poisson isomorphisms. Similarly, there is a natural one-to-one correspondence between infinitesimal automorphisms of a Lie algebroid A and fiber-wise linear Poisson vector fields on A^*.

Example 8.2.5. Let ξ be a section of a Lie algebroid A. Then ξ may be viewed as a fiber-wise linear function on A^*. Its corresponding Hamiltonian vector field X_ξ is fiber-wise linear, and generates a (local) fiber-wise linear flow of Poisson isomorphisms of A^*, and hence a flow (of automorphisms) on A by duality. More generally, ξ may be a time-dependent section (i.e., a family of sections dependent on a time parameter), then it still generates a flow of automorphisms on A.

One may also be tempted to define morphisms (and not just isomorphisms) from a Lie algebroid A_1 to a Lie algebroid A_2 via fiber-wise linear Poisson morphisms from A_2^* to A_1^*. This actually leads to the notion of *comorphisms* of Lie algebroids [174]. Note that in general these comorphisms are *not* the infinitesimal version of Lie groupoid morphisms.

8.3   Lie algebroid morphisms

Definition 8.3.1. A *Lie algebroid morphism* from a Lie algebroid $(A \to M, [,], \sharp)$ to a Lie algebroid $(A' \to M', [,]', \sharp')$ is a vector bundle morphism $\phi : A \to A'$ (i.e., a smooth fiber-wise linear map), which is compatible with the anchor maps and the Lie brackets.

Denote the projection of ϕ to M by the same letter $\phi : M \to M'$. The compatibility of ϕ with the anchor maps means that

$$\sharp(\phi(\alpha)) = (\phi)_*(\sharp\alpha) \ \forall \ \alpha \in A. \tag{8.16}$$

The compatibility of ϕ with the Lie brackets means the following, according to Higgins and Mackenzie [174]: for any smooth sections α, β of A with decompositions

$$\phi \circ \alpha = \sum_i f_i(\alpha_i' \circ \phi), \ \phi \circ \beta = \sum_i g_i(\beta_i' \circ \phi), \tag{8.17}$$

where f_i, g_i are functions on M and α_i', β_i' are sections of A', we have

$$\phi \circ [\alpha, \beta] = \sum_{ij} f_i g_j([\alpha_i', \beta_j'] \circ \phi) + \sum_j (\sharp\alpha(g_j))(\beta_j' \circ \phi) - \sum_j (\sharp\beta(f_j))(\alpha_j' \circ \phi). \tag{8.18}$$

In particular, if there are smooth sections α', β' of A' such that

$$\phi \circ \alpha = \alpha' \circ \phi, \ \phi \circ \beta = \beta' \circ \phi, \tag{8.19}$$

then we also have

$$\phi \circ [\alpha, \beta] = [\alpha', \beta'] \circ \phi. \tag{8.20}$$

Exercise 8.3.2. Show that, if α and β are given, then the right-hand side of (8.18) does not depend on the choice of decompositions of $\phi \circ \alpha$ and $\phi \circ \beta$ in (8.17).

Theorem 8.3.3 ([174]). *There is a natural functor from the category of Lie groupoids to the category of Lie algebroids, which associates to each Lie groupoid morphism a Lie algebroid morphism.*

Proof. Let $\phi : (\Gamma_1 \rightrightarrows M_1) \to (\Gamma_2 \rightrightarrows M_2)$ be a Lie groupoid morphism. For $i = 1, 2$, identify the Lie algebroid $A_i = Lie(\Gamma_i)$ of Γ_i with the vector bundle over M_i whose fiber over $x \in M_i$ is the tangent space to $t^{-1}(x)$ at x (see Subsection 8.1.2). Then the map $\phi_* : T\Gamma_1 \to T\Gamma_2$ induces a map, denoted by the same symbol ϕ_*, from $Lie(\Gamma_1)$ to $Lie(\Gamma_2)$. One verifies directly that

$$\phi_* : Lie(\Gamma_1) \to Lie(\Gamma_2)$$

satisfies the conditions of a Lie algebroid morphism. The functoriality of the construction is obvious. □

Example 8.3.4. If $(A, [,], \sharp)$ is a Lie algebroid over M, then the anchor map $\sharp : A \to TM$ is a Lie algebroid morphism from A to the tangent algebroid of M.

Example 8.3.5. Let $(A, [,], \sharp)$ be a Lie algebroid over M and z a point of M. Then the inclusion map from the isotropy algebra $\ker \sharp_z$ of z to A is a Lie algebroid morphism.

Example 8.3.6. Let L be a leaf of the characteristic foliation of a Lie algebroid A. Then there is a short exact sequence of Lie algebroid morphisms

$$0 \to \ker_L \to A_L \to TL \to 0, \tag{8.21}$$

where A_L is the restriction of A to L; $\ker_L$ denotes the totally intransitive Lie algebroid consisting of isotropy Lie algebras of A over L (ker means the kernel of the anchor map).

Example 8.3.7. Given two Lie algebroids A_1 over M_1 and A_2 over M_2, we can define their *direct product* to be the Lie algebroid $A_1 \times A_2$ which is dual to the direct product $A_1^* \times A_2^*$ of fiber-wise linear Poisson manifolds. The natural projections from $A_1 \times A_2$ to A_1 and A_2 are Lie algebroid morphisms.

Example 8.3.8. A smooth A-*path* of a Lie algebroid A is a Lie algebroid morphism from TI to A, where TI means the tangent algebroid of the interval $I = [0,1]$. It can be characterized as a path $\gamma : I \to A$ in A, which projects to a base path $\pi \circ \gamma : I \to M$ in M, where $\pi : A \to M$ denotes the projection, such that

$$\sharp(\gamma(t)) = \frac{\mathrm{d}}{\mathrm{d}t}\pi(\gamma(t)) \ \ \forall \, t \in I. \tag{8.22}$$

Exercise 8.3.9. Show that a Lie algebroid morphism between two tangent algebroids TN_1 and TN_2 is the lifting of a smooth map between the manifolds N_1 and N_2.

Remark 8.3.10. There is also a notion of *comorphisms* of Lie groupoids, which is a global version of comorphisms of Lie algebroids, see [175].

8.4 Lie algebroid actions and connections

Actions of Lie algebroids of manifolds are defined similarly to actions of Lie groupoids.

Definition 8.4.1. A *Lie algebroid action* of a Lie algebroid $(A \to M, [,], \sharp)$ on a manifold N consists of a map $\mu : N \to M$ called the *moment map* (or *momentum map*), and a Lie algebra homomorphism $\rho : \Gamma(A) \to \mathcal{V}^1(N)$ from the Lie algebra of sections of A to the Lie algebra of vector fields on N, which satisfies the following compatibility conditions: if f is a function on M and α is a section of A then

$$\rho(f\alpha) = (\mu^* f)\rho(\alpha) \tag{8.23}$$

and

$$\sharp\alpha(\mu(y)) = \mu_*(\rho(\alpha)(y)) \ \forall \ y \in N. \tag{8.24}$$

Exercise 8.4.2. Show that if a groupoid $\mathcal{G}$ acts on a manifold N then it induces an action of the Lie algebroid $A = \text{Lie}(\mathcal{G})$ on N.

Example 8.4.3. $(A \to M, [,], \sharp)$ acts on M in an obvious way: the moment map is the identity map, and the action map $\rho : \Gamma(A) \to \mathcal{V}^1(M)$ is the anchor.

Exercise 8.4.4. (See [174].) Let $\rho : \Gamma(A) \to \mathcal{V}^1(N)$ be an action of A on N with moment map $\mu : N \to M$. Show that the pull-back $\mu^* A$ of the bundle $A \to M$ by the map μ has a natural Lie algebroid structure over N.

A special case of Lie algebroid actions is when $N = E$ is a vector bundle over M, the moment map $\mu : E \to M$ is the projection map, and the vector fields $\rho(\alpha)$, $\alpha \in \Gamma(A)$, are fiber-wise linear (i.e., they generate local linear isomorphisms of the vector bundle E). In this case we say that we have a *linear representation* of A, or a *linear A-module*. It is clear that a linear module of a Lie groupoid $\mathcal{G}$ is also a linear module of its Lie algebroid $Lie(\mathcal{G})$.

Exercise 8.4.5. Let $E \to M$ be a vector bundle, and denote by $gl(E) = \text{Lie}(GL(E))$ the Lie algebroid of the Lie groupoid of linear isomorphisms among the fibers of E (see Exercise 7.1.21). Show that a linear action of a Lie algebroid $(A \to M, [,], \sharp)$ on E is the same as a Lie algebroid morphism from A to $gl(E)$ which projects to the identity map on M.

Exercise 8.4.6. Show that a linear representation of a tangent algebroid TM on a vector bundle E over M is the same as a flat linear connection on E.

The above exercise suggests that linear representations of general Lie algebroids may be viewed as flat connections in an appropriate sense.

Definition 8.4.7 ([131]). If E is a vector bundle over the base space M of a Lie algebroid A, then by a *A-connection* on E we mean a linear map (the *covariant derivative* map)

$$\Gamma(A) \times \Gamma(E) \to \Gamma(E), \ (\alpha, \xi) \mapsto \nabla_\alpha \xi \tag{8.25}$$

such that $\nabla_{f\alpha}\xi = f\nabla_\alpha \xi$ and $\nabla_\alpha(f\xi) = f\nabla_\alpha \xi + \sharp\alpha(f)\xi$ for any function f on M, section α of A and section ξ of E.

Remark 8.4.8. The above notion of A-connections generalizes the usual notion of linear connections on vector bundles (the case $A = TM$). When $A = T^*M$ is the cotangent algebroid of a Poisson manifold, a T^*M-connection is also known as a *contravariant connection*, see, e.g., [333, 130]. One can also define A-connections on principal bundles in a similar way [131].

Similarly to the case of usual linear connections, each A-connection is determined by a corresponding A-*horizontal lifting*: it is a $C^\infty(M)$-linear map

$$\text{hor} : \Gamma(A) \to \mathcal{V}^1_{\text{linear}}(E) \tag{8.26}$$

from the space of sections of A to the space of fiber-wise linear vector fields on E such that

$$\pi_* \, \text{hor}(\alpha) = \sharp\alpha \tag{8.27}$$

for any $\alpha \in \Gamma(A)$. The word *horizontal* is not totally correct, because if an element $p \in A$ lies in the kernel of $\sharp$, then hor(p) is actually a vertical vector in E which may be nontrivial, but we will use it anyway. Given an A-horizontal lifting, the corresponding A-covariant derivative can be defined by the usual formula

$$\nabla_\alpha \xi = \lim_{\delta \to 0} \frac{\phi^\delta_{\text{hor}(\alpha)}\xi - \xi}{\delta}, \tag{8.28}$$

where $\phi^\delta_{\text{hor}(\alpha)}$ is the (local) fiber-wise linear flow of time δ generated by the vector field hor(α).

Another equivalent way to define a connection is via parallel transport. Given an A-path a with base bath $\gamma : I \to M$ (see Example 8.3.8), and a point $u \in E_{\gamma(0)}$, the *parallel transport* of u along a is the path $u(t) \in E_{\gamma(t)}$ defined by the formula

$$u(t) = \phi^t_{\text{hor}(\hat{\alpha}_t)} u \,, \tag{8.29}$$

where $\hat{\alpha}_t$ is a time-dependent section of A such that $\hat{\alpha}_t(\gamma_t) = \alpha(t) \, \forall \, t$, and $\phi^t_{\text{hor}(\hat{\alpha}_t)}$ is the flow of the time-dependent vector field hor$(\hat{\alpha}_t)$. Equivalently, $u(t)$ is the solution of the ordinary differential equation

$$\nabla_\alpha u(t) = 0 \tag{8.30}$$

with initial boundary value $u(0) = u$, where $\nabla_\alpha u(t)$ is the covariant derivative of $u(t)$ along α, which can be defined by the formula

$$\nabla_\alpha u(t) = \nabla_\alpha \xi_t(x) + \frac{d\xi_t}{dt}(x) \quad \text{at } x = \gamma(t), \tag{8.31}$$

where ξ_t is a time-dependent section of E such that $\xi_t(\gamma(t)) = u(t)$. (The above formulas do not depend on the choice of $\hat{\alpha}_t$ and ξ_t.)

Given an A-connection, its *curvature* is a 2-form R_∇ on A with values in $E^* \otimes E$, given by the usual formula

$$R_\nabla(\alpha, \beta) = \nabla_\alpha \nabla_\beta - \nabla_\beta \nabla_\alpha - \nabla_{[\alpha,\beta]}. \tag{8.32}$$

(One verifies directly that $R_\nabla(\alpha, \beta)\xi$ is $C^\infty(M)$-linear in α, β and ξ.) When the curvature vanishes, i.e., $R_\nabla \equiv 0$, one says that ∇ is a *flat connection*. In terms of horizontal lifting, the flatness condition can be written as

$$[\text{hor}(\alpha), \text{hor}(\beta)] = \text{hor}([\alpha, \beta]) \ \ \forall \ \alpha, \beta \in \Gamma(A), \tag{8.33}$$

which means that the map hor is a linear representation of A. In other words, *a flat linear A-connection is nothing but a linear representation of A.*

Example 8.4.9. Let $\mathcal{O} \subset M$ be a leaf of the characteristic foliation of a Lie algebroid $(A, [,], \sharp)$ over M. Then $\ker \sharp_\mathcal{O}$ is a vector bundle over $\mathcal{O}$, and the formula $\nabla_\alpha \beta := [\alpha, \beta]$, $\alpha \in \Gamma(A_\mathcal{O})$, $\beta \in \Gamma(\ker \sharp_\mathcal{O})$ defines a flat $A_\mathcal{O}$-connection on $\ker \sharp_\mathcal{O}$ (which is called the *Bott connection*). The holonomy of this flat connection, i.e., the linear maps defined by parallel transport along $A_\mathcal{O}$-paths, is called the *linear holonomy* of A [146, 131].

8.5   Splitting theorem and transverse structures

The following *splitting theorem* for Lie algebroids first appeared in a weaker form in the work of Fernandes [131] and Weinstein [352], and was then refined in [112].

Theorem 8.5.1. *Let $(A, [,], \sharp)$ be a (smooth or analytic) Lie algebroid over a manifold M. Then in a neighborhood of each point of M the algebroid A can be decomposed locally into the direct product of a tangent algebroid and a Lie algebroid whose rank at the origin is 0. More precisely, if the rank of $\sharp$ at a point z of M is q, then there is a system of local coordinates $(x_1, \ldots, x_q, y_1, \ldots, y_s)$ on M centered at z, and a basis of local sections $(\alpha_1, \ldots, \alpha_q, \beta_1, \ldots, \beta_r)$ of A over a neighborhood of z, such that the following conditions are satisfied for all possible indices:*

$$[\alpha_i, \alpha_j] = 0, \quad \sharp\alpha_i = \frac{\partial}{\partial x_i}, \tag{8.34}$$

$$[\beta_i, \alpha_j] = 0, \quad \sharp\beta_i(x_j) = 0, \quad \mathcal{L}_{\partial/\partial x_j}\sharp\beta_i = 0, \tag{8.35}$$

$$[\beta_i, \beta_j] = \sum_k f_{ij}^k(y)\beta_k, \tag{8.36}$$

where $f_{ij}^k(y)$ are functions which depend only on the variables $y = (y_1, \ldots, y_s)$.

Proof. We will prove by induction on the rank $q = \text{rank}\,\sharp_z$. When $q = 0$ there is nothing to prove, so we will assume $q > 0$. Suppose by induction that, for an integer d with $0 \leq d < q$, we have found a local coordinate system $(x_1, \ldots, x_d, y_1, \ldots, y_{s+q-d})$ of M, and a basis of local sections $(\alpha_1, \ldots, \alpha_d, \beta_1, \ldots, \beta_{r+q-d})$ of A, which satisfy Equations (8.34), (8.35), (8.36). We will now try to replace d by $d + 1$.

Since $d < q$, there must be an index i, $1 \leq i \leq r+q-d$, such that $\sharp\beta_i(z) \neq 0$. We may assume that $\sharp\beta_1(z) \neq 0$. By a change of variables (which leaves Equations

(8.34), (8.35), (8.36) intact), we may assume that

$$\sharp\beta_1 = \frac{\partial}{\partial y_1}. \tag{8.37}$$

For each $i = 2, \ldots, r + q - 1$, we will replace β_i by $\widetilde{\beta}_i = \sum_{j=1}^{r+q-1} g_i^j \beta_j$, where (g_i^j) is an invertible matrix whose entries are functions of variables $(y_1, \ldots, y_{s+q-d})$. By such a change, we want to achieve the following condition:

$$[\beta_1, \widetilde{\beta}_i] = 0 \;\; \forall\, i = 2, \ldots, r + q - d. \tag{8.38}$$

By the Leibniz rule, the above condition is equivalent to the following system of ordinary differential equations:

$$\frac{\partial(g_i^j)}{\partial y_1} + (g_i^k)(f_{1k}^j) = 0, \tag{8.39}$$

where (f_{1k}^j) is a matrix of functions of variables $(y_1, \ldots, y_{s+q-r})$ which appear in Equation (8.14), i.e.,

$$[\beta_1, \beta_k] = \sum_{j=1}^{r+q-d} f_{1k}^j \beta_j. \tag{8.40}$$

Equation (8.39) is just an ordinary differential equation, which can be solved locally, say by fixing the following boundary condition: when $y_1 = 0$ we pose $g_i^i = 1$ and $g_i^j = 0$ if $i \neq j$. Thus we may assume that Condition (8.38) is satisfied. We may rename $\widetilde{\beta}_i$ by β_i, and forget about the tilde, so we have

$$[\beta_1, \beta_i] = 0 \;\; \forall\, i = 2, \ldots, r + q - d. \tag{8.41}$$

This last equation implies in particular that $\sharp\beta_i$ is invariant with respect to $\partial/\partial y_1$. Indeed, we have

$$\mathcal{L}_{\partial/\partial y_1}\sharp\beta_i = [\sharp\beta_1, \sharp\beta_i] = \sharp[\beta_1, \beta_i] = 0. \tag{8.42}$$

Note that $[\beta_1, (\sharp\beta_i(y_1))\beta_1] = 0$ because $\sharp\beta_i(y_1)$ is invariant with respect to y_1. Replacing β_i by $\beta_i - (\sharp\beta_i(y_1))\beta_1$ for each $i \geq 2$, we may also assume that

$$\sharp\beta_i(y_1) = 0 \;\; \forall\, i = 2, \ldots, r + q - d. \tag{8.43}$$

Finally, look at the equation

$$[\beta_i, \beta_j] = f_{ij}^1 \beta_1 + \sum_{k=2}^{r+q-d} f_{ij}^k \beta_k \tag{8.44}$$

for $i, j \geq 2$. We want to show that $f_{ij}^1 = 0$. Applying the anchor map to the above equation, and valuing it on y_1, we have $0 = [\sharp\beta_i, \sharp\beta_j](y_1) = \sharp(f_{ij}^1\beta_1 +$

$\sum_{k=2}^{r+q-d} f_{ij}^k \beta_k)(y_1) = f_{ij}^1$, whence $f_{ij}^1 = 0$. Thus we have

$$[\beta_i, \beta_j] = \sum_{k=2}^{r+q-d} f_{ij}^k \beta_k. \tag{8.45}$$

Now we can rename y_1 by x_{d+1} and β_1 by α_{d+1} to achieve our induction procedure. $\square$

Remark 8.5.2. The above splitting theorem implies that, similarly to the case of Poisson manifolds, to study the local structure of a Lie algebroid, it is enough to consider the case when the anchor map vanishes at a point.

Remark 8.5.3. A similar splitting theorem for the so-called Koszul–Vinberg algebroids was obtained by Nguiffo Boyom and Wolak in [278].

Corollary 8.5.4. *If the characteristic distribution* Im$\sharp$ *of a Lie algebroid A has constant rank q near a point $z \in M$, then A is locally isomorphic to the product of the tangent algebroid $T\mathbb{K}^q$ with a bundle of Lie algebras.*

Proof. The constant rank condition means that the vector fields $\sharp\beta_i$, where β_i are sections of A given by the splitting theorem 8.5.1, must vanish. This leads to the result. $\square$

Example 8.5.5. If A is a transitive Lie algebroid, then locally A is isomorphic to the direct product of a tangent algebroid with a finite-dimensional Lie algebra. In particular, the isotropy Lie algebras of a transitive Lie algebroid are isomorphic. More generally, if A is a Lie algebroid, then the isotopy algebras of points on a same orbit of A are isomorphic.

Remark 8.5.6. Similarly to the case of Poisson manifolds (see Remark 1.5.10), a direct consequence of Theorem 8.5.1 is that to each Lie algebroid $(A, [,], \sharp)$ over a manifold M there is a natural associated singular foliation on M whose tangent distribution is the characteristic distribution of A.

Let $(A, [.,.], \sharp)$ be a Lie algebroid over a manifold M, and N be a submanifold of M which is transversal to the characteristic foliation, i.e., for any point $x \in N$ we have

$$T_x N + \text{Im}\sharp(x) = T_x M. \tag{8.46}$$

(We don't require that $T_x N \cap \text{Im}\sharp(x) = 0$.) Then we can define the algebroid restriction A_N of A to N as follows:

$$A_N = \{\alpha \in A \mid \sharp\alpha \in TN\}. \tag{8.47}$$

Due to the transversality of N to the characteristic foliation, A_N is a subbundle of $A|_N$ (whose codimension in $A|_N$ is equal to the codimension of N in M). There is a unique natural Lie algebroid structure on A_N, for which the inclusion map $A_N \hookrightarrow A$ is a Lie algebroid morphism. The Lie bracket on $\Gamma(A_N)$ can be defined by

$$[\alpha, \beta] = [\widetilde{\alpha}, \widetilde{\beta}]|_N, \tag{8.48}$$

where α, β are sections of A_N, and $\tilde{\alpha}, \tilde{\beta}$ are sections of A which extend them. One verifies directly that the above formula does not depend on the choice of $\tilde{\alpha}$ and $\tilde{\beta}$.

In particular, when N is transversal to the characteristic foliation at a point $x \in M$, and $T_N \cap \text{Im}\sharp(x) = 0$, then the germ of the Lie algebroid A_N at x is called the *transverse Lie algebroid* of A at x. Given any such a transversal submanifold N, by repeating the proof of Theorem 8.5.1 but taking N into account, we can choose a local coordinate system $(x_1, \ldots, x_q, y_1, \ldots, y_s)$ on M centered at x, and a basis of local sections $(\alpha_1, \ldots, \alpha_q, \beta_1, \ldots, \beta_r)$ of A, which satisfy the conditions of Theorem 8.5.1, and such that locally N is given by $\{x_1 = \cdots = x_q = 0\}$, and A_N is spanned by $\beta_1, \ldots, \beta_r$. Theorem 8.5.1 says that the germ of A at x can be written as the direct product of the germ of A_N at x with the germ of a tangent algebroid.

Let N' be another submanifold transversal to the characteristic foliation at a point x' which lies on the same orbit as x, such that $T'_N \cap \text{Im}\sharp(x') = 0$. To justify the words *transverse Lie algebroid*, we will show that the germ of A_N at x is isomorphic to the germ of A'_N at x'. It is a direct consequence of the following:

Proposition 8.5.7 ([131]). *With the above notations, there is an isomorphism of A which sends a neighborhood of x in N to a neighborhood of x' in N'.*

Proof. We can assume that $x \neq x'$. There is a smooth section α of A with compact support, such that the time-1 flow $\exp(\sharp\alpha)$ of $\sharp\alpha$ sends x to x' and a neighborhood of x in N to a neighborhood of x' in N'. The time-1 flow $\exp(X_\alpha)$ of the Hamiltonian vector field X_α of α on A^* is then a fiber-wise linear Poisson isomorphism, which projects to the map $\exp(\sharp\alpha)$ on N. By duality, this fiber-wise linear Poisson isomorphism of A^* corresponds to a Lie algebroid isomorphism of A. $\square$

8.6 Cohomology of Lie algebroids

Associated to each Lie algebroid $(A \to M, [,], \sharp)$ there is a natural differential complex, called the *de Rham complex* of A:

$$\cdots \longrightarrow \Gamma(\wedge^{p-1}A^*) \xrightarrow{\text{d}_A} \Gamma(\wedge^p A^*) \xrightarrow{\text{d}_A} \Gamma(\wedge^{p-1}A^*) \longrightarrow \cdots, \qquad (8.49)$$

where the space $\Gamma(\wedge^p A^*)$ of p-cochains is the space of smooth p-forms on A, i.e., $C^\infty(M)$-multilinear antisymmetric maps

$$\Gamma(A) \times \cdots \times \Gamma(A) \longrightarrow C^\infty(M),$$

and the differential operator d_A is given by Cartan's formula:

$$\text{d}_A \omega(s_1, \ldots, s_{p+1}) = \sum_i (-1)^{i+1} \sharp(s_i)\omega(s_0, \ldots \hat{s}_i \ldots, s_p)$$

$$+ \sum_{i<j}(-1)^{i+j}\omega([s_i, s_j], s_1, \ldots \hat{s}_i \ldots \hat{s}_j \ldots, s_p). \qquad (8.50)$$

Exercise 8.6.1. Show that the anchor map and the Lie bracket of the Lie algebroid A are completely determined by the above differential operator d_A, and $d_A \circ d_A = 0$.

Definition 8.6.2. The cohomology of the above de Rham complex of a Lie algebroid A is called the *algebroid cohomology* of A and denoted by $H^\star(A)$.

For example, if $A = TM$ is the tangent algebroid of a manifold M, then $H^\star(A)$ is nothing but the de Rham cohomology $H^\star_{dR}(M)$ of M. If $A = T^*M$ is the cotangent algebroid of a Poisson manifold (M, Π), then the above de Rham complex coincides with the Lichnerowicz complex of (M, Π), and $H^\star(A) = H^\star(M, \Pi)$ is the Poisson cohomology of A.

Remark 8.6.3. One may view $\Gamma(\wedge^\star A^*) = \oplus_p \Gamma(\wedge^p A^*)$ as the space of superfunctions on a supermanifold $\mathcal{A}$ associated to the vector bundle $A \to M$, and the differential d_A of the de Rham complex of A as a vector field of degree 1 on $\mathcal{A}$. This leads to an equivalent definition of a Lie algebroid as a supermanifold with a vector field d_A of degree 1 such that $(d_A)^2 = 0$ [331, 10].

More generally, given a linear module E of a Lie algebroid A, one has a natural differential complex

$$\cdots \longrightarrow \Gamma(\wedge^{p-1}A^* \otimes E) \xrightarrow{\ d\ } \Gamma(\wedge^p A^* \otimes E) \xrightarrow{\ d\ } \Gamma(\wedge^{p-1}A^* \otimes E) \longrightarrow \cdots \qquad (8.51)$$

which generalizes the Chevalley–Eilenberg complex. The differential d is again given by Cartan's formula

$$d\omega(s_1, \ldots, s_{p+1}) = \sum_i (-1)^{i+1} \nabla_{s_i} \omega(s_0, \ldots \hat{s}_i \ldots, s_p)$$

$$+ \sum_{i<j} (-1)^{i+j} \omega([s_i, s_j], s_1, \ldots \hat{s}_i \ldots \hat{s}_j \ldots, s_p), \qquad (8.52)$$

where ∇ denotes the covariant derivative of the corresponding flat A-connection on E. The corresponding cohomology is called the *algebroid cohomology* of A with coefficients in E. See, e.g., [85, 127, 131, 153, 181, 202, 203, 360] and references therein for some results on Lie algebroid cohomology and its relations with homology, characteristic classes of Lie algebroids, and differentiable cohomology of Lie groupoids.

Here we want to discuss another cohomology of Lie algebroids, called *deformation cohomology* [88], which is more directly related to the problems of normal forms and deformations of Lie algebroids.

Let $E \to M$ be a vector bundle over a manifold M. A vector field X on E is called *fiber-wise constant vertical* (resp., *fiber-wise linear*) if the derivation by X of any fiber-wise linear function is basic (resp., fiber-wise linear) and the derivation of any basic function vanishes (resp., is basic). In a fibered local system of coordinates $(x_1, \ldots, x_n, y_1, \ldots, y_q)$, where x_i are coordinates on M and y_j are linear coordinates on the fiber, a fiber-wise constant vertical vector field has the form

$$\sum_j X_j(x)\partial y_j,$$

and a fiber-wise linear vector field has the form

$$\sum_{kj} X_j^k(x)y_k\partial y_j + \sum_i B_i(x)\partial x_i.$$

For $p > 0$, we denote by $\mathcal{V}_{\mathrm{lin}}^p(E)$ the $C^\infty(M)$-module of p-vectors on E generated by the exterior products $X_1 \wedge \cdots \wedge X_p$ where $X_1,\ldots,X_{p-1}$ are fiber-wise constant vertical vector fields and X_p is fiber-wise linear. Elements of $\mathcal{V}_{\mathrm{lin}}^p(E)$ are called *fiber-wise linear p-vector fields*; in the fibered coordinates $(x_1,\ldots,x_n,$ $y_1,\ldots,y_q)$, they have the local form

$$\sum_{i_1\ldots i_p i} a_{i_1\ldots i_p}^i(x)y_i\partial y_{i_1} \wedge \cdots \wedge \partial y_{i_p} + \sum_{j_1\ldots j_{p-1}k} b_{j_1\ldots j_{p-1}}^k(x)\partial y_{j_1} \wedge \cdots \wedge \partial y_{j_{p-1}} \wedge \partial x_k.$$

In particular, $\mathcal{V}_{\mathrm{lin}}^1(E)$ is the set of fiber-wise linear vector fields on E. We denote by $\mathcal{V}_{\mathrm{lin}}^0(E)$ the $C^\infty(M)$-module of functions on E which is generated by basic functions and fiber-wise linear functions. The space $\mathcal{V}_{\mathrm{lin}}^\star(E) = \oplus_{p\geq 0}\mathcal{V}_{\mathrm{lin}}^p(E)$ is called the space of fiber-wise linear multi-vectors on E.

For example, if A is a Lie algebroid, then the corresponding Poisson structure on A^* is a fiber-wise linear 2-vector field on A^*.

The following fundamental lemma, whose proof is straightforward, allows us to specialize Poisson cohomology to the fiber-wise linear world.

Lemma 8.6.4. *The Schouten bracket of two fiber-wise linear multi-vector fields on a vector bundle is again a fiber-wise linear multi-vector field.*

Let Π be a fiber-wise linear Poisson tensor (i.e., an element of $\mathcal{V}_{\mathrm{lin}}^2(E)$ with $[\Pi,\Pi] = 0$). Then the differential operator $\delta : A \mapsto [\Pi, A]$ of the Lichnerowicz complex (see Section 2.1) gives, by restriction, a differential complex $(\mathcal{V}_{\mathrm{lin}}(E),\delta)$, i.e.,

$$\cdots \xrightarrow{\delta} \mathcal{V}_{\mathrm{lin}}^{p-1}(E) \xrightarrow{\delta} \mathcal{V}_{\mathrm{lin}}^p(E) \xrightarrow{\delta} \mathcal{V}_{\mathrm{lin}}^{p+1}(E) \xrightarrow{\delta} \cdots. \tag{8.53}$$

We will denote the cohomology of this complex by $H_{\mathrm{lin}}^\star(E,\Pi)$ and call it the *fiber-wise linear Poisson cohomology* of (E,Π). If A is a Lie algebroid and (A^*,Π) its dual fiber-wise linear Poisson manifold, then $H_{\mathrm{lin}}^\star(A^*,\Pi)$ is also called the *deformation cohomology* of A, see Crainic and Moerdijk [88] who also gave some relations between usual algebroid cohomology (with coefficients in A-modules) and deformation cohomology.

It is clear that, just as Poisson cohomology governs deformations of Poisson structures, fiber-wise linear Poisson cohomology governs fiber-wise linear deformations of fiber-wise linear Poisson structures. In other words, deformation cohomology of Lie algebroids govern their deformations. In particular, similarly to

Section 2.2, deformation cohomology may be used to study local normal forms of Lie algebroids.

Due to the local splitting theorem for Lie algebroids, in the study of local structure of a Lie algebroid $(A \to M, [,], \sharp)$ near a point $0 \in M$, we may assume that the rank of $\#_0 : A_0 \to T_0 M$ is zero. Fix a local coordinate system $(x_1, \ldots, x_m, y_1, \ldots, y_n)$ on the dual bundle A^* of A, where $y_1, \ldots, y_n$ are fiber-wise linear functions and $x_1, \ldots, x_m$ are basic function which vanish at $0 \in M$. Suppose that the Taylor expansion of the associated fiber-wise linear Poisson structure Π on A^* begin with terms of degree k,

$$\Pi = \Pi^{(k)} + \Pi^{(k+1)} + \cdots , \tag{8.54}$$

where, for every $r > 0$, $\Pi^{(r)}$ is homogeneous of degree r; it is a fiber-wise linear 2-vector field of the form $\sum_{jli} a^i_{jl}(x) y_i \partial y_j \wedge \partial y_l + \sum_{su} b^u_s(x) \partial y_s \wedge \partial x_u$, where the coefficients $a^i_{jl}(x)$ are homogeneous polynomials of degree $r - 1$ in variables $x = (x_1, \ldots, x_m)$ and $b^u_s(x)$ are homogeneous polynomials of degree r. In particular, $\Pi^{(k)}$ is the "principal part" of Π, i.e., the non-zero homogeneous part of lowest degree. When $k = 1$, we recover the linear part. The lowest degree terms of the equation $[\Pi, \Pi] = 0$ give $[\Pi^{(k)}, \Pi^{(k)}] = 0$, so the principal part $\Pi^{(k)}$ is again a fiber-wise Poisson structure. In the language of Lie algebroids, $\Pi^{(k)}$ corresponds to a *homogeneous Lie algebroid* of degree k, which will be called the *homogeneous part* of Π at 0, and which doesn't depend on the choice of local coordinates. Similarly, one can define *quasi-homogeneous* Lie algebroids, exactly as in Section 2.2. Moreover, similarly to the Poisson case, the deformation cohomology also has quasi-homogeneous versions $H^p_{\mathrm{lin},(r)}(A^*, \Pi^{(k)})$ when we restrict our attention to fiber-wise linear quasi-homogeneous multi-vector fields of degree r. So Proposition 2.2.1 and Theorem 2.2.2 have versions with the suffix lin, i.e., in the fiber-wise linear world. This gives the theoretic framework for the study of (formal) normal formal norms and, in particular, homogenization of Lie algebroids near a singular point. The problem of linearization of Lie algebroids will be considered in detail in the next section.

8.7   Linearization of Lie algebroids

Recall that, due to the local splitting theorem for Lie algebroids, in the study of local structure of a Lie algebroid $(A, [,], \sharp)$ near a point x, we may assume that the rank of $\#_x : A_x \to T_x M$ is zero.

Consider a Lie algebroid $(A, [,], \sharp)$, over a neighborhood U of the origin in $\mathbb{K}^N$ with a local system of coordinates $x = (x_1, \ldots, x_N)$, such that $\sharp_0 = 0$. Denote by $(\alpha_1, \ldots, \alpha_n)$ a basis of sections of A over U. Write

$$[\alpha_i, \alpha_j] = \sum_{k=1}^n c^k_{ij} \alpha_k + h.o.t. \tag{8.55}$$

and

$$\sharp \alpha_i = \sum_{j,k} b_{ij}^k x_k \frac{\partial}{\partial x_j} + h.o.t., \tag{8.56}$$

where c_{ij}^k and b_{ij} are constants, and $h.o.t.$ means higher-order terms. If we forget about the higher-order terms in the above expressions, then we get an n-dimensional Lie algebra (namely the isotropy algebra of A at 0) with structural coefficients c_{ij}^k, which acts on the linear space $V = \mathbb{K}^N$ via linear vector fields $\sum_{j,k} b_{ij}^k x_k \partial / \partial x_j$. The transformation algebroid of this linear Lie algebra action is called the *linear part* of A and denoted by $A^{(1)}$.

Of course, $A^{(1)}$ does not depend on the choice of local coordinates and local sections of A. It is the transformation algebroid $\mathfrak{l} \dot{\ltimes} V$ of a linear action of $\mathfrak{l}$ on $V = \mathbb{K}^N$, where $\mathfrak{l} = A_0$ is the isotropy algebra of A at 0.

The local linearization problem for Lie algebroids is: does there exist a local (formal, analytic or smooth) isomorphism from A to $A^{(1)}$. If it is the case, then we say that A is locally *linearizable* (formally, analytically or smoothly).

Similarly to the problem of formal linearization of Poisson structures, the formal linearization of a Lie algebroid A whose anchor vanishes at a point is governed by a cohomology group, namely the group

$$H^2\big(\mathfrak{l}, \oplus_{k\geq 1}\mathcal{S}^k(V^*) \otimes \mathfrak{l}\big) \oplus H^1\big(\mathfrak{l}, \oplus_{k\geq 2}\mathcal{S}^k(V^*) \otimes V\big),$$

where $\mathcal{S}^k(V^*)$ denotes the k-symmetric power of the $\mathfrak{l}$-module V^* (the dual of $\mathfrak{l}$-module V), and $\mathfrak{l}$ acts on itself by the adjoint action. In other words, we have the following result, first mentioned in [353] (in a not very precise way).

Theorem 8.7.1. *With the above notations, if*

$$H^2(\mathfrak{l}, \oplus_{p\geq 1}\mathcal{S}^p(V^*) \otimes \mathfrak{l}) = 0, \tag{8.57}$$

then A is formally isomorphic to the transformation algebroid of an action of $\mathfrak{l}$ with a fixed point. If, moreover,

$$H^1(\mathfrak{l}, \oplus_{p\geq 2}\mathcal{S}^p(V^*) \otimes V) = 0, \tag{8.58}$$

then A is formally linearizable.

Proof. Suppose that $H^2(\mathfrak{l}, \oplus_{p\geq 1}\mathcal{S}^p(V^*) \otimes \mathfrak{l}) = 0$. We will find a formal basis of sections $\alpha_1^\infty, \ldots, \alpha_n^\infty$ of A such that

$$[\alpha_i^\infty, \alpha_j^\infty] = \sum_{k=1}^n c_{ij}^k \alpha_k^\infty. \tag{8.59}$$

If such a formal basis of sections can be found, then A is formally isomorphic to the transformation algebroid of the formal (nonlinear) action of $\mathfrak{l}$ on V generated by $\sharp \alpha_i^\infty$. We will do it by induction. Suppose that, for an integer $p \geq 0$, we found a basis of sections $\alpha_1^p, \ldots, \alpha_n^p$ of A such that

$$[\alpha_i^p, \alpha_j^p] = \sum_{k=1}^n c_{ij}^k \alpha_k^p + \beta_{ij}^{(p)} + h.o.t., \tag{8.60}$$

where $\beta_{ij}^{(p)} = \sum_{k=1}^n b_{ij}^{k(p)} \alpha_k^p$ are sections whose coefficients $b_{ij}^{k(p)}$ are homogeneous polynomials of degree p with respect to a given system of coordinates $(x_1, \ldots, x_N)$ on $V = \mathbb{K}^N$. (When $p = 0$, such a basis of sections exists, because Equation (8.60) is the same as Equation (8.55).) We can identify $\beta_{ij}^{(p)}$ with an element of $\mathcal{S}^p(V^*) \otimes \mathfrak{l}$, by mapping (α_i^p) to a basis (ξ_i) of $\mathfrak{l}$ such that $[\xi_i, \xi_j] = \sum c_{ij}^k \xi_k$. One checks directly that, due to the Jacobi identity for the sections of A, $\beta^p : \xi_i \wedge \xi_j \mapsto \sum_{k=1}^n b_{ij}^{k(p)} \xi_k$ is a 2-cocycle of the $\mathfrak{l}$-module $\mathcal{S}^p(V^*) \otimes \mathfrak{l}$. By our assumptions, it is a coboundary. Denote by $\zeta^p : \xi_i \mapsto \sum_{k=1}^n z_i^{k(p)} \xi_k$ a 1-cochain whose coboundary is β^p. Put $\alpha_i^{p+1} = \alpha_i^p - \sum_{k=1}^n z_i^{k(p)} \alpha_k^p$. Then we have

$$[\alpha_i^{p+1}, \alpha_j^{p+1}] = \sum_{k=1}^n c_{ij}^k \alpha_k^{p+1} + O(p+1), \tag{8.61}$$

where $O(p+1)$ means terms of degree $\geq p+1$ in variables $(x_1, \ldots, x_n)$.

Repeating the above process for each $p \in \mathbb{N}$, we get a series of bases α_i^p, which converges formally. The formal limit $\alpha_i^\infty = \lim_{p \to \infty} \alpha_i^p$ satisfies Equation (8.59). The first part of Theorem 8.7.1 is proved.

The second part of Theorem 8.7.1 follows from the first part and the following linearization result for formal Lie algebra actions.

Theorem 8.7.2 ([172]). *Suppose that a finite-dimensional Lie algebra $\mathfrak{l}$ acts formally on a manifold $\mathbb{K}^N$ with the origin 0 as a fixed point, and that the linear part of the action of $\mathfrak{l}$ at 0 corresponds to an $\mathfrak{l}$-module $V = \mathbb{K}^N$ such that $H^1(\mathfrak{l}, \oplus_{p \geq 2} \mathcal{S}^p(V^*) \otimes V) = 0$. Then this action of $\mathfrak{l}$ is formally linearizable.*

We will leave the proof of Theorem 8.7.2, which is implicitly given in [172], as an exercise (see the proof of Theorem 3.1.6). It is very similar to the proof of the first part of Theorem 8.7.1, except for the fact that we will have to deal with some 1-cocycles instead of 2-cocycles.   □

Exercise 8.7.3. Write down the relation between the cohomology groups that appear in Theorem 8.7.1 and the deformation cohomology of the transformation algebroid $\mathfrak{l} \ltimes V$.

A particular case of Theorem 8.7.1 is when the isotropy algebra $\mathfrak{l}$ is semisimple. In that case, due to Whitehead's lemma, the cohomological conditions in Theorem 8.7.1 are automatically satisfied, and we get:

Theorem 8.7.4 ([353, 112]). *Let A be a formal Lie algebroid over a space $\mathbb{K}^N$ ($\mathbb{K} = \mathbb{R}$ or $\mathbb{C}$), whose anchor vanishes at the origin 0 and whose isotropic algebra at 0 is semisimple, then A is formally linearizable.*

In analogy with Poisson structure, one may say that a finite-dimensional module V of a finite-dimensional Lie algebra $\mathfrak{l}$ is formally (resp. analytically, resp. smoothly) *nondegenerate*, if the corresponding linear transformation algebroid $\mathfrak{l} \ltimes V$ is formally (resp. analytically, resp. smoothly) nondegenerate in the sense that any Lie algebroid, whose anchor vanishes at a point and whose linear part at that point is isomorphic to $\mathfrak{l} \ltimes V$, is formally (resp. analytically, resp. smoothly) linearizable. Theorem 8.7.4 says that any finite-dimensional module of a semisimple Lie algebra is formally nondegenerate.

In fact, Theorem 8.7.4 also holds in the analytic case (i.e., any finite-dimensional module of a semisimple Lie algebra is *analytically* nondegenerate), and it is a special case of the following Theorem 8.7.5, which is the formal and analytic Levi decomposition theorem for Lie algebroids.

Theorem 8.7.5 ([369]). *Let A be a local analytic (resp. formal) Lie algebroid over $(\mathbb{K}^N, 0)$, whose anchor map $\#_x : A_x \to T_x\mathbb{K}^N$ vanishes at 0: $\#_0 = 0$. Denote by $\mathfrak{l} = A_0$ the isotropy algebra A at 0, and by $\mathfrak{l} = \mathfrak{g} \ltimes \mathfrak{r}$ its Levi decomposition. Then there exists a local analytic (resp. formal) system of coordinates $(x_1^\infty, \ldots, x_N^\infty)$ of $(\mathbb{K}^N, 0)$, and a local analytic (resp. formal) basis of sections $(\alpha_1^\infty, \ldots, \alpha_m^\infty, \beta_1^\infty, \ldots, \beta_{n-m}^\infty)$ of A, where $n = \dim \mathfrak{l}$ and $m = \dim \mathfrak{g}$, such that we have:*

$$
\begin{aligned}
[\alpha_i^\infty, \alpha_j^\infty] &= \sum_k c_{ij}^k \alpha_k^\infty \ , \\
[\alpha_i^\infty, \beta_j^\infty] &= \sum_k a_{ij}^k \beta_k^\infty \ , \\
\#\alpha_i^\infty &= \sum_{j,k} b_{ij}^k x_k^\infty \partial/\partial x_j^\infty \ ,
\end{aligned}
\tag{8.62}
$$

where $c_{ij}^k, a_{ij}^k, b_{ij}^k$ are constants, with c_{ij}^k being the structural coefficients of the semi-simple Lie algebra $\mathfrak{g}$.

The above form (8.62) is called a *Levi normal form* or *Levi decomposition* of A.

Proof. Theorem 8.7.5 may be viewed as a special case of Theorems 3.2.3 and 3.2.6, due to the characterization of Lie algebroids as fiber-wise linear Poisson structures. We will indicate here how to get a proof of Theorem 8.7.5 from the proof of Theorems 3.2.3 and 3.2.6 given in Chapter 3.

Denote by $(\alpha_1, \ldots, \alpha_n)$ a basis of local analytic (resp. formal) sections on A, and $(x_1, \ldots, x_N)$ a local system of coordinates on $\mathbb{K}^N$. Together they form a local fiber-wise linear coordinate system on the dual bundle A^* of A. Recall from Section 8.2 that A^* is equipped with the corresponding fiber-wise linear Poisson structure Π whose Poisson bracket is given as follows:

$$
\begin{aligned}
\{\alpha_i, \alpha_j\} &= [\alpha_i, \alpha_j], \\
\{\alpha_i, x_j\} &= \sharp\alpha_i(x_j), \\
\{x_i, x_j\} &= 0.
\end{aligned}
\tag{8.63}
$$

Since $\natural_0 = 0$, Π also vanishes at the origin of A^* (i.e., the null point in the fiber A_0^* of A^* over 0). The linear part of Π at the origin is $\mathfrak{l} \ltimes V$, and its Levi factor is the same as the Levi factor of $\mathfrak{l}$.

It is easy to see that, to find a Levi decomposition of A is the same as to find a Levi decomposition of Π, given by a coordinate system

$$(\alpha_1^\infty, \ldots, \alpha_m^\infty, \beta_1^\infty, \ldots, \beta_{n-m}^\infty, x_1^\infty, \ldots, x_N^\infty)$$

on A^*, such that $(\alpha_1^\infty, \ldots, \alpha_m^\infty)$ form a Levi factor, $\alpha_i^\infty, \beta_i^\infty$ are fiber-wise linear functions, and x_i^∞ are base functions. Such a Levi decomposition of Π on A^* will be called a fiber-wise linear Levi decomposition.

The existence of a Levi decomposition of Π on A^* is provided by (the proof of) Theorem 3.2.6 given in Chapter 3 (which includes a proof of Theorem 3.2.3). But we have to modify that proof a little bit to make sure that this Levi decomposition, in the case of fiber-wise Poisson structures, can be chosen fiber-wise linear. Here are the few modifications to be made to the construction of Levi decomposition given in Section 3.3:

i) After Step l ($l \geq 0$), we will get a local coordinate system

$$(\alpha_1^l, \ldots, \alpha_m^l, \beta_1^l, \ldots, \beta_{n-m}^l, x_1^l, \ldots, x_N^l)$$

of A^* with the following properties: $x_1^l, \ldots, x_n^l$ are base functions (i.e., functions on $(\mathbb{K}^N, 0)$); $\alpha_1^l, \ldots, \alpha_m^l, \beta_1^l, \ldots, \beta_{n-m}^l$ are fiber-wise linear functions (i.e., they are sections of A); $\{\alpha_i^l, \alpha_j^l\} - \sum_k c_{ij}^k \alpha_k^l = O(|x|^{2^l})$; $\{\alpha_i^l, \beta_j^l\} - \sum_k a_{ij}^k \beta_k^l = O(|x|^{2^l})$; $\{\alpha_i^l, x_j^l\} - \sum_k b_{ij}^k x_k^l = O(|x|^{2^l+1})$. Here $c_{ij}^k, a_{ij}^k, b_{ij}^k$ are structural constants as appeared in the statement of Theorem 8.7.5.

ii) Replace the space $\mathcal{O}$ of all local analytic functions by the subspace of local analytic functions which are fiber-wise linear. Similarly, replace the space $\mathcal{O}_l$ of local analytic functions without terms of order $\leq 2^l$ by the subspace of fiber-wise linear analytic functions without terms of order $\leq 2^l$.

iii) Replace $\mathcal{Y}^l$ by the subspace of vector fields of the following form:

$$\sum_{i=1}^{n-m} p_i \partial/\partial v_i^l + \sum_{i=1}^{N} q_i \partial/\partial x_i^l$$

where p_i are fiber-wise linear functions and q_i are base functions. For the replacement of $\mathcal{Y}_k^l$, we require that p_i do not contain terms of order $\leq 2^k - 1$ in variables $(x_1, \ldots, x_N)$, and q_i do not contain terms of order $\leq 2^k$.

One checks that the above subspaces are invariant under the $\mathfrak{g}$-actions introduced in Section 3.3, and the cocycles introduced there will also live in the corresponding quotient spaces of these subspaces. Details are left to the reader. $\square$

Similarly, a slight modification of the proof of the smooth Levi decomposition Theorem 3.2.9 for Poisson structures leads to the following smooth Levi decomposition theorem for Lie algebroids.

Theorem 8.7.6 ([263]). *For each $p \in \mathbb{N} \cup \{\infty\}$ and $n, N \in \mathbb{N}$ there is $p' \in \mathbb{N} \cup \{\infty\}$, $p' < \infty$ if $p < \infty$, such that the following statement holds: Let A be an n-dimensional, $C^{p'}$-smooth Lie algebroid over a neighborhood of the origin 0 in $\mathbb{R}^N$ with anchor map $\# : A \to T\mathbb{R}^N$, such that $\#_0 = 0$. Denote by $\mathfrak{l}$ the isotropy Lie algebra of A at 0, and by $\mathfrak{l} = \mathfrak{g} + \mathfrak{r}$ a decomposition of $\mathfrak{L}$ into a direct sum of a semisimple compact Lie algebra $\mathfrak{g}$ and a linear subspace $\mathfrak{r}$ which is invariant under the adjoint action of $\mathfrak{g}$. Then there exists a local C^p-smooth system of coordinates $(x_1^\infty, \ldots, x_N^\infty)$ of $(\mathbb{R}^N, 0)$, and a local C^p-smooth basis of sections $(s_1^\infty, s_2^\infty, \ldots, s_m^\infty, v_1^\infty, \ldots, v_{n-m}^\infty)$ of A, where $m = \dim \mathfrak{g}$, such that we have:*

$$
\begin{aligned}
[s_i^\infty, s_j^\infty] &= \sum_k c_{ij}^k s_k^\infty \ , \\
[s_i^\infty, v_j^\infty] &= \sum_k a_{ij}^k v_k^\infty \ , \\
\# s_i^\infty &= \sum_{j,k} b_{ij}^k x_k^\infty \partial/\partial x_j^\infty \ ,
\end{aligned}
\tag{8.64}
$$

where $c_{ij}^k, a_{ij}^k, b_{ij}^k$ are constants, with c_{ij}^k being the structural constants of the compact semisimple Lie algebra $\mathfrak{g}$.

In particular, when $\mathfrak{r} = 0$, Theorem 8.7.6 gives a local smooth linearization of a Lie algebroid whose isotropy algebra is semisimple compact. This may be viewed as the infinitesimal version of Theorem 7.4.7 in the compact semisimple case.

8.8 Integrability of Lie brackets

The so-called *Lie's third theorem* says that any (finite-dimensional) Lie algebra can be integrated into a Lie group. A natural question arises: does Lie's third theorem hold for Lie algebroids, i.e., is it true that any (smooth real) Lie algebroid can be integrated into a Lie groupoid? This question was studied by many people, starting with Pradines [295] who claimed, erroneously, that the answer is positive. What Pradines actually obtained is that any Lie algebroid can be integrated into a *local groupoid*: the product map is not necessarily globally defined on $\Gamma_{(2)}$, but only locally defined near the identity section. In the case of transitive Lie algebroids, an abstract theory of homological obstructions to integrability was developed by Mackenzie [226], and a concrete non-integrable example was found by Almeida and Molino [11]. Recently, a necessary and sufficient condition for smooth Lie algebroids (and Poisson manifolds) to be integrable was found by Crainic and Fernandes [86, 87]. We will briefly present their results here, referring the reader to [86, 87] for the details.

8.8.1 Reconstruction of groupoids from their algebroids

Assume that a Lie algebroid A over a connected manifold M is integrable, i.e., it is the Lie algebroid of a Lie groupoid $\mathcal{G} \rightrightarrows M$. Similarly to the case of Lie groups, by taking a covering of the identity component of $\mathcal{G}$, we may assume that $\mathcal{G}$ is connected and its s-fibers are simply-connected, and then $\mathcal{G}$ is uniquely determined by A [257]. Remark that the inversion map of $\mathcal{G}$ sends s-fibers to t-fibers, so the t-fibers of G are also simply-connected.

We will try to reconstruct $\mathcal{G}$ from A. By a $\mathcal{G}$-*path*, we will mean a map

$$g : I = [0,1] \to \mathcal{G} \qquad (8.65)$$

of smoothness class C^2, such that $g(0) = \varepsilon(x)$, the identity element of x for some $x \in M$, and $t(g(\tau)) = x \; \forall \; \tau \in I$. (In [86] it is written as $s(g(\tau)) = x$, because they adopt a different convention: they use *right*-invariant vector fields on a Lie groupoid to define its Lie algebroid.)

Differentiating a $\mathcal{G}$-path g, we will get a A-path $a : [0,1] \to A$ (i.e., a Lie algebra homomorphism from TI to A, where $I = [0,1]$, see Example 8.3.8) of class C^1 by the formula

$$a(\tau) = g(\tau)^{-1} \frac{\mathrm{d}g}{\mathrm{d}\tau}(\tau). \qquad (8.66)$$

Conversely, any A-path of class C^1 can be integrated to a $\mathcal{G}$-path.

If two $\mathcal{G}$-paths g and g' have the same end point $q = g(1) = g'(1)$, then they are homotopic in the corresponding t-fiber by a homotopy g_ρ ($g_0 = g, g_1 = g'$) of $\mathcal{G}$-paths with fixed end points, because we have assumed that the t-fibers are simply-connected. We may assume that this homotopy is of class C^2. By differentiating this homotopy, we get a family of A-paths a_ρ which are *equivalent* in the sense that their corresponding $\mathcal{G}$-paths have the same end point. We can write

$$\mathcal{G} = P(A)/ \sim , \qquad (8.67)$$

where $P(A)$ is the space of A-paths of class C^1 and $\sim$ is the above equivalence. This equivalence can be characterized infinitesimally, i.e., in terms of A, as follows:

Let a_ρ be a family of A-paths of class C^1 which depends on a parameter $\rho \in [0,1]$ in a C^2 fashion. Denote by $\gamma_\rho = \pi \circ a_\rho$ the corresponding family of base paths, where $\pi : A \to M$ is the projection. If a_ρ is the differentiation of g_ρ then $\gamma_\rho(\tau) = s(g_\rho(\tau))$. We will assume that these base paths have the same end points: $\gamma_\rho(0) = \gamma_0(0)$ and $\gamma_\rho(1) = \gamma_0(1)$ for all ρ. Let ξ_ρ be a family of time-dependent sections of A such that

$$\xi_\rho(\tau, \gamma_\rho(\tau)) = a_\rho(\tau). \qquad (8.68)$$

Put

$$b(\rho,\tau) = \int_0^\tau \phi_{\xi_\rho}^{\tau,r} \frac{\mathrm{d}\xi_\rho}{\mathrm{d}\rho}(r, \gamma_\rho(r))dr, \qquad (8.69)$$

where $\phi_{\xi_\rho}^{\tau,r}$ denotes the flow in A from time r to time τ generated by the time-dependent section ξ_ρ (see Example 8.2.5). One verifies directly that b depends only on the family a_ρ but not on the choice of time-dependent sections ξ_ρ. Moreover, if g_ρ is the family of $\mathcal{G}$-paths integrating a_ρ, then we have

$$b(\rho,\tau) = g_\rho(\tau)^{-1} \frac{\mathrm{d}g_\rho(\tau)}{\mathrm{d}\rho}. \qquad (8.70)$$

In other words, $b(\rho, \tau)$ is the differentiation of $g(\rho, \tau) = g_\rho(\tau)$ in the ρ-direction. We stress here the fact that $b(\rho, \tau)$ is well defined by Formula (8.69) even if the algebroid A is not integrable. In the integrable case, the condition that the family of $\mathcal{G}$-paths g_ρ (starting from a common point $\varepsilon(x)$) have the same end point may be expressed infinitesimally as

$$b(\rho, 1) = 0 \; \forall \; \rho \in [0, 1]. \tag{8.71}$$

Definition 8.8.1. We say that two A-paths a and a' of class C^1, whose base paths γ and γ' have the same end points, are *equivalent*, and write $a \sim a'$, if they can be connected by a C^2-family of A-paths a_ρ whose base paths have the same end points and such that $b(\rho, \tau)$ defined by Formula (8.69) satisfies the equation $\beta(\rho, 1) = 0$ for all $\rho \in [0, 1]$.

Clearly, this definition of equivalence coincides with the previous equivalence notion in the integrable case. Even if A is not integrable, the following natural fact still holds true:

Proposition 8.8.2 ([86]). *Let a_1 and a_1 be two equivalent A-paths from x to y, where A is a Lie algebroid over a manifold M and $x, y \in M$. Then for any flat A-connection on a vector bundle E over M, the parallel transports $E_x \to E_y$ along a_0 and a_1 coincide.*

In general, for any Lie algebroid A, the quotient

$$\mathcal{G}(A) = P(A)/\sim \tag{8.72}$$

is called the *Weinstein groupoid* of A (the construction is due to Weinstein and is inspired by a similar construction for the case of Lie algebras [121]). It coincides with the connected t-simply connected groupoid integrating A in the integrable case. In general, it is a topological groupoid: the groupoid structure comes from the concatenation of A-paths. (The concatenation of two A-paths of class C^1 is only piecewise differentiable in general, but it can be made C^1 by a homotopy using a cut-off function trick, see [86].) The algebroid A is integrable if and only if $\mathcal{G}$ is smooth. In fact, $P(A)$ is a Banach manifold, the equivalence relation $\sim$ defines a smooth foliation on $P(A)$ of finite codimension, and $\mathcal{G}(A)$ is the leaf space of this foliation. It is a smooth (not necessarily Haussdorff) manifold if and only if each leaf admits a transversal section which intersects every leaf at at most one point.

8.8.2 Integrability criteria

For any point $x \in M$, denote by $\mathfrak{g}_x = \ker \sharp$ the isotropy Lie algebra of A at x, considered as a Lie algebroid over x, and by $\mathcal{G}(\mathfrak{g}_x)$ the simply-connected Lie group integrating $\mathfrak{g}_x$, considered as the group of $\mathfrak{g}_x$-paths modulo equivalence. Then there is a natural homomorphism from $\mathcal{G}(\mathfrak{g}_x)$ to the isotropy group $\mathcal{G}(A)_x$ of x in $\mathcal{G}(A)$ (consider $\mathfrak{g}_x$-paths as A-paths). Denote by $\tilde{N}_x(A)$ the kernel of this homomorphism: it is the subgroup of $\mathcal{G}(\mathfrak{g}_x)$ consisting of equivalence classes of $\mathfrak{g}_x$-paths which are equivalent in A to the trivial A-path over x. If $\mathcal{G}(A)$ is smooth then in particular $\tilde{N}_x(A)$ must be discrete in $\mathcal{G}(\mathfrak{g}_x)$, i.e., the quotient $\mathcal{G}(\mathfrak{g}_x)/\tilde{N}_x(A)$ must be a Lie group of Lie algebra $\mathfrak{g}_x$.

We can characterize $\tilde{N}_x(A)$ as the image of the *second monodromy* of the leaf $\mathcal{O}(x)$ through x in M as follows: Let $[\gamma] \in \pi_2(\mathcal{O}(x), x)$ be represented by a smooth map $\gamma : I \times I \to \mathcal{O}(x)$ which maps the boundary into x. We can choose a morphism of Lie algebroid

$$a\mathrm{d}\tau + b\mathrm{d}\rho : TI \times TI \to A \tag{8.73}$$

(which maps $u(\rho,\tau)\partial/\partial\tau + v(\rho,\tau)\partial/\partial\rho$ to $u(\rho,\tau)a(\rho,\tau) + v(\rho,\tau)b(\rho,\tau) \in A$), which lifts $\mathrm{d}\gamma : TI \times TI \to T\mathcal{O}(x)$ via the anchor, and such that $a(0,\tau), b(\rho,0)$ and $b(\rho,1)$ vanish. Since γ is constant on the boundary, $a_1(\tau) = a(1,\tau)$ stays inside $\mathfrak{g}_x$, i.e., a_1 is a $\mathfrak{g}_x$-path. Denote the equivalence class of a_1 in $\mathcal{G}(\mathfrak{g}_x)$ by $\partial(\gamma)$.

Lemma 8.8.3. *The map $\gamma \mapsto \partial(\gamma)$ is well defined and depends only on the homotopy class $[\gamma]$ of γ. It induces a homomorphism*

$$\partial : \pi_2(\mathcal{O}(x), x) \longrightarrow \mathcal{G}(\mathfrak{g}_x), \tag{8.74}$$

whose image is exactly $\tilde{N}_x(A)$.

In fact, the image $\partial([\gamma])$ of any $[\gamma] \in \pi_2(\mathcal{O}(x), x)$ lies in the center of $\mathcal{G}(\mathfrak{g}_x)$. To see it, recall that there is a natural flat $A_{\mathcal{O}}$-connection on the isotropy bundle $\ker \sharp_{\mathcal{O}}$ where $\mathcal{O}$ is the leaf through x (see Example 8.4.9). By construction, the A-path a_1 which represents $\partial(\gamma)$ is equivalent in A (hence in $A_{\mathcal{O}}$) to the trivial path, so by Proposition 8.8.2 the parallel transport in $\ker \sharp_{\mathcal{O}}$ along a_1 with respect to this flat $A_{\mathcal{O}}$-connection, $T_{a_1} : \mathfrak{g}_x \to \mathfrak{g}_x$, is the identity map of $\mathfrak{g}_x$. On the other hand, since $a_1(\tau) \in \mathfrak{g}_x \ \forall \ \tau$, it is easy to see that this parallel transport map T_{a_1} is nothing but the adjoint action map $\mathrm{Ad}_{[a_1]} : \mathfrak{g}_x \to \mathfrak{g}_x$, where $[a_1]$ is the element in $\mathcal{G}(\mathfrak{g}_x)$ represented by a_1. Hence $\mathrm{Ad}_{[a_1]} = Id$, which means that $\partial([\gamma]) = [a_1]$ lies in the center of $\mathcal{G}(\mathfrak{g}_x)$.

Denote by $N_x(A)$ the subset of $\mathfrak{g}_x$ formed by those elements v of $\mathfrak{g}_x$ with the property that the constant A-path v is equivalent to the trivial A-path over x. Then we have:

Lemma 8.8.4. *For any Lie algebra A over M and any $x \in M$, the group $\tilde{N}_x(A)$ lies in the center $Z(\mathcal{G}(\mathfrak{g}_x))$ of $\mathcal{G}(\mathfrak{g}_x)$, and its intersection with the connected component $Z(\mathcal{G}(\mathfrak{g}_x))^0$ of the center is isomorphic to $N_x(A)$. In particular, $\tilde{N}_x(A)$ is discrete in $\mathcal{G}(\mathfrak{g}_x)$ if and only if $N_x(A)$ is discrete in $\mathfrak{g}_x$.*

In particular, $\mathcal{G}(A)_x$ is a Lie group of Lie algebra $\mathfrak{g}_x$ if and only if $N_x(A)$ is discrete. For $\mathcal{G}(A)$ to be integrable, we will need a stronger condition, namely that $N_x(A)$ is locally "uniformly" discrete when x varies. More precisely, fix an arbitrary continuous fiber-wise metric d on A, and put $r(x) = \infty$ if $N_x(A) = \{0\}$ and

$$r(x) = d(0, N_x(A) \setminus \{0\}) \tag{8.75}$$

otherwise. Then the condition that $N_x(A)$ is discrete is equivalent to the condition $r(x) > 0$, and the local uniform discreteness means that $\liminf_{y \to x} r(y) > 0$. A careful analysis of the smoothness of $\mathcal{G}(A)$ leads to the following theorem.

Theorem 8.8.5 (Crainic–Fernandes [86]). *A Lie algebroid A over a manifold M is integrable if and only if for every $x \in A$, the group $N_x(A)$ is discrete and $\liminf_{y \to x} r(y) > 0$.*

For example, when the Lie algebroid is totally intransitive (i.e., the anchor map is trivial), then $r(x) = \infty$ for all $x \in M$, and we recover from Theorem 8.8.5 the following integrability result of Douady and Lazard [106]: any Lie algebra bundle can be integrated into a Lie group bundle.

What makes Theorem 8.8.5 effective is the fact that, in many cases, the sets $N_x(A)$ can be computed explicitly via Proposition 8.8.6 below. Suppose that $\mathcal{O} \ni x$ is a leaf of A on M, and $\sigma : TL \to A_{\mathcal{O}}$ is a splitting of the surjective linear map $\sharp_{\mathcal{O}} : A_{\mathcal{O}} \to T\mathcal{O}$. The *curvature* of σ is a 2-form on $\mathcal{O}$ with values in $\mathfrak{g}_{\mathcal{O}} = \ker \sharp_{\mathcal{O}}$, defined by the formula

$$\Omega_\sigma(X, Y) = [\sigma(X), \sigma(Y)] - \sigma([X, Y]). \tag{8.76}$$

The vector bundle $Z(\mathfrak{g}_{\mathcal{O}})$, whose fiber over a point $y \in \mathcal{O}$ is the center $Z(\mathfrak{g}_y)$ of the isotropy algebra $\mathfrak{g}_y$, admits a canonical flat linear connection given by $\nabla_X \alpha = [\sigma(X), \alpha]$ for $X \in \mathcal{V}^1(\mathcal{O}), \alpha \in \Gamma(Z(\mathfrak{g}_{\mathcal{O}}))$ (this connection does not depend on the choice of σ).

Proposition 8.8.6 ([86]). *With the above notations, suppose that the image of the curvature form Ω_σ lies in $Z(\mathfrak{g}_{\mathcal{O}})$, then we have*

$$N_x(A) = \left\{ \int_\gamma \Omega_\sigma \mid \gamma \in \pi_2(\mathcal{O}(x), x) \right\}. \tag{8.77}$$

In the above proposition, the values of Ω_σ are translated to $Z(\mathfrak{g}_x)$ by the canonical flat connection on $Z(\mathfrak{g}_{\mathcal{O}})$ (via parallel transport along the paths lying in a representative of γ) before taking the integral.

Example 8.8.7. (Almeida–Molino's non-integrable example [11].) If ω is an arbitrary closed 2-form on a connected manifold M, then it induces a transitive Lie algebroid $A_\omega = TM \oplus \mathbb{L}$, where $\mathbb{L}$ is the trivial line bundle, with anchor $(X, \lambda) \mapsto X$ and Lie bracket

$$[(X, f), (Y, g)] = ([X, Y], X(g) - Y(f) + \omega(X, Y)). \tag{8.78}$$

(See, e.g., [226].) Applying Proposition 8.8.6 to the obvious splitting of A_ω, we obtain that

$$N_x(A_\omega) = \left\{ \int_\gamma \omega \mid \gamma \in \pi_2(M, x) \right\} \subset \mathbb{R} \tag{8.79}$$

is the group of periods of ω [86]. If for example, this group of periods is dense in $\mathbb{R}$ (e.g., it has two incommensurable generators), then the corresponding Lie algebroid is not integrable [11].

Exercise 8.8.8. Recover from Theorem 8.8.5 the following result of Dazord [96]: every transformation Lie algebroid (of a smooth action of a Lie algebra on a manifold) is integrable.

8.8.3   Integrability of Poisson manifolds

Recall that a Poisson manifold (M, Π) is *integrable* if there is a symplectic groupoid over it. The problem of integrability of Poisson manifolds is very similar to the problem of integrability of Lie algebroids, due to the following result:

Theorem 8.8.9 (Mackenzie–Xu [231]). *A Poisson manifold is integrable if and only if its cotangent Lie algebroid is integrable.*

Conversely, a Lie algebroid A is integrable if and only if A^* with the corresponding fiber-wise linear symplectic structure is an integrable Poisson manifold (see, e.g., [228]).

In particular, one can use Theorem 8.8.5 and Proposition 8.8.6 (rewritten in terms of Poisson structures) to verify integrability and non-integrability of Poisson manifolds. See [87] and references therein for details and other results concerning integrability of Poisson structures. In particular, let us mention the following result:

A symplectic realization (M, ω) of a Poisson manifold (P, Π) is called *complete* if for any function f on P such that the Hamiltonian vector field of f on P is complete, the Hamiltonian vector field of the lifting of f on M is also complete.

Theorem 8.8.10 ([87]). *A Poisson manifold admits a complete symplectic realization if any only if it is integrable.*

Example 8.8.11. Any two-dimensional Poisson manifold is integrable: if the Poisson structure is nondegenerate then it is obviously integrable, if not then the corresponding cotangent algebra is integrable by Theorem 8.8.5, because the second homotopy groups of the symplectic leaves are trivial.

Remark 8.8.12. The problem of integrability of regular Poisson manifolds was studied by Alcalde-Cuesta and Hector [4]. In the regular case, the groups $N_x(A)$ which control integrability can be given in terms of variations of symplectic areas of 2-spheres on symplectic leaves, see [4, 87].

Remark 8.8.13. In the case when A is the cotangent algebroid of a Poisson manifold, the Weinstein groupoid $\mathcal{G}(A)$ was also constructed by Cattaneo and Felder [68] by a symplectic reduction process. Their original motivation was not the integrability problem, but rather Poisson sigma-models and quantization of Poisson manifolds (see Appendix A.9).

Appendix

A.1 Moser's path method

A smooth time-dependent vector field $X = (X_t)_{t \in]a,b[}$ on a manifold M is a smooth path of smooth vector fields X_t on M, parametrized by a parameter t (the *time parameter*) taken in some interval $]a, b[$. Any smooth path $(\phi_t)_{t \in]a,b[}$ of diffeomorphisms determines a time-dependent vector field X by the formula

$$X_t(\phi_t(x)) = \frac{\partial \phi_t}{\partial t}(x). \tag{A.1}$$

Conversely, the classical theory of ordinary differential equations says that, given a smooth time-dependent vector field $X = (X_t)_{t \in]a,b[}$, in a neighborhood of any (x_0, t_0) in $M \times]a, b[$, we can define a unique smooth map $(x, t) \mapsto \phi_t(x)$, called the *flow* of X starting at time t_0, with $\phi_{t_0}(x) \equiv x$ and which satisfies Equation (A.1). Because ϕ_{t_0} is locally the identity, ϕ_t are local diffeomorphisms for t sufficiently near t_0. In some circumstances, for example when X has compact support, these local diffeomorphisms extend to global diffeomorphisms. Taking then $t_0 = 0$, we can get by this procedure a path of (local) diffeomorphisms $(\phi_t)_t$ with $\phi_0 = \mathrm{Id}$.

Suppose that we want to prove that two tensors Λ and Λ', on a manifold M, are isomorphic, i.e., we want to show the existence of a diffeomorphism ϕ of the ambient manifold such that

$$\phi^*(\Lambda') = \Lambda. \tag{A.2}$$

Sometimes, this isomorphism problem can be solved with the help of *Moser's path method*, which consists of the following:

- First, construct an adapted smooth path $(\Lambda_t)_{t \in [0,1]}$ of such tensors, with $\Lambda_0 = \Lambda$ and $\Lambda_1 = \Lambda'$.

- Second, try to construct a smooth path $(\phi_t)_{t \in [0,1]}$ of diffeomorphisms of M with $\phi_0 = \mathrm{Id}$ and

$$\phi_t^*(\Lambda_t) = \Lambda_0 \quad \forall\, t \in [0, 1], \tag{A.3}$$

  or equivalently,

$$\partial(\phi_t^*(\Lambda_t))/\partial t = 0 \quad \forall\, t \in [0, 1]. \tag{A.4}$$

One tries to construct a time-dependent vector field $X = (X_t)$ whose flow (starting at time 0) gives ϕ_t. Equation (A.4) is then translated to the following equation on (X_t):

$$\mathcal{L}_{X_t}(\Lambda_t) = -\frac{\partial \Lambda_t}{\partial t}. \tag{A.5}$$

If this last equation can be solved, then one can define ϕ_t by integrating X, and $\phi = \phi_1$ will solve Equation (A.2).

Moser's path method works best for local problems because time-dependent vector fields can always be integrated locally. In global problems, one usually needs an additional compactness condition to assure that $\phi = \phi_1$ is globally defined.

The method is named after Jürgen Moser, who first used it to prove the following result:

Theorem A.1.1 ([265]). *Let ω and ω' be two volume forms on a manifold M which coincide everywhere except on a compact subset K. If we have $\omega - \omega' = d\alpha$ where the form α has its support in K, then there is a diffeomorphism ϕ of M, which is identity on $M \setminus K$, such that $\phi^*(\omega') = \omega$.*

Proof. We have $\omega' = f\omega$ where f is a strictly positive function on M ($f = 1$ on $M \setminus K$). Then $\omega_t = f_t\omega$, with $f_t = tf + 1 - t$, $t \in [0, 1]$, is a path of volume forms on M. Now Equation (A.5) reduces to

$$d i_{X_t}\omega_t = \omega - \omega'. \tag{A.6}$$

The hypothesis of the theorem allows us to replace Equation (A.6) by

$$i_{X_t}\omega_t = \alpha. \tag{A.7}$$

But this last equation has a unique solution X_t. Since the support of X_t lies in K, we can integrate this time-dependent vector field to a path of diffeomorphisms ϕ_t on M which is identity on $M \setminus K$, and such that $\phi_1^*(\omega') = \omega$. □

In particular, when $M = K$ we get the following corollary:

Corollary A.1.2 ([265]). *Two volume forms on a compact manifold are isomorphic if and only if they have the same total volume.*

It was pointed out by Weinstein [344, 345] that the path method works very well in the local study of symplectic manifolds. A basic result in that direction is the following.

Theorem A.1.3. *Let $(\omega_t)_{t \in [0,1]}$ be a smooth path of symplectic forms on a manifold M. If we have $\partial \omega_t / \partial t = d\gamma_t$ for a smooth path γ_t of 1-forms with compact support, then there is a diffeomorphism ϕ of M with $\phi^*\omega_1 = \omega_0$.*

Proof. Equation (A.5) follows from the equation $i_{X_t}\omega_t = -\gamma_t$, which has a solution $(X_t)_t$ because ω_t is nondegenerate, so the path method works in this case. □

Theorem A.1.4 (see [243]). *Let K be a compact submanifold of a manifold M. Suppose that ω_0 and ω_1 are two symplectic forms on M which coincide at each point of K. Then there exist neighborhoods N_0 and N_1 of K and a diffeomorphism $\phi : N_0 \longrightarrow N_1$, which fixes K, such that $\phi^*\omega_1 = \omega_0$.*

Proof. Consider the path $\omega_t = (1-t)\omega_0 + t\omega_1$ of symplectic forms in a neighborhood of K. Similarly to the proof of the previous theorem, it is sufficient to find a 1-form γ such that $\gamma(x) = 0$ for any $x \in K$ and

$$d\gamma = \partial\omega_t/\partial t = \omega_1 - \omega_0. \tag{A.8}$$

The existence of such a γ is a generalization of Poincaré's lemma which says that a closed form is exact on any contractible open subset of a manifold. It can be proved by the following method, inspired by the Moser path method. Choose a sufficiently small tubular neighborhood T of K in M and denote by ψ_t the mapping from T into T which is the linear contraction $v \mapsto tv$ along the fibers of T. With $\theta = \omega_1 - \omega_0$, we have

$$\theta = \psi_1^*(\theta) - \psi_0^*(\theta) = \int_0^1 \frac{\partial\psi_t^*\theta}{\partial t} dt, \tag{A.9}$$

because ψ_1 is the identity and ψ_0 has its range in K. Now, if we denote by Y_t the time-dependent vector field associated by Formula (A.1) to the path $(\psi_t)_{t\in]0,1]}$, we get

$$\frac{\partial\psi_t^*\theta}{\partial t} = \psi_t^*\mathcal{L}_{Y_t}\theta = d\gamma_t, \tag{A.10}$$

with $\gamma_t = \psi_t^*\iota_{Y_t}\theta$. The path γ_t is, a priori, defined only for $t > 0$, but it extends clearly to $t = 0$; also it vanishes on K. So Equation (A.9) gives $\theta = d\gamma$ with $\gamma = \int_0^1 \gamma_t dt$, and leads to the conclusion. $\quad\square$

A direct corollary of Theorem A.1.4 is the following result of Weinstein:

Theorem A.1.5 ([344]). *Let L be a compact Lagrangian submanifold of the symplectic manifold (M, ω). There is a neighborhood N_1 of L in M, a neighborhood N_0 of L (identified with the zero section) in T^*L and a diffeomorphism $\phi : N_0 \longrightarrow N_1$, which fixes L, such that $\phi^*\omega$ is the canonical symplectic form on T^*L.*

Proof (sketch). Choose a Lagrangian complement E_x to each T_xL in T_xM in order to get a fiber bundle E over L complement to TL in $TM|_L$. We construct a fiber bundle isomorphism $f : E \longrightarrow T^*L$ by $f(v)(w) = \omega(v, w)$. As E realizes a normal bundle to L, we can consider that f gives a diffeomorphism from a tubular neighborhood of L in M to a neighborhood of L in T^*L which sends $\omega|_L$ to the canonical symplectic form of T^*L (restricted to L). So we can suppose that ω is defined on a neighborhood of L in T^*L and is equal to the canonical symplectic form at every point on L. Then we achieve our goal using Theorem A.1.4. $\quad\square$

When the compact submanifold K is just one point, we recover from Theorem A.1.4 the classical Darboux's theorem:

Theorem A.1.6 (Darboux). *Every point of a symplectic manifold admits a neighborhood with a local system of coordinates* $(p_1, \ldots, p_n, q_1, \ldots, q_n)$ (*called* Darboux coordinates *or* canonical coordinates) *in which the symplectic form has the standard form* $\omega = \sum_{i=1}^{n} \mathrm{d}p_i \wedge \mathrm{d}q_i$.

Proof. We can suppose that we work with a symplectic form ω near the origin 0 in $\mathbb{R}^{2n}$. Moreover we can suppose, up to a linear change, that a first system of coordinates is chosen such that $\omega(0) = \omega_0(0)$ where $\omega_0 = \sum_{i=1}^{n} \mathrm{d}p_i \wedge \mathrm{d}q_i$. Then we apply Theorem A.1.4 to the case $K = \{0\}$. $\qquad\square$

In fact, Moser's path method gives a simple proof of the following *equivariant* version of Darboux's theorem [345], which would be very hard (if not impossible) to prove by the classical method of coordinate-by-coordinate construction.

Theorem A.1.7 (Equivariant Darboux theorem). *Let* G *be a compact Lie group which acts symplectically on a symplectic manifold* (M, ω) *and which fixes a point* $z \in M$. *Then there is a local canonical system of coordinates* $(p_1, \ldots, p_n, q_1, \ldots, q_n)$ *in a neighborhood of* z *in* M, *with respect to which the action of* G *is linear.*

Proof. One first linearizes the action of G near z using Bochner's Theorem 7.4.1. Then, after a linear change, one arrives at a system of coordinates $(p_1, \ldots, p_n, q_1, \ldots, q_n)$ in which the action of G is linear, and such that $\omega(0) = \omega_0(0)$ where $\omega_0 = \sum_{i=1}^{n} \mathrm{d}p_i \wedge \mathrm{d}q_i$. One then uses the path method to move ω to ω_0 in a G-equivariant way. To do this, one must find a 1-form γ such that $\omega - \omega_0 = \mathrm{d}\gamma$ as in the proof of Theorem A.1.3, and moreover γ must be G-invariant in order to assure that the resulting flow ϕ_t is G-invariant (i.e., it preserves the action of G). Note that both ω and ω_0 are G-invariant. In order to find such a G-invariant 1-form γ, one starts with an arbitrary 1-form $\hat{\gamma}$ such that $\omega - \omega_0 = \mathrm{d}\hat{\gamma}$, and then averages it by the action of G:

$$\gamma = \int_G \rho(g)^* \hat{\gamma} \mathrm{d}\mu, \tag{A.11}$$

where ρ denotes the action of G, and $\mathrm{d}\mu$ denotes the Haar measure on G. $\qquad\square$

Moser's path method has also become an essential tool in the study of singularities of smooth maps. In that domain we often have to construct equivalences between two maps f_0 and f_1, e.g., relations $g \circ \phi = f$ (right equivalence), or $g \circ \phi = \psi \circ f$ (right-left equivalence), or more general equivalences (contact equivalence, etc.), which are given by some diffeomorphisms (ϕ, ψ, etc.). To use the path method we first construct an appropriate path (f_t) which connects f_0 to f_1. Then we try to find a path of diffeomorphisms which gives the corresponding equivalence between f_t and f_0. Differentiating the equation with respect to t, we get a version of Equation (A.5). For example, in the case of right-left equivalence we find the equation

$$\mathrm{d}f_t(X_t(x)) + Y_t(f_t(x)) = -\frac{\partial f_t(x)}{\partial t}, \tag{A.12}$$

where the unknowns are time-dependent vector fields X_t and Y_t on the source and the target spaces respectively. The singularists have developed various methods to solve these equations, such as the celebrated preparation theorem (see, e.g., [150, 16]). The Tougeron theorem that we present below is typical of the use of the path method in this domain.

Let $C_0^\infty(\mathbb{R}^n)$ be the algebra of germs at the origin of smooth functions $f : \mathbb{R}^n \to \mathbb{R}$. We denote by the same letter such a function and its germ at the origin. Let $\Delta(f) \subset C_0^\infty(\mathbb{R}^n)$ be the ideal generated by the partial derivatives $\frac{\partial f}{\partial x_1}, \ldots,$ $\frac{\partial f}{\partial x_n}$ of f. The *codimension* of $f \in C_0^\infty(\mathbb{R}^n)$ is, by definition, the dimension of the real vector space $C_0^\infty(\mathbb{R}^n)/\Delta(f)$.

We say that $f \in C_0^\infty(\mathbb{R}^n)$ is *k-determinant* if every $g \in C_0^\infty(\mathbb{R}^n)$, such that f and g have the same Taylor expansion at the origin up to order k, is right equivalent to f.

Theorem A.1.8 (Tougeron [330]). *If $f \in C_0^\infty(\mathbb{R}^n)$ has finite codimension k, then it is $(k+1)$-determinant.*

Proof. Denote by $\mathfrak{M} \subset C_0^\infty(\mathbb{R}^n)$ the ideal of germs of functions vanishing at the origin. Consider the following sequence of inequalities:

$$\dim C_0^\infty(\mathbb{R}^n)/(\mathfrak{M} + \Delta(f)) \leq \dim C_0^\infty(\mathbb{R}^n)/(\mathfrak{M}^2 + \Delta(f)) \leq \cdots$$
$$\leq \dim C_0^\infty(\mathbb{R}^n)/(\mathfrak{M}^m + \Delta(f)) \leq \dim C_0^\infty(\mathbb{R}^n)/(\mathfrak{M}^{m+1} + \Delta(f)) \leq \cdots \leq k. \tag{A.13}$$

It follows that there is a number $q \in \mathbb{Z}, 0 \leq q \leq k$, such that $\mathfrak{M}^q + \Delta(f) = \mathfrak{M}^{q+1} + \Delta(f)$, and hence $\mathfrak{M}^q \subset \mathfrak{M}^{q+1} + \Delta(f)$. Applying Nakayama's lemma (Lemma A.1.9) to this relation, we get $\mathfrak{M}^q \subset \Delta(f)$, which implies that

$$\mathfrak{M}^{k+2} \subset \mathfrak{M}^2 \Delta(f). \tag{A.14}$$

Let $g \in C_0^\infty(\mathbb{R}^n)$ be a function with the same $(k+1)$-Taylor expansion as f. We consider the path $f_t = (1-t)f + tg$ and, following the path method, try to construct a path of local diffeomorphisms $(\phi_t)_t$ which fixes the origin and such that $f_t \circ \phi_t = f$. Here the equation to solve (Equation (A.5)) becomes

$$\sum_{i=1}^n \frac{\partial f_t}{\partial x_i}(x) X_i(t, x) = (f - g)(x), \tag{A.15}$$

where the unknown functions $X_i(t, x)$ must be such that $X_i(t, 0) = 0$. In fact, we will try to find X_i such that $X_i(t, .) \in \mathfrak{M}^2$ for any t, so the differential of the resulting diffeomorphisms ϕ_t at 0 will be equal to identity.

By compactness of the interval $[0, 1]$, we need only be able to construct $(\phi_t)_t$ for t near any fixed $t_0 \in [0, 1]$,, i.e., we need to solve Equation (A.15) only for t near t_0 (and x near 0). We denote by A the ring of germs at $(t_0, 0)$ of smooth functions of $(t, x) \in \mathbb{R} \times \mathbb{R}^n$; $C_0^\infty(\mathbb{R}^n)$ is considered naturally as a subring of A.

Then Equation (A.15) (with $X_i \in \mathfrak{M}^2$, where $\mathfrak{M}$ is now the ideal of A generated by germs of functions f such that $f(t,0) = 0$ for any t near t_0) can be replaced by

$$\mathfrak{M}^{k+2} \subset \mathfrak{M}^2 \Delta(f_t), \tag{A.16}$$

where $\Delta(f_t)$ is now the ideal of A generated by the functions $h(t,x) = \frac{\partial f_t(x)}{\partial x_i}$ (recall that $f - g$ belongs to $\mathfrak{M}^{k+2}$).

Now, because $\frac{\partial f_t}{\partial x_i} = \frac{\partial f}{\partial x_i}$ modulo $\mathfrak{M}^{k+1}A$, we have $\Delta(f) \subset \Delta(f_t) + \mathfrak{M}^{k+1}A$, and Relation (A.14) leads to

$$\mathfrak{M}^{k+2}A \subset \mathfrak{M}^2\Delta(f_t) + \mathfrak{M}^{k+3}A, \tag{A.17}$$

or

$$\mathfrak{M}^{k+2}A \subset \mathfrak{M}^2\Delta(f_t) + I \cdot \mathfrak{M}^{k+2}A, \tag{A.18}$$

where I is the ideal of A consisting of germs of functions vanishing at $(t_0, 0)$. Applying again Nakayama's lemma, we get $\mathfrak{M}^{k+2}A \subset \mathfrak{M}^2\Delta(f_t)$, i.e., Relation (A.16), which leads to the result. $\qquad\qquad\square$

Lemma A.1.9 (Nakayama's lemma). *Let R be a commutative ring with unit and I an ideal of R such that, for any x in I, $1 + x$ is invertible. If M and N are two R-modules, M being finitely generated, then the relation $M \subset N + IM$ implies that $M \subset N$.*

Proof. Choose a system of generators $m_1, \dots, m_q$ for M. Then $M \subset N + IM$ gives a linear system $\sum_{j=1,\dots,q} a_i^j m_j = n_i$ where n_i are elements of N, $i = 1, \dots, q$ and a_i^j are elements of R of the form $a_i^j = \delta_i^j + \nu_i^j$, where δ_i^j is the Kronecker symbol and ν_i^j are in I. As the matrix (a_i^j) is invertible, we can write the m_i as linear combinations of the n_j and obtain $M \subset N$. $\qquad\qquad\square$

A special case of the preceding Tougeron's theorem with $k = 1$ is the Morse lemma which says that, near any singular point with invertible Hessian matrix, a smooth function is right equivalent to $(x_1, \dots, x_n) \mapsto c + \sum_{i=1,\dots,n} \pm x_i^2$.

Moser's path method is also useful in the study of contravariant tensors, e.g., vector fields and Poisson structures. In the case of vector fields, we find the equation

$$[X_t, Z_t] = -\frac{\partial Z_t}{\partial t}, \tag{A.19}$$

where $(Z_t)_t$ is a given path of vector fields and $(X_t)_t$ is the unknown (see, e.g., [304]).

For Poisson structures, we get a similar equation:

$$[X_t, \Pi_t] = -\frac{\partial \Pi_t}{\partial t}, \tag{A.20}$$

where $(\Pi_t)_t$ is a path of Poisson structures. However, the use of Moser's path method in the study of Poisson structures is rather tricky, because it is not easy to find a path of Poisson structures which connects two given Poisson structures, even locally, due to the Jacobi condition. One needs to do some preparatory work first, for example to make the two original Poisson structures have the same characteristic foliation (then it will become easier to find a path connecting the two structures). The *equivariant splitting theorem* for Poisson structures can be proved using this approach, see [251].

A.2 Division theorems

We will give here a version of Saito's division theorem [308] which works not only for germs of analytic forms but also for germs of smooth forms.

Let R be a commutative ring with unit. A *regular sequence* of R is a sequence $(a_1, a_2, \ldots, a_q)$ of elements of R such that

 i) a_1 is not a zero divisor of R, and
 ii) the class of a_i is not a zero divisor in $R/(a_1 R + \cdots + a_{i-1} R)$ for $i = 2, \ldots, q$.

Let M be a free module over R of finite rank n. Let $\omega_1, \ldots, \omega_k$ be given elements of M and $e_1, \ldots, e_n$ a basis of M,

$$\omega_1 \wedge \cdots \wedge \omega_k = \sum_{1 \leq i_1 < \cdots < i_k \leq n} a_{i_1 \cdots i_k} e_{i_1} \wedge \cdots \wedge e_{i_k}.$$

We call A the ideal generated by the family of coefficients $a_{i_1 \cdots i_k}$.

Theorem A.2.1. *With the above notations, let ω be an element of the exterior product $\bigwedge^p M$ with p strictly smaller than the maximal length of regular sequences of R with all elements lying in A. Then the relation*

$$\omega \wedge \omega_1 \wedge \cdots \wedge \omega_k = 0 \tag{A.21}$$

implies the existence of k elements $\eta_1, \ldots, \eta_k$ of $\bigwedge^{p-1} M$ such that

$$\omega = \sum_{i=1}^{k} \eta_i \wedge \omega_i. \tag{A.22}$$

When $k = 1$ this theorem is de Rham's division theorem [98].

Proof. We will use a double induction on (p, k).

For $p = 0$, ω is just an element of R and the hypothesis implies that it is 0. So the result is proved in that case.

We suppose that either $k = 1$ and the result is proved for $(p - 1, 1)$, or $k > 1$ and the result is proved for $(p, k - 1)$ and $(p - 1, k)$.

Let $(a_1, a_2, \ldots, a_q)$ be a regular sequence of R with all elements in A. Fix an element $\omega \in \bigwedge^p M$ with $0 < p < q$. We extend R to R_{a_1}, the localization of R by the powers of a_1; this means that we add $1/a_1$ (the inverse of a_1). We can consider that ω and the ω_i have coefficients in R_{a_1}. With this trick we obtain that the ideal generated by the coefficients of $\omega_1 \wedge \cdots \wedge \omega_k$ contains an invertible element. It means that $\omega_1, \ldots, \omega_k$ are independent and can be seen as a part of a basis of M over R_{a_1}. The hypothesis $\omega \wedge \omega_1 \wedge \cdots \wedge \omega_k = 0$ implies that $\omega = \sum_{i=1}^{k} \eta_i \wedge \omega_i$ over R_{a_1}. Since this relation is valid only after the preceding localization, we get

$$a_1^m \omega = \sum_{i=1}^{k} \eta_i \wedge \omega_i \tag{A.23}$$

for an appropriate natural number m, when we return to the initial ring R.

Now, when we tensorize with $R/a_1^m R$, this last relation gives simply

$$0 = \sum_{i=1}^{k} \overline{\eta}_i \wedge \overline{\omega}_i, \tag{A.24}$$

where the image of $\alpha \in \bigwedge^r M$ in $\bigwedge^r M \otimes R/a_1^m R$ is written $\overline{\alpha}$. For any $j = 1, \ldots, k$ we have

$$\overline{\eta}_j \wedge \overline{\omega}_1 \wedge \cdots \wedge \overline{\omega}_k = \pm \Big(\sum_{i=1}^{k} \overline{\eta}_i \wedge \overline{\omega}_i \Big) \wedge \overline{\omega}_1 \wedge \cdots \wedge \overline{\omega}_{j-1} \wedge \overline{\omega}_{j+1} \wedge \cdots \wedge \overline{\omega}_k = 0. \tag{A.25}$$

Denote by $\overline{a}$ the class of an element a of R modulo a_1^m. Then the sequence $(\overline{a}_2, \overline{a}_3, \ldots, \overline{a}_q)$ is a regular sequence in $R/a_1^m R$ (see Exercise 11 of Chapter X of [44]). So we can apply to $\overline{\eta}_j$ the induction hypothesis: we obtain $\overline{\xi}_j^i \in \bigwedge^{p-2} M \otimes R/a_1^m R$ $(i = 1, \ldots, k)$ such that $\overline{\eta}_j = \sum_{i=1}^{k} \overline{\xi}_j^i \wedge \overline{\omega}_i$. Returning to R, these relations give

$$\eta_j = \sum_{i=1}^{k} \xi_j^i \wedge \omega_i + a_1^m \nu_j. \tag{A.26}$$

So Equation (A.23) becomes

$$a_1^m \Big(\omega - \sum_{j=1}^{k} \nu_j \wedge \omega_j \Big) = \sum_{i,j=1}^{k} \xi_j^i \wedge \omega_i \wedge \omega_j. \tag{A.27}$$

Multiplying by $\omega_2 \wedge \cdots \wedge \omega_k$, we get

$$a_1^m \Big(\omega - \sum_{j=1}^{k} \nu_j \wedge \omega_j \Big) \omega_2 \wedge \cdots \wedge \omega_k = 0 \tag{A.28}$$

(in the case $k = 1$ we skip this step). As a_1 is not a zero divisor, we can erase a_1^m in this last equation. In the case $k = 1$ this gives the result. If $k > 1$ we first remark that the ideal generated by the coefficients of $\omega_2 \wedge \cdots \wedge \omega_k$ contains A. Hence we can apply our induction hypothesis (case $(p, k-1)$) to get the equation

$$\omega - \sum_{j=1}^{k} \nu_j \wedge \omega_j = \sum_{i=2}^{k} \beta_i \wedge \omega_i \tag{A.29}$$

which leads to the result. $\qquad\square$

Let the 1-form $\alpha = \sum_{i=1}^{n} a_i dx_i$ be analytic (resp. smooth) on $\mathbb{R}^n$ or holomorphic on $\mathbb{C}^n$. We say that the origin is an algebraically isolated singularity of α if the ideal generated by the germs at 0 of $a_1, \ldots, a_n$ has finite $\mathbb{R}$-codimension in the algebra of germs at 0 of analytic (resp. smooth) or holomorphic functions on $\mathbb{R}^n$ or on $\mathbb{C}^n$.

We say that such a 1-form has the *division property* at 0 if the germ relation $\omega \wedge \alpha = 0$ implies a germ relation $\omega = \beta \wedge \alpha$ for any germ of p-form ω with $p < n$ (β being the germ of a $p-1$-form).

The following theorem is a consequence of de Rham's division theorem (i.e., the case $k = 1$ of Theorem A.2.1).

Theorem A.2.2. *Let the 1-form $\alpha = \sum_{i=1}^{n} a_i dx_i$ be analytic or smooth on a neighborhood of 0 in $\mathbb{R}^n$, or holomorphic on a neighborhood of 0 in $\mathbb{C}^n$. If 0 is an algebraically isolated singularity of α, then α has the division property at 0.*

More generally, if α is a holomorphic 1-form on $(\mathbb{C}^n, 0)$ such that the singular set $S = \{x \in \mathbb{C}^n \mid \alpha(x) = 0\}$ has codimension $m \geq 2$, then any germ of holomorphic p-form ω such that $p < m$ and $\omega \wedge \alpha = 0$, is dividable by α. See, e.g., [267, 233].

A.3 Reeb stability

Recall that the *holonomy* of a leaf L of regular foliation $\mathcal{F}$ on a manifold M at a point $x \in L$ is a homomorphism from the fundamental group $\pi_1(L, x)$ to the group $\mathrm{Diff}(\mathbb{R}^p, 0)$ of germs of diffeomorphisms of $(\mathbb{R}^p, 0)$, where p is the codimension of $\mathcal{F}$. It can be defined as follows: Let $\gamma : [0, 1] \to L$ be a loop in L, $\gamma(0) = \gamma(1) = x$. Denote by $\mathcal{U}$ a tubular neighborhood of γ in M, and $\pi : \mathcal{U} \to L$ a projection (which is identity on $L \cap \mathcal{U}$), such that the preimage of each point of $y \in L \cap \mathcal{U}$ is a small p-dimensional disk D_y^p. If $z \in D_x$ is sufficiently close to x, then there is a unique loop $\gamma_z : [0, 1] \to L_z \cap \mathcal{U}$, where L_z is the leaf through z, such that $\gamma_z(0) = z$ and $\pi(\gamma_z(t)) = \gamma(t) \ \forall \ t \in [0, 1]$. The map $\phi_\gamma : z \mapsto \gamma_z(1)$ is a local diffeomorphism of (D_x^p, x). Its germ at x is called the holonomy around γ and depends only on the homotopy class of the loop γ, so we can assign to each element $[\gamma]$ of $\pi_1(L, x)$ a germ of diffeomorphism ϕ_γ of $(D_x^p, 0)$. The concatenation of two loops clearly leads to the composition of two corresponding germs of diffeomorphisms, so we get a group homomorphism from $\pi_1(L, x)$ to $\mathrm{Diff}(D_x^p, 0)$. This is, by definition, the holonomy of $\mathcal{F}$ at x. Its image in $\mathrm{Diff}(D_x^p, 0)$ is called the holonomy group of $\mathcal{F}$ at x. Up to isomorphisms, this holonomy depends only on $\mathcal{F}$ and L, and not on the choice of x, $\mathcal{U}$ and π. We will say that a leaf L has finite (resp. trivial) holonomy if its holonomy group is finite (resp., trivial). For example, when L is simply connected, then it automatically has trivial holonomy.

The following theorem, called the *Reeb stability* theorem, says that if the holonomy of a leaf is trivial, then the foliation is "stable" near that leaf.

Theorem A.3.1 (Reeb [356]). *Let L be a compact leaf of a regular foliation $\mathcal{F}$ of codimension p on a manifold M. Assume that L has trivial holonomy (this condition is automatically satisfied if L is simply-connected). Then there is a neighborhood $\mathcal{U}$ of L in M in which the foliation $\mathcal{F}$ is diffeomorphic to a trivial fibration $\mathcal{F} \times \mathbb{R}^p \to \mathbb{R}^p$.*

Proof (sketch). Since L is compact, there is a tubular neighborhood $\mathcal{U}$ of L in M and a projection $\pi : \mathcal{U} \to L$, such that the preimage of a point $x \in L$ is a p-dimensional disk D_x^p. Since the holonomy of L is trivial, for each point $y \in \mathcal{U}$ close enough to L, there is a unique point $\rho(y) \in D_x^p$ such that y can be connected to $\rho(y)$ by a path which lies on a leaf of $\mathcal{F}$ and which is close to L. The map

$$y \mapsto (\pi(y), \rho(y)) \tag{A.30}$$

is a diffeomorphism from a neighborhood of L in $\mathcal{U}$ to a neighborhood of $L \times \{x\}$ in $L \times D_x^p$. Under this map, leaves of $\mathcal{F}$ are sent to fibers of the trivial fibration $L \times D_x^p \to D_x^p$, hence the conclusion of the theorem. $\square$

Reeb's stability theorem can be used to give a simple proof of the following folklore theorem, which is somehow difficult to find in the literature (see, e.g., [132]):

Theorem A.3.2. *Let $\mathfrak{g}$ be the Lie algebra of a connected simply-connected compact Lie group G. Then any smooth action of $\mathfrak{g}$ on a manifold M with a fixed point $p \in M$ can be integrated, in a neighborhood of p, to an action of G.*

Proof (sketch). Following Palais [290], let us consider the action of $\mathfrak{g}$ on $G \times M$ given by $x \mapsto x^+ + \xi_x$, where x^+ denotes the left-invariant vector field on G generated by $x \in \mathfrak{g}$, and ξ_x is the vector field on M generated by x via the action of $\mathfrak{g}$. The image of this action spans a regular involutive distribution on $G \times M$ (of rank equal to the dimension of G), so by Frobenius' theorem this distribution is integrable. Since p is a fixed point of the action of $\mathfrak{g}$, the submanifold $G \times \{p\}$ is a leaf of the corresponding foliation on $G \times M$. Since this leaf is compact simply-connected, the nearby leaves are also diffeomorphic to G by Reeb's stability theorem. This implies that the action of $\mathfrak{g}$ on M integrates to a free (right) action of G in a neighborhood of $G \times \{p\}$ in $G \times M$ (which projects to the action of G on itself by multiplication from the right). By projecting the action of G to the second component of the product, one gets an action of G in a neighborhood of p in M. $\square$

A.4 Action-angle variables

There is a huge amount of literature on integrable systems. See, e.g., [19, 38, 20, 102, 2] for an introduction to the subject and many concrete examples. In this section, we will discuss mainly only one particular aspect of integrable Hamiltonian systems, namely the existence of action-angle variables, with an emphasis on action variables. We need these action variables in the study of proper symplectic groupoids (Section 7.5) and convexity of momentum maps (see [370]).

Recall that a function H (or the corresponding Hamiltonian vector field X_H) on a $2n$-dimensional symplectic manifold (M^{2n}, ω) is called *Liouville integrable*, if it admits n functionally independent first integrals in involution. In other words, there are n functions $F_1 = H, F_2, \ldots, F_n$ on M^{2n} such that $dF_1 \wedge \cdots \wedge dF_n \neq 0$ almost everywhere and $\{F_i, F_j\} = 0 \ \forall \ i, j$. The map

$$\mathbf{F} = (F_1, \ldots, F_n) : (M^{2n}, \omega) \to \mathbb{K}^n \qquad (A.31)$$

(where $\mathbb{K} = \mathbb{R}$ or $\mathbb{C}$) is called the *momentum map* of the system. It is actually the momentum map of a Hamiltonian action of the Abelian Lie algebra of dimension n on (M^{2n}, ω). The (regular) level sets of $\mathbf{F}$ are Lagrangian submanifolds of (M^{2n}, ω), and they are invariant with respect to X_H. A classical result attributed to Liouville [215] says that, in the smooth case, if a connected level set N is compact and does not intersect with the boundary of M^{2n} (of if M^{2n} has no boundary), then it is diffeomorphic to a standard torus $\mathbb{T}^n$ (= quotient of $\mathbb{R}^n$ by a co-compact lattice), and the Hamiltonian system X_H is quasi-periodic on N. In other words, there is a periodic coordinate system $(q_1, \ldots, q_n)$ on N with respect to which the restriction of X_H to N has constant coefficients: $X_H = \sum \gamma_i \partial/\partial q_i$, γ_i being constants. For this reason, N is called a *Liouville torus*.

In practice, one often deals with Hamiltonian systems which admit a non-Abelian algebra of symmetries. For such systems, it is more convenient to work with a generalized notion of Liouville integrability. The following definition is essentially due to Nekhoroshev [276] and Mischenko and Fomenko [252]:

Definition A.4.1. A Hamiltonian vector field X_H on a Poisson manifold (M, Π) is called *integrable in generalized Liouville sense* if there are positive integers p, q, a q-tuple $\mathbf{F} = (F_1, \ldots, F_q)$ of functions on M and a p-tuple $\mathbf{X} = (X_{G_1}, \ldots, X_{G_p})$ of Hamiltonian vector fields on (M, Π), such that the following conditions are satisfied:

   i) $\{F_i, G_j\} = \{G_i, G_j\} = \{F_i, H\} = \{G_i, H\} = 0 \ \forall \ i, j$,
   ii) $dF_1 \wedge \cdots \wedge dF_q \neq 0$ and $X_{G_1} \wedge \cdots \wedge X_{G_p} \neq 0$ almost everywhere,
   iii) $p + q = \dim M$.

We may put $G_1 = H$ in the above definition. When (M, Π) is symplectic and $p = q = \dim M/2$, the functions in $\mathbf{F}$ automatically Poisson-commute and we get back to the usual Liouville integrability. The above notion of integrability is also called *noncommutative integrability* due to the fact that the components of $\mathbf{F}$ do not Poisson-commute in general, and in many cases one may choose them from a finite-dimensional non-Abelian Lie algebra of functions under the Poisson bracket.

By abuse of language, we will also say that $(\mathbf{X}, \mathbf{F})$ is an *integrable Hamiltonian system* in generalized Liouville sense, and call (p, q) the *bi-degree* of the system. We will say that the system $(\mathbf{X}, \mathbf{F})$ is *proper* if the generalized momentum map $\mathbf{F} : M \to \mathbb{R}^q$ is a proper map from M to its image, and the image of the singular set $\{x \in M, X_{G_1} \wedge \cdots \wedge X_{G_p}(x) = 0\}$ of the commuting Hamiltonian vector fields under the map $\mathbf{F} : M \to \mathbb{R}^q$ is nowhere dense in $\mathbb{R}^q$. A connected component N of a level set of $\mathbf{F}$ will be called a *regular* if it is regular with respect to $\mathbf{F}$, and the vector fields of $\mathbf{X}$ are independent *everywhere* on N. We have the following obvious generalization of Liouville's theorem: a regular connected level set N of a proper noncommutatively integrable Hamiltonian system $(\mathbf{X}, \mathbf{F})$ is diffeomorphic to a torus $\mathbb{T}^p$ on which the flow of any vector field X_H such that $\{H, F_i\} = 0 \; \forall i$ is quasi-periodic; in analogy with the Liouville-integrable case, we will call such N a *Liouville torus*.

We will restrict our attention to the case when (M, Π) is a symplectic manifold, i.e., Π is nondegenerate. There is a very close relation between noncommutatively integrable Hamiltonian systems and symplectically complete foliations in the sense of Libermann (see Section 1.9). Indeed, if $(\mathbf{X}, \mathbf{F})$ is an integrable system on a symplectic manifold (M, ω), then the regular part of M (consisting of points x such that $dF_1 \wedge \cdots \wedge dF_q(x) \neq 0$ and $X_{G_1} \wedge \cdots \wedge X_{G_p}(x) \neq 0$) is a symplectically complete foliation by isotropic submanifolds (= level sets of $\mathbf{F}$): the dual foliation is given by the level set of the map $\mathbf{G} = (G_1, \ldots, G_p)$. (The pair $(\mathbf{F}, \mathbf{G})$ is a so-called *dual pair*, see [288].) Conversely, assume that $\mathcal{W}$ is a symplectically complete foliation on M. Then the set of points $x \in M$ such that $\dim(T_x \mathcal{W} \cap (T_x \mathcal{W})^\perp)$ is locally constant is a dense open subset of M on which $T_x \mathcal{W} \cap (T_x \mathcal{W})^\perp$ is an integrable distribution. The corresponding foliation $\mathcal{F}$ is isotropic symplectically complete, its dual foliation $\mathcal{G}$ is given by the coisotropic distribution $T_x \mathcal{W} \cup (T_x \mathcal{W})^\perp$. Assume that we can realize the leaves of $\mathcal{F}$ (resp., $\mathcal{G}$) as the connected components of the level set of a map $\mathbf{F} = (F_1, \ldots, F_q)$ from M to $\mathbb{R}^q$ (resp., of a map $\mathbf{G} = (F_1, \ldots, F_p)$ from M to $\mathbb{R}^p$) almost everywhere. Then $(X_{G_1}, \ldots, X_{G_p}, F_1, \ldots, F_q)$ is an integrable Hamiltonian system of bi-degree (p, q).

Example A.4.2. A Hamiltonian $\mathbb{T}^p$-action on a Poisson manifold can be seen as a proper integrable system: $\mathbf{X}$ is generated by the components of the momentum map of the $\mathbb{T}^p$-action, and $\mathbf{F}$ consists of an appropriate number $\mathbb{T}^p$-invariant function. One can associate to each Hamiltonian action of a compact Lie group G on a Poisson manifold a proper integrable system: H is the composition of the momentum map $\mu \colon M \to \mathfrak{g}^*$ with a generic Ad^*-invariant function $h \colon \mathfrak{g}^* \to \mathbb{R}$ (see [371]).

Example A.4.3. If $(\Gamma, \omega) \rightrightarrows (P, \Pi)$ is a proper symplectic groupoid, then it gives rise to a natural proper integrable Hamiltonian system $(\mathbf{X}, \mathbf{F})$ on (Γ, ω): $\mathbf{F}$ consists of functions of the types $s^* f$ and $t^* f$, $f : P \to \mathbb{R}$, and $\mathbf{X}$ consists of Hamiltonian vector fields of the type $X_{s^* f}$, where f is a Casimir function on P. The corresponding symplectically complete isotropic foliation is given by the submanifolds $s^{-1}(x) \cap t^{-1}(y)$, $x, y \in P$.

Remark A.4.4. Proper integrable Hamiltonian systems in generalized Liouville sense can also be made into Liouville integrable systems (by changing the set of first integrals), see [137].

The description of a Hamiltonian system near a Liouville torus is given by the following theorem, called the *Liouville–Mineur–Arnold theorem*, about the existence of action-angle variables. It was obtained by Henri Mineur in 1935 [249, 250] for Liouville-integrable systems, was forgotten, then rediscovered by V.I. Arnold and other people in 1960s, and extended in a straightforward way to the noncommutatively integrable case by Nekhoroshev [276].

Theorem A.4.5 (Liouville–Mineur–Arnold). *Let N be a Liouville torus of a proper Hamiltonian system $(\mathbf{X}, \mathbf{F})$ of bi-degree (p, q) on a symplectic manifold (M, ω). Then there is a neighborhood $\mathcal{U}(N)$ of N and a symplectomorphism*

$$\Psi : (\mathcal{U}(N), \omega) \to (D^{2r} \times D^p \times \mathbb{T}^p, \sum_1^r \mathrm{d}x_i \wedge \mathrm{d}y_i + \sum_1^p \mathrm{d}\nu_i \wedge \mathrm{d}\phi_i), \qquad (A.32)$$

where $2r = q - p = \dim M - 2p$, (x_i, y_i) are coordinates on D^{2r}, ν_i are coordinates of D^p, ϕ_i (mod 1) are periodic coordinates of $\mathbb{T}^p$, such that $\mathbf{F}$ does not depend on $\theta_i = \Psi^* \phi_i$, and the functions which generate the Hamiltonian vector fields in $\mathbf{X}$ depend only on $I_1, \ldots, I_p$, where $I_i = \Psi^* \nu_i$.

The variables (I_i, θ_i) in the above theorem are called *action-angle variables*. In the Liouville-integrable case $(p = \dim M/2)$, they form a complete system of variables in a neighborhood of N. If $p < \dim M/2$, we need $\dim M - 2p$ additional variables. The map

$$(I_1, \ldots, I_n) : (\mathcal{U}(N), \omega) \to \mathbb{R}^p \qquad (A.33)$$

is the momentum map of a free Hamiltonian torus $\mathbb{T}^p$-action on $(\mathcal{U}(N), \omega)$ which preserves $\mathbf{F}$. The existence of this Hamiltonian torus action is essentially equivalent to the Liouville–Mineur–Arnold theorem. Indeed, once the action variables $(I_1, \ldots, I_p)$ which generate a free torus action are found, the corresponding angle variables $(\theta_1, \ldots, \theta_p)$ can also be found easily by the following method: fix a coisotropic section to the foliation by Liouville tori in $\mathcal{U}(N)$, put $\theta_i = 0$ on this coisotropic section, and extend them to the rest of $\mathcal{U}(N)$ in a unique way such that $X_{I_i}(\theta_j) = 0$ if $i \neq j$ and $X_{I_i}(\theta_i) = 1$. The fact that the section is coisotropic will ensure that $\{\theta_i, \theta_j\} = 0$. When $p < \dim M/2$, we can find additional variables in a way similar to the classical proof of Darboux's theorem. Remark that the quasi-periodicity of the system on N also follows immediately from the existence of this torus action.

The existence of action-angle variables is very important, both for the theory of near-integrable systems (K.A.M. theory), and for the quantization of integrable systems (Bohr–Sommerfeld rule). Actually, Mineur was an astrophysicist, and Bohr–Sommerfeld quantization was his motivation for finding action-angle variables.

Mineur [250] also wrote down the following simple formula, which we will call the *Mineur–Arnold formula*, for action functions:

$$I_i(z) = \int_{\Gamma_i(z)} \beta, \tag{A.34}$$

where z is a point in $\mathcal{U}(N)$, β is a primitive of the symplectic form ω, i.e., $d\beta = \omega$, and $\Gamma_i(z)$ is a 1-cycle on the Liouville torus which contains z and which depends on z continuously (the cycle given by the orbits of the periodic vector field $X_{I_i} = \frac{\partial}{\partial \theta_i}$). Of course, this formula is an immediate consequence of Theorem A.4.5.

Since Formula (A.34) depends on the choice of Γ_i and β, the momentum map $(I_1, \ldots, I_p)$ is not unique, but it is unique only up to an integral affine transformation. As a consequence, they induce a well-defined integral affine structure. This affine structure lives on the base space (= space of Liouville tori) in the Liouville-integrable case, and on the "reduced" base space (= leaf space of the coisotropic foliation dual to the foliation by Liouville tori) in the noncommutatively integrable case.

Remark A.4.6. To a large extent, torus actions (of appropriate dimensions) and (partial) action-angle variables also exist near *singularities* of proper integrable Hamiltonian systems, see [371] and references therein.

A.5 Normal forms of vector fields

A.5.1 Poincaré–Dulac normal forms

Let X be an analytic or formal vector field in a neighborhood of 0 in $\mathbb{K}^m$, where $\mathbb{K} = \mathbb{R}$ or $\mathbb{C}$, with $X(0) = 0$. When $\mathbb{K} = \mathbb{R}$, we may also view X as a holomorphic (i.e., complex analytic) vector field by complexifying it. Denote by

$$X = X^{(1)} + X^{(2)} + X^{(3)} + \cdots \tag{A.35}$$

the Taylor expansion of X in some local system of coordinates, where $X^{(k)}$ is a homogeneous vector field of degree k for each $k \geq 1$. The algebra of linear vector fields on $\mathbb{K}^m$, under the standard Lie bracket, is nothing but the reductive algebra $gl(m, \mathbb{K}) = sl(m, \mathbb{K}) \oplus \mathbb{K}$. In particular, we have

$$X^{(1)} = X^s + X^{nil}, \tag{A.36}$$

where X^s (resp., X^{nil}) denotes the semisimple (resp., nilpotent) part of $X^{(1)}$. There is a complex linear system of coordinates (x_j) in $\mathbb{C}^m$ which puts X^s into diagonal form:

$$X^s = \sum_{j=1}^{m} \gamma_j x_j \partial/\partial x_j, \tag{A.37}$$

where γ_j are complex coefficients, called *eigenvalues* of X (or $X^{(1)}$) at 0.

Definition A.5.1. The vector field X is said to be in *Poincaré–Dulac normal form*, or *normal form* for short, if it commutes with the semisimple part of its linear part:

$$[X, X^s] = 0. \tag{A.38}$$

A transformation of coordinates which puts X in Poincaré–Dulac normal form is called a *Poincaré–Dulac normalization*.

For each natural number $k \geq 1$, the vector field X^s acts linearly on the space of homogeneous vector fields of degree k by the Lie bracket, and the monomial vector fields are the eigenvectors of this action:

$$\Big[\sum_{j=1}^{m} \gamma_j x_j \partial / \partial x_j, x_1^{b_1} x_2^{b_2} \dots x_n^{b_n} \partial / \partial x_l \Big] = \Big(\sum_{j=1}^{n} b_j \gamma_j - \gamma_l \Big) x_1^{b_1} x_2^{b_2} \dots x_n^{b_n} \partial / \partial x_l. \tag{A.39}$$

When an equality of the type

$$\sum_{j=1}^{m} b_j \gamma_j - \gamma_l = 0 \tag{A.40}$$

holds for some nonnegative integer m-tuple (b_j) with $\sum b_j \geq 2$, we will say that the monomial vector field $x_1^{b_1} x_2^{b_2} \dots x_m^{b_m} \partial / \partial x_l$ is a resonant term, and that the m-tuple $(b_1, \dots, b_l - 1, \dots, b_m)$ is a *resonance relation* for the eigenvalues (γ_i). Equation (A.38) means that a vector field is in Poincaré–Dulac normal form if and only if its Taylor expansion does not contain any nonresonant nonlinear term. In particular, if X is nonresonant, i.e., its eigenvalues do not satisfy any resonance relation, then X becomes linear when it is normalized. In general, Poincaré–Dulac normalization may be viewed as a partial linearization (see Subsection A.5.3). Of course, when the semisimple linear part X^s of X is trivial or very resonant, then the Poicaré–Dulac normal form does not give much information, and more refined normal forms are be needed.

Theorem A.5.2 (Poincaré–Dulac). *Any analytic or formal vector field admits a formal Poincaré–Dulac normalization.*

The proof of the above theorem is rather simple and is given by the (formal) composition of an (infinite) sequence of local coordinate transformations which kill all the nonresonant nonlinear terms, term by term. Each of these coordinate transformations is provided by the following simple lemma (which is similar to Proposition 2.2.1 of Chapter 2):

Lemma A.5.3. *Let $X = X^{(1)} + X^{(2)} + \cdots$ be a vector field such as above. Suppose that we have $X^{(q)} = [Z, X^{(1)}] + T^{(q)}$, where Z is a homogeneous vector field of degree q with $q > 1$. Then the time-1 flow ϕ_Z^1 of Z transforms X to a vector field admitting a Taylor expansion*

$$(\phi_Z^1)_* X = X^{(1)} + \cdots + X^{(q-1)} + T^{(q)} + \cdots, \tag{A.41}$$

i.e., this transformation eliminates the term $[Z, X^{(1)}]$ in X without changing the terms of degree smaller than q.

It is much more difficult to show the existence, or lack thereof, of a *convergent* (i.e., local analytic) normalization for a given analytic vector field. One of the best results in this direction is a theorem of A. Bruno, which we use several times in this book:

Theorem A.5.4 (Bruno [48]). *If a local holomorphic vector field X on $(\mathbb{C}^m, 0)$ with $X(0) = 0$ satisfies the ω-condition, and admits a formal normal form of the type $\hat{a}(x) \sum_{j=1}^{m} \gamma_i x_i \partial/\partial x_j$ where $\hat{a}(x)$ is a formal first integral of $\sum_{j=1}^{m} \gamma_i x_i \partial/\partial x_j$, then it admits a local analytic normalization. In particular, if X is nonresonant and satisfies the ω-condition, then it is locally analytically linearizable.*

The ω-*condition* in the above theorem is the following Diophantine condition, invented by Bruno, on the eigenvalues of X:

$$\sum_{k=1}^{\infty} \frac{-\log \omega_k}{2^k} < +\infty, \tag{A.42}$$

where

$$\omega_k = \min \left\{ \left| \sum_{j=1}^{m} c_j \gamma_j - \gamma_l \right| \; ; \; (c_i) \in \mathbb{Z}_+^n, 2 \le \sum c_j \le 2^k, l = 1, \dots, m \right\}. \tag{A.43}$$

Remark that the ω-condition is weaker than the following Diophantine condition of Siegel: there are positive constants $C, q > 0$ such that

$$\left| \sum_i c_i \gamma_i - \gamma_l \right| \le C \left(\sum_i |c_i| \right)^{-q} \tag{A.44}$$

for all $(c_i) \in \mathbb{Z}_+^n$ such that $2 \le \sum c_j \le 2^k$ and all $l = 1, \dots, m$.

The proof of Theorem A.5.4 is based on Kolmogorov's fast convergence method. See, e.g., [49, 122, 239, 323] for details and some generalizations.

A.5.2   Birkhoff normal forms

Consider now a local analytic or formal Hamiltonian vector field $X = X_H$ on a symplectic space $(\mathbb{K}^{2n}, 0)$ with a standard symplectic structure, such that $X(0) = 0$. We may assume that $H(0) = 0$ and $dH(0) = 0$. Then the Taylor expansion of X corresponds to the Taylor expansion

$$H = H^{(2)} + H^{(3)} + H^{(4)} + \cdots \tag{A.45}$$

of H, $X^{(j)} = X_{H^{(j+1)}}$. In particular, $X^{(1)} = X_{H^{(2)}} \in sp(2n, \mathbb{K})$, which is a simple Lie algebra, and the decomposition $X^{(1)} = X^s + X^{\mathrm{nil}}$ lives in this algebra and corresponds to a decomposition of $H^{(2)}$,

$$H^{(2)} = H^s + H^{\mathrm{nil}}. \tag{A.46}$$

There is a complex canonical linear system of coordinates (x_j, y_j) in $\mathbb{C}^{2n}$ in which H^s has diagonal form:

$$H^s = \sum_{j=1}^{n} \lambda_j x_j y_j, \qquad (A.47)$$

where λ_j are complex coefficients, called *frequencies* of H (or X_H) at 0. These frequencies are uniquely determined by H only up to a sign (i.e., up to multiplication by ± 1), and the eigenvalues of X_H are $\lambda_1, -\lambda_1, \ldots, \lambda_n, -\lambda_n$. In particular, a Hamiltonian vector field will have *auto-resonance* relations of the type $\lambda_i + (-\lambda_i) = 0$.

Equation (A.38) in the Hamiltonian case can also be written as

$$\{H, H^s\} = 0, \qquad (A.48)$$

and one says that H (or X_H) is in *Birkhoff normal form* if H satisfies the above equation. A symplectic transformation of coordinates which puts X_H in Birkhoff normal form is called a *Birkhoff normalization*. In other words, a Birkhoff normalization is a Poincaré–Dulac normalization with the additional property of being symplectic.

The theory of Birkhoff normal forms is very similar to the theory of Poincaré–Dulac normal forms, and one may regroup the two theories together and call it the theory of *Poincaré–Birkhoff normal forms*. In particular, the proof of the following theorem is absolutely similar to the proof of Theorem A.5.2.

Theorem A.5.5 (Birkhoff et al.). *Any analytic or formal Hamiltonian vector field on a symplectic manifold admits a formal Birkhoff normalization.*

Moreover, it follows from the toric characterization of Poincaré–Birkhoff normal forms (see Subsection A.5.3) that we have:

Proposition A.5.6. *An analytic Hamiltonian vector field X_H on a symplectic manifold which vanishes at a point admits a local analytic Birkhoff normalization if and only if it admits a local analytic Poincaré–Dulac normalization (when one forgets about the symplectic structure).*

For example, Theorem A.5.4 together with Proposition A.5.6 imply the following positive result of Rüssmann [307] and Bruno [48]: if H is an analytic Hamiltonian function on a symplectic manifold which admits a formal normal form of the type $\hat{h}(\sum_{j=1}^{n} \lambda_j x_j y_j)$ for some formal function $\hat{h}$ of one variable, and the numbers $\pm \lambda_1, \ldots, \pm \lambda_n$ satisfy Bruno's ω-condition, then H admits a local analytic Birkhoff normalization. (However, this result is not very applicable in practice, because it is very rare for a Hamiltonian to have a normal form of the type $\hat{h}(\sum_{j=1}^{n} \lambda_j x_j y_j)$.)

A *resonance relation* in the Hamiltonian case is a relation of the type

$$\sum_{j=1}^{n} c_j \lambda_j = 0 \text{ with } (c_j) \in \mathbb{Z}^n, \qquad (A.49)$$

where λ_j are the frequencies. They correspond to monomial functions (resonance terms) $\prod_{j=1}^n x_j^{a_j} y_j^{b_j}$ such that $\{H^s, \prod_{j=1}^n x_j^{a_j} y_j^{b_j}\} = 0$. Here $c_j = a_j - b_j$. In particular, the terms $\prod_{j=1}^n (x_j y_j)^{a_j}$ are always resonant (we will call them *auto-resonant*). A Hamiltonian vector field X_H is called *nonresonant* at 0 if its frequencies do not admit any nontrivial resonance relation $\sum_{j=1}^n c_j \lambda_j = 0$ with $(c_j) \neq 0$. Due to auto-resonant terms, a nonresonant Hamiltonian vector field is not formally linearizable in general, and the best one can do is to put a nonresonant Hamiltonian function in the form $H = h(x_1 y_1, \ldots, x_n y_n)$ via a Birkhoff normalization.

Due to auto-resonances, it is even more difficult for a Hamiltonian vector field to admit a *convergent* normalization than a general vector field. Nevertheless, in dimension 4 there is a positive result by Moser [266]: if λ_1, λ_2 are two non-zero complex numbers such that $\lambda_1 / \lambda_2 \notin \mathbb{R}$, then any local holomorphic Hamiltonian vector field in $\mathbb{C}^4$ with frequencies λ_1, λ_2 at 0 admits a convergent Birkhoff normalization. If $\lambda_1 / \lambda_2 \in \mathbb{R}$ then a convergent normalization doesn't exist in general [316].

A.5.3   Toric characterization of normal forms

In this subsection we will explain a simple but important fact about Poincaré–Birkhoff normal forms, namely that they are governed by torus actions.

Denote by $\mathcal{R}$ the set of resonance relations for the eigenvalues $(\gamma_1, \ldots, \gamma_m)$ of a given vector field X which vanishes at 0. In other words, $(c_j) \in \mathcal{R}$ if and only if $(c_j) \in \mathbb{Z}^n$, $\sum_j c_j \gamma_j = 0$, $\sum_j c_j \geq 1$, $c_j \geq -1$ for all j, and at most one of the c_j is negative. The number

$$r = \dim_{\mathbb{Z}}(\mathcal{R} \otimes \mathbb{Z}) \tag{A.50}$$

is called the *degree of resonance* of X. If $X = X_H$ is a Hamiltonian vector field on a symplectic space of dimension $m = 2n$, then we always have $r \geq n$ due to n auto-resonance relations $\lambda_j + (-\lambda_j) = 0$, and we will call $r' = r - n$ the *Hamiltonian* degree of resonance of X_H.

Denote by $\mathcal{Q} \subset \mathbb{Z}^m$ the integral sublattice of $\mathbb{Z}^m$ consisting of m-dimensional vectors $(\rho_j) \in \mathbb{Z}^m$ which satisfy the following properties:

$$\sum_{j=1}^m \rho_j c_j = 0 \; \forall \; (c_j) \in \mathcal{R} \;, \quad \text{and} \quad \rho_j = \rho_k \;\; \text{if} \;\; \gamma_j = \gamma_k. \tag{A.51}$$

The number

$$d = \dim_{\mathbb{Z}} \mathcal{Q} \tag{A.52}$$

is called the *toric degree* of X at 0. In general, we have $d + r \leq m$. In the Hamiltonian case, we have $d + r = m = 2n$, or $d + r' = n$.

Let $(\rho_j^1), \ldots, (\rho_j^d)$ be a basis of $\mathcal{Q}$. For each $k = 1, \ldots, d$ define the following diagonal linear vector field Z_k:

$$Z_k = 2\pi\sqrt{-1} \sum_{j=1}^m \rho_j^k x_j \partial/\partial x_j. \tag{A.53}$$

The above vector fields $Z_1, \ldots, Z_d$ have the following remarkable properties: They commute pairwise and commute with X^s and X^{nil}. They are linearly independent almost everywhere. Each Z_k, $k = 1, \ldots, d$, is periodic of real period 1, and together they generate a linear action of the real torus $\mathbb{T}^d$ in $\mathbb{C}^m$, which preserves X^s and X^{nil}. In the Hamiltonian case, this action also preserves the standard symplectic structure. The semisimple linear part X^s is a linear combination of $Z_1, \ldots, Z_d$. A simple calculation shows that X is in Poincaré–Birkhoff normal form, i.e., $[X, X^s] = 0$, if and only if we have

$$[X, Z_k] = 0 \quad \forall\, k = 1, \ldots, d. \tag{A.54}$$

The above commutation relations mean that if X is in normal form, then it is preserved by the effective d-dimensional torus action generated by $Z_1, \ldots, Z_d$. Conversely, if there is a torus action which preserves X, then because the torus is a compact group we can linearize this torus action (using Bochner's linearization Theorem 7.4.1 in the non-Hamiltonian case, and the equivariant Darboux theorem A.1.7 in the Hamiltonian case), leading to a normalization of X. In other words, we have the following simple result, probably first written down explicitly in [372, 368], which says that a Poincaré–Birkhoff normalization for a vector field is nothing but a linearization of a corresponding torus action:

Theorem A.5.7. *A holomorphic (Hamiltonian) vector field X in a neighborhood of 0 in $\mathbb{C}^m$ (or $\mathbb{C}^{2n}$ with a standard symplectic form), which vanishes at 0, admits a convergent Poincaré–Birkhoff normalization if and only if it is preserved by an effective holomorphic (Hamiltonian) action of a real torus of dimension d, where d is the toric degree of X at 0, in a neighborhood of 0 in $\mathbb{C}^m$ (or $\mathbb{C}^{2n}$), which has 0 as a fixed point and whose linear part at 0 has appropriate weights (given by the lattice $\mathcal{Q}$ defined in (A.51)).*

The above theorem is true in the formal category as well. But of course, any vector field admits a formal Poincaré–Birkhoff normalization, and a formal torus action. Theorem A.5.7 has many direct implications. One of them is that, a real analytic vector field admits a local real analytic Poincaré–Birkhoff normalization if and only if its complexification admits a local holomorphic normalization, see [372, 368]. Another one is Proposition A.5.6 mentioned in the previous subsection. In fact, if there is a torus action which preserves a Hamiltonian vector field X_H and whose linear part is generated by the linear vector fields $Z_1, \ldots, Z_d$ defined above, then one can show easily that this torus action must automatically preserve the symplectic structure.

Theorem A.5.7 leads to another method for finding a convergent Poincaré–Birkhoff normalization: find a torus action. This method works well for integrable systems, due to Liouville–Mineur–Arnold Theorem A.4.5 and its generalizations about the existence of torus actions in integrable systems. In particular, we have:

Theorem A.5.8. *Any integrable analytic vector field which vanishes at a point admits a convergent Poincaré–Birkhoff normalization.*

In the above theorem, *integrable* means integrable in generalized Liouville sense in the Hamiltonian case, or a similar notion of integrability in the general case (existence of a complete set of analytic commuting vector fields and common first integrals, see [368]). For the proof, see [372, 368]. A particular important case of the above theorem, when X is a nonresonant Hamiltonian vector field, was obtained by Ito [183] (Ito's proof is based on the fast convergence method and is very technical). Some other partial cases of Theorem A.5.8 were obtained in [336, 337, 50, 188], using different methods.

If a dynamical system near an equilibrium point is invariant under a compact group action which fixes the equilibrium point, then this compact group action commutes with the (formal) torus action of the Poincaré–Birkhoff normalization. Together, they form a bigger compact group action, whose linearization leads to a simultaneous Poincaré–Birkhoff normalization and linearization of the compact symmetry group, i.e., we can perform the Poincaré–Birkhoff normalization in an invariant way. This is a known result in dynamical systems, see, e.g., [367], but the toric point of view gives a simple proof of it. The case of equivariant vector fields is similar. For example, one can speak about Poincaré–Dulac normal forms for time-reversible vector fields, see, e.g., [206].

One can probably use the toric point of view to study normal forms of Hamiltonian vector fields on *Poisson* manifolds as well. For example, let $\mathfrak{g}^*$ be the dual of a semisimple Lie algebra, equipped with the standard linear Poisson structure, and let $H : \mathfrak{g}^* \to \mathbb{K}$ be a regular function near the origin 0 of $\mathfrak{g}^*$. The corresponding Hamiltonian vector field X_H will vanish at 0, because the Poisson structure itself vanishes at 0. Applying Poincaré–Birkhoff normalization techniques, we can kill the "nonresonant terms" in H (with respect to the linear part of H, or $dH(0)$). The normalized Hamiltonian will be invariant under the coadjoint action of a subtorus of a Cartan torus of the (complexified) Lie group of $\mathfrak{g}$. In the "nonresonant" case, we have a Cartan torus action which preserves the system.

A.5.4   Smooth normal forms

Let X be a C^∞ vector field on $\mathbb{R}^n$, vanishing at 0. We decompose its linear part at 0 in the form $X^{(1)} = X^s + X^{nil}$ like in the analytic case (semisimple plus nilpotent part), and say that X is in normal form if we have $[X^s, X] = 0$. A smooth local transformation of coordinates which puts X in normal form is called a smooth normalization of X. The smooth normalization is substantially different, and in a sense simpler, than the analytic normalization problem. In particular, according to a famous theorem of Sternberg [321] generalized by Chen [76], if X is a hyperbolic vector field (i.e., its eigenvalues have non-zero real parts), then X admits a smooth normalization. See, e.g., [304, 73, 182, 46, 22, 47, 26, 74, 75] for results and methods on problems of smooth normalization and C^k normalization of vector fields. The usual methods for proving the existence of smooth normalization of a vector field extend readily to the case with parameters, i.e., families of vector fields. For example, let $X(u)$ be a smooth vector field on $\mathbb{R}^n$, depending smoothly on a parameter u belonging to a compact manifold P. We suppose that $X(u)$ vanishes at the origin of $\mathbb{R}^n$, for every u and has a linear part $X^{(1)}$ which is independent of u. If the eigenvalues $\lambda_1, \ldots, \lambda_n$ of $X^{(1)}$ have all non-zero real parts and have no resonance relations, then there is a smooth family $\phi(u)$ of local diffeomorphisms of $\mathbb{R}^n$, fixing the origin and all defined in a fixed neighborhood of the origin such that $\phi(u)_* X(u) = X^{(1)}$ for every u.

A.6 Normal forms along a singular curve

Let (M, Π) be a C^∞-smooth Poisson manifold, and m be a point where Π vanishes. We will assume that the curl vector field $D_\omega \Pi$ of Π with respect to some smooth density ω (see Section 2.6) doesn't vanish at m; this requirement is independent of the chosen density, and the Poisson structure has to vanish along the whole orbit of $D_\omega \Pi$ through m. In this section, we are interested in the case where this orbit is closed, and we denote it by Γ. As we want to study the germ of Π along Γ we may consider that M is a neighborhood of $\Gamma = S^1 \times \{0\}$ in $S^1 \times \mathbb{R}^n$ with coordinates $(\theta, x_1, \ldots, x_n)$, and that Π vanishes on Γ.

The first invariant attached to this situation is the period c of the curl vector field $D_\omega \Pi$ on Γ: it doesn't depend on the choice of ω because Π vanishes on Γ.

The second thing we have to do is to take care of the linear part of Π at the points of Γ. If we choose local coordinates $(\widetilde{x}_0, \widetilde{x}_1, \ldots, \widetilde{x}_n)$ vanishing at m, and such that $D_\omega \Pi$ is $\partial/\partial \widetilde{x}_0$, then the linear part $\{,\}_{(1)}$ of the Poisson bracket satisfies relations of the type

$$
\begin{aligned}
\{\widetilde{x}_0, \widetilde{x}_i\}_{(1)} &= \sum_{j=1}^n d_{ij} \widetilde{x}_j, \\
\{\widetilde{x}_r, \widetilde{x}_i\}_{(1)} &= \sum_{j=1}^n c_{ri}^j \widetilde{x}_j,
\end{aligned}
\tag{A.55}
$$

for r and i varying from 1 to n. This means that it corresponds to a semi-direct product $\mathbb{R} \ltimes L$ of $\mathbb{R}$ with an n-dimensional Lie algebra L. Since $D_\omega \Pi$ is an infinitesimal automorphism of Π, the isomorphism class of this Lie algebra $\mathbb{R} \ltimes L$ depends on Γ but does not depend on the choice of the point m on Γ, so it is also an invariant of Π along Γ.

We will impose the following additional hypothesis on the matrix (d_{ij}): its eigenvalues, denoted by $\lambda_1, \ldots, \lambda_n$ are real and do not satisfy any resonance relation of the types

$$
\begin{aligned}
\lambda_i &= p_1 \lambda_1 + \cdots + p_n \lambda_n \\
\text{or} \qquad \lambda_i + \lambda_j &= p_1 \lambda_1 + \cdots + p_n \lambda_n \,,
\end{aligned}
\tag{A.56}
$$

where $(p_1, \ldots, p_n)$ is a non-zero multi-index with $p_k \geq 0$ and $\sum_{i=1}^n p_i \geq 2$, except trivial relations $\lambda_i + \lambda_j = \lambda_i + \lambda_j$. In particular, the relations $\lambda_i + \lambda_j = 2\lambda_i$, with $i \neq j$, i.e., $\lambda_i = \lambda_j$, are forbidden, so the matrix (d_{ij}) is diagonalizable, so by a linear change of coordinates, we may assume that (d_{ij}) is diagonal: $d_{ij} = 0$ if $i \neq j$, and $d_{ii} = \lambda_i$. It follows from the relations $\{\widetilde{x}_0, \widetilde{x}_i\}_{(1)} = \lambda_i \widetilde{x}_i$ and the Jacobi identity that we have $\{\widetilde{x}_0, \{\widetilde{x}_i, \widetilde{x}_j\}_{(1)}\}_{(1)} = (\lambda_i + \lambda_j)\{\widetilde{x}_i, \widetilde{x}_j\}_{(1)}$. Since $\lambda_i + \lambda_j$ is not an eigenvalue due to the nonresonance condition, we have $\{\widetilde{x}_i, \widetilde{x}_j\}_{(1)} = 0$, i.e., L is an Abelian algebra.

Using the standard normalization procedure, one can show that, in this situation, Π is formally nonhomogeneously quadratizable near every point of Γ, i.e., we can find formal coordinates in which our Poisson structure has only linear and quadratic terms. Here we will present a result of Brahic [45] which says that this quadratization extends to a neighborhood of the curve Γ. Moreover we will describe the invariants attached to the isomorphism class of the germ of Π along Γ.

Theorem A.6.1 ([45]). *Under the above hypotheses, up to a double covering, there is a smooth system of coordinates* $(\theta, x_1, \ldots, x_n)$ *in a neighborhood of* Γ*, where* $\theta \in S^1 = \mathbb{R}/\mathbb{Z}$ *is a periodic coordinate, such that*

$$\{x_i, x_j\} = a_{ij} x_i x_j + o_\infty(x) \qquad i, j = 1, \ldots, n,$$
$$\{\theta, x_i\} = \mu_i x_i \qquad i = 1, \ldots, n,$$

where $o_\infty(x)$ *stands for a smooth function flat along* Γ*, and* μ_i *and* a_{ij} *are constant.*

Proof. We will divide the proof into several small steps.

Step 1. The brackets induced by Π can, in a neighborhood of Γ, be written as

$$\{x_i, x_j\} = O_2(x) ,$$
$$\{\theta, x_i\} = \sum_{j=1}^n h_{i,j}(\theta) x_j + O_2(x) ,$$

where $h_{i,j}$ are smooth functions on $S^1 = \mathbb{R}/\mathbb{Z}$, $O_2(x)$ denotes smooth functions on $S^1 \times \mathbb{R}^n$ which are of order 2 in the variables $x_1, \ldots, x_n$, and the matrix $H_\theta = (h_{i,j}(\theta))_{i,j=1,\ldots,n}$ has eigenvalues $k(\theta)\lambda_i$, $i = 1, \ldots, n$ for some $k \in C^\infty(S^1)$.

Step 2. Up to a double covering of a neighborhood of Γ, in an appropriate coordinate system we have:

$$\{x_i, x_j\} = O_2(x) ,$$
$$\{\theta, x_i\} = k(\theta)\lambda_i x_i + O_2(x).$$

Indeed, since for each $\theta \in S^1$, the eigenvalues $k(\theta)\lambda_i$, $i = 1, \ldots, n$, of H_θ are different, the space $\mathbb{R}^n = \{\theta\} \times \mathbb{R}^n$ decomposes into a direct sum of n one-dimensional eigenspaces E_i of H. By varying θ, we get n line bundles over S^1. By taking a double covering of the circle if necessary, we may assume that these line bundles are trivial. Then we can diagonalize H_θ by a smooth change of variables which is linear on each $\{\theta\} \times \mathbb{R}^n$. (It is easy to construct examples where a double covering is really necessary.)

Step 3. Reparametrizing S^1 by the formula

$$\theta^{new} = \frac{\int_0^\theta k^{-1}(t) dt}{\int_0^1 k^{-1}(t) dt},$$

we can write

$$\{x_i, x_j\} = O_2(x), \quad \{\theta, x_i\} = \mu_i x_i + O_2(x),$$

where μ_i are non-zero constants.

Step 4. The Hamiltonian vector field X_θ is tangent to each subspace $\{\theta\} \times \mathbb{R}^n$ and has the type $X_\theta = \sum_{i=1}^{n}(\mu_i x_i + p_i)\frac{\partial}{\partial x_i}$, where p_i are $O_2(x)$. Since, by our hypotheses, there is no resonance relation among the eigenvalues μ_i, the parametrized version of Sternberg's smooth linearization theorem (see [304]) implies that there is a smooth coordinate transformation which fixes θ and which linearizes X_θ. Applying this coordinate transformation, we get

$$\{x_i, x_j\} = O_2(x), \quad \{\theta, x_i\} = \mu_i x_i.$$

Step 5. Denote $u_{ij} = \{x_i, x_j\}$. The Jacobi identity $\oint_{\theta, x_i, x_j} \{\theta, \{x_i, x_j\}\} = 0$ now becomes

$$(\mu_i + \mu_j)u_{ij} = X_\theta(u_{ij}).$$

Denote by $\sum_I u_{ijI}(\theta)x^I$ the Taylor expansion of u_{ij} in the variables $(x_1, \ldots, x_n)$. Then the above equation implies

$$(\mu_i + \mu_j)u_{ijI}(\theta) = \Big(\sum_{j=1}^{n} I_j \mu_j\Big)u_{ijI}(\theta)$$

for every multi-index $I = (I_1, \ldots, I_n)$ with $\sum_{j=1}^{n} I_j \geq 2$. The non-resonance hypothesis says that $u_{ijI} = 0$ unless $I_i = I_j = 1$ and $I_k = 0 \ \forall \ k \neq i, j$. In other words, we have $u_{ij} = v_{ij}(\theta)x_i x_j + o_\infty(x)$. Finally, to turn $v_{ij}(\theta)$ into constants, we may apply the diffeomorphism

$$(\theta, x_1, \ldots, x_n) \longmapsto (\theta, x_1, \chi_2(\theta)x_2, \ldots, \chi_n(\theta)x_n),$$

where

$$\chi_j(\theta) = \exp\left(\frac{1}{\mu_1}\int_0^\theta (u_{1,j}(t) - \overline{u}_{1,j})\mathrm{d}t\right), \quad j = 2, \ldots, n,$$

with $\overline{u}_{1,j} = \int_0^1 u_{1,j}(t)\mathrm{d}t$. $\square$

Remark A.6.2. In the case $n = 1$ the same proof shows that we have a *smooth* normal form

$$\Pi = \mu x \frac{\partial}{\partial\theta} \wedge \frac{\partial}{\partial x},$$

where μ is a constant, in a neighborhood of the curve Γ. A more elaborate analysis [45] shows that we also have a smooth nonhomogeneous quadratic normal form without the flat term in the case when $n = 2$ and when $n > 2$ but all the μ_i have the same sign. When Π is analytic and the eigenvalues satisfy some Diophantine conditions, then there is also an analytic normalization.

The constants μ_i and a_{ij}, which appear in Theorem A.6.1, are invariants of the Poisson structure. Moreover, we can give a geometrical meaning to these invariants as follows. The sum of the μ_i is the inverse of the period of the curl vector field along the singular curve; the n-tuple $(\mu_1, \ldots, \mu_n)$, up to multiplication by a scalar, is determined by the Lie algebra $\mathbb{R} \ltimes L$ which corresponds to the linear part of Π at a point on Γ; and the numbers a_{ij} can be read off the holonomy of singular symplectic foliation of Π along Γ.

To see the geometrical meaning of a_{ij}, let us first look at what happens in the case $n = 2$: in that case, the normal form is

$$\Pi = \frac{\partial}{\partial \theta} \wedge \left(\mu_1 x_1 \frac{\partial}{\partial x_1} + \mu_2 x_2 \frac{\partial}{\partial x_1} \right) + a x_1 x_2 \frac{\partial}{\partial x_1} \wedge \frac{\partial}{\partial x_1},$$

and the regular symplectic leaves are surfaces which intersect each plane $\{\theta = \text{constant}\}$ along trajectories of the vector field $X_\theta = \mu_1 x_1 \partial/\partial x_1 + \mu_2 x_2 \partial/\partial x_1$; after going around Γ once, a symplectic leaf which intersects $\{\theta = 0\}$ at the X_θ-trajectory of a point $(x_1, x_2) \in \mathbb{R}^2$ will intersect $\{\theta = 0\}$ again at the X_θ-trajectory of the point $(x_1, x_2 \exp(-a/\mu_1))$.

When $n > 2$, remark that restriction of the normal form

$$\Pi = \frac{\partial}{\partial \theta} \wedge \left(\sum_{i=1}^n \mu_i x_i \frac{\partial}{\partial x_i} \right) + \sum_{i,j=1}^n a_{i,j} x_i x_j \frac{\partial}{\partial x_i} \wedge \frac{\partial}{\partial x_j}$$

to any (foliation invariant) subspace $\{x_k = 0 \; \forall \; k \neq i, j\}$ is again a Poisson structure in normal form with $n = 2$ as above but with (i, j) in place of $(1, 2)$ and a_{ij} in place of a. So the preceding paragraph shows that we can read a_{ij} off the behavior of the symplectic leaves. This works also in the smooth case of Theorem A.6.1, when there are flat terms (a nontrivial exercise).

A.7   The neighborhood of a symplectic leaf

In this section, following Vorobjev [340], we will give a description of a Poisson structure in the neighborhood of a symplectic leaf in terms of *geometric data*, and then use these geometric data to study the problem of linearization of Poisson structures along a symplectic leaf.

A.7.1   Geometric data and coupling tensors

First let us recall the notion of an Ehresmann (nonlinear) connection. Let $p : E \longrightarrow S$ be a submersion over a manifold S. Denote by $T_V E$ the vertical subbundle of the tangent bundle TE of E, and by $\mathcal{V}_V^1(E)$ the space of vertical tangent vector fields (i.e., vector fields tangent to the fibers of the fibration) of E. An *Ehresmann connection* on E is a splitting of TE into the direct sum of $T_V E$ and another tangent subbundle $T_H E$, called the *horizontal subbundle* of E. It can be defined by a $\mathcal{V}_V^1(E)$-valued 1-form $\Gamma \in \Omega^1(E) \otimes \mathcal{V}_V^1(E)$ on E such that $\Gamma(Z) = Z$ for every $Z \in T_V E$. Then the horizontal subbundle is the kernel of Γ: $T_H E := \{X \in TE, \Gamma(X) = 0\}$. For every vector field $u \in \mathcal{V}^1(S)$ on S, there is a unique lifting of u to a horizontal vector field $\text{Hor}(u) \in \mathcal{V}_H^1(E)$ on E (whose projection to S is u). The *curvature* of an Ehresmann connection is a $\mathcal{V}_V^1(E)$-valued 2-form on S,

$\mathrm{Curv}_\Gamma \in \Omega^2(S) \otimes \mathcal{V}_V^1(E)$, defined by

$$\mathrm{Curv}_\Gamma(u, v) := [\mathrm{Hor}(u), \mathrm{Hor}(v)] - \mathrm{Hor}([u, v]), \ \ u, v \in \mathcal{V}^1(S), \qquad (A.57)$$

and the associated covariant derivative $\partial_\Gamma : \Omega^i(S) \otimes C^\infty(E) \longrightarrow \Omega^{i+1}(S) \otimes C^\infty(E)$
is defined by an analog of Cartan's formula:

$$\partial_\Gamma K(u_1, \ldots, u_{k+1}) = \sum_i (-1)^{i+1} \mathcal{L}_{\mathrm{Hor}(u_i)}(K(u_1, \ldots, \widehat{u_i}, \ldots, u_{k+1}))$$
$$+ \sum_{i<j} (-1)^{i+j} K([u_i, u_j], u_1, \ldots, \widehat{u_i}, \ldots, \widehat{u_j}, \ldots, u_{k+1}). \qquad (A.58)$$

Remark that $\partial_\Gamma \circ \partial_\Gamma = 0$ if and only if Γ is a flat connection, i.e., $\mathrm{Curv}_\Gamma = 0$.

Suppose now that S is a symplectic leaf in a Poisson manifold (M, Π), and E is a small tubular neighborhood of S with a projection $p : E \longrightarrow S$. Then there is a natural Ehresmann connection $\Gamma \in \Omega^1(E) \otimes \mathcal{V}_V^1(E)$ on E, whose horizontal subbundle is spanned by the Hamiltonian vector fields $X_{f \circ p}$, $f \in C^\infty(S)$. The Poisson structure Π splits into the sum of its horizontal part and its vertical part,

$$\Pi = \mathcal{V} + \mathcal{H}, \qquad (A.59)$$

where $\mathcal{V} = \Pi_V \in \mathcal{V}_V^2(E)$ and $\mathcal{H} = \Pi_H \in \mathcal{V}_H^2(E)$ (there is no mixed part). The horizontal 2-vector field $\mathcal{H}$ is nondegenerate on $T_H E$. Denote by $\mathbb{F}$ its dual 2-form; it is a section of $\wedge^2 T_H^* E$ which can be defined by the following formula:

$$\mathbb{F}(X_{f \circ p}, X_{g \circ p}) = \langle \Pi, p^* \mathrm{d}f \wedge p^* \mathrm{d}g \rangle, \ \ f, g \in C^\infty(S), \qquad (A.60)$$

(recall that $X_{f \circ p}, X_{g \circ p} \in \mathcal{V}_H^1(E)$). Via the horizontal lifting of vector fields, $\mathbb{F}$ may be viewed as a nondegenerate $C^\infty(E)$-valued 2-form on S, $\mathbb{F} \in \Omega^2(S) \otimes C^\infty(E)$.

The above triple $(\mathcal{V}, \Gamma, \mathbb{F})$ is called a set of *geometric data* for (M, Π) in a neighborhood of S.

Conversely, given a set of geometric data $(\mathcal{V}, \Gamma, \mathbb{F})$, one can define a 2-vector field Π on E by the formula $\Pi = \mathcal{V} + \mathcal{H}$, where $\mathcal{H}$ is the horizontal 2-vector field dual to $\mathbb{F}$. A natural question arises: how to express the condition $[\Pi, \Pi] = 0$, i.e., Π is a Poisson structure, in terms of geometric data $(\mathcal{V}, \Gamma, \mathbb{F})$? The answer to this question is given by the following theorem:

Theorem A.7.1 (Vorobjev [340]). *A triple of geometric data $(\mathcal{V}, \Gamma, \mathbb{F})$ on a fibration $p : E \longrightarrow S$, where Γ is an Ehresmann connection on E, $\mathcal{V} \in \mathcal{V}_V^2(E)$ is a vertical 2-vector field, and $\mathbb{F} \in \Omega^2(S) \otimes C^\infty(E)$ is a nondegenerate $C^\infty(E)$-valued 2-form on S, determines a Poisson structure on E (by the above formulas) if and only if it satisfies the following four compatibility conditions:*

$$[\mathcal{V}, \mathcal{V}] = 0, \qquad (A.61)$$

$$\mathcal{L}_{\mathrm{Hor}(u)} \mathcal{V} = 0 \qquad \qquad \forall \ u \in \mathcal{V}^1(S), \qquad (A.62)$$

$$\partial_\Gamma \mathbb{F} = 0, \qquad (A.63)$$

$$\mathrm{Curv}_\Gamma(u, v) = \mathcal{V}^\sharp(\mathrm{d}(\mathbb{F}(u, v))) \quad \forall \ u, v \in \mathcal{V}^1(S), \qquad (A.64)$$

where $\mathcal{V}^\sharp$ means the map from $T^* E$ to TE defined by $\langle \mathcal{V}^\sharp(\alpha), \beta \rangle = \langle \mathcal{V}, \alpha \wedge \beta \rangle$.

Remark A.7.2. Equations (A.61) and (A.62) mean that the vertical part $\mathcal{V}$ of Π is a Poisson structure (on each fiber of E) which is preserved under parallel transport. This gives another proof of Theorem 1.6.1 which says that the transverse Poisson structure to a symplectic leaf is unique up to local isomorphisms.

Remark A.7.3. In the above theorem, E is not necessarily a tubular neighborhood of S. The symplectic case (E is a symplectic manifold) of the above theorem was obtained by Guillemin, Lerman and Sternberg in [159]. In fact, the proof of the symplectic case can be easily adapted to the Poisson case because a Poisson manifold is just a singular foliation by symplectic manifolds. The Poisson structure Π is called the *coupling tensor* of $(\mathcal{V}, \Gamma, \mathbb{F})$ (it couples a horizontal tensor with a vertical tensor via a connection).

Proof. Consider a local system of coordinates $(x_1, \ldots, x_m, y_1, \ldots, y_{n-m})$ on E, where $y_1, \ldots, y_{n-m}$ are local functions on a fiber and $x_1, \ldots, x_m$ are local functions on S ($m = \dim S$ is even). Denote the horizontal lifting of the vector field $\partial x_i := \partial/\partial x_i$ from S to E by $\overline{\partial x_i}$. Then we can write $\Pi = \mathcal{V} + \mathcal{H}$, where

$$\mathcal{V} = \frac{1}{2}\sum_{ij} a_{ij}\partial y_i \wedge \partial y_j \quad (a_{ij} = -a_{ji}), \tag{A.65}$$

and

$$\mathcal{H} = \frac{1}{2}\sum_{ij} b_{ij}\overline{\partial x_i} \wedge \overline{\partial x_j} \quad (b_{ij} = -b_{ji}) \tag{A.66}$$

is the dual horizontal 2-vector field of $\mathbb{F}$.

The condition $[\Pi, \Pi] = 0$ is equivalent to

$$0 = [\mathcal{V}, \mathcal{V}] + 2[\mathcal{V}, \mathcal{H}] + [\mathcal{H}, \mathcal{H}] = A + B + C + D, \tag{A.67}$$

where

$$A = [\mathcal{V}, \mathcal{V}], \tag{A.68}$$

$$B = 2\sum_i [\mathcal{V}, \overline{\partial x_i}] \wedge X_i, \text{ where } X_i = \sum_j b_{ij}\overline{\partial x_j}, \tag{A.69}$$

$$C = \sum_{ij} [\mathcal{V}, b_{ij}] \wedge \overline{\partial x_i} \wedge \overline{\partial x_j} + \sum_{ij} \overline{\partial x_i} \wedge \overline{\partial x_j} \wedge \left(\sum_{kl} b_{ik}b_{jl}[\overline{\partial x_k}, \overline{\partial x_l}]\right), \tag{A.70}$$

$$D = \sum_{ijkl} b_{ij}\overline{\partial x_j}(b_{kl}) \overline{\partial x_i} \wedge \overline{\partial x_k} \wedge \overline{\partial x_l}. \tag{A.71}$$

Notice that A, B, C, D belong to complementary subspaces of $\mathcal{V}^3(E)$, so the condition $A + B + C + D = 0$ means that $A = B = C = D = 0$.

The equation $A = 0$ is nothing but Condition (A.61): $[\mathcal{V}, \mathcal{V}] = 0$.

The equation $B = 0$ means that $[\mathcal{V}, \overline{\partial x_i}] = 0 \ \forall \ i$, i.e., $\mathcal{L}_{\overline{\partial x_i}}\mathcal{V} = 0 \ \forall \ i$, which is equivalent to Condition (A.62).

The equation $D = 0$ means that $\oint_{ikl} \sum_j b_{ij} \overline{\partial x_j}(b_{kl}) = 0$ $\forall\, i, k, l$, where $\oint_{ikl}$ denotes the cyclic sum. Let us show that this condition is equivalent to Condition (A.63). Notice that $\mathbb{F}(\partial x_i, \partial x_j) = c_{ij}$, where (c_{ij}) is the inverse matrix of (b_{ij}), and $\partial_\Gamma \mathbb{F}(\partial x_i, \partial x_j, \partial x_k) = \oint_{ijk} \overline{\partial x_i}(c_{jk})$. By direct computations, we have

$$\partial_\Gamma \mathbb{F}\Big(\sum_\alpha b_{i\alpha}\partial x_\alpha, \sum_\beta b_{j\beta}\partial x_\beta, \sum_\gamma b_{k\gamma}\partial x_\gamma\Big) = 2\oint_{ijk} \sum_l b_{il}\overline{\partial x_l}(b_{jk}). \qquad (A.72)$$

Thus the condition $D = 0$ is equivalent to the condition

$$\partial_\Gamma \mathbb{F}\Big(\sum_\alpha b_{i\alpha}\partial x_\alpha, \sum_\beta b_{j\beta}\partial x_\beta, \sum_\gamma b_{k\gamma}\partial x_\gamma\Big) = 0 \quad \forall\, i, j, k. \qquad (A.73)$$

Since the matrix (b_{ij}) is invertible, the last condition is equivalent to $\partial_\Gamma \mathbb{F} = 0$.

Similarly, by direct computations, one can show that the condition $C = 0$ is equivalent to Condition (A.64). $\qquad\square$

Theorem A.7.4 (Vorobjev [340]). *Let E be a sufficiently small neighborhood E of a symplectic leaf S of a Poisson manifold (M, Π), together with a given projection $p : E \longrightarrow S$. Denote by $(\mathcal{V}, \Gamma, \mathbb{F})$ the associated geometric data in E. Consider an arbitrary tensor field $\phi \in \Omega^1(S) \otimes C^\infty(E)$ whose restriction to $\Omega^1(S) = \Omega^1(S) \otimes C^\infty(S)$ via the inclusion $S \hookrightarrow E$ is trivial, and the following new set of geometric data:*

$$\mathcal{V}' = \mathcal{V}, \qquad (A.74)$$

$$\Gamma' = \Gamma - \mathcal{V}^\sharp(\mathrm{d}p^*\phi), \qquad (A.75)$$

$$\mathbb{F}' = \mathbb{F} - \partial_\Gamma\phi - \{\phi, \phi\}_\mathcal{V}. \qquad (A.76)$$

Then the coupling tensor Π' of $(\mathcal{V}', \Gamma', \mathbb{F}')$ is also a Poisson tensor, and there is a diffeomorphism f between neighborhoods of S, which fixes every point of S and such that $f_*\Pi = \Pi'$.

In the above theorem, $\mathcal{V}^\sharp(\mathrm{d}p^*\phi)$ means an element of $\Omega^1(E) \otimes \mathcal{V}^1_V(E)$ defined by the formula $\mathcal{V}^\sharp(\mathrm{d}p^*\phi)(w) = \mathcal{V}^\sharp(\mathrm{d}(\phi(p_*w)))$, $w \in TE$, where p_*w is the projection of w to S, $\phi(p_*w)$ is viewed as a function on the fiber T_xE over the origin x of p_*w, and $\mathcal{V}^\sharp(\mathrm{d}(\phi(p_*w)))$ is the Hamiltonian vector field with respect to $\mathcal{V}$ on T_xE of the function $\phi(p_*w)$. Similarly, $\{\phi, \phi\}_\mathcal{V}$ means an element of $\Omega^2(S) \otimes C^\infty(E)$ defined by $\{\phi, \phi\}_\mathcal{V}(u, v) = \{\phi(u), \phi(v)\}_\mathcal{V}$, where $u, v \in \mathcal{V}^1(S)$, and the bracket is taken with respect to $\mathcal{V}$.

Proof (sketch). We will use Moser's path method. Consider the following family of geometric data,

$$\mathcal{V}_t = \mathcal{V},$$

$$\Gamma_t = \Gamma - t\mathcal{V}^\sharp(\mathrm{d}p^*\phi),$$

$$\mathbb{F}_t = \mathbb{F} - t\partial_\Gamma\phi - t^2\{\phi, \phi\}_\mathcal{V},$$

and the corresponding family of coupling tensors Π_t, $t \in [0,1]$. Define a time-dependent vector field $X = (X_t)_{t \in [0,1]}$ as follows: X_t is the unique horizontal vector field with respect to Γ_t which satisfies the equation

$$X_t \lrcorner \mathbb{F}_t = -\phi$$

(where ϕ and $\mathbb{F}_t$ are considered as differential forms on E by lifting). One verifies directly that we have

$$[X_t, \Pi_t] = -\frac{\partial \Pi_t}{\partial t}. \qquad (A.77)$$

It implies that the time-1 flow φ_X^1 of $X = (X_t)$ moves $\Pi = \Pi_0$ to $\Pi' = \Pi_1$. As a consequence, Π' is automatically a Poisson tensor. Note that X vanishes on S, so φ_X^1 fixes every point of S. $\qquad\qquad\qquad\qquad\qquad\qquad\qquad\qquad\qquad\square$

Remark A.7.5. The flow φ_X^t in the above proof preserves the symplectic leaves of Π (so Π and Π' have the same foliation though not the same symplectic forms on the leaves). What the flow does is to change the projection map p. It also allows us to compare different geometric data of the same Poisson structure but with respect to different projection maps.

A.7.2   Linear models

Consider the geometric data $(\mathcal{V}, \Gamma, \mathbb{F})$ of a Poisson structure in a neighborhood E of a symplectic leaf S with respect to a projection $p : E \to S$. We will embed E in the normal bundle NS of S by a fiber-wise embedding which maps S to the zero section in NS and which projects to the identity map on S. Then we can view $(\mathcal{V}, \Gamma, \mathbb{F})$ as geometric data in a neighborhood of S (identified with the zero section) in NS.

Denote by $\mathcal{V}^{(1)}$ the fiber-wise linear part of $\mathcal{V}$, $\Gamma^{(1)}$ the fiber-wise linear part of Γ, and $\mathbb{F}^{(1)}$ the fiber-wise affine part of $\mathbb{F}$ in NS. For example, if $X, Y \in T_x S$, then $\mathbb{F}(X, Y)$ is a function on a neighborhood of zero in $N_x S$, and $\mathbb{F}^{(1)}(X, Y)$ is the sum of the constant part and the linear part of $\mathbb{F}(X, Y)$ on N_x. By looking at the fiber-wise linear terms of the equations in Theorem A.7.1, we obtain immediately that $(\mathcal{V}^{(1)}, \Gamma^{(1)}, \mathbb{F}^{(1)})$ also satisfy these equations, which implies that the coupling tensor $\Pi^{(1)}$ of $(\mathcal{V}^{(1)}, \Gamma^{(1)}, \mathbb{F}^{(1)})$ is also a Poisson structure, defined in a neighborhood of S in NS. We will call $\Pi^{(1)}$ the *Vorobjev linear model* of Π along the symplectic leaf S.

Theorem A.7.6. *Up to isomorphisms, the Vorobjev linear model of a Poisson structure Π along a symplectic leaf S is uniquely determined by Π and S (and does not depend on the choice of the projection).*

Proof. We will fix a projection $p : E \to S$, and linearize the fibers of E by an embedding from E to NS which is compatible with p. This way we may consider the linear model $\Pi^{(1)}$ of Π with respect to p as living in E. Consider now another arbitrary projection p_1. We can find a smooth path of projections p_t with $p_0 = p$ and $p_1 = p_1$. There is a unique time-dependent vector field $Y = (Y_t)$ in a neighborhood of S which satisfies the following properties: Y_t is tangent to the symplectic leaves of Π, is symplectically orthogonal to the intersections of the fibers of p_t with the symplectic leaves, vanishes on S, and the flow φ_Y^t of Y moves the fibers of p_0 to the fibers of p_t: $p_t \circ \varphi_Y^t = p_0$. Denote $\Pi_t = (\varphi_Y^t)^{-1}\Pi$. Denote by $\Pi_t^{(1)}$ the Vorobjev linear model of Π_t with respect to the projection p ($\Pi_t^{(1)}$ also lives in E via the fixed linearization of E). To prove the theorem, it is sufficient to show that $\Pi_1^{(1)}$ is isomorphic to $\Pi^{(1)}$ by a diffeomorphism in a neighborhood of S.

Denote by $(\mathcal{V}_t, \Gamma_t, \mathbb{F}_t)$ the geometric data of Π_t with respect to p. Note that $\frac{\partial \Pi_t}{\partial t} = -[Y_t, \Pi_t]$ by construction. Similar to the proof of Theorem A.7.4, we have

$$\mathcal{V}_t = \mathcal{V}, \quad \frac{\partial \Gamma_t}{\partial t} = -\mathcal{V}^\sharp(dp^* \phi_t), \quad \frac{\partial \mathbb{F}_t}{\partial t} = -\partial_{\Gamma_t} \phi_t,$$

where ϕ_t is a family of elements of $\Omega^1(S) \otimes \mathcal{C}^\infty(E)$ defined by $\phi_t = -Y_t \lrcorner \mathbb{F}_t$. Looking only at the fiber-wise linear terms of the above equations, we get

$$\mathcal{V}_t^{(1)} = \mathcal{V}^{(1)}, \quad \frac{\partial \Gamma_t^{(1)}}{\partial t} = -(\mathcal{V}^{(1)})^\sharp(dp^* \phi_t^{(1)}), \quad \frac{\partial \mathbb{F}_t^{(1)}}{\partial t} = -\partial_{\Gamma_t^{(1)}} \phi_t^{(1)},$$

which implies that

$$\frac{\partial \Pi_t^{(1)}}{\partial t} = -[Z_t, \Pi_t^{(1)}],$$

where $Z = (Z_t)$ is the time-dependent vector field defined by the formula $\phi_t^{(1)} = -Z_t \lrcorner \mathbb{F}_t^{(1)}$. As a consequence, the time-1 flow of (Z_t) moves $\Pi^{(1)}$ to $\Pi_1^{(1)}$. $\qquad \square$

Remark A.7.7. The linear model of a Poisson structure along a symplectic leaf can also be constructed from the transitive Lie algebroid which is the restriction of the cotangent algebroid to the symplectic leaf, see [340]. We will leave it as an exercise for the reader to show that Vorobjev's original construction via transitive Lie algebroids is equivalent to the above construction.

The following simple example shows that, in general, one can't hope to find a local isomorphism between a Poisson structure and its Vorobjev linear model along a symplectic leaf; even if the leaf is simply-connected, the normal bundle is trivial and the transverse Poisson structure is linearizable.

Example A.7.8. Put $M = S^2 \times \mathbb{R}^3$ with Poisson structure $\Pi = f\Pi_1 + \Pi_2$, where Π_1 is a nondegenerate Poisson structure on S^2, $\Pi_2 = x\partial y \wedge \partial z + y\partial z \wedge \partial x + z\partial x \wedge \partial y$ is the Lie–Poisson structure on $\mathbb{R}^3$ corresponding to $so(3)$, and $f = f(x^2 + y^2 + z^2)$ is a Casimir function on $(\mathbb{R}^3, \Pi_2)$. Since the linear part of f on $\mathbb{R}^3$ is trivial, the linear model of Π is $f(0)\Pi_1 + \Pi_2$. If f is not a constant then Π can't be isomorphic to $\Pi^{(1)}$ near S for homological reasons: the regular symplectic leaves are $S^2 \times S^2$, the integral of the symplectic form over the first component S^2 is a constant (does not depend on the leaf) in the linear model $\Pi^{(1)}$, but is not a constant when the symplectic form comes from Π.

If one wants to linearize only $\mathcal{V}$ and Γ but not $\mathbb{F}$, then the situation becomes more reasonable. See [45] for some results in that direction.

A.8   Dirac structures

An *almost Dirac structure* on a manifold M is a subbundle L of the bundle $TM \oplus T^*M$, which is isotropic with respect to the natural indefinite symmetric scalar product on $TM \oplus T^*M$,

$$\langle (X_1, \alpha_1), (X_2, \alpha_2) \rangle := \frac{1}{2}(\langle \alpha_1, X_2 \rangle + \langle \alpha_2, X_1 \rangle) \qquad (\text{A.78})$$

for $(X_1, \alpha_1), (X_2, \alpha_2) \in \Gamma(TM \oplus T^*M)$, and such that the rank of L is maximal possible, i.e., equal to the dimension of M.

For example, if ω is an arbitrary differential 2-form on M, then its graph $L_\omega = \{(X, i_X \omega) \mid X \in TM\}$ is an almost Dirac structure. Furthermore, an almost Dirac structure L is the graph of a 2-form if and only if $L_x \cap (\{0\} \oplus T_x^*M) = \{0\}$ for any $x \in M$. Similarly, if Λ is an arbitrary 2-vector field on M, then the set $L_\Lambda = \{(i_\alpha \Lambda, \alpha) \mid \alpha \in T^*M\}$ is also an almost Dirac structure.

A Dirac structure is an almost Dirac structure plus an integrability condition. To formulate this condition, consider the following bracket on $\Gamma(TM \oplus T^*M)$, called the *Courant bracket* [84]:

$$[(X_1, \alpha_1), (X_2, \alpha_2)]_C = ([X_1, X_2], \mathcal{L}_{X_1} \alpha_2 - i_{X_2} d\alpha_1). \qquad (\text{A.79})$$

An almost Dirac structure L is called a *Dirac structure* if it is closed under the Courant bracket, i.e., $[(X_1, \alpha_1), (X_2, \alpha_2)]_C \in \Gamma(L)$ for any $(X_1, \alpha_1), (X_2, \alpha_2) \in \Gamma(L)$. In this case, the pair (M, L) is called a *Dirac manifold*.

Example A.8.1. If ω is a 2-form on M, then the almost Dirac structure $L_\omega = \{(X, X \lrcorner \omega) \mid X \in TM\}$ is a Dirac structure if and only if ω is closed. Similarly, if Λ is a 2-vector field on M, then the almost Dirac structure $L_\Lambda = \{(\alpha \lrcorner \Lambda, \alpha) \mid \alpha \in T^*M\}$ is a Dirac structure if and only if Λ is a Poisson structure. In other words, Dirac structures generalize both presymplectic structures and Poisson structures.

Example A.8.2. If L is a Dirac structure on M such that its canonical projection $pr_1 : L \to M$ to M vanishes at a point $x_0 \in M$, then for x near x_0 we have $L_x \cap (T_xM \oplus \{0\}) = \{0\}$, which implies that locally $L = L_\Lambda$ is the graph of a 2-vector field Λ, and the integrability of L means that Λ satisfies the Jacobi identity. Thus locally a Dirac structure whose projection to TM vanishes at a point is the same as a Poisson structure which vanishes at that point.

The Courant bracket (A.79) is not anti-symmetric nor does it satisfy the Jacobi identity on $\Gamma(TM \oplus T^*M)$. But if L is a Dirac structure, then one can verify easily that the restriction of the Courant bracket to $\Gamma(L)$ is anti-symmetric and satisfies the Jacobi identity, and it turns L into a Lie algebroid over M whose anchor map is the canonical projection $pr_1 : L \to TM$ from L to TM, see [84]. For example, when $L = L_\Lambda$ comes from a Poisson structure, then this Lie algebroid is naturally isomorphic to the cotangent algebroid associated to Λ.

In particular, if L is a Dirac structure, then its associated distribution $\mathcal{D}_L$ on M, $(\mathcal{D}_L)_x = pr_1(L_x)$, is integrable and gives rise to the associated foliation $\mathcal{F}_L$ on M. Furthermore, there is a 2-form ω_L defined on each leaf of this foliation by the formula

$$\Omega_L(X,Y) = \langle \alpha, Y \rangle \quad \forall\, (X,\alpha), (Y,\beta) \in L_x. \tag{A.80}$$

The fact that L is isotropic assures that Ω_L is well defined and skew-symmetric. Moreover, we have:

Theorem A.8.3 ([84]). *If L is a Dirac structure on M, then* $\mathrm{d}\Omega_L = 0$ *on any leaf of the associated singular foliation $\mathcal{F}_L$ of L.*

The meaning of the above proposition is that, roughly speaking, a Dirac structure is a singular foliation by *presymplectic leaves*. Its proof is a straightforward verification similar to the Poisson case. Note that L is completely determined by $\mathcal{D}_L$ and Ω_L.

A submanifold of a Poisson manifold is not a Poisson manifold in general, but is a Dirac manifold under some mild assumptions. More generally, we have:

Proposition A.8.4 ([84]). *Let Q be a submanifold of a Dirac manifold (M, L). If $L_q \cap (T_q Q \oplus T_q^* M)$ has constant dimension (i.e., its dimension does not depend on $q \in Q$), then there is a natural induced Dirac structure L_Q on Q defined by the formula*

$$(L_Q)_q = \frac{L_q \cap (T_q Q \oplus T_q^* M)}{L_q \cap (\{0\} \oplus (T_Q)^0)}. \tag{A.81}$$

A special case of the above proposition is when $Q = N$ is a slice, i.e., a local transversal to a presymplectic leaf $\mathcal{O}$ at a point x_0. Then the condition of the theorem is satisfied, so N admits an induced Dirac structure, whose projection to TN vanishes at x_0, thus in fact N admits a Poisson structure which vanishes at x_0, and one can talk about the transverse Poisson structure to a presymplectic leaf in a Dirac manifold – provided that it is unique up to local isomorphisms.

Vorobjev's (semi)local description of Poisson structures via coupling tensors (see Subsection A.7.1) can be naturally extended to the case of Dirac structures. More precisely, given a triple of geometric data $(\mathcal{V}, \Gamma, \mathbb{F})$ on a manifold E with a submersion $p : E \to S$, where Γ is an Ehresmann connection, $\mathcal{V}$ is a vertical 2-vector field, and $\mathbb{F}$ is a (maybe degenerate) $C^\infty(E)$-valued 2-form on S, denote by $L = L(\mathcal{V}, \Gamma, \mathbb{F})$ the associated subbundle of $TE \oplus T^* E$, which is generated by sections of the types $(\alpha, i_\alpha \mathcal{V})$ and $(X, i_X \mathbb{F})$, where $X \in \mathcal{V}_H^1 E$ is a horizontal vector field and α is a vertical 1-form, i.e., $\alpha|_{T_H E} = 0$. Here $i_X \mathbb{F}$ means the contraction of $\mathbb{F}$, considered as a horizontal 2-form on E, with X. Then L is an almost Dirac structure on E.

Theorem A.8.5 ([115]). *Given a set of geometric data $(\mathcal{V}, \Gamma, \mathbb{F})$ for a submersion $p : E \to S$ such as above, the corresponding almost Dirac structure $L(\mathcal{V}, \Gamma, \mathbb{F})$ is a Dirac structure if and only if the following four conditions (the same as in*

Theorem A.7.1) are satisfied:

$$[\mathcal{V}, \mathcal{V}] = 0, \tag{A.82}$$

$$\mathcal{L}_{\mathrm{Hor}(u)}\mathcal{V} = 0 \qquad\qquad \forall\, u \in \mathcal{V}^1(S), \tag{A.83}$$

$$\partial_\Gamma \mathbb{F} = 0, \tag{A.84}$$

$$\mathrm{Curv}_\Gamma(u, v) = \mathcal{V}^\sharp(\mathrm{d}(\mathbb{F}(u, v))) \quad \forall\, u, v \in \mathcal{V}^1(S). \tag{A.85}$$

Conversely, if E is a sufficiently small tubular neighborhood of a presymplectic leaf S with a projection map $p : E \to S$ in a Dirac manifold (M, L), then there is a unique triple of geometric data $(\mathcal{V}, \Gamma, \mathbb{F})$ on (E, p) such that $L = L(\mathcal{V}, \Gamma, \mathbb{F})$ on E. Moreover, the vertical Poisson structure $\mathcal{V}$ vanishes on S, and the restriction of $\mathbb{F}$ to S is the presymplectic form of S induced from L.

The proof of Theorem A.8.5 is absolutely similar to the Poisson case. (The only difference between the Dirac case and the Poisson case is that the horizontal 2-form $\mathbb{F}$ may be degenerate in the Dirac case.) A direct consequence of Theorem A.8.5 is that, similarly to the Poisson case, the transverse Poisson structure to a presymplectic leaf in a Dirac manifold is well defined, i.e., up to local isomorphisms it does not depend on the choice of the slice. Another simple consequence is that the dimensions of the presymplectic leaves of a Dirac structure have the same parity.

Dirac and almost Dirac structures provide a convenient setting in which to study dynamical systems with constraints (holonomic and non-holonomic) and control theory, and there is a theory of symmetry and reduction of (almost) Dirac structures, which generalizes the theory for symplectic and Poisson structures. See, e.g., [84, 105, 91, 32, 33, 34] and references therein.

For a generalization of the notion of Dirac structures to Lie algebroids, see [216]. In a different development, the complex version of Dirac structures (L is a complex subbundle of $(TM \oplus T^*M) \otimes \mathbb{C}$ which satisfies the same conditions as in the real case) leads to *generalized complex structures*, see, e.g., [176, 156].

A.9 Deformation quantization

A product operation

$$\star : C^\infty(M)[[\nu]] \times C^\infty(M)[[\nu]] \longrightarrow C^\infty(M)[[\nu]], \tag{A.86}$$

on the space $C^\infty(M)[[\nu]]$ of formal series in ν with coefficients in the space $C^\infty(M)$ of smooth functions on a smooth manifold M, which turns $C^\infty(M)[[\nu]]$ into an *associative* algebra, is called a *differential star product* on M, or *star product* for short, if it satisfies the following conditions:

i) $1 \star F = F \star 1 = F$ for any $F \in C^\infty(M)[[\nu]]$,

ii) $\star$ is ν-linear, i.e.,

$$(\nu F) \star G = F \star (\nu G) = \nu(F \star G) \quad \forall\, F, G \in C^\infty(M)[[\nu]] , \tag{A.87}$$

iii) there are bi-differential operators $C_n(.,.)$ on M with smooth coefficients such that for any $f, g \in C^\infty(M)$ we have

$$f \star g = fg + \sum_{n=1}^{\infty} \nu^n C_n(f, g). \qquad (A.88)$$

Given a star product $\star$ on M with Taylor expansion (A.88), put

$$\{f, g\} = \frac{1}{2}(C_1(f, g) - C_1(g, f)) \qquad (A.89)$$

for $f, g \in C^\infty(M)$. Then $\{.,.\}$ is a Poisson bracket on M. Indeed, denote $[f, g]_\star = \frac{1}{2\nu}(f \star g - g \star f)$. It follows from the associativity of $\star$ that $[.,.]_\star$ is a Lie bracket, i.e., it satisfies the Jacobi identity. Since $\{f, g\}$ is the zeroth term of $[f, g]_\star$ in its Taylor expansion with respect to ν, the zeroth term of the Taylor expansion of the Jacobi identity for $[.,.]_\star$ is nothing but the Jacobi identity for $\{.,.\}$. Similarly, by associativity of $\star$ we have $[f_1 \star f_2]_\star \star g = f_1 \star [f_2, g]_\star + [f_1, g]_\star \star f_2$, and the zeroth term of this equality yields the Leibniz equality $\{f_1 f_2, g\} = f_1\{f_2, g\} + \{f_1, g\}f_2$ for $\{.,.\}$. The anti-commutativity of $\{.,.\}$ is obvious.

Conversely, given a Poisson manifold (M, Π) one is interested in finding a star product on M, which satisfies Equation (A.89), where the Poisson bracket comes from Π. If such a star product exists, then one says that (M, Π) admits a *deformation quantization*.

Deformation quantization was proposed by Bayen, Flato, Frønsdal, Lichnerowicz and Sternheimer [24] as a tool to study quantum physics, based on the philosophy that a higher level (quantum) physical theory is a deformation of a lower level (classical) one. The first example comes from the *Weyl quantization* rule, which associates to each symbol[1] $a(x_1, \ldots, x_n, \xi_1, \ldots, \xi_n)$ on $\mathbb{R}^{2n} = T^*\mathbb{R}^n$ a pseudo-differential operator $W(a)$ on $\mathbb{R}^n$ defined by the oscillatory integral formula

$$(W(a)u)(x) = \frac{1}{(2\pi\hbar)^n} \int_{\mathbb{R}^{2n}} e^{i\langle x-y, \xi\rangle/\hbar} a\left(\frac{x+y}{2}, \xi\right) u(y)\mathrm{d}y\mathrm{d}\xi. \qquad (A.90)$$

For example, $W(\xi_j)u = \frac{\hbar}{i}\frac{\partial}{\partial x_j}u$, $W(x_j)u = x_j.u$, and one has the usual Heisenberg commutation relations $[P_j, Q_k] = \frac{\hbar}{i}\delta_{jk}Id$ where $P_j = W(\xi_j), Q_k = W(x_k)$. If a and b are two symbols on $\mathbb{R}^{2n}$ then one may write

$$W(a)W(b) = W(c), \qquad (A.91)$$

---

[1] A *symbol* a on $\mathbb{R}^{2n} = T^*\mathbb{R}^n$ is a function which satisfies the following asymptotic condition: there is a real number N such that for any compact subset $K \subset \mathbb{R}^n$ and any multi-indices $\alpha, \beta \in \mathbb{Z}_+^n$ there is a constant $C_{K,\alpha,\beta}$ such that $\sup_{x \in K} |\partial_x^\alpha \partial_\xi^\beta a(x, \xi)| \leq C_{K,\alpha,\beta}(1 + |\xi|)^{N-|\alpha|}$.

where c has the following asymptotic expansion when $\hbar \to 0$ (see, e.g., [179]):

$$c(x,\xi) \sim \sum_{\alpha,\beta \in \mathbb{Z}_+^n} (-1)^{|\beta|} \frac{(\hbar/2i)^{|\alpha+\beta|}}{\alpha!\beta!} (\partial_\xi^\alpha \partial_x^\beta a(x,\xi))(\partial_x^\alpha \partial_\xi^\beta b(x,\xi))$$

$$= \exp\left(\frac{\hbar}{2i}\sum_j(\partial_{\xi_j}.\partial_{y_j} - \partial_{x_j}.\partial_{\zeta_j})\right) a(x,\xi)b(y,\zeta)\Big|_{y=x,\zeta=\xi}. \quad \text{(A.92)}$$

Considering $\nu = \hbar/i$ in the above formula as a formal variable and writing $c = a \star b$, one gets the following star product on the standard symplectic space $(\mathbb{R}^{2n}, \sum_j \partial\xi_j \wedge \partial x_j)$, called the *Moyal product* [270]:

$$f \star g(x,\xi) = \exp\left(\frac{\nu}{2}\sum_j(\partial_{\xi_j}.\partial_{y_j} - \partial_{x_j}.\partial_{\zeta_j})\right) f(x,\xi)g(y,\zeta)\Big|_{y=x,\zeta=\xi}. \quad \text{(A.93)}$$

The existence of a star product for an arbitrary symplectic manifold was first established by De Wilde and Lecomte [99] in 1983 using algebraic methods. A more geometric proof, using Weyl quantization and symplectic connections, was found by Fedosov [128] and Omori–Maeda–Yoshioka [286]. The problem of classification of star products on symplectic manifolds was studied by Nest–Tsygan [277], Deligne [100], Bertelson–Cahen–Gutt [28], and other people. The result is that, given a symplectic manifold (M,ω), there is a one-to-one correspondence between the set of equivalence classes of star products for (M,ω) and $H^2(M,\mathbb{R})[[\nu]]$. See [163] for a very nice exposition of this result.

The existence and classification up to equivalence of star products on an arbitrary Poisson manifold was obtained in 1997 by Kontsevich:

Theorem A.9.1 (Kontsevich [195]). *For any smooth Poisson manifold (M,Π), the set of equivalence classes of differential star products on (M,Π) can be naturally identified with the set of equivalence classes of formal Poisson deformation of Π:*

$$\Pi_\nu = \Pi + \nu\Pi_1 + \nu^2\Pi_2 + \cdots \in \mathcal{V}^2(M)[[\nu]], \ [\Pi_\nu, \Pi_\nu] = 0. \quad \text{(A.94)}$$

In particular, one can take $\Pi_\nu = \Pi$, so a deformation quantization for (M,Π) always exists. If the Poisson structure is rigid (i.e., does not admit any nontrivial formal deformation), then the corresponding star product is unique up to equivalence, and vice versa.

Kontsevich obtained the above theorem as a corollary of another very deep result, called the *formality theorem*. Consider the algebra $\mathcal{D}_{\text{poly}}(M)$ of poly-differential operators on a manifold M, equipped with the Hochschild differential and the Gerstenhaber bracket coming from the associative algebra of linear operators on $C^\infty(M)$ (see [142]), and the algebra $\mathcal{T}_{\text{poly}}(M) = \bigoplus_k \mathcal{V}^k(M)$ of multi-vector fields on M with the trivial differential and the Schouten bracket. According to a smooth version of Hochschild–Kostant–Rosenberg's theorem [177], $\mathcal{T}_{\text{poly}}(M)$ is the Hochschild cohomology of $\mathcal{D}_{\text{poly}}(M)$. Kontsevich's *formality theorem* says that

$\mathcal{D}_{\mathrm{poly}}(M)$ is formal, i.e., it is quasi-isomorphic, as a differential graded Lie algebra (or more precisely, as a L_∞-algebra), to its cohomology $\mathcal{T}_{\mathrm{poly}}(M)$. We will not try to explain what it means here, referring the reader to [195, 196, 103, 64] instead. As a particular case of an explicit construction of a quasi-isomorphism from $\mathcal{T}_{\mathrm{poly}}(\mathbb{R}^n)$ to $\mathcal{D}_{\mathrm{poly}}(\mathbb{R}^n)$, Kontsevich [195] gave the following explicit formula for a star product on $(\mathbb{R}^n, \Pi)$:

$$f \star g = \sum_{k=0}^{\infty} \nu^k \sum_{\Gamma \in G_k} w_\Gamma C_\Gamma(\Pi)(f,g), \qquad (A.95)$$

where:

- G_k is the set of oriented graphs Γ such that Γ has $k+2$ vertices $V_\Gamma = \{1,\ldots,k,L,R\}$ and $2k$ labelled edges (with no multiple edges and no edges of the form (v,v) for $v \in V_\Gamma$), $E_\Gamma = (e_1^1, e^2, 1, \ldots, e_k^1, e_k^2)$ where e_j^1 and e_j^2 start at vertex j.

- $C_\Gamma(\Pi)$ is a bi-differential operator defined by the following formula, where $\Pi_{ij} = \{x_i, x_j\}$ are the coefficients of Π on $\mathbb{R}^n$:

$$C_\Gamma(\Pi)(f,g) = \sum_{I:E_\Gamma \to \{1,\ldots,m\}} \Big[\prod_{j=1}^{k} \Big(\prod_{e \in E_\Gamma;\ e=(*,j)} \partial_{x_{I(e)}} \Big) \Pi_{I(e_j^1)I(e_j^2)} \Big]$$

$$\times \Big(\prod_{e \in E_\Gamma;\ e=(*,L)} \partial_{x_{I(e)}} \Big) f \times \Big(\prod_{e \in E_\Gamma;\ e=(*,R)} \partial_{x_{I(e)}} \Big) g. \qquad (A.96)$$

- w_Γ is a real number defined as follows. Denote by $\mathcal{H} = \{p \in \mathbb{C};\ Im(p) > 0\}$ the upper half-plane, and $\mathcal{H}_k = \{(p_1,\ldots,p_k) \mid p_j \in \mathcal{H}, p_i \neq p_j \ \forall\, i \neq j\}$. For $p \in \mathcal{H}, q \in \mathcal{H} \cup \mathbb{R}$, define $\phi(p,q) = \frac{1}{2i} \log((q-p)(\bar{q}-p)(q-\bar{p})^{-1}(\bar{q}-\bar{p})^{-1})$. Assign a point $p_j \in \mathcal{H}$ to each vertex j of Γ, $1 \leq j \leq k$, point $0 \in \mathbb{R} \subset \mathbb{C}$ to the vertex L, and point $1 \in \mathbb{R} \subset \mathbb{C}$ to the vertex R. Every edge $e \in E_\Gamma$ defines an ordered pair (p,q) of points on $\mathcal{H} \cup \mathbb{R}$, thus a function $\phi_e = \phi(p,q)$ on $\mathcal{H}$ with values in $\mathbb{R}/2\pi\mathbb{Z}$. Now put

$$w_\Gamma = \frac{1}{k!(2\pi)^{2k}} \int_{\mathcal{H}_k} \wedge_{j=1}^{k} (d\phi_{e_j^1} \wedge d\phi_{e_j^2}). \qquad (A.97)$$

Kontsevich [195] obtained the formality theorem (and hence the existence of deformation quantization) for a general manifold from the formality theorem for $\mathbb{R}^n$ and abstract arguments. A more explicit globalization of Kontsevich's star product formula (A.95) from $\mathbb{R}^n$ to an arbitrary Poisson manifold (M, Π) was obtained by Cattaneo–Felder–Tomassini [69, 67] by a method similar to Fedosov's method [128] for the symplectic case.

For people familiar with quantum field theory, Formula (A.95) looks like the expansion of a Feynman path integral. In fact, it seems that a field theory that gives rise to star products on Poisson manifolds exists and is known as the *Poisson* *sigma model*, first studied by Shaller–Strobl [310] and other physicists. A field in a Poisson sigma model is a pair (X, η), where $X : \Sigma \to M$ is a map from a two-dimensional surface Σ to a given Poisson manifold (M, Π), and η is a section of $T^*\Sigma \otimes X^*(T^*M)$ with the boundary condition that $\eta(u)v = 0 \; \forall \; u \in \partial\Sigma, v \in T_u(\partial\Sigma)$. The action functional is

$$S(X, \eta) = \int_{\Sigma} \langle \eta, \mathrm{d}X \rangle + \frac{1}{2} \langle \eta, (\Pi \circ X)\eta \rangle. \tag{A.98}$$

Cattaneo and Felder [66] found the following path integral formula for a star product on an arbitrary given Poisson manifold (M, Π):

$$f \star g = \int_{X(\infty)=x} f(X(1))g(X(0))e^{iS(X,\eta)/\hbar}\mathrm{d}X\mathrm{d}\eta, \tag{A.99}$$

where Σ is now a two-dimensional disk, $0, 1, \infty$ are three distinct points on the boundary of Σ, S is the above functional, and f, g and functions on (M, Π). The interested reader may consult, e.g., [66, 67, 68, 69] for details and relations to formal symplectic groupoids. Unfortunately, the authors of this book are not familiar with quantum field theory, and the above formulas look mysterious to them.

Deformation quantization has many other aspects and is related to many other subjects, e.g., quantum groups, representation theory, index theory, other quantization theories, etc. See, e.g., the survey articles [322, 162] and references therein.

Bibliography

[1] A. Abouqateb and M. Boucetta, *The modular class of a regular Poisson manifold and the Reeb class of its symplectic foliation*, C. R. Math. Acad. Sci. Paris **337** (2003), no. 1, 61–66.

[2] M. Adler, P. van Moerbeke, and P. Vanhaecke, *Algebraic completely integrable systems, Painlevé architecture and Lie algebras*, Springer, 2004.

[3] Claude Albert and Pierre Dazord, *Groupoïdes de Lie et groupoïdes symplectiques*, Symplectic geometry, groupoids, and integrable systems (Berkeley, CA, 1989), Math. Sci. Res. Inst. Publ., vol. 20, Springer, New York, 1991, pp. 1–11.

[4] F. Alcalde-Cuesta and G. Hector, *Intégration symplectique des variétés de Poisson régulières*, Israel J. Math. **90** (1995), no. 1-3, 125–165.

[5] Anton Alekseev, *On Poisson actions of compact Lie groups on symplectic manifolds*, J. Differential Geom. **45** (1997), no. 2, 241–256.

[6] Anton Alekseev and Yvette Kosmann-Schwarzbach, *Manin pairs and moment maps*, J. Differential Geom. **56** (2000), no. 1, 133–165.

[7] Anton Alekseev, Yvette Kosmann-Schwarzbach, and Eckhard Meinrenken, *Quasi-Poisson manifolds*, Canad. J. Math. **54** (2002), no. 1, 3–29.

[8] Anton Alekseev, Anton Malkin, and Eckhard Meinrenken, *Lie group valued moment maps*, J. Differ. Geom. **48** (1998), no. 3, 445–495.

[9] D. Alekseevsky and P. Guha, *On decomposability of Nambu-Poisson tensor*, Acta Math. Univ. Comenian. (N.S.) **65** (1996), no. 1, 1–9.

[10] M. Alexandrov, A. Schwarz, O. Zaboronsky, and M. Kontsevich, *The geometry of the master equation and topological quantum field theory*, Internat. J. Modern Phys. A **12** (1997), no. 7, 1405–1429.

[11] Rui Almeida and Pierre Molino, *Suites d'Atiyah et feuilletages transversalement complets*, C. R. Acad. Sci. Paris Sér. I Math. **300** (1985), no. 1, 13–15.

[12] D. Arnal, M. Cahen, and J. Ludwig, *Lie groups whose coadjoint orbits are of dimension smaller or equal to two.*, Lett. Math. Phys. **33** (1995), no. 2, 183–186.

[13] V. I. Arnold, *Normal forms of functions near degenerate critical points, the Weyl groups A_k, D_k, E_k and Lagrangian singularities*, Funkcional. Anal. i Priložen. **6** (1972), no. 4, 3–25.

[14] ———, *Geometrical methods in the theory of ordinary differential equations*, Springer-Verlag, New York, 1988.

[15] ———, *Poisson structures on the plane and other powers of volume forms*, J. Soviet. Math. **47** (1989), 2509–2516.

[16] V. I. Arnold, S. M. Gusejn-Zade, and A. N. Varchenko, *Singularities of differentiable maps. Volume I: The classification of critical points, caustics and wave fronts.*, Monographs in Mathematics, Vol. 82., 382 p., 1985.

[17] V. I. Arnold and Yu. S. Ilyashenko, *Ordinary differential equations. In: Dynamical systems I. Transl. from the Russian.*, Encyclopaedia of Mathematical Sciences, 1. Berlin etc.: Springer-Verlag., 1988.

[18] M. Artin, *On the solutions of analytic equations*, Invent. Math. **5** (1968), 277–291.

[19] Michèle Audin, *Spinning tops*, Cambridge University Press, Cambridge, 1996.

[20] Olivier Babelon, Denis Bernard, and Michel Talon, *Introduction to classical integrable systems*, Cambridge Monographs on Mathematical Physics, Cambridge University Press, 2003.

[21] A. Balinsky and Yu. Burman, *Quadratic Poisson brackets and the Drinfeld theory for associative algebras*, Lett. Math. Phys. **38** (1996), no. 1, 63–75.

[22] A. Banyaga, R. de la Llave, and C. E. Wayne, *Cohomology equations near hyperbolic points and geometric versions of Sternberg linearization theorem*, J. Geom. Anal. **6** (1996), no. 4, 613–649.

[23] L. Bates and E. Lerman, *Proper group actions and symplectic stratified spaces*, Pacific J. Math. **181** (1997), no. 2, 201–229.

[24] F. Bayen, M. Flato, Ch. Frønsdal, A. Lichnerowicz, and D. Sternheimer, *Deformation theory and quantization. I and II*, Ann. Phys. **111** (1978), 61–151.

[25] A. A. Belavin and V. G. Drinfeld, *Triangle equations and simple Lie algebras*, Sov. Sci. Rev. Math. **4** (1984), 93–165.

[26] G. Belitskii, C^∞-*normal forms of local vector fields*, Acta Appl. Math. **70** (2002), no. 1-3, 23–41.

[27] Moulay-Tahar Benameur and Victor Nistor, *Homology of algebras of families of pseudodifferential operators*, J. Funct. Anal. **205** (2003), no. 1, 1–36.

[28] Mélanie Bertelson, Michel Cahen, and Simone Gutt, *Equivalence of star products*, Classical Quantum Gravity **14** (1997), no. 1A, A93–A107.

[29] K. H. Bhaskara and K. Rama, *Quadratic Poisson structures*, J. Math. Phys. **32** (1991), no. 9, 2319–2322.

[30] K. H. Bhaskara and K. Viswanath, *Calculus on Poisson manifolds*, Bull. London Math. Soc. **20** (1988), 68–72.

[31] _____, *Poisson algebras and Poisson manifolds*, Pitman Research Notes in Mathematics Series, vol. 174, Longman Scientific & Technical, Harlow, 1988.

[32] G. Blankenstein and A. J. van der Schaft, *Symmetry and reduction in implicit generalized Hamiltonian systems*, Rep. Math. Phys. **47** (2001), no. 1, 57–100.

[33] Guido Blankenstein and Tudor S. Ratiu, *Singular reduction of implicit Hamiltonian systems*, Rep. Math. Phys. **53** (2004), no. 2, 211–260.

[34] A. M. Bloch, *Nonholonomic mechanics and control*, Interdisciplinary Applied Mathematics, vol. 24, Springer-Verlag, New York, 2003.

[35] Philip Boalch, *Stockes matrices and Poisson-Lie groups*, Invent. Math. **146** (2001), 479–506.

[36] Salomon Bochner, *Compact groups of differentiable transformations*, Annals of Math. (2) **46** (1945), 372–381.

[37] A.V. Bolsinov, *Compatible Poisson brackets on Lie algebras and completeness of families of functions in involution*, Math. USSR Izvestiya **38** (1992), 69–90.

[38] A.V. Bolsinov and A.T. Fomenko, *Integrable Hamiltonian Systems. Geometry, Topology, Classification. Vol. 1 and 2 (in Russian)*, 1999.

[39] M. Bordemann, A. Makhlouf, and T. Petit, *Déformation par quantification et rigidité des algèbres enveloppantes*, preprint math.RA/0211416 (2002).

[40] A. Borel and N. Wallach, *Continuous cohomology, discrete subgroups, and representations of reductive groups*, Mathematical Surveys and Monographs, vol. 67, American Mathematical Society, Providence, RI, 2000.

[41] Raoul Bott and Loring W. Tu, *Differential forms in algebraic topology*, Graduate Texts in Mathematics, vol. 82, Springer-Verlag, New York, 1982.

[42] N. Bourbaki, *Elements de mathematique. Fasc. XXVI: Groupes et algebres de Lie. Chap. 1: Algebres de Lie*, Paris: Hermann & Cie. 142 p., 1960.

[43] _____, *Eléments de mathématique. Algèbre. Chapitre 3*, Paris: Hermann, 1970.

[44] _____, *Elements of mathematics. Commutative algebra*, Hermann, Paris, 1972.

[45] Olivier Brahic, *Formes normales semi-locales de structures de Poisson*, Thèse de doctorat, Université Montpellier II (2004).

[46] I. U. Bronstein and A. Ya. Kopanskii, *Smooth invariant manifolds and normal forms*, World Scientific Series on Nonlinear Science. Series A: Monographs and Treatises, vol. 7, World Scientific Publishing Co. Inc., River Edge, NJ, 1994.

[47] _____, *Normal forms of vector fields satisfying certain geometric conditions*, Nonlinear dynamical systems and chaos (Groningen, 1995), Progr. Nonlinear Differential Equations Appl., vol. 19, Birkhäuser, Basel, 1996, pp. 79–101.

[48] A. D. Bruno, *Analytic form of differential equations*, Trans. Moscow Math. Soc. **25** (1971), 131–288.

[49] _____, *Local methods in nonlinear differential equations*, Springer Series in Soviet Mathematics, Springer-Verlag, Berlin, 1989.

[50] A.D. Bruno and S. Walcher, *Symmetries and convergence of normalizing transformations*, J. Math. Anal. Appl. **183** (1994), 571–576.

[51] Jean-Luc Brylinski, *A differential complex for Poisson manifolds*, J. Differential Geom. **28** (1988), no. 1, 93–114.

[52] H. Bursztyn and M. Crainic, *Dirac structures, moment maps and quasi-Poisson manifolds*, preprint math.DG/0310445 v.2 (2004).

[53] H. Bursztyn, M. Crainic, A. Weinstein, and C. Zhu, *Integration of twisted Dirac brackets*, Duke Math. J. **123** (2004), no. 3, 549–607.

[54] H. Bursztyn and O. Radko, *Gauge equivalence of Dirac structures and symplectic groupoids*, Ann. Inst. Fourier (Grenoble) **53** (2003), no. 1, 309–337.

[55] Michel Cahen, Simone Gutt, and John Rawnsley, *Nonlinearizability of the Iwasawa Poisson Lie structure*, Lett. Math. Phys. **24** (1992), no. 1, 79–83.

[56] _____, *Some remarks on the classification of Poisson Lie groups*, Symplectic geometry and quantization (Sanda and Yokohama, 1993), Contemp. Math., 179, Amer. Math. Soc., Providence, 1994, pp. 1–16.

[57] Grant Cairns and Étienne Ghys, *The local linearization problem for smooth SL(n)-actions*, Enseign. Math. **43** (1997), no. 1-2, 133–171.

[58] César Camacho, *Structural stability theorems for integrable differential forms on 3-manifolds*, Topology **17** (1978), no. 2, 143–155.

[59] César Camacho and Alcides Lins Neto, *The topology of integrable differential forms near a singularity*, Inst. Hautes Études Sci. Publ. Math. (1982), no. 55, 5–35.

[60] Ana Cannas da Silva and Alan Weinstein, *Geometric models for noncommutative algebras*, American Mathematical Society, Providence, RI, 1999.

[61] J. F. Cariñena, A. Ibort, G. Marmo, and A. Perelomov, *On the geometry of Lie algebras and Poisson tensors*, J. Phys. A **27** (1994), no. 22, 7425–7449.

[62] Roger Carles, *Sur la structure des algèbres de Lie rigides*, Ann. Inst. Fourier (Grenoble) **34** (1984), no. 3, 65–82.

[63] ———, *Déformations et éléments nilpotents dans les schémas définis par les identités de Jacobi*, C. R. Acad. Sci. Paris I **312** (1991), 671–674.

[64] Alberto Cattaneo and Davide Indelicato, *Formality and star products*, preprint math.QA/0403135 (2004).

[65] Alberto Cattaneo and Ping Xu, *Integration of twisted Poisson structures*, J. Geom. Phys. **49** (2004), no. 2, 187–196.

[66] Alberto S. Cattaneo and Giovanni Felder, *A path integral approach to the Kontsevich quantization formula*, Commun. Math. Phys. **212** (2000), no. 3, 591–611.

[67] ———, *On the globalization of Kontsevich's star product and the perturbative Poisson sigma model*, Progr. Theoret. Phys. Suppl. (2001), no. 144, 38–53.

[68] ———, *Poisson sigma models and symplectic groupoids*, Landsman, N. P. et al. (ed.), Quantization of singular symplectic quotients. Progress in Mathematics, Vol. 198, 2001, pp. 61–93.

[69] Alberto S. Cattaneo, Giovanni Felder, and Lorenzo Tomassini, *From local to global deformation quantization of Poisson manifolds.*, Duke Math. J. **115** (2002), no. 2, 329–352.

[70] D. Cerveau and J.-F. Mattei, *Formes intégrables holomorphes singulières*, Astérisque, vol. 97, Société Mathématique de France, Paris, 1982.

[71] Dominique Cerveau, *Distributions involutives singulières*, Ann. Inst. Fourier **29** (1979), no. 3, 261–294.

[72] Dominique Cerveau and Alcides Lins Neto, *Formes tangentes à des actions commutatives*, Ann. Fac. Sci. Toulouse Math. (5) **6** (1984), no. 1, 51–85.

[73] Marc Chaperon, *Géométrie différentielle et singularités de systèmes dynamiques*, Astérisque (1986), no. 138-139, 440.

[74] ———, *Invariant manifolds revisited*, Tr. Mat. Inst. Steklova **236** (2002), no. Differ. Uravn. i Din. Sist., 428–446.

[75] ———, *Stable manifolds and the Perron-Irwin method*, Ergodic Theory Dynam. Systems **24** (2004), no. 5, 1359–1394.

[76] Kuo-Tsai Chen, *Equivalence and decomposition of vector fields about an elementary critical point*, Amer. J. Math. **85** (1963), 693–722.

[77] Claude Chevalley and Samuel Eilenberg, *Cohomology theory of lie groups and lie algebras*, Trans. Am. Math. Soc. **63** (1948), 85–124.

[78] Véronique Chloup, *Bialgebra structures on a real semisimple Lie algebra*, Bull. Belg. Math. Soc. Simon Stevin **2** (1995), no. 3, 265–278.

[79] ———, *Linearization of some Poisson-Lie tensor*, J. Geometry and Physics **24** (1997), 46–52.

[80] Jack F. Conn, *Normal forms for analytic Poisson structures*, Annals of Math. (2) **119** (1984), no. 3, 577–601.

[81] ———, *Normal forms for smooth Poisson structures*, Annals of Math. (2) **121** (1985), no. 3, 565–593.

[82] Alain Connes, *Noncommutative geometry*, Academic Press, San Diego, 1994.

[83] A. Coste, P. Dazord, and A. Weinstein, *Groupoïdes symplectiques*, Publications du Département de Mathématiques. Nouvelle Série. A, Vol. 2, Publ. Dép. Math. Nouvelle Sér. A, vol. 87, Univ. Claude-Bernard, Lyon, 1987, pp. i–ii, 1–62.

[84] T. J. Courant, *Dirac manifolds*, Trans. Amer. Math. Soc. **319** (1990), no. 2, 631–661.

[85] Marius Crainic, *Differentiable and algebroid cohomology, van Est isomorphisms, and characteristic classes*, Comment. Math. Helv. **78** (2003), no. 4, 681–721.

[86] Marius Crainic and Rui L. Fernandes, *Integrability of Lie brackets*, Ann. of Math. (2) **157** (2003), no. 2, 575–620.

[87] _____, *Integrability of Poisson brackets*, J. Diff. Geom. **66** (2004), no. 1, 71–137.

[88] Marius Crainic and Ieke Moerdijk, *Deformations of Lie brackets: Cohomological aspects*, preprint math.DG/0403434 (2004).

[89] Richard Cushman and Mark Roberts, *Poisson structures transverse to coadjoint orbits*, Bull. Sci. Math. **126** (2002), no. 7, 525–534.

[90] Richard Cushman and Jedrzej Sniatycki, *Singular reduction for proper actions*, preprint Utrecht No. 1133 (2000).

[91] Morten Dalsmo and Arjan van der Schaft, *On representations and integrability of mathematical structures in energy-conserving physical systems*, SIAM J. Control Optim. **37** (1999), no. 1, 54–91 (electronic).

[92] Pantelis A. Damianou, *Transverse Poisson structures of coadjoint orbits*, Bull. Sci. Math. **120** (1996), no. 2, 195–214.

[93] Pierre Dazord, *Feuilletages à singularités*, Indag. Math. **47** (1985), 21–39.

[94] _____, *Stabilité et linéarisation dans les variétés de Poisson*, Séminaire Sud-Rhodanien Symplectic geometry and mechanics, Balaruc 1983, 1985, pp. 59–75.

[95] _____, *Groupoïdes symplectiques et troisième théorème de Lie "non linéaire"*, Géométrie symplectique et mécanique (La Grande Motte, 1988), Lecture Notes in Math., vol. 1416, Springer, Berlin, 1990, pp. 39–74.

[96] _____, *Groupoïde d'holonomie et géométrie globale*, C. R. Acad. Sci. Paris Sér. I Math. **324** (1997), no. 1, 77–80.

[97] Pierre Dazord and D. Sondaz, *Groupes de Poisson affines*, Symplectic geometry, groupoids, and integrable systems (Berkeley, CA, 1989), Math. Sci. Res. Inst. Publ., vol. 20, Springer, New York, 1991, pp. 99–128.

[98] Georges de Rham, *Sur la division de formes et de courants par une forme linéaire*, Comment. Math. Helv. **28** (1954), 346–352.

[99] Marc De Wilde and Pierre B. A. Lecomte, *Existence of star-products and of formal deformations of the Poisson Lie algebra of arbitrary symplectic manifolds*, Lett. Math. Phys. **7** (1983), no. 6, 487–496.

[100] P. Deligne, *Déformations de l'algèbre des fonctions d'une variété symplectique: comparaison entre Fedosov et De Wilde, Lecomte*, Selecta Math. (N.S.) **1** (1995), no. 4, 667–697.

[101] André Diatta and Alberto Medina, *Classical Yang-Baxter equation and left invariant affine geometry on Lie groups*, Manuscripta Math. **114** (2004), no. 4, 477–486.

[102] L.A. Dickey, *Soliton equations and Hamiltonian systems. 2nd ed.*, Advanced Series in Mathematical Physics 26, World Scientific, Singapore, 2003.

[103] Vasiliy Dolgushev, *Covariant and equivariant formality theorems*, preprint math.QA/0307212 (2003).

[104] J. Donin and D. Gurevich, *Some Poisson structures associated to Drinfel'd-Jimbo R-matrices and their quantization*, Israel J. Math. **92** (1995), no. 1-3, 23–32.

[105] Irene Dorfman, *Dirac structures and integrability of nonlinear evolution equations*, Nonlinear Science: Theory and Applications, John Wiley & Sons Ltd., Chichester, 1993.

[106] A. Douady and M. Lazard, *Espaces fibres en algèbres de Lie et en groupes*, Invent. Math. **1** (1966), 133–151.

[107] V. G. Drinfeld, *Hamiltonian structures on Lie groups, Lie bialgebras and geometric meaning of classical Yang-Baxter equations*, Dokl. Akad. Nauk. SSSR **268** (1983), 285–287.

[108] ———, *Quantum groups*, Proceedings ICM, Berkeley 1986, Vol. 1, 1986, pp. 789–820.

[109] ———, *On Poisson homogeneous spaces of Poisson-Lie groups*, Theoret. and Math. Phys. **95** (1993), no. 2, 524–525.

[110] J.-P. Dufour, *Linéarisation de certaines structures de Poisson*, J. Differential Geom. **32** (1990), no. 2, 415–428.

[111] ———, *Hyperbolic actions of R^p on Poisson manifolds.*, Symplectic geometry, groupoids, and integrable systems, Sémin. Sud- Rhodan. Geom. VI, Berkeley/CA (USA) 1989, Math. Sci. Res. Inst. Publ. 20, 137-150 , 1991.

[112] ———, *Normal forms of Lie algebroids*, Banach Center Publications **54** (2001), 35–41.

[113] J.-P. Dufour and J.-Ch. Molinier, *Une nouvelle famille d'algèbres de Lie non dégénérées*, Indag. Math. (N.S.) **6** (1995), no. 1, 67–82.

[114] J.-P. Dufour and A. Wade, *Formes normales de structures de Poisson ayant un 1-jet nul en un point*, J. Geom. Phys. **26** (1998), no. 1-2, 79–96.

[115] ———, *On the local structure of Dirac manifolds*, preprint math.SG/0405257 (2004).

[116] J.-P. Dufour and M. Zhitomirskii, *Classification of non-resonant Poisson structures*, J. London Math. Soc. **(2) 60** (1999), no. 3, 935–950.

[117] ———, *Singularities and bifurcations of 3-dimensional Poisson structures*, Israel J. Math. **121** (2001), 199–220.

[118] Jean-Paul Dufour and Abdeljalil Haraki, *Rotationnels et structures de Poisson quadratiques*, C. R. Acad. Sci. Paris Sér. I Math. **312** (1991), no. 1, 137–140.

[119] Jean-Paul Dufour and Nguyen Tien Zung, *Linearization of Nambu structures*, Compositio Math. **117** (1999), no. 1, 77–98.

[120] ———, *Nondegeneracy of the Lie algebra $\mathfrak{aff}(n)$*, C. R. Math. Acad. Sci. Paris **335** (2002), no. 12, 1043–1046.

[121] Hans Duistermaat and J. A. C. Kolk, *Lie groups*, Universitext, Springer-Verlag, Berlin, 2000.

[122] Jean Écalle, *Singularités non abordables par la géométrie*, Ann. Inst. Fourier (Grenoble) **42** (1992), no. 1-2, 73–164.

[123] Charles Ehresmann, *Œuvres complètes et commentées. I-1,2. Topologie algébrique et géométrie différentielle*, Cahiers Topologie Géom. Différentielle **24** (1983), no. suppl. 1, xxix+601 pp.

[124] M. El Galiou, *Structures de Poisson homogènes. Déformations. Star-Produits*, Thèse de la Faculté des Sciences-Semlalia (1996).

[125] Pavel Etingof and Olivier Schiffmann, *Lectures on the dynamical Yang-Baxter equations*, Quantum groups and Lie theory (Durham, 1999), London Math. Soc. Lecture Note Ser., vol. 290, Cambridge Univ. Press, Cambridge, 2001, pp. 89–129.

[126] Sam Evens and Jiang-Hua Lu, *On the variety of Lagrangian subalgebras. I*, Ann. Sci. École Norm. Sup. (4) **34** (2001), no. 5, 631–668.

[127] Sam Evens, Jiang-Hua Lu, and Alan Weinstein, *Transverse measures, the modular class and a cohomology pairing for Lie algebroids*, Quart. J. Math. Oxford Ser. (2) **50** (1999), no. 200, 417–436.

[128] Boris V. Fedosov, *A simple geometrical construction of deformation quantization*, J. Differential Geom. **40** (1994), no. 2, 213–238.

[129] Ping Feng and Boris Tsygan, *Hochschild and cyclic homology of quantum groups*, Comm. Math. Phys. **140** (1991), no. 3, 481–521.

[130] Rui L. Fernandes, *Connections in Poisson geometry. I. Holonomy and invariants*, J. Differential Geom. **54** (2000), no. 2, 303–365.

[131] _____, *Lie algebroids, holonomy and characteristic classes*, Adv. in Math. **70** (2002), 119–179.

[132] Rui Loja Fernandes and Philippe Monnier, *Linearization of Poisson brackets*, Lett. Math. Phys. **69** (2004), 89–114.

[133] M. Fernández, R Ibáñez, and M. de León, *On a Brylinski conjecture for compact symplectic manifolds*, Proceedings of the meeting on quaternionic structures in mathematics and physics, SISSA, Trieste, 1994.

[134] V.T. Filippov, *n-ary Lie algebras*, Sibirskii Math. J. **26** (1985), 879–891.

[135] M. Flato and J. Simon, *On a linearization program of nonlinear field equations*, Phys. Lett. B **94** (1980), no. 4, 518–522.

[136] A. S. Fokas and I. M. Gel'fand, *Quadratic Poisson algebras and their infinite-dimensional extensions*, J. Math. Phys. **35** (1994), no. 6, 3117–3131.

[137] A. T. Fomenko, *Integrability and nonintegrability in geometry and mechanics*, Kluwer, Dordrecht, 1988.

[138] A. T. Fomenko and D. B. Fuks, *Kurs gomotopicheskoi topologii*, "Nauka", Moscow, 1989.

[139] B. Fuchssteiner, *The Lie algebra structure of degenerate Hamiltonian and bi-Hamiltonian systems*, Progr. Theoret. Phys. **68** (1982), 1082–1104.

[140] Philippe Gautheron, *Some remarks concerning Nambu mechanics.*, Lett. Math. Phys. **37** (1996), no. 1, 103–116.

[141] I. M. Gel'fand and I. Ya. Dorfman, *Hamiltonian operators and the classical Yang-Baxter equation*, Functional Anal. Appl. **16** (1982), no. 4, 241–248.

[142] Murray Gerstenhaber, *On the deformation of rings and algebras*, Ann. of Math. (2) **79** (1964), 59–103.

[143] Viktor L. Ginzburg, *Momentum mappings and Poisson cohomology*, Internat. J. Math. **7** (1996), no. 3, 329–358.

[144] _____, *Equivariant Poisson cohomology and a spectral sequence associated with a moment map*, Internat. J. Math. **10** (1999), no. 8, 977–1010.

[145] _____, *Grothendieck groups of Poisson vector bundles*, J. Symplectic Geom. **1** (2001), no. 1, 121–169.

[146] Viktor L. Ginzburg and Alex Golubev, *Holonomy on Poisson manifolds and the modular class*, Israel J. Math. **122** (2001), 221–242.

[147] Viktor L. Ginzburg and Jiang-Hua Lu, *Poisson cohomology of Morita-equivalent Poisson manifolds*, Internat. Math. Res. Notices (1992), no. 10, 199–205.

[148] Viktor L. Ginzburg and Alan Weinstein, *Lie-Poisson structure on some Poisson Lie groups.*, J. Am. Math. Soc. **5** (1992), no. 2, 445–453.

[149] Claude Godbillon and Jacques Vey, *Un invariant des feuilletages de codimension 1*, C. R. Acad. Sci. Paris Sér. A-B **273** (1971), A92–A95.

[150] M. Golubitsky and V. Guillemin, *Stable mappings and their singularities. 2nd corr. printing.*, Graduate Texts in Mathematics, 14. New York - Heidelberg - Berlin: Springer-Verlag. XI, 209 p., 1980.

[151] M. Goze and J. M. Ancochea Bermudez, *On the classification of rigid Lie algebras*, J. Algebra **245** (2001), no. 1, 68–91.

[152] J. Grabowski and G. Marmo, *On Filippov algebroids and multiplicative Nambu-Poisson structures*, Differential Geom. Appl. **12** (2000), no. 1, 35–50.

[153] J. Grabowski, G. Marmo, and P. Michor, *Homology and modular classes of lie algebroids*, preprint math.DG/0310072 (2003).

[154] Janusz Grabowski, *Quasi-derivations and QD-algebroids*, Rep. Math. Phys. **52** (2003), no. 3, 445–451.

[155] Karsten Grove, Hermann Karcher, and Ernst A. Ruh, *Group actions and curvature*, Invent. Math. **23** (1974), 31–48.

[156] Marco Gualtieri, *Generalized complex geometry*, preprint math.DG/0401221 (2004).

[157] A. Guichardet, *Cohomologie des groupes topologiques et des algèbres de Lie*, Textes Mathématiques, vol. 2, CEDIC, Paris, 1980.

[158] Victor Guillemin, Viktor Ginzburg, and Yael Karshon, *Moment maps, cobordisms, and Hamiltonian group actions*, Mathematical Surveys and Monographs, vol. 98, American Mathematical Society, Providence, RI, 2002.

[159] Victor Guillemin, Eugene Lerman, and Shlomo Sternberg, *Symplectic fibrations and multiplicity diagrams*, Cambridge: Cambridge Univ. Press, 1996.

[160] Victor W. Guillemin and Shlomo Sternberg, *Remarks on a paper of Hermann*, Trans. Am. Math. Soc. **130** (1968), 110–116.

[161] D. I. Gurevich, *Poisson brackets associated with the classical Yang-Baxter equation*, Funct. Anal. Appl. **23** (1989), no. 1, 57–59.

[162] Simone Gutt, *Variations on deformation quantization*, Conférence Moshé Flato 1999, Vol. I (Dijon), Math. Phys. Stud., vol. 21, Kluwer Acad. Publ., Dordrecht, 2000, pp. 217–254.

[163] Simone Gutt and John Rawnsley, *Equivalence of star products on a symplectic manifold; an introduction to Deligne's Čech cohomology classes*, J. Geom. Phys. **29** (1999), no. 4, 347–392.

[164] André Haefliger, *Structures feuilletées et cohomologie à valeur dans un faisceau de groupoïdes.*, Comment. Math. Helv. **32** (1958), 248–329.

[165] _____, *Groupoïdes d'holonomie et classifiants*, Astérisque (1984), no. 116, 70–97.

[166] Richard S. Hamilton, *Deformation of complex structures on manifolds with boundary. I. The stable case*, J. Differential Geometry **12** (1977), no. 1, 1–45.

[167] _____, *The inverse function theorem of Nash and Moser*, Bull. Amer. Math. Soc. (N.S.) **7** (1982), no. 1, 65–222.

[168] William R. Hamilton, *Second essay on a general mthod in dynamics*, Philosophical Transactions of the Royal Society **I** (1835), 95–144.

[169] A. Haraki, *Quadratisation de certaines structures de Poisson*, J. London Math. Soc. (2) **56** (1997), no. 2, 384–394.

[170] Sigurdur Helgason, *Differential geometry, Lie groups, and symmetric spaces*, Graduate Studies in Mathematics, vol. 34, American Mathematical Society, Providence, RI, 2001.

[171] Robert Hermann, *On the accessibility problem in control theory*, Internat. Sympos. Nonlinear Differential Equations and Nonlinear Mechanics, Academic Press, New York, 1963, pp. 325–332.

[172] _____, *The formal linearization of a semisimple Lie algebra of vector fields about a singular point*, Trans. Amer. Math. Soc. **130** (1968), 105–109.

[173] Jean-Claude Herz, *Pseudo-algèbres de Lie. I*, C. R. Acad. Sci. Paris **236** (1953), 1935–1937 and 2289–2291.

[174] Philip Higgins and Kirill Mackenzie, *Algebraic constructions in the category of Lie algebroids*, Journal of Algebra **129** (1990), 194–230.

[175] _____, *Duality for base-changing morphisms of vector bundles, modules, Lie algebroids and Poisson structures*, Math. Proc. Cambridge Philos. Soc. **114** (1993), no. 3, 471–488.

[176] Nigel Hitchin, *Generalized Calabi-Yau manifolds*, Quarterly J. Math. **54** (2003), no. 3, 281–308.

[177] G. Hochschild, Bertram Kostant, and Alex Rosenberg, *Differential forms on regular affine algebras*, Trans. Amer. Math. Soc. **102** (1962), 383–408.

[178] G. Hochschild and J.-P. Serre, *Cohomology of Lie algebras*, Ann. Math. **57** (1953), 591–603.

[179] L. Hörmander, *The Weyl calculus of pseudodifferential operators*, Comm. Pure Appl. Math. **32** (1979), no. 3, 360–444.

[180] Johannes Huebschmann, *Poisson cohomology and quantization*, J. für die Reine und Angew. Math. **408** (1990), 57–113.

[181] _____, *Duality for Lie-Rinehart algebras and the modular class*, J. Reine Angew. Math. **510** (1999), 103–159.

[182] Yu. S. Ilyashenko and S. Yu. Yakovenko, *Finitely smooth normal forms of local families of diffeomorphisms and vector fields*, Russian Math. Surveys **46** (1991), no. 1, 1–43.

[183] Hidekazu Ito, *Convergence of Birkhoff normal forms for integrable systems*, Comment. Math. Helv. **64** (1989), no. 3, 412–461.

[184] V. Itskov, M. Karasev and Yu. Vorobjev, *Infinitesimal Poisson cohomology*, Amer. Math. Soc. Transl. **187** (1998), 327–360.

[185] C. G. J. Jacobi, *Vorlesungen über Dynamik* (1842–1843). *Gesammelte Werke. Bände I–VIII*, Herausgegeben auf Veranlassung der Königlich Preussischen Akademie der Wissenschaften. Zweite Ausgabe, Chelsea Publishing Co., New York, 1969.

[186] Nathan Jacobson, *Lie algebras*, Interscience Publishers, 1962.

[187] Michio Jimbo, *Yang-Baxter equation in integrable systems (edited by M. Jimbo)*, Advanced Series in Mathematical Physics, vol. 10, World Scientific Publishing, 1989.

[188] T. Kappeler, Y. Kodama, and A. Némethi, *On the Birkhoff normal form of a completely integrable Hamiltonian system near a fixed point with resonance*, Ann. Scuola Norm. Sup. Pisa Cl. Sci. **XXVI** (1998), no. 4, 623–661.

[189] M. V. Karasev, *Analogues of objects of Lie group theory for nonlinear Poisson brackets*, Math. USSR Izvestiya **28** (1987), 497–527.

[190] M. V. Karasev and V. P. Maslov, *Nonlinear Poisson brackets: geometry and quantization*, Translations of Mathematical Monographs, Vol. 119, AMS, Providence, 1993.

[191] Eugene Karolinsky, *A classification of Poisson homogeneous spaces of complex reductive Poisson-Lie groups*, Poisson geometry (Warsaw, 1998), Banach Center Publ., vol. 51, Polish Acad. Sci., Warsaw, 2000, pp. 103–108.

[192] D. Kastler and R. Stora, *Lie-Cartan pairs*, J. Geom. Phys. **2** (1985), no. 3, 1–31.

[193] A. A. Kirillov, *Lectures on the orbit method*, Graduate Studies in Mathematics, vol. 64, American Mathematical Society, Providence, RI, 2004.

[194] Anthony W. Knapp, *Lie groups beyond an introduction*, Progress in Mathematics, vol. 140, Birkhäuser Boston Inc., Boston, MA, 2002.

[195] Maxim Kontsevich, *Deformation quantization of Poisson manifolds, I*, preprint q-alg/9709040 (1997).

[196] ———, *Operads and motives in deformation quantization*, Lett. Math. Phys. **48** (1999), no. 1, 35–72.

[197] Yvette Kosmann-Schwarzbach, *Quasi, twisted and all that ... in Poisson geometry and Lie algebroid theory*, preprint math.SG/0310359 (2003).

[198] Yvette Kosmann-Schwarzbach and Franco Magri, *Poisson-Lie groups and complete integrability. I. Drinfel'd bialgebras, dual extensions and their canonical representations*, Ann. Inst. H. Poincaré Phys. Théor. **49** (1988), no. 4, 433–460.

[199] ———, *Poisson-Nijenhuis structures*, Ann. Inst. H. Poincaré Phys. Théor. **53** (1990), no. 1, 35–81.

[200] Jean-Louis Koszul, *Sur certains groupes de transformations de Lie*, Géométrie différentielle. Colloques Internationaux du CNRS, Strasbourg, 1953, CNRS, Paris, 1953, pp. 137–141.

[201] ———, *Crochet de Schouten-Nijenhuis et cohomologie*, Astérisque (1985), no. Numero Hors Serie, 257–271.

[202] Jan Kubarski, *Characteristic classes of regular Lie algebroids—a sketch*, Rend. Circ. Mat. Palermo (2) Suppl. (1993), no. 30, 71–94.

[203] ———, *Bott's vanishing theorem for regular Lie algebroids.*, Trans. Am. Math. Soc. **348** (1996), no. 6, 2151–2167.

[204] I. Kupka, *The singularities of integrable structurally stable Pfaffian forms*, Proc. Natl. Acad. Sci. USA **52** (1964), 1431–1432.

[205] A.G. Kushnirenko, *Linear-equivalent action of a semisimple Lie group in the neighborhood of a stationary point*, Funkts. Anal. Prilozh. **1** (1967), no. 1, 103–104.

[206] Jeroen S. W. Lamb and John A. G. Roberts, *Time-reversal symmetry in dynamical systems: a survey*, Phys. D **112** (1998), no. 1-2, 1–39.

[207] N. P. Landsman, *Mathematical topics between classical and quantum mechanics*, Springer Monographs in Mathematics, Springer-Verlag, New York, 1998.

[208] Luen Chau Li and Serge Parmentier, *Nonlinear Poisson structures and r-matrices*, Comm. Math. Phys. **125** (1989), no. 4, 545–563.

[209] Paulette Libermann, *Problèmes d'équivalence et géométrie symplectique*, Astérisque, vol. 107, Soc. Math. France, Paris, 1983, pp. 43–68.

[210] Paulette Libermann and Charles-Michel Marle, *Symplectic geometry and analytical mechanics*, D. Reidel Publishing Co., Dordrecht, 1987.

[211] André Lichnerowicz, *Les variétés de Poisson et leurs algebrès de Lie associées*, J. Differential Geometry **12** (1977), no. 2, 253–300.

[212] André Lichnerowicz and Alberto Medina, *On Lie groups with left-invariant symplectic or Kählerian structures*, Lett. Math. Phys. **16** (1988), no. 3, 225–235.

[213] Sophus Lie, *Theorie des transformationgrupen*, Teubner, Leipzig, 1890.

[214] Alcides Lins Neto, *Local structural stability of C^2 integrable 1-forms*, Ann. Inst. Fourier (Grenoble) **27** (1977), no. 2, vi, 197–225.

[215] Joseph Liouville, *Note sur l'intégration des équations differentielles de la dynamique, présentée au bureau des longitudes le 29 juin 1853*, Journal de Math. Pures et Appl. **20** (1855), 137–138.

[216] Zhang Ju Liu, Alan Weinstein, and Ping Xu, *Dirac structures and Poisson homogeneous spaces*, Comm. Math. Phys. **192** (1998), no. 1, 121–144.

[217] Zhang Ju Liu and Ping Xu, *On quadratic Poisson structures*, Lett. Math. Phys. **26** (1992), no. 1, 33–42.

[218] Frank Loray, *Analytic normal forms for nondegenerate singularities of planar vector fields*, preprint CRM Barcelona (2003).

[219] Philipp Lorhmann, *Formes normales de structures de Poisson ayant 1-jet nul*, preprint (2005).

[220] Jiang-Hua Lu, *Multiplicative and affine Poisson structures on Lie groups*, Thesis, Berkeley (1990).

[221] _____, *Momentum mappings and reduction of Poisson actions*, Symplectic geometry, groupoids, and integrable systems (Berkeley, CA, 1989), Math. Sci. Res. Inst. Publ., vol. 20, Springer, New York, 1991, pp. 209–226.

[222] _____, *Poisson homogeneous spaces and Lie algebroids associated to Poisson actions*, Duke Math. J. **86** (1997), no. 2, 261–304.

[223] Jiang-Hua Lu and Alan Weinstein, *Groupoïdes symplectiques doubles des groupes de Lie-Poisson*, C. R. Acad. Sci. Paris Sér. I Math. **309** (1989), no. 18, 951–954.

[224] _____, *Poisson Lie groups, dressing transformations, and Bruhat decompositions*, J. Differ. Geom. **31** (1990), no. 2, 501–526.

[225] O. V. Lychagina, *Degenerate Poisson structures in dimension 3*, Mat. Zametki **63** (1998), no. 4, 579–592.

[226] Kirill Mackenzie, *Lie groupoids and Lie algebroids in differential geometry*, London Math. Soc. Lecture Notes Series No. 124, Cambridge University Press, 1987.

[227] _____, *Lie algebroids and Lie pseudoalgebras*, Bull. London Math. Soc. **27** (1995), no. 2, 97–147.

[228] _____, *General theory of Lie groupoids and Lie algebroids*, Cambridge University Press, 2005.

[229] Kirill Mackenzie and Ping Xu, *Lie bialgebroids and Poisson groupoids*, Duke Math. J. **73** (1994), no. 2, 415–452.

[230] _____, *Classical lifting processes and multiplicative vector fields*, Quart. J. Math. Oxford II. Ser. **49** (1998), no. 193, 59–85.

[231] _____, *Integration of Lie bialgebroids*, Topology **39** (2000), no. 3, 445–467.

[232] Franco Magri, *A simple model of the integrable Hamiltonian equation.*, J. Math. Phys. **19** (1978), 1156–1162.

[233] Bernard Malgrange, *Frobenius avec singularités. I. Codimension un*, Inst. Hautes Études Sci. Publ. Math. (1976), no. 46, 163–173.

[234] _____, *Frobenius avec singularites. II: Le cas general.*, Invent. Math. **39** (1977), 67–89.

[235] Marí Beffa, Gloria, *A transverse structure for the Lie-Poisson bracket on the dual of the Virasoro algebra*, Pacific J. Math. **163** (1994), no. 1, 43–72.

[236] _____, *Transverse sections for the second Hamiltonian KdV structure*, J. Geom. Anal. **8** (1998), no. 3, 385–407.

[237] Jerrold Marsden and Alan Weinstein, *Reduction of symplectic manifolds with symmetry*, Rep. Mathematical Phys. **5** (1974), no. 1, 121–130.

[238] Jerrold E. Marsden and Tudor Ratiu, *Reduction of Poisson manifolds*, Lett. Math. Phys. **11** (1986), no. 2, 161–169.

[239] Jean Martinet, *Normalisation des champs de vecteurs holomorphes (d'après A. D. Brujno*, Séminaire Bourbaki **564** (1980).

[240] David Martinez Torres, *Global classification of generic multi-vector fields of top degree*, J. London Math. Soc. (2) **69** (2004), no. 3, 751–766.

[241] Olivier Mathieu, *Harmonic cohomology classes of symplectic manifolds*, Comment. Math. Helvetici **70** (1995), 1–9.

[242] J.-F. Mattei and R. Moussu, *Holonomie et intégrales premières*, Ann. Sci. École Norm. Sup. (4) **13** (1980), no. 4, 469–523.

[243] Dusa McDuff and Dietmar Salamon, *Introduction to symplectic topology*, Oxford Mathematical Monographs, The Clarendon Press Oxford University Press, New York, 1998.

[244] A. S. de Medeiros, *Structural stability of integrable differential forms.*, Geom. Topol., III. Lat. Am. Sch. Math., Proc., Rio de Janeiro 1976, Lect. Notes Math. 597, 395-428 , 1977.

[245] _____, *Singular foliations and differential p-forms*, Ann. Fac. Sci. Toulouse Math. (6) **9** (2000), no. 3, 451–466.

[246] Richard Melrose, *The Atiyah–Patodi–Singer index theorem*, Research Notes In Mathematics 4, A K Peters Ltd., Wellesley, 1993.

[247] Kenneth R. Meyer, *Symmetries and integrals in mechanics*, Dynamical systems (Proc. Sympos., Univ. Bahia, Salvador, 1971), Academic Press, New York, 1973, pp. 259–272.

[248] Kentaro Mikami and Alan Weinstein, *Moments and reduction for symplectic groupoids.*, Publ. Res. Inst. Math. Sci. **24** (1988), no. 1, 121–140.

[249] Henri Mineur, *Sur les systemes mecaniques admettant n integrales premieres uniformes et l'extension a ces systemes de la methode de quantification de Sommerfeld*, C. R. Acad. Sci., Paris **200** (1935), 1571–1573 (French).

[250] _____, *Sur les systemes mecaniques dans lesquels figurent des parametres fonctions du temps. Etude des systemes admettant n integrales premieres uniformes en involution. Extension a ces systemes des conditions de quantification de Bohr-Sommerfeld.*, Journal de l'Ecole Polytechnique, Série III, 143ème année (1937), 173–191 and 237–270.

[251] Eva Miranda and Nguyen Tien Zung, *A note on equivariant normal forms of Poisson structures*, in preparation (2005).

[252] A. S. Mischenko and A. T. Fomenko, *A generalized method for Liouville integration of Hamiltonian sytems*, Funct. Anal. Appl. **12** (1978), 46–56.

[253] _____, *Euler equation on finite-dimensional Lie groups*, Izv. Akad. Nauk SSSR Ser. Mat. **42** (1978), no. 2, 396–415, 471.

[254] I. Moerdijk and J. Mrčun, *Introduction to foliations and Lie groupoids*, Cambridge Studies in Advanced Mathematics, vol. 91, Cambridge University Press, 2003.

[255] I. Moerdijk and D. A. Pronk, *Orbifolds, sheaves and groupoids*, K-Theory **12** (1997), no. 1, 3–21.

[256] Ieke Moerdijk, *Orbifolds as groupoids: an introduction*, Orbifolds in mathematics and physics (Madison, WI, 2001), Contemp. Math., vol. 310, Amer. Math. Soc., 2002, pp. 205–222.

[257] Ieke Moerdijk and Janez Mrcun, *On integrability of infinitesimal actions.*, Am. J. Math. **124** (2002), no. 3, 567–593.

[258] J.-C. Molinier, *Linéarisation de structures de Poisson*, Thèse, Montpellier (1993).

[259] Pierre Molino, *Structure transverse aux orbites de la représentation coadjointe: le cas des orbites réductives*, Séminaire Gaston Darboux de Géométrie Différentielle à Montpellier (1983–1984), 55–62.

[260] _____, *Riemannian foliations*, Progress in Mathematics, vol. 73, Birkhäuser Boston Inc., Boston, MA, 1988.

[261] _____, *Orbit-like foliations*, Geometric study of foliations (Tokyo, 1993), World Sci. Publishing, River Edge, NJ, 1994, pp. 97–119.

[262] Philippe Monnier, *Poisson cohomology in dimension two*, Israel J. Math. **129** (2002), 189–207.

[263] Philippe Monnier and Nguyen Tien Zung, *Levi decomposition for smooth Poisson structures*, J. Diff. Geom. **68** (2004), no. 2, 347–395.

[264] Kiiti Morita, *Duality for modules and its applications to the theory of rings with minimum condition*, Sci. Rep. Tokyo Kyoiku Daigaku Sect. A **6** (1958), 83–142.

[265] Jürgen Moser, *On the volume elements on a manifold*, Trans. Am. Math. Soc. **120** (1965), 286–294.

[266] Jürgen Moser, *On the generalization of a theorem of A. Liapounoff*, Comm. Pure Appl. Math. **11** (1958), 257–271.

[267] Robert Moussu, *Sur l'existence d'intégrales premières pour un germe de forme de Pfaff*, Ann. Inst. Fourier **26** (1976), no. 2, 170–220.

[268] _____, *Classification C∞ des équations de Pfaff intégrables à singularités isolées*, Invent. Math. **73** (1983), 419–436.

[269] _____, *Sur l'existence d'intégrales premières holomorphes*, Ann. Scuola Norm. Sup. Pisa Cl. Sci. (4) **26** (1998), no. 4, 709–717.

[270] J. E. Moyal, *Quantum mechanics as a statistical theory*, Proc. Cambridge Philos. Soc. **45** (1949), 99–124.

[271] Paul S. Muhly, Jean N. Renault, and Dana P. Williams, *Equivalence and isomorphism for groupoid C^*-algebras*, J. Operator Theory **17** (1987), no. 1, 3–22.

[272] Shingo Murakami, *Sur la classification des algèbres de Lie réelles et simples*, Osaka J. Math. **2** (1965), 291–307.

[273] Nobutada Nakanishi, *Poisson cohomology of plane quadratic Poisson structures*, Publ. Res. Inst. Math. Sci. **33** (1997), no. 1, 73–89.

[274] ———, *On Nambu-Poisson manifolds*, Rev. Math. Phys. **10** (1998), no. 4, 499–510.

[275] Y. Nambu, *Generalized Hamiltonian dynamics*, Phys. Rev. D **7** (1973), 2405–2412.

[276] N.N. Nekhoroshev, *Action-angle variables and their generalizations*, Trans. Moscow Math. Soc. **26** (1972), 180–198.

[277] Ryszard Nest and Boris Tsygan, *Algebraic index theorem for families*, Adv. Math. **113** (1995), no. 2, 151–205.

[278] Michel Nguiffo Boyom and Robert Wolak, *Local structure of Koszul-Vinberg and of Lie algebroids*, Bull. Sci. Math. **128** (2004), no. 6, 467–479.

[279] A. Nijenhuis and R. W. Richardson, *Cohomology and deformations in graded Lie algebras*, Bull. Amer. Math. Ssoc. **72** (1966), no. 1, 1–29.

[280] Albert Nijenhuis, *Jacobi-type identities for bilinear differential concomitants of certain tensor fields*, Indag. Math. **17** (1955), 390–403.

[281] Anatol Odzijewicz and Tudor S. Ratiu, *Banach Lie-Poisson spaces and reduction*, Comm. Math. Phys. **243** (2003), no. 1, 1–54.

[282] W. Oevel, *Poisson brackets for integrable lattice systems*, Algebraic aspects of integrable systems, Progr. Nonlinear Differential Equations Appl., vol. 26, Birkhäuser Boston, Boston, MA, 1997, pp. 261–283.

[283] Walter Oevel and Orlando Ragnisco, *R-matrices and higher Poisson brackets for integrable systems*, Phys. A **161** (1989), no. 1, 181–220.

[284] Yong-Geun Oh, *Some remarks on the transverse Poisson structures of coadjoint orbits*, Lett. Math. Phys. **12** (1986), no. 2, 87–91.

[285] Peter J. Olver, *Applications of Lie groups to differential equations*, Graduate Texts in Mathematics, vol. 107, Springer-Verlag, New York, 1993.

[286] Hideki Omori, Yoshiaki Maeda, and Akira Yoshioka, *Weyl manifolds and deformation quantization*, Adv. Math. **85** (1991), no. 2, 224–255.

[287] Juan-Pablo Ortega and Tudor S. Ratiu, *Singular reduction of Poisson manifolds*, Lett. Math. Phys. **46** (1998), no. 4, 359–372.

[288] ———, *Momentum maps and Hamiltonian reduction*, Progress in Mathematics, Vol. 222, Birkhäuser, 2004.

[289] V. Yu. Ovsienko and B. A. Khesin, *Symplectic leaves of the Gel'fand-Dikii brackets and homotopy classes of nonflattening curves*, Funkt. Anal. Pril. **24** (1990), no. 1, 38–47.

[290] Richard S. Palais, *A global formulation of the Lie theory of transformation groups*, Mem. Amer. Math. Soc. No. **22** (1957), iii+123.

[291] ———, *On the existence of slices for actions of non-compact Lie groups*, Ann. of Math. (2) **73** (1961), 295–323.

[292] J. Patera, R. T. Sharp, P. Winternitz, and H. Zassenhaus, *Invariants of real low dimension Lie algebras*, J. Mathematical Phys. **17** (1976), no. 6, 986–994.

[293] S.-D. Poisson, *Sur la variation des constantes arbitraires dans les questions de mécanique*, J. Ecole Polytechnique **8** (1809), Cah. 15, 266–344.

[294] Jean Pradines, *Théorie de Lie pour les groupoides différentiables*, C. R. Acad. Sci. Paris Sér. A Math. **264** (1967), 245–248.

[295] _____, *Troisième théorème de Lie pour les groupoïdes différentiables*, C. R. Acad. Sci. Paris Sér. A-B **267** (1968), A21–A23.

[296] Olga Radko, *A classification of topologically stable Poisson structures on a compact oriented surface*, J. Symplectic Geom. **1** (2002), no. 3, 523–542.

[297] K. Rama, K. H. Bhaskara, and John V. Leahy, *Poisson structures due to Lie algebra representations*, Internat. J. Theoret. Phys. **34** (1995), no. 10, 2031–2037.

[298] Bruce L. Reinhart, *Foliated manifolds with bundle-like metrics*, Ann. of Math. (2) **69** (1959), 119–132.

[299] Jean Renault, *A groupoid approach to C^*-algebras*, Lecture Notes in Mathematics, vol. 793, Springer, Berlin, 1980.

[300] R. W. Richardson, *On the rigidity of semi-direct products of Lie algebras*, Pacific J. of Math. **22** (1967), 339–344.

[301] G. Rinehart, *Differential forms for general commutative algebras*, Trans. Amer. Math. Soc. **108** (1963), 195–222.

[302] Claude Roger and Pol Vanhaecke, *Poisson cohomology of the affine plane*, J. Algebra **251** (2002), no. 1, 448–460.

[303] Julien Roger, *Cohomologie des variétés de poisson de dimension deux*, Mémoire DEA, Université Toulouse III (2004).

[304] Robert Roussarie, *Modèles locaux de champs et de formes*, Astérisque, Société Mathématique de France, Paris, 1975.

[305] Dmitry Roytenberg, *Poisson cohomology of SU(2)-covariant "necklace" Poisson structures on S^2*, J. Nonlinear Math. Phys. **9** (2002), no. 3, 347–356.

[306] Walter Rudin, *Function theory in the unit ball of $\mathbf{C}^n$*, Grundlehren der Mathematischen Wissenschaften, vol. 241, Springer-Verlag, New York, 1980.

[307] Helmut Rüssmann, *Über die Normalform analytischer Hamiltonscher Differentialgleichungen in der Nähe einer Gleichgewichtslösung*, Math. Ann. **169** (1967), 55–72.

[308] Kyoji Saito, *On a generalization of de-Rham lemma*, Ann. Inst. Fourier (Grenoble) **26** (1976), no. 2, 165–170.

[309] Ichirô Satake, *The Gauss-Bonnet theorem for V-manifolds*, J. Math. Soc. Japan **9** (1957), 464–492.

[310] Peter Schaller and Thomas Strobl, *Poisson structure induced (topological) field theories*, Modern Phys. Lett. A **9** (1994), no. 33, 3129–3136.

[311] J. A. Schouten, *Über Differentialkonkomitanten zweier kontravarianter Größen*, Indag. Math. **2** (1940), 449–452.

[312] _____, *On the differential operators of first order in tensor calculus*, Convegno Int. Geom. Diff. Italia, 1953, Ed. Cremonese, Roma, 1954, pp. 1–7.

[313] Michael A. Semenov-Tian-Shansky, *What is a classical r-matrix ?*, Funct. Anal. Appl. **17** (1983), no. 4, 259–272.

[314] _____, *Dressing transformations and Poisson group actions*, Publ. Res. Inst. Math. Sci. **21** (1985), no. 6, 1237–1260.

[315] Pavol Ševera and Alan Weinstein, *Poisson geometry with a 3-form background*, Progr. Theoret. Phys. Suppl. (2001), no. 144, 145–154.

[316] Carl Ludwig Siegel, *Über die Existenz einer Normalform analytischer Hamiltonscher Differentialgleichungen in der Nähe einer Gleichgewichtslösung*, Math. Ann. **128** (1954), 144–170.

[317] Reyer Sjamaar and Eugene Lerman, *Stratified symplectic spaces and reduction*, Ann. of Math. (2) **134** (1991), no. 2, 375–422.

[318] E. K. Sklyanin, *Some algebraic structures connected with the Yang-Baxter equation*, Functional Anal. Appl. **16** (1982), no. 4, 263–270.

[319] J.-M. Souriau, *Structure des systèmes dynamiques*, Maîtrises de mathématiques, Dunod, Paris, 1970.

[320] P. Stefan, *Accessible sets, orbits, and foliations with singularities*, Proc. London Math. Soc. (3) **29** (1974), 699–713.

[321] Shlomo Sternberg, *On the structure of local homeomorphisms of euclidean n-space. II.*, Am. J. Math. **80** (1958), 623–631.

[322] Daniel Sternheimer, *Deformation quantization: twenty years after*, Particles, fields, and gravitation (Łódź, 1998), AIP Conf. Proc., vol. 453, Amer. Inst. Phys., pp. 107–145.

[323] Laurent Stolovitch, *Singular complete integrability*, Publications IHES **91** (2003), 134–210.

[324] _____, *Sur les structures de Poisson singulières*, Ergodic Theory Dyn. Sys. **24** (2004), no. 5, 1833–1863.

[325] Ewa Stróżyna and Henryk Żoładek, *The analytic and formal normal form for the nilpotent singularity*, J. Differential Equations **179** (2002), no. 2, 479–537.

[326] Yu. B. Suris, *On the bi-Hamiltonian structure of Toda and relativistic Toda lattices*, Phys. Lett. A **180** (1993), no. 6, 419–429.

[327] Héctor J. Sussmann, *Orbits of families of vector fields and integrability of distributions*, Trans. Amer. Math. Soc. **180** (1973), 171–188.

[328] L. Takhtajan, *On foundation of the generalized Nambu mechanics*, Comm. Math. Phys. **160** (1994), 295–315.

[329] William P. Thurston, *Three-dimensional geometry and topology. Vol. 1*, Princeton Mathematical Series, vol. 35, Princeton University Press, Princeton, NJ, 1997.

[330] Jean-Claude Tougeron, *Idéaux de fonctions différentiables. I*, Ann. Inst. Fourier (Grenoble) **18** (1968), no. fasc. 1, 177–240.

[331] A. Yu. Vaintrob, *Lie algebroids and homological vector fields*, Russian Math. Surveys **52** (1997), no. 2, 428–429.

[332] Izu Vaisman, *Remarks on the Lichnerowicz-Poisson cohomology*, Ann. Inst. Fourier (Grenoble) **40** (1990), no. 4, 951–963.

[333] _____, *Lectures on the geometry of Poisson manifolds*, Progress in Mathematic, Vol. 118, 1994.

[334] _____, *Nambu-Lie groups*, J. Lie Theory **10** (2000), no. 1, 181–194.

[335] V. S. Varadarajan, *Lie groups, Lie algebras, and their representations*, Graduate Texts in Mathematics, vol. 102, Springer-Verlag, New York, 1984, Reprint of the 1974 edition.

[336] Jacques Vey, *Sur certains systèmes dynamiques séparables*, Amer. J. Math. **100** (1978), no. 3, 591–614.